Handbook of Geochemistry

Vol. I

Executive Editor: K. H. Wedepohl

Springer-Verlag Berlin · Heidelberg · New York 1969

ISBN 3-540-06577-6 Springer-Verlag Berlin · Heidelberg · New York
ISBN 0-387-06577-6 Springer-Verlag New York · Heidelberg · Berlin

PREFACE

Geochemistry is concerned with the laws governing the distribution of the chemical elements and their isotopes throughout the Earth. As a concept it has been recognized for 130 years but it has grown into a separate Earth science during this century. Geochemistry has mutual links with many neighbouring disciplines. Its present field of activity is determined by many problems of broad interest and by the availability of methods. Several exterior influences have recently developed. Thus, nuclear physics and its specific measuring techniques made isotope geochemistry possible, while space research has stimulated the development of cosmochemistry. Except a few "standard" materials as G 1 and W 1 there is no other rock on earth whose composition is as well known as that of meteorites colliding with our planet on their cosmic course. Biochemistry is linked with the rapidly developing new branch of organic geochemistry.

Our discipline has moved forward in step with the advancement of analytical chemistry. When optical and X-ray spectrochemical analysis came into use and with the discovery of natural and artificial radio-activity, many new elements were identified. With the development of spectrophotometers, radiation counters and nuclear sources over the last 20 years, a flood of analytical data on geological subjects has been released, and we ought to make use of it.

Parallel with the advance in analytical geochemistry there has been a big improvement in the facilities for synthesizing minerals, due to progress in high-pressure techniques and the production of high-temperature materials. Many chemical reactions analogous to processes occurring within the Earth's crust and mantle can today be observed under controlled conditions in the laboratory. The individual scientist cannot benefit from the mass of data until it has been arranged in a readily accessible form. The system of data compilation may be classified by geological and petrological units, or by chemical elements in atomic number sequence. F. W. CLARKE (1847—1931) used the first system in his "Data of Geochemistry" (5th edn 1924), currently under revision by M. FLEISCHER and a large group of coworkers. V. M. GOLDSCHMIDT (1888—1947) in "Geochemistry" edited after his death by A. MUIR in 1954 classified geochemical facts and data according to both their geological-petrological and their chemical features.

GOLDSCHMIDT's main emphasis was the consideration of crystal-chemical relationships in natural systems. "Geochemistry" by RANKAMA and SAHAMA, published in 1950, presents the data available before the recent introduction of greatly improved and rationalized analytical procedures. Several specialized journals have been launched since then, but there is still a mass of valuable data strewn throughout the literature of science which is not readily accessible to the individual scientist.

The idea of extracting important geochemical data and facts from the exponentially expanding literature was born in the 1960s, but even then it was clear that one or two authors could not possibly prepare such a book within a reasonably short space of time. A group of editors with a realistic appreciation of the amount of work involved thought it might be possible to compile this urgently needed book by dividing it up among some 70 authors. Splitting the work up like this does, of course, entail some loss of consistency. It will be impossible to compete with the design of CLARKE or GOLDSCHMIDT. For optimum uniformity, the chapters allotted to the various authors must fit into a pre-arranged scheme, even though this must unfortunately restrict their individuality.

The editors, with the experience of their predecessors did not assume that some 70 authors can work at the same rate and deliver their contributions on the same date. Anticipating the natural phasing in the preparation of the various chapters, the editors decided on a loose leaf insertion method. Volume II of the handbook will therefore be published in 4 instalments. The binders will contain a total of about 3,300 pages. There are undeniable drawbacks to this method, but it does have the advantages mentioned above and, moreover, permits the subsequent replacement of any pages in particular need of revision.

Each element, with the exception of the noble gases, the lanthanides, and the platinum elements will be dealt with in a separate chapter of Volume II of the handbook. The chapters are numbered with the atomic numbers of the elements, and each is subdivided in sections as follows:

A. Crystal chemistry
B. Isotopes in nature
C. Abundance in cosmos, meteorites, tektites and lunar materials
D. Abundance in rock-forming minerals (phase equilibria), minerals
E. Abundance in common igneous rock types
F. Behavior in magmatogenic processes (pegmatites, gas transport, ore deposition etc.)
G. Behavior during weathering and alteration of rocks
H. Solubilities of compounds which control concentrations of the element in natural waters; adsorption processes; valence states in natural environments
I. Abundance in natural waters and in the atmosphere
K. Abundance in common sediments and sedimentary rock types
L. Biogeochemistry
M. Abundance in common metamorphic rock types
N. Behavior in metamorphic reactions
O. Relations to other elements, economic importance etc.

Each section is numbered as an independent unit. For example, the second page of the section on the crystal chemistry of copper will be 29-A-2. Tables and diagrams are also numbered consecutively within the sections. In general, each chapter, with the exception of section A (crystal chemistry) and possibly B (natural isotopes) was prepared by a single author. Sub-sections of sections are designated by Roman numerals and additional subdivisions by small letters. To give the work international currency, results have wherever possible been presented in table and graph form. Data on concentrations are normally given as follows:

0.1—100% wt
0.1—1000 ppm (0.00001—0.1% wt)
X ppb (0.00X ppm).

In a few cases atomic percentages are also given in brackets. It is often possible to assess the value of the data from the method of analysis, therefore the tables contain symbols indicating the method used (see Preface of Volume II).

Each chapter has its own bibliography. Section A (crystal chemistry) often carries its own references and so, in certain cases, does Section B (isotopes in nature). Technical difficulties prevent any stricter application of the separate section principle.

Volume I contains reviews on fundamental facts of geochemistry, geophysics and cosmochemistry, together with definitions, dimensions, methods of evaluation etc. Controversial names of rocks or geological concepts etc. are used throughout Volume II in the sense in which they were introduced and defined in Volume I. The editors' purpose in so doing is to maintain consistancy in nomenclature throughout; they do not want to compete with international commissions on nomenclature.

K. H. Wedepohl

CONTENTS

Chapter 12

A. Heydemann

CONTRIBUTORS

C. W. Correns	Geochemisches Institut der Universität, Göttingen, Goldschmidtstr. 1 (Germany)
G. Goles	Center for Volcanology, Department of Geology, University of Oregon, Eugene, Oregon (U.S.A.)
A. Heydemann	Geochemisches Institut der Universität, Göttingen, Goldschmidtstr. 1 (Germany)
K. Keil	Institute of Meteoritics, Department of Geology and Institute of Meteoritics, The University of New Mexico Albuquerque, New Mexico (U.S.A.)
K. B. Krauskopf	Department of Geology, School of Earth Sciences, Stanford University, Stanford, Calif. (U.S.A.)
K. R. Mehnert	Mineralogisches Institut der Universität, Berlin-Lichterfelde, Holbeinstr. 48 (Germany)
U. Schmucker	Göttingen, Herzberger Landstr. 42 (Germany)
D. M. Shaw	Department of Geology, McMaster University Hamilton, Ontario (Canada)
K. K. Turekian	Department of Geology, Yale University, Box 2161 Yale Station New Haven, Conn. (U.S.A.)
K. H. Wedepohl	Geochemisches Institut der Universität, Göttingen, Goldschmidtstr. 1 (Germany)
J. Zemann	Mineralogisches Institut der Universität Wien, Dr.-Karl-Lueger-Ring 1, A-1010 Wien (Austria)

CHAPTER 1

C. W. CORRENS

THE DISCOVERY OF THE CHEMICAL ELEMENTS THE HISTORY OF GEOCHEMISTRY DEFINITIONS OF GEOCHEMISTRY

I. The Discovery of the Chemical Elements

The relations between chemistry and mineralogical objects go back a long way even in prehistoric times, minerals were used in the making of metals, pottery, dyes etc. But a scientific understanding of these relations first begins with the development of chemistry as a science.

From the long history of chemistry we will here consider only the discovery of the elements. *Gold, silver, copper, iron, tin, mercury, lead, sulfur and carbon* were already known in antiquity, but were not regarded as elements in the modern sense. As recently as 1600, the school of ARISTOTLE (384—322 B.C.) still reigned: matter was constituted of the elements: earth, water, air and fire. These were regarded as symbolic of the four pairs of the main qualities, warm, cold, dry and wet, the elements being transformable, e.g. water (cold and wet) can be transformed into air (warm and wet) if the heat of the fire overcomes the coldness of the water. In the seventeenth century the classic concept of the elements begins with DANIEL SENNERT (1572—1637), JOACHIM JUNGIUS (1587—1637) and ROBERT BOYLE (1627—1691). But even BOYLE considered white lead, $Pb_3(OH)_2(CO_3)_2$, and red lead, (Pb_3O_4), to be one substance (HOOYKAAS, 1957).

Besides the elements known in antiquity, some were discovered in the Middle Ages. In ancient times the sulfides of arsenic, orpiment and realgar (sandarek), were called "arsenic"; the element *arsenic* was probably first isolated by ALBERTUS MAGNUS (1193—1280) who obtained it by heating orpiment with soap. *Antimony* was also known as sulfide to the Ancients and was used to darken and beautify the eyebrows, but it is possible that the element was also used as a metal. G. AGRICOLA (1494—1555) was familiar with metallic antimony and its use. Metallic *bismuth* was in use at least as early as 1480. These elements play an important part in the literature of the alchemists.

Phosphorus was discovered in Hamburg about 1670 by HENNIG BRAND, who prepared it from urine. His "cold fire" roused the interest of many scientists, among them the famous philosopher, G. W. LEIBNIZ, who visited BRAND in 1678.

With the advance of analytical chemistry in the 18th century, the number of new elements begins to increase. The history of their discovery is sometimes rather complicated. It frequently happened that two or more men discovered the same element indepedently. In other cases various observers recognized the existence of a new element long before it was actually isolated, so it is not possible to give a detailed account here. For instance: *silicon* was probably prepared in 1811 as an impure substance by Gay-Lussac and Thenard; Berzelius, generally credited with its discovery, succeeded in preparing silicon in 1824. It was not until 1854 that Deville prepared crystalline silicon. Mary Elvira Weeks has given these often very interesting details in her book, Discovery of the Elements. There are also some very amusing stories in the correspondence between Berzelius and Wöhler (1823—1848) (ed. Wallach): for instance, the claimed discovery of the elements apollon, dian and ostran by J. F. Breithaupt in 1826.

Fig. 1-1 shows the elements discovered from 1720—1961, arranged in decades. We observe three periods of great activity: the first begins with the discovery of

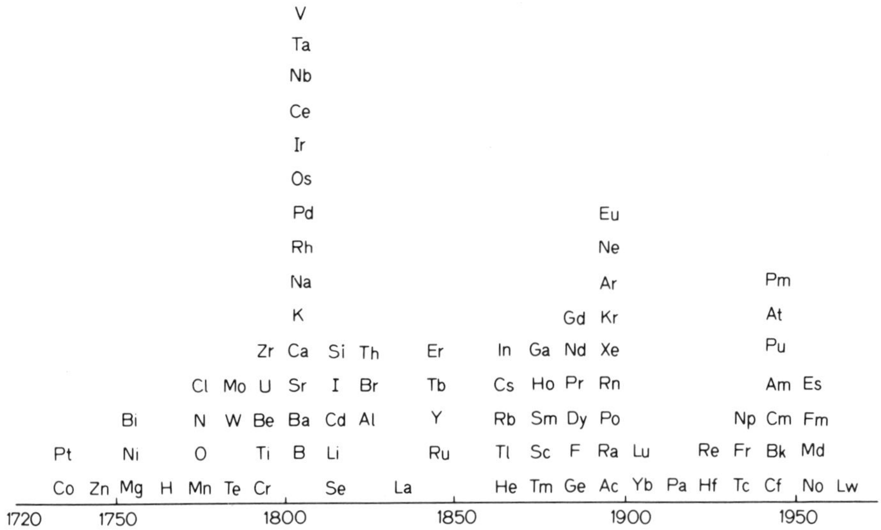

Fig. 1-1. Elements discovered from 1720—1961 arranged in decades

hydrogen (Cavendish, 1766), *oxygen* (Priestley, 1774; Scheele) and *nitrogen* (Daniel Rutherford, 1772) and ends about 1850. During this time 40 new elements were discovered, including some of the most abundant such as oxygen, *silicon, aluminium,* the *alkalis, alkaline earths* and *halogens.* Usually these elements were found in minerals, although *iodine* was detected 1811 by Courtois in brown algae (Fucus, Laminaria). This was the period during which wet methods in analytical chemistry were developed. A new period began with the development of optical emission spectrography by Bunsen and Kirchhoff. In 1860 they detected *cesium* in the mineral waters of Bad Dürckheim and in 1861 *rubidium* in the mineral lepidolite. In the same year Sir William Crookes discovered *thallium* in the seleniferous residues from the sulfuric acid factory at Tilkerode in the Harz Mountains. In 1863, F. Reich and H. T. Richter detected *indium* in sphalerite from the Himmelfürst mine at Freiberg, Saxony. Most of the rare earth elements were discovered spectroscopically, also the socalled inert gases. *Helium* was 1868 first observed in the sun by Janssen and, not until 27 years later, in the mineral cleveite by Sir William Ramsay.

Table 1-1. *Year of discovery and discoverers of the elements*

Atom-No.	Name	Symbol	Discovery year	by
1	Hydrogen	H	1766	CAVENDISH
2	Helium	He	1868	JANSSEN
3	Lithium	Li	1817	ARFWEDSON
4	Beryllium	Be	1798	VAUQUELIN
5	Boron	B	1808	DAVY; GAY-LUSSAC and THENARD
6	Carbon	C	antiquity	
7	Nitrogen	N	1772	DAN RUTHERFORD
8	Oxygen	O	1774	PRIESTLEY (SCHEELE, publ. 1777)
9	Fluorine	F	1886	MOISSAN
10	Neon	Ne	1898	RAMSAY and TRAVERS
11	Sodium	Na	1807	DAVY
12	Magnesium	Mg	1755	BLACK (DAVY, 1808)
13	Aluminium	Al	1827	WÖHLER
14	Silicon	Si	1811	GAY-LUSSAC and THENARD, 1824 BERZELIUS, 1854 DEVILLE
15	Phosphorus	P	1669	BRAND
16	Sulfur	S	antiquity	
17	Chlorine	Cl	1774	SCHEELE
18	Argon	Ar	1894	RAYLEIGH and RAMSAY
19	Potassium	K	1807	DAVY } Sir HUMPHRY
20	Calcium	Ca	1808	DAVY } Sir HUMPHRY
21	Scandium	Sc	1879	NILSON
22	Titanium	Ti	1791	GREGOR, named by KLAPROTH 1795
23	Vanadium	V	1801	DEL RIO
24	Chromium	Cr	1797	VAUQUELIN
25	Manganese	Mn	1774	GAHN
26	Iron	Fe	antiquity	
27	Cobalt	Co	1735	BRAND
28	Nickel	Ni	1751	CRONSTEDT
29	Copper	Cu	antiquity	
30	Zinc	Zn	1746	MARGGRAF
31	Gallium	Ga	1857	LECOQ DE BOISBAUDRAN
32	Germanium	Ge	1886	WINKLER
33	Arsenic	As	antiquity	
34	Selenium	Se	1817	BERZELIUS
35	Bromine	Br	1826	BALARD
36	Krypton	Kr	1898	RAMSAY and TRAVERS
37	Rubidium	Rb	1861	BUNSEN and KIRCHHOFF
38	Strontium	Sr	1808	DAVY
39	Yttrium	Y	1843	MOSANDER
40	Zirconium	Zr	1789	KLAPROTH, BERZELIUS 1824
41	Niobium	Nb	1801	HATCHETT, 1864 BLOMSTRAND
42	Molybdenum	Mo	1778	SCHEELE, 1782 HJELM
43	Technetium	Te	1937	PERRIER and SEGRÉ
44	Ruthenium	Ru	1844	KLAUS
45	Rhodium	Rh	1803/4	WOLLASTON
46	Palladium	Pd	1803	WOLLASTON
47	Silver	Ag	antiquity	
48	Cadmium	Cd	1817	STROHMEYER
49	Indium	In	1863	REICH and RICHTER
50	Tin	Sn	antiquity	
51	Antimony	Sb	antiquity	

Table 1-1. (Continuation)

Atom-No.	Name	Symbol	Discovery year	Discovery by
52	Tellurium	Te	1782	MÜLLER v. REICHENSTEIN, named by KLAPROTH 1798
53	Iodine	I	1811	COURTOIS
54	Xenon	Xe	1898	RAMSAY and TRAVERS
55	Cesium	Cs	1860	BUNSEN and KIRCHHOFF
56	Barium	Ba	1808	DAVY
57	Lanthanum	La	1839	MOSANDER
58	Cerium	Ce	1803	KLAPROTH, BERZELIUS and HISINGER
59	Praseodymium	Pr	1885	AUER v. WELSBACH
60	Neodymium	Nd	1885	AUER v. WELSBACH
61	Promethium	Pm	1945	MARINSKY, GLENDENIN and CORYELL
62	Samarium	Sm	1879	LECOQ DE BOISBAUDRAN
63	Europium	Eu	1896	DEMARÇAY
64	Gadolinium	Gd	1880	MARIGNAC
65	Terbium	Tb	1843	MOSANDER
66	Dysprosium	Dy	1886	LECOQ DE BOISBAUDRAN
67	Holmium	Ho	1878	DELAFONTAINE and SORET, CLEVE
68	Erbium	Er	1843	MOSANDER
69	Thulium	Tm	1879	CLEVE
70	Ytterbium	Yb	1907	URBAIN
71	Lutetium	Lu	1907	URBAIN
72	Hafnium	Hf	1923	COSTER and VON HEVESY
73	Tantalum	Ta	1802	EKEBERG, ROSE 1844, MARIGNAC 1866
74	Tungsten	W	1781	SCHEELE
75	Rhenium	Re	1925	NODDACK, TACKE and BERG
76	Osmium	Os	1803	TENNANT
77	Iridium	Ir	1803	TENNANT
78	Platinum	Pt	1735	ULLOA, 1741 WOOD
79	Gold	Au	antiquity	
80	Mercury	Hg	antiquity	
81	Thallium	Tl	1861	CROOKES
82	Lead	Pb	antiquity	
83	Bismuth	Bi	1753	GEOFFROY (the younger)
84	Polonium	Po	1892	M. CURIE
85	Astatine	At	1940	CORSON, MACKENZIE and SEGRÉ
86	Radon	Rn	1900	DORN, 1908 RAMSAY and GRAY
87	Francium	Fr	1939	PEREY, MLLE MARGUERITE
88	Radium	Ra	1898	P. and M. CURIE
89	Actinium	Ac	1889	DEBIERNE (GIESEL, 1902)
90	Thorium	Th	1828	BERZELIUS
91	Protactinium	Pa	1918	HAHN and MEITNER
92	Uranium	U	1789	KLAPROTH
93	Neptunium	Np	1940	MCMILLAN and ABELSON
94	Plutonium	Pu	1948	SEABORG, MCMILLAN, KENNEDY, WAHL
95	Americium	Am	1944	SEABORG et al., JAMES, MORGAN and GHIORSO
96	Curium	Cm	1944	SEABORG, JAMES and GHIORSO
97	Berkelium	Bk	1949	THOMSON, GHIORSO and SEABORG
98	Californium	Cf	1950	THOMPSON, STREET, GHIORSO and SEABORG
99	Einsteinium	Es	1952	GHIORSO and coworkers
100	Fermium	Fm	1953	GHIORSO and coworkers
101	Mendelevium	Md	1955	GHIORSO and coworkers
102	Nobelium	No	1958	GHIORSO and coworkers
103	Lawrencium	Lw	1961	GHIORSO and coworkers

The periodic system of the elements, developed in 1868 by LOTHAR MEYER and independently by D. J. MENDELEEV in 1869, enabled the latter to predict the properties of some then undiscovered elements and their compounds. Three of these elements were discovered soon after: *scandium* (eka-bor) in 1879 by NILSON in euxenite and gadolinite, *gallium* (eka aluminium) in 1875 by LECOQ DE BOISBAUDRAN in sphalerite and *germanium* (eka silicium) in 1886 by WINKLER in argyrodite from Freiberg, Saxony.

The discovery of the natural radioactive elements began in 1898 when P. CURIE and his wife, MARIE SKLODOWSKA-CURIE, found *radium* and *polonium* in pitchblende from Joachimstal, Bohemia. Their study, which initiated the development of the modern science of nuclear physics, is described by OTTO HAHN in his autobiography, especially the discovery of *protactinium* with LISE MEITNER in 1918. X-ray spectroscopy, a new physical method, based on laws discovered by MOSELEY in 1913, enabled D. COSTER and G. v. HEVESY in 1923 to discover element 72, *hafnium,* in zircon, and I. TACKE (later Mrs. NODDACK) and W. NODDACK in 1925 *rhenium* in columbite. These scientists reported erroneously in the same year that they had discovered element 43, which they named masurium. Element 43 was actually discovered by PERRIER and SEGRÉ in Italy in 1937 and was named *technetium*. It was found in a sample of molybdenum which had been bombarded by deuterons in the Berkeley cyclotron. Searches have been made for the element in terrestrial materials without success, but it has been reported from the spectrum of stars.

The rare earth element 61 was also erroneously reported discovered (illinium, florentium). This active element was isolated by MARINSKY, GLENDENIN and CORYELL in 1945, also by the use of ion-exchange chromatography, from fission products of the neutron bombardment of neodymium and was named *promethium*. Searches for the element in nature have been fruitless. Later on other artificial elements were found. They have the atomic numbers 93—103. This third period in the discovery of the elements we owe to nuclear physics. New elements synthesized in the laboratory by atomic bombardment filled the last gaps in the periodic system and extended it beyond uranium (92). Table 1-1 shows the years of discovery and the discoverers of the elements from 1 to 103.

Man used some elements from the very earliest times. With the exception of iron these metals were not the most abundant but those which could easily be extracted. The most abundant elements were discovered during and after the development of chemistry as a science: 1774 O, 1807 K, Na, 1808 Ca, 1824 Si, 1827 Al. It is evident that a history of geochemistry in the modern sense must begin at this time.

II. The History of Geochemistry

As early as 1838 CHRISTIAN FRIEDRICH SCHÖNBEIN (1799—1869), then professor of chemistry at Basle, used the word "Geochemie". Earlier still in 1821, the eminent Swedish chemist, JÖRN JACOB BERZELIUS (1779—1848), called mineralogy the chemistry of the Earth's crust. From 1845 to 1854 GUSTAV BISCHOF (1792—1870) published his "Lehrbuch der chemischen und physikalischen Geologie" in 3 volumes. A revised English edition appeared under the title: "Elements of Chemical and Physical Geology" (1854—59). The second German edition was published in 1863—1866. It is largely thanks to BISCHOF that chemistry was introduced into geology. He was a neptunist and tried to explain all problems by the influence of water at low temperature, and he was the first to consider the influence of living organisms on the distribution of the elements.

On the other hand, the geochemistry of volcanic products was promoted by ELIE DE BEAUMONT (1798—1874), who emphasized high-temperature processes under the influence of magma. In his "Note sur les Emanations Volcaniques et Métallifères", 1847, he treats the abundance of elements in different rocks of the Earth's crust and in meteorites. His table of the distribution of elements (Fig. 1-2) gives only the presence of elements in different rocks, but no quantitative data. At this time chemical analysis was not sufficiently advanced to detect, for example, B, Ni and Zr in basic rocks nor Sr and Ba in granites.

In the following decade the quarrel between neptunists and plutonists subsided. This was the time (1851—1858) when the microscopic study of thin sections was introduced and a wealth of new observations stimulated petrographic studies and theories. In the petrographical work of this time, e.g. in the books by ROSENBUSCH and ZIRKEL, problems of geochemistry are scarcely discussed.

The famous chemist, R. W. BUNSEN (1811—1899), already mentioned as one of the founders of spectrography, was also interested in the chemistry of rocks. In 1851 he spent three and a half months in Iceland with the German mineralogist, W. SARTORIUS VON WALTERSHAUSEN, and the French mineralogist, A. L. O. DES-CLOISEAUX. The fruits of BUNSEN's stay on the volcanic island were his theory of geysers, the term "pneumatolytic" and a theory now obsolete explaining volcanic rocks by the mixing of two melts: "normalpyroxenitic" and "normaltrachytic". CLARKE (1925) writes: "Although BUNSEN was the first to show that a magma is really a solution, little attention was paid to this consideration until Lagorio ... (1887)". The old argument, still revived sometimes, that granite could not have crystallized from a melt, because quartz, which has the highest melting point, crystallized after feldspar and mica, which have lower melting points, was already disproved in 1860 by BUNSEN.

Improvements in chemical analysis furnished new data on elements in rocks. We owe, for instance, to FRIDOLIN VON SANDBERGER (1826—1898) much information on trace elements in minerals and sediments accompanying ore veins. He was a pupil of BISCHOF and defended the theory of lateral secretion, i.e. that the contents of a vein or lode are derived from the adjacent wall rock. G. FORCHHAMMER, professor of chemistry and mineralogy at Copenhagen, 1784—1865, tested 1835 rocks for trace elements and began in 1845 to analyse seawater (final publication 1865). He found only very slight variations in the interelemental ratios, if analyses for certain marginal seas, such as the Baltic, were excluded. He mentioned also already the elements Br, B, Co, Ni, Fe, Mn, Al, Sr, Ba in seawater and Ag, Cu, Pb, Zn, B, Sr, Ba, I in marine organisms.

Another branch of geochemistry also developed at this time. Already LEIBNIZ (1646—1716) had suggested in his "Protogaea" (published 1749) that chemical experiments should be applied in Earth science, but he was in advance of his time. The synthesis of minerals and rocks did not flourish until the middle of the 19th century. A comprehensive report was given by FOUQUÉ and MICHEL LEVY in 1888; it shows that numerous minerals and some rocks had been made artificially. It may

Tableau de la distribution des corps simples dans la nature.

	1 Corps les plus repandus sur la surface du globe.	2 Roches volcaniques actuelles.	3 Roches volcaniques anciennes.	4 Roches basiques.	5 Granites.	6 Filons stannifères.	7 Filons ordinaires et géodes.	8 Sources minérales.	9 Émanations volcaniques.	10 Radicaux natifs.	11 Aérolithes.	12 Corps organisés.
1 Potassium	*	*	*	*	*	*	*	*	*	...	*	*
2 Sodium	*	*	*	*	*	*	*	*	*	...	*	*
3 Lithium	...	...	...	...	*	*	...	*				
4 Barium	...	...	...	...	...	*	*	*				
5 Strontium	...	...	...	...	...	...	*	*				
6 Calcium	*	*	*	*	*	*	*	*	*	...	*	*
7 Magnesium	*	*	*	*	*	*	*	*	...	...	*	*
8 Yttrium	...	...	...	...	*	*						
9 Glucinium	...	...	...	...	*	*	*					
10 Aluminium	*	*	*	*	*	*	*	*	*	...	*	*
11 Zirconium	...	...	...	...	*	*						
12 Thorium	...	...	...	...	*							
13 Cerium	...	...	...	...	*	*						
14 Lanthane	...	...	...	...	*	*						
15 Didymium	...	...	...	...	*	*						
16 Urane	...	...	...	...	*	*	*					
17 Manganèse	*	*	*	*	*	*	*	*	*	...	*	*
18 Fer	*	*	*	*	*	*	*	*	*	...	*	*
19 Nickel	...	...	...	...	...	*	*	...	...	...	*	
20 Cobalt	...	...	...	*	*	*	*	...	*	...	*	
21 Zinc	...	...	...	*	*	*	*					
22 Cadmium	...	...	...	...	...	*	*					
23 Étain	...	...	...	...	*	*	*					
24 Plomb	...	...	...	*	*	*	*	...	*	*	*	
25 Bismuth	...	...	...	*	*	*	*	...	...	*		
26 Cuivre	...	...	...	*	*	*	*	* ?	*	*	*	
27 Mercure	...	...	...	...	...	...	*	...	...	*		
28 Argent	...	...	...	*	*	*	*	...	...	*		
29 Palladium	...	...	...	*	* ?	*	*	...	...	*		
30 Rhodium	...	...	...	*	...	...	...	...	...	*		
31 Ruthenium	...	...	...	*	...	...	...	...	...	*		
32 Iridium	...	...	...	*	...	...	...	...	...	*		
33 Platine	...	...	...	*	...	...	*	...	...	*		
34 Osmium	...	...	...	*	...	...	...	...	...	*		
35 Or	...	...	...	*	*	*	*	...	...	*		
36 Hydrogène	*	*	*	*	*	*	*	*	*	...	*	*
37 Silicium	*	*	*	*	*	*	*	*	*	...	*	*
38 Carbone	*	...	...	...	*	*	*	*	*	*	*	*
39 Bore	...	...	...	...	*	*	*	*	*			
40 Titane	...	*	*	*	*	*	*					
41 Tantale	...	...	...	...	*	*						
42 Nobium	...	...	...	...	*	*						
43 Pelopium	...	...	...	...	*	*						
44 Tungstène	...	...	...	...	*	*						
45 Molybdène	...	...	...	...	*	*	*					
46 Vanadium	...	...	...	...	...	*	*					
47 Chrome	...	...	...	*	*	*	*	...	...	...	*	
48 Tellure	...	...	...	...	...	*	*	...	...	*		
49 Antimoine	...	...	...	...	...	*	*	...	...	*		
50 Arsenic	...	...	...	*	*	*	*	*	*	*		
51 Phosphore	*	...	*	*	*	*	*	*	...	...	*	*
52 Azote	*	...	...	...	...	...	...	*	*	*	*	*
53 Sélénium	...	...	...	...	...	*	*	...	*	*	*	
54 Soufre	*	*	*	*	*	*	*	*	*	*	*	*
55 Oxygène	*	*	*	*	*	*	*	*	*	*	*	*
56 Iode	...	...	...	...	...	...	*	*				
57 Brome	...	...	...	...	...	...	*	*				
58 Chlore	*	*	*	*	*	*	*	*	*	...	*	*
59 Fluor	*	*	*	*	*	*	*	*	...	...	...	*
	16	14	15	30	42	48	43	24	19	20	21	16

Fig. 1-2. ELIE de BEAUMONT's (1847) table of the frequency of elements in different rocks and in meteorites

be noted that orthoclase was synthesized as early as 1877 from melts with tungstic acid and sodium or potassium phosphates (1880) and hydrothermally by FRIEDEL and SARASIN in 1879. But, of course, these syntheses were made without benefit of the laws of physical chemistry, so the application of the experiments was in many cases doubtful or misleading.

Cosmochemistry, too, begins about this time. Already in 1794 CHLADNI had explained meteorites as solid debris, reaching the Earth from outer space. The development of spectral analysis by KIRCHHOFF and BUNSEN in 1859 pointed the way to an exploration of the abundance of the elements in the stars.

Spectrographic analyses were applied to the quantitative determination of the content of elements in rocks. In 1897 Sir WALTER N. HARTLEY and H. RAMAGE showed that gallium, indium, thallium, rubidium and silver are widely diffused in rocks and minerals. G. EBERHARD (1908, 1910) emphasized the fact that Si, La and Y are very common constituents of minerals and rocks, even though they occur in low concentrations. Also the Russian scientist, W. J. VERNADSKY (1863—1945), one of the founders of modern geochemistry, demonstrated the worldwide distribution of In, Tl, Ga, Rb, Cs. A German edition of his book: "Geochemie in ausgewählten Kapiteln", translated from the Russian by E. KORDES, appeared in 1930. The chemical study of the biosphere was among of objects main interest to VERNADSKY, whereas his younger contemporary and compatriot, A. E. FERSMAN (1883—1945), focussed his activity on the geochemical investigation of the lithosphere. He tried to find the ultimate causes of the distribution of the elements in their atomic structure.

FRANK W. CLARKE (1847—1931) published in 1908 "The Data of Geochemistry". This book is a critical compilation of analyses of rocks and minerals which he evaluated to explain the chemical processes occuring in the crust of the Earth. He was the first since SCHÖNBEIN to use the word *geochemistry*, apparently without knowing of his predecessor. The fifth edition of his book appeared in 1924, the sixth, with M. FLEISCHER as editor, is beginning to appear in parts from 1962 on.

Starting in 1923, V. M. GOLDSCHMIDT (1888—1947) studied the quantitative distribution of the elements with particular attention to the rare elements in numerous investigations in Oslo and later in Göttingen. His book: "Geochemistry" appeared in 1954, seven years after his death, and was edited by ALEX. MUIR.

Most of the older data were obtained by wet chemical analysis. From the turn of the century, new methods brought new and more accurate data, e.g. optical spectrometry (spark, arc, flame inclusive atomic absorption) X-ray spectrometry, neutron activation and γ-spectrometry, isotope dilution and mass-spectrometry microprobe-analysis.

The scope of geochemistry was enlarged by the development of crystalchemistry, mainly by V. M. GOLDSCHMIDT (1888—1947) in Oslo and Göttingen. C. E. TILLEY wrote in 1948 about his geochemical work: "In this work GOLDSCHMIDT had succeeded in solving the basic geochemical problem that he had set himself at the start of the investigation, namely — What are the characteristics of atoms that determine their distribution in the crust of the Earth? He showed that for the most part these were ionic size, charge and polarizability. This work, combined with the knowledge of silicate structures derived from the Bragg school, explained with the utmost elegance and simplicity the puzzling rules of ionic substitution in minerals. The older mineral chemists had used the analogies of organic chemistry in putting forward formulae for the complex minerals such as the micas and feldspars. GOLDSCHMIDT showed that they were on the wrong track and in fact a much simpler explanation was that of the piling of spheres in different size. The possibility of substituting any one atom by another even of different charge was possible so long as ionic sizes did not diverge too much."

Another approach to an understanding of the laws of distribution of elements in the Earth is physical chemistry. The famous physical chemist, J. H. VAN T'HOFF (1852—1911) initiated a new period of geochemistry when in 1898 with numerous coworkers he began to investigate in Berlin the processes by which salt deposits are formed. These studies stimulated further work, e.g. in 1908 H. E. BOEKE investigated the substitution of Br in various salt minerals and in 1911 the substitution of Fe.

The systematic experimental study of the physical chemistry of igneous rocks began with the foundation of the Geophysical Laboratory of the Carnegie Institution in Washington in 1902. "Invaluable information has been gained concerning equilibria in silicate systems at elevated temperatures" (V. M. GOLDSCHMIDT, 1954). The earlier results of these important investigations were summarized by N. L. BOWEN in his book: The Evolution of Igneous Rocks, 1928[1]. In later years the laboratory also worked on hydrothermal and high pressure systems.

An entirely new branch of geochemistry came from nuclear physics after H. BEQUEREL detected the radiation of uranium in 1896 as well as SODDY in 1910, I. I. THOMPSON in 1913 and F. W. ASTON in 1919 demonstrated the existence of isotopes. Already in 1907 BOLTWOOD used the different Pb/U ratios of minerals to estimate their age. A. HOLMES made the first suggestion about the use of stable isotopes in geological problems. In 1932 he proposed that the possible variations in the isotopic constitution of calcium be used in petrogenic research. In the same year, H. UREY discovered deuterium. He too demonstrated in 1947 the feasibility of using the O^{16}/O^{18} ratio to determine palaeotemperatures.

III. Definitions of Geochemistry

SCHÖNBEIN (1838) not only coined the word "Geochemie" but also mapped out the program for this science. He writes: "We have to study in the greatest detail the properties of all geological formations, we have to find out the relationships of their chemical and physical qualities and their chronological sequence as exactly as possible and we have at the same time to compare carefully the products of the chemical forces active at the present time with the organic substance of the past ('Urwelt')." SCHÖNBEIN did not influence the science of his time.

The next definition to be published is that of F. W. CLARKE. He writes in the 5th edition (1925) of his book: "Each rock may be regarded, for present purposes, as a chemical system in which, by various agencies, chemical changes can be brought about. Every such change implies a disturbance of equilibrium, with the ultimate formation of a new system which, under the new conditions, is itself stable in turn. The study of these changes is the province of geochemistry. To determine what changes are possible, how and when they occur, to observe the phenomena which

[1] Most of which was contributed by himself.

attend them, and to note their final results are the functions of the geochemist. Analysis and synthesis are his two chief instruments of research, but they become effective only when guided by a broad knowledge of chemical principles which correlate the data obtained and extract from the evidence its full meaning."

V. I. Vernadsky wrote in the 1924 French and in the 1930 German edition: "Geochemistry studies the chemical elements, i.e. the atoms, in the crust of the Earth and as far as possible in the entire Earth. It studies their history, their distribution in space in the present and the past. Geochemistry can be distinguished strictly from mineralogy because the latter considers only the development of the compounds of the atoms, the molecules and crystals."

A. E. Fersman's 1922 definition in the translation by Tomkeieff (1944), cited by Rankama and Sahama states: "The purpose of geochemistry is the study of the element-atom in the conditions prevailing in the Earth's crust (as well as in the parts of the Cosmos accessible to our exact observations). Geochemistry studies: (a) the quantitative distribution of the chemical elements in the Earth's crust and their dispersion and local concentration; (b) the combinations of different elements in the different parts of the Earth's crust and their distribution in space and time under the influence of different chemical processes; (c) the migration of elements and the laws of such migration as determined by the different thermodynamic conditions of their environment; and (d) the behavior of chemical elements either in the environment of the Earth's crust or as compounds and particularly as crystals. This may be expressed even more simply: geochemistry studies the history of chemical elements in the Earth's crust and their behavior under different thermodynamic and physico-chemical natural conditions."

Goldschmidt's definition of 1933, translated by B. Mason in 1960, reads as follows: "The primary purpose of geochemistry is on the one hand to determine quantitatively the composition of the Earth and its parts, and on the other to discover the laws which control the distribution of the individual elements. To solve these problems the geochemist requires a comprehensive collection of analytical data on terrestrial material, such as rocks, waters, and the atmosphere; he also uses analyses of meteorites, astrophysical data on the composition of other cosmic bodies, and geophysical data on the nature of the Earth's interior. Much valuable information has also been derived from the laboratory synthesis of minerals and the investigation of their mode of formation and their stability conditions." In 1954 he completed his definition by the indication to the significance of the isotopes.

Without the study of the distribution of the elements, there will be no recognition of laws governing the distribution. So we can shorten Goldschmidt's definition as follows: "Geochemistry is concerned with the laws governing the distribution of the chemical elements and their isotopes throughout the Earth."

Rankama and Sahama (1950) agree with Goldschmidt's definition but propose to distinguish between geochemistry and chemical geology. They wrote about the geochemist: "His methods and, as may be especially emphasized, his problems are those of a chemist or a physical chemist . . . , conversely, the chemical geologist examines his problems from the viewpoint of a geologist." This distinction has not found many followers.

On the other hand there can sometimes be heard the claim that petrology and geochemistry are identical. There is no doubt that the fields of petrology and geochemistry overlap in many cases. But studies of the fabric and mechanical behavior of rocks are tasks for the petrologist, while the study of isotopes is mostly non-petrological. It may be mentioned incidentally, that the term "petrology", was first coined by the French geologist, DUROCHER, in 1857.

Nature itself is indivisible, it has no provinces or territories; man has made the demarcations and has had to do so for practical purposes. So expediency should determine the demarcation lines between the various fields of science.

References

ADAMS, FR. D.: The birth and development of the geological sciences. Baltimore: Williams & Wilkins Co., 1938.

BEAUMONT, ELIE DE: Note sur les émanations volcaniques et métallifères. Bull. soc. géol. France, II. Ser., 1249—1333 (1847).

BISCHOF, G.: Lehrbuch der chemischen und physikalischen Geologie, II. Aufl. Bonn: Marcus 1863—1866.

BUNSEN, R.: Über die Processe der vulkanischen Gesteinsbildungen Islands. Poggendorffs Ann. physik. Chem. **83**, 197 (1851).

— Über die Bildung des Granites. Z. deut. geol. Ges. **13**, 61 (1860).

CLARKE, F. W.: Data of geochemistry, V. e. U.S. Survey Bull. 770 (1924).

CORRENS, C. W.: Über die Entwicklung und Situation in der Geochemie in der Bundesrepublik. Fortschr. Mineral. **41**, 92—98 (1963).

DUROCHER, J.: Essai de pétrologie comparée. Ann. mines, 5th ser. **11**, 217 (1857).

FLEISCHER, M. (ed.): Data of geochemistry (sixth ed.) U.S. Geol. Survey Profess. Papers 440 seit 1962.

FORCHHAMMER, G.: On the composition of sea-water in the different parts of the ocean. Phil. Trans. Roy. Soc. London **155** (1865).

FOUQUÉ, F., et M. LÉVY: Synthèse des minéraux et des roches. Paris: Masson & Cie. 1882.

GOLDSCHMIDT, V. M.: Grundlagen der quantitativen Geochemie. Fortschr. Mineral. Krist. Petrog. **17**, 112 (1933).

— Geochemistry. Oxford: Clarendon Press 1954.

HAHN, O.: Vom Radiothor zur Uranspaltung. Braunschweig: Fr. Vieweg & Sohn 1962.

HODGMAN, CH. D. (ed.): Handbook of chemistry and physics. 46th edit. Cleveland, Ohio: The Chemical Rubber Co. 1965/66.

HOOYKAAS, B.: Elementenlehre und Atomistik im 17. Jahrhundert. In: Die Entfaltung der Wissenschaft. Veröff. d. Joachim Jungius-Ges. der Wissenschaften, Hamburg. 1957.

LEIBNIZ, G. W.: Übers. von W. v. ENGELHARDT 1949, Protogaea. Stuttgart: Kohlhammer

MASON, B.: Principles of geochemistry, III. ed. New York: John Wiley & Sons 1966.

RANKAMA, K., and TH. G. SAHAMA: Geochemistry. Chicago: Chicago University Press 1950.

SANDBERGER, FR.: Untersuchungen über Erzgänge. Wiesbaden: Kreidel 1882.

SCHÖNBEIN, C. F.: Über die Ursache der Farbenveränderung, welche manche Körper unter dem Einflusse der Wärme erleiden. Poggendorffs Ann. physik. Chem. **45**, 263 (1838).

TILLEY, C. E.: Victor Moritz Goldschmidt. Obituary Notices of Fellows of the Royal Society **6** (1948).

VERNADSKY, W. J.: Geochemie in ausgewählten Kapiteln. Aut. Übersetzung von Dr. E. KORDES. Leipzig: Akad. Verlagsges. 1930.

WALLACH, O. (ed.): Briefwechsel zwischen J. BERZELIUS und F. WÖHLER. Leipzig: W. Engelmann 1901.

WEDEPOHL, K. H.: Geochemie. Samml. Göschen. Berlin: W. de Gruyter & Co. 1967.

WEEKS, M. E.: Discovery of the elements (sixth edit.). 910 pp. Easton, Pa.: J. Chem. Educ. 1956. Sec. print 1960.

Chapter 2

Josef Zemann

CRYSTAL CHEMISTRY

Introduction

There exist many good introductions and textbooks in the field of crystal chemistry (see below), so it was considered more useful not to write one more text of this kind, but rather to stress in relatively more detail some chapters which offer pitfalls or which tend to be neglected. For the many phenomena in crystal chemistry which have now become general knowledge, no literature references are given. The papers quoted in the text therefore represent a highly individual choice and serve as a rule only to illustrate the examples.

The Classics

Bragg, W. L.: Atomic structure of minerals. London 1937: Humphrey Milford. New edition: Bragg, W. L., and G. F. Claringbull: Crystal structures of minerals (The crystalline state, Vol. IV.). London: Bell & Sons 1965.

— The crystalline state. Vol. I. A general survey. London: Bell & Sons 1933. Reprinted with corrections 1949.

Goldschmidt, V. M.: Geochemische Verteilungsgesetze der Elemente. VII.: Die Gesetze der Krystallochemie. VIII.: Untersuchungen über Bau und Eigenschaften von Kristallen. Skrifter Norske Videnskaps-Akad. Oslo, Math.-Naturv. Kl. 1926, Nr. 2 und 1927, Nr. 8.

— Kristallchemie. In „Handwörterbuch der Naturwissenschaften", Bd. 5. Jena: Fischer 1934.

Hassel, O.: Kristallchemie. Dresden u. Leipzig: Steinkopff 1934.

Some Textbooks on Crystal Chemistry

Addison, W. E.: Structural principles in inorganic compounds. New York: Wiley & Sons 1961.

Brandenberger, E., u. W. Epprecht: Röntgenographische Chemie. Basel u. Stuttgart: Birkhäuser, 2nd edition 1960.

Evans, R. C.: An introduction to crystal chemistry. Cambridge: University Press, 2nd edition 1964 and First paperback edition 1966.

Fyfe, W. S.: Geochemistry of solids. New York: McGraw-Hill 1964.

Garner, E.: Chemistry of the solid state. London: Butterworths Scientific Publications 1955.

Halla, F.: Kristallchemie und Kristallphysik metallischer Werkstoffe. Leipzig: Barth, 3rd edition 1957.

Hiller, J.-E.: Grundriß der Kristallchemie. Berlin: De Gruyter 1952.

Hume-Rothery, W.: Atomic theory for students of metallurgy. London: Institute of Metals (Monograph 3), 4th reprinted edition with corrections.

—, and G. V. Raynor: The structure of metals and alloys. London: Institute of Metals (Monograph 1), 4th edition 1962.

Kleber, W.: Kristallchemie. Leipzig: Teubner 1963.
Krebs, H.: Grundzüge der anorganischen Kristallchemie. Stuttgart: Enke 1968.
Machatschki, F.: Grundlagen der allgemeinen Mineralogie und Kristallchemie. Wien: Springer 1946.
Niggli, P.: Grundlagen der Stereochemie. Basel: Birkhäuser 1945.
Stillwell, Ch. W.: Crystal chemistry. New York and London: McGraw-Hill 1938.
Wells, A. F.: Structural inorganic chemistry. Oxford: Clarendon, 3rd edition 1962.
Winkler, H. G. F.: Struktur und Eigenschaften der Kristalle. Berlin: Springer, 2nd edition 1955.
Zemann, J.: Kristallchemie. Sammlung Göschen, Bd. 1220/1220a. Berlin: de Gruyter 1966.

Textbooks on the Chemical Bond

Coulson, C. A.: Valence. London: Oxford Univ. Press, 2nd edition 1961.
Orgel, L. E.: An introduction to transition-metal chemistry: ligand-field theory. New York: John Wiley 1960.
Pauling, L.: The nature of the chemical bond. New York: Cornell Univ. Press, 3rd edition 1960.
Schläfer, H. L., u. G. Gliemann: Einführung in die Ligandenfeldtheorie. Frankfurt am Main: Akademische Verlagsgesellschaft 1967.

Tables and Reference Series

Donnay, J. D. H. (editor): Crystal data. Washington: Amer. Cryst. Ass. (Monograph.-Nr. 5), 2nd edition 1963.
"International tables for X-ray crystallography", Vol. 1: Symmetry groups. Birmingham: Kynoch 1952 (Geometry of space groups).
"Landolt-Börnstein" Zahlenwerte und Funktionen, Vol. I, Part IV: Kristalle. Berlin-Göttingen-Heidelberg: Springer 1955.
Pearson, W. B.: A handbook of lattice spacings and structures of metals and alloys. London: Pergamon 1958.
Strunz, H.: Mineralogische Tabellen. Leipzig: Akademische Verlagsgesellschaft, 4th edition 1966. (Lattice parameters and space groups of minerals.)
"Strukturbericht", Vol. 1—7. Leipzig: Akademische Verlagsgesellschaft 1931—1943. Continued by "Structure reports"[2], Vol. 8—23. Utrechts: Oosthoek. (References of all determinations of structures of crystals. At time up to the literature of 1959; will be continued.)
"Tables of interatomic distances and configuration in molecules and ions." London: The Chemical Society, Special publication No. 11, 1958, and Supplement, London: The Chemical Society, Special publications No. 18, 1965.
Wyckoff, R. W. G.: Crystal structures. New York and London: Interscience Publishers, 5 volumes 1948—1960. (Vol. 1—3 and 5 of the new edition already published; will be continued.)

Like many branches of science which are in a state of rapid development, the field of crystal chemistry is not altogether clearly defined. One of its main tasks is to find correlations between the atomic arrangement of a crystal species and its chemical composition — the ultimate aim being the prediction of the crystal structure from the known chemical composition at given temperature and pressure. But crystal chemistry is not restricted to this special field of stereochemistry. Other topics of interest include the phenomenon of analogous crystal structures of compounds with different chemical composition (isotypism; example: NaCl, KF, MgO, PbS, WC ... which all have the rocksalt-type of structure), the phenomenon of different crystal structures of one and the same chemical substance (polymorphism[1]; example: calcite—

[1] Polytypism is a special kind of polymorphism; example: the different forms of wurtzite.
[2] Abbreviation "SR" used in sections A of volume II of this handbook.

aragonite — vaterite) the mechanism of polymorphic transitions, solid solutions (mixed crystals), diffusion in the solid state, reactions in the solid state, etc.

The importance of crystal chemistry for an understanding of the chemistry of the crust of the Earth was clearly recognized in the third decade of this century by V. M. GOLDSCHMIDT and others. A great deal has been done since in this field, but it is still in active development. In this short survey only those parts of crystal chemistry which are of special importance to geochemistry will be dealt with briefly.

I. The Geometrical Framework of Crystal Structures

An "ideal crystal" is defined as a body in which the atoms (ions) form a pattern exactly periodic in the three dimensions of space. Although this definition implies abstractions from the crystal as a physical body (e.g. a crystal according to this definition cannot have boundaries), it provides a very important geometrical framework for the atomic arrangement in crystals. This derives from the fact that the deviations from exact periodicity which are caused by strain deformations in solid solutions and by thermal vibrations are very small compared with the identity periods. More serious deviations, like dislocations, usually concern only a small fraction of the crystal.

The *description of a crystal structure* comprises:

a) The Lattice Constants of the Unit Cell and the Space Group

To describe the whole periodic pattern of the ideal crystal structure, it is sufficient to give the description of a parallelepiped of minimum size having the property that the whole structure can be thought of as being built from such cells by translations which coincide in directions with, and have the same lengths as the edges of the cell. For reasons of symmetry sometimes the chosen parallelepiped is not the smallest possible, but a larger, so-called multiple cell (one-face centred, all-faces centred, body-centred cell; rhombohedral centred hexagonal cell).

For crystals with low symmetry, the unit cell is not uniquely defined and additional conventions are necessary to enable different workers to choose the same cell, which of course is highly desirable. The general designations of the lengths of the edges of the unit cell are: a_0, b_0, c_0 (also only a, b, c), and of its angles: α, β, γ. The unit of length is the Ångström (1 Å $= 10^{-8}$ cm). The lattice constants of minerals measure as a rule between 3 and 25 Å; larger values are rather rare.

Any structure of an "ideal" crystal belongs to one of the 230 periodic combinations of symmetry elements in 3-dimensional space, i.e. to one of the 230 space groups[1]. Among them there are 11 pairs which differ only in the "hand" of the screw-axes. This causes that in some cases the two kinds of a crystal species which belong to an enantiomorphic class have different space groups (e.g. right- and left-

[1] Especially for magnetically ordered crystal structures, an extension of the usual space groups to "black-white space groups", and generally "coloured space groups" has proved to be useful (cf. SHUBNIKOV and BELOV: Coloured Symmetry, 1964).

quartz), while in other cases the right and the left crystal belong to the same space group (example: right- and left-epsomite, $MgSO_4 \cdot 7H_2O$). An extensive description of the space groups is given in the *International Tables for X-Ray Crystallography*.

The designation of a space group is given by the Hermann-Mauguin symbol, the Schoenflies symbol, or sometimes also simply by the number of the space group in the arrangement of the International Tables. Of these, the Hermann-Mauguin symbol gives the most detailed information, and with some few exceptions it is possible to derive from it the whole set of symmetry elements of the space group. The non-crystallographer should, however, remember that this derivation is not always easy, and he is advised to consult the International Tables. It is also important to realize that for the lower symmetries the symbol for the space group varies with the chosen orientation of the unit cell. E.g. the Hermann-Mauguin symbol of the space group in which olivine crystallizes can be *Pnma*, *Pbnm*, *Pmcn*, *Pcmn*, *Pmnb*, and *Pnam*, corresponding to the six possible orientations of the *X*-, *Y*- and *Z*-axes parallel to the axes of the unit cell. Of these six symbols, only *Pnam* is given in the main table of the International Tables; the other symbols are found only in the auxiliary tables of the same book.

The Schoenflies symbol combines the crystal class of the space group under consideration and the number of the space group in this class in a short symbol. For example, the Schoenflies symbol of the space group of olivine is D_{2h}^{16}. The Schoenflies symbol does *not* give the orientation of the axes, and therefore it does not allow a complete description of the symmetry elements for crystals which belong to crystal classes of low symmetry. But as this symbol readily enables the full description of the space group to be found in the International Tables, it is often given in addition to the Hermann-Mauguin symbol. The Schoenflies symbol is also appropriate to designate a space group if the orientation of the symmetry elements to the crystallographic axes is of no interest.

b) The Content of the Unit Cell

In well ordered crystals, the material content of the unit cell is a multiple of the formula unit. It is usually given either in the form: $Z=n$, or with a chemical formula which gives the content of the unit-cell, e.g. for rocksalt; NaCl: $Z=4$, or cell content: Na_4Cl_4.

c) The Positions of the Atoms and their Parameters

The designation of the occupied positions follows exclusively the nomenclature of the International Tables. The parameters of the atoms (x, y, z) are given in fractions of the corresponding cell edges (a_0, b_0, c_0). Standard deviations for the parameters are now given in the majority of papers. If the position, as it usually does, contains several symmetrically equivalent atoms, the parameters are given for only one of them. — Modern crystal structure papers very often give individual isotropic or anisotropic "temperature factors". The non-crystallographic reader should realize that the number quoted has this meaning only, if absorption and anomalous dispersion have been carefully corrected for, and if the crystal does not contain disorder (e.g. local strains in solid solutions).

Of the many deviations from the "ideal crystal" which are found in real crystals, one should be mentioned here, because it leads to atomic arrangements in space the geometry of which is closely related to but not identical with that of space groups. There are cases in which sheets of a structure (which are periodic in two dimensions and belong to one of the 80 "plane space groups") are not stacked in such a way that a 3-dimensional periodic pattern arises, but according to less restrictive rules. A typical case is illustrated in Fig. 2-1. All pairs of neighbouring sheets are congruent, but when going from sheet n to sheet $n+1$ one has two possibilities. There

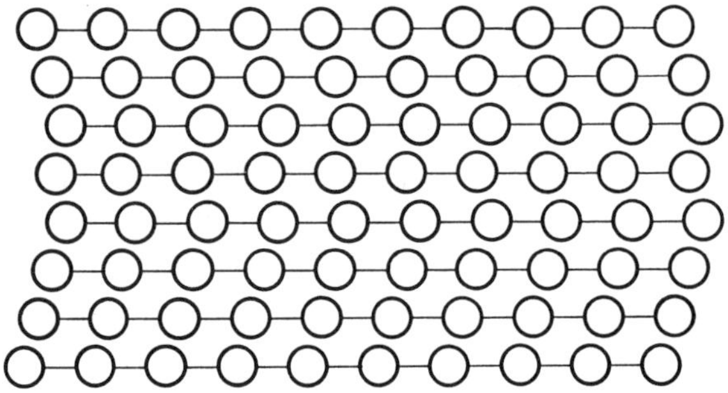

Fig. 2-1. 2-dimensional example of stacking-disorder

may occur some disorder in the "sequence of stacking", and then one speaks of 1-dimensional disorder or of (disordered) OD-structures of sheets. Such atomic arrangements do not belong to one of the usual space groups — but the repetition follows the axioms of groupoids (cf. DORNBERGER-SCHIFF, 1964). Mineralogical examples are found e.g. in the group of the micas. Cases where rods (fully ordered in one dimension) are similarly packed in two dimensions — structures with 2-dimensional disorder — are much rarer; an example is artinite (JAGODZINSKI, 1965).

II. Some Geometrical Problems of Crystal Structures not Related to Space Group Symmetry

In many inorganic structures, the constituent particles and the forces between them have very nearly spherical symmetry. This is true e.g. for part of the intermetallic compounds and for crystals built up from ions, the electron configuration of which corresponds to that of inert gases. In such cases, consideration of the possible packings of spheres is of great value. Unfortunately, relatively few contributions to this problem have been made from general geometrical aspects; for example, it has not even been proved that it is impossible to find in 3-dimensional space a periodic packing of equal spheres which is denser than the well known varieties of "closest packing".

Also many other geometrical problems of crystal structures are far from being solved. As an example, consider the difficulties in the definition and nomenclature of coordination polyhedra. There is general agreement that the concept of the coordination polyhedron (or the coordination figure) is very important in stereochemistry because it gives the geometry of the "surroundings" of the atoms (ions) in the crystal. A clear, but formal definition would be: The coordination polyhedron around an atom is the convex polyhedron which can be constructed by connecting the centres of the nearest neighbours by planes. One thereby obtains for rocksalt the coordination number 6 around Na^+ and an octahedron as the coordination polyhedron, for sphalerite the coordination number 4 around Zn and as the coordination polyhedron a tetrahedron etc. But if one adheres to this definition too strictly, the coordination number around Ti in rutile is only four, while it seems to be natural to designate the coordination as (distorted) octahedral with 6 "equal" Ti-O distances. The difference of the Ti-O bond lengths is, in fact, only 2% (4 at 1.944 ± 0.004 Å, 2 at 1.988 ± 0.006 Å; BAUR, 1956); the deviations of the O-Ti-O bond angles compared with those of the ideal octahedron do not exceed 10°. — Another, more complicated example is $PbCl_2$ (cotunnite). The nearest Pb-Cl distances measure (in Å): 2.86 (1 ×), 2.90 (2 ×), 3.06 (1 ×), 3.08 (3 ×), and 3.64 (2 ×) — the next larger Pb-Cl distance is 4.56 Å, which is larger than the smallest Pb-Pb distance (4.54 Å). The standard deviation of the bond lengths is ca. 0.05 Å (cf. SAHL, 1963). What is now the coordination number around Pb? Certainly, it is not larger than nine — but one can also choose 7 (perhaps even 3!). In such cases the coordination number depends upon the decision up to which limit deviations from the minimum bond length are considered to be non-essential; often ca. 15% is used for this purpose.

Not only the coordination number, but also the designation of the coordination figures presents problems. Clear cases occur only when all the corners of the coordination polyhedron are equivalent from symmetry (and there are no further neighbours at almost the same distance). This case has been treated exhaustively by DONNAY et al. (1964). But very often it seems to be natural to consider the coordination figure as composed not of only one set of symmetrically equivalent atoms, but of two or more, for which the distances to the coordinated atom are very similar. One example, i.e. rutile, has already been mentioned. Another very typical example is provided by the crystal structure of grossularite, $Ca_3Al_2[SiO_4]_3$, in which Ca^{2+} occupies the position 24(c) with point symmetry 222 in space group Ia3d—O_h^{10}. Each Ca^{2+}-ion is surrounded by two sets of oxygens — the four smaller Ca-O distances measure 2.326 Å, the four larger ones 2.483 Å — both determined with an error of ± 0.003 Å (cf. the refinement of this structure with X-rays and neutrons by PRANDL, 1966). Both sets form a rhombic disphenoid — one is steep, the other flat. But as the difference in the Ca-O distances is smaller than 7%, it seems to be justified to consider the Ca^{2+}-ions to be 8-coordinated. The coordination polyhedron is then bounded by 12 irregular triangles; but of the 18 edges of this body, 12 are definitely shorter than the rest, and as these form the edges of an distorted cube, it is also justified to call the coordination distorted cubic.

Some highly symmetrical coordinations are given in Fig. 2-2.

The problems which arise in the definition of coordination numbers and coordination polyhedra are among the greatest obstacles for a well established symbolism of crystal chemical formulas. It is now fairly common to put in such formulas the coordination number in square brackets above the chemical symbol of the element. The Commission on Mineral Data of the International Mineralogical Association

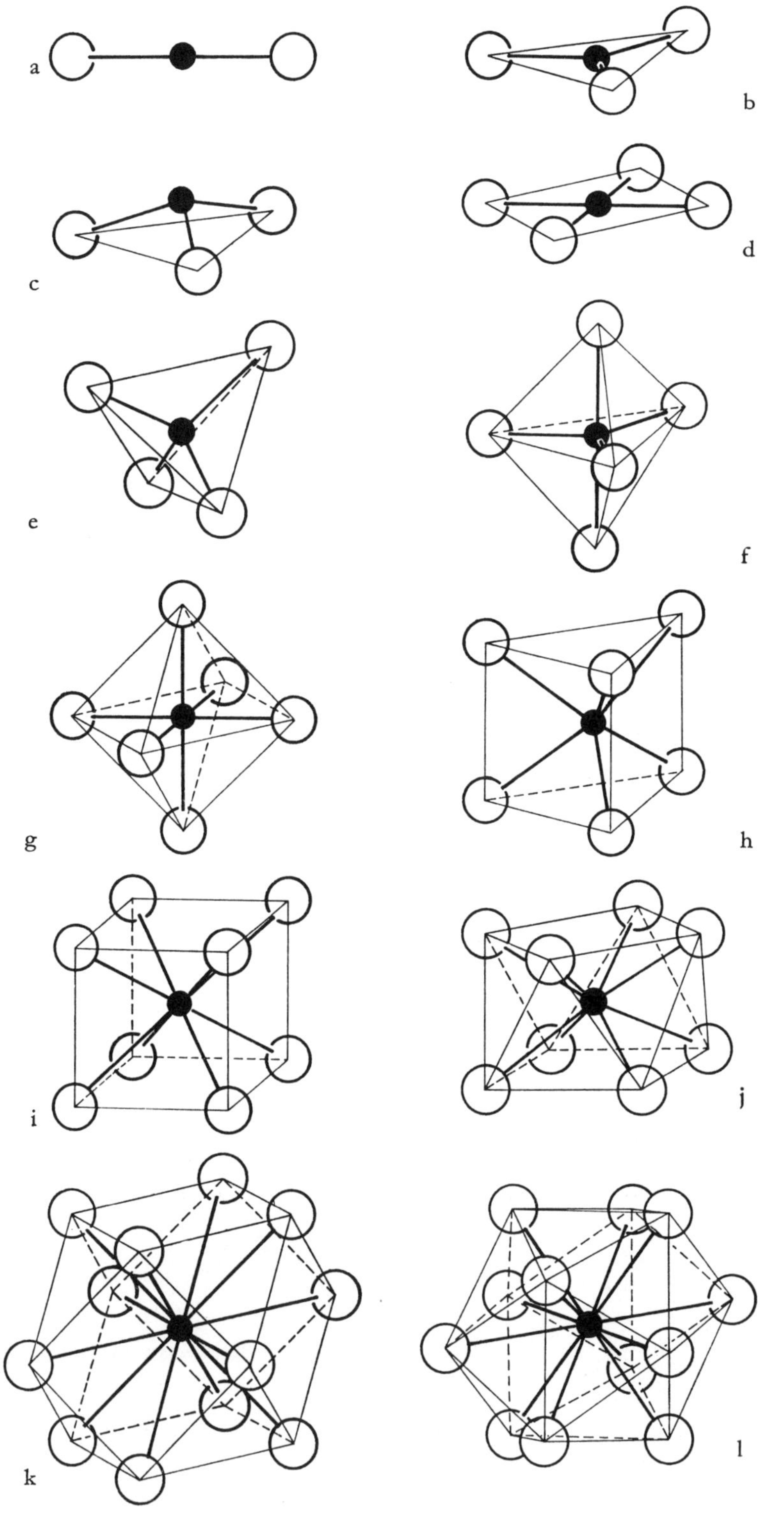
a
b
c
d
e
f
g
h
i
j
k
l

(IMA) has made preliminary recommendations for crystal chemical formulas in 1960, according to which the following notation should be used[1]:

Mineral	Structural formula
Calcite	$Ca^{[6]}[CO_3]$
Aragonite	$Ca^{[9]}[CO_3]$
Grossularite	$Ca_3^{[8]}Al_2^{[6]}[SiO_4]_3$
Topaz	$Al_2^{[2+4]}[F_2\|SiO_4]$
Malachite	$Cu^{[2+4]}Cu^{[4+2]}[(OH)_2\|CO_3]$

For the Handbook of Geochemistry a variation of this notation is recommended, the main differences being:

1. If a crystal contains different electronegative radicals (OH^-, CO_3^{2-}, SO_4^{2-}, SiO_4^{4-}, ...) they are not embraced together by square brackets. This seems to be justified because usually there are no chemical bonds between the electronegative groups.

2. In the notation of the IMA-recommendations the coordination number $[m+n+\cdots]$ means that the atom (ion) under consideration is surrounded by m particles of the first anion, by n particles of the second anion, etc. For instance, in topaz, $Al^{[2+4]}[F_2|SiO_4]$, aluminium is coordinated by two F^--ions and two oxygens of silicate groups. In this Handbook a notation is intended in which the coordination number $[m+n+\cdots]$ means that there are m ligands at (approximately) equal distances, n ligands at a larger distance, etc. As in topaz the AlF- and AlO-distances do not differ significantly in length, the formula is written: $Al^{[6]}\,F_2[SiO_4]$. In malachite, however, both (crystallographically different) Cu-atoms are surrounded by four O and OH resp. at 2.0 Å in the form of a slightly distorted square, but additional oxygens (and hydroxyl groups) at distances between 2.35 and 2.63 Å complete a strongly distorted octahedron (Wells, 1951; Süsse, 1967); the structural formula is therefore written: $Cu_2^{[4+2]}(OH)_2[CO_3]$.

If it seems necessary to indicate the chemical nature of the ligand, the chemical symbol is given; in case that there are different ligands at approximately the same distance they are separated by commas. More detailed formulas for topaz and malachite accordingly read:

$Al_2^{[4\,O,2\,F]}\,F_2[SiO_4]$ and $Cu^{[2\,O,2\,OH+2\,O]}\,Cu^{[2\,O,2\,OH+2\,OH]}\,(OH)_2[CO_3]$.

3. In the IMA-recommendations only the coordination number is given. It is, however, easily possible to indicate the main geometric features of the coordination figure for coordination numbers up to six. If the coordination number is given without

[1] The IMA-notation as well as the one recommended in this article are rather similar to that of Machatschki (1947).

Fig. 2-2a—l. Some important coordination figures of high symmetry. (a) linear 2-coordination, (b) planar triangular coordination, (c) trigonal pyramidal coordination, (d) planar square coordination, (e) tetrahedral coordination, (f) trigonal dipyramidal coordination, (g) octahedral coordination, (h) trigonal prismatic coordination, (i) cubic coordination, (j) tetragonal antiprismatic coordination, (k) cubooctahedral coordination, (l) regular ikosahedral coordination

an additional symbol, the coordination figure shall correspond to the electrostatic most favourable arrangement, i.e. for 2 the linear coordination, for 3 the plane triangular coordination, for 4 the tetrahedron, for 5 the trigonal dipyramid, and for 6 the octahedron. To indicate a one-sided coordination, the symbol po (for *po*lar) is recommended: [2 po] means the bent 2-coordination, [3 po] the one-sided triangular coordination (like in the AsS_3-group), [4 po] the one sided 4-coordination. — For the planar 4-coordination the symbol [4 pl] will be used, for the 5-coordination in which the ligands occupy the corners of a tetragonal pyramid the symbol [5 py] (after *py*ramidal). — For the coordination numbers 7 and larger, the shape of the coordination polyhedron is often complicated, and no symbolism is therefore intended to be used here.

4. To indicate chain- and layer-(sheet-)structures, $\frac{1}{\infty}$ or $\frac{2}{\infty}$ is written before the formula, e.g. $\frac{1}{\infty}\,Se^{[2]}$ for metallic selenium, $\frac{2}{\infty}\,C^{[3]}$ for graphite, etc.

Neither the IMA-recommendations nor the modifications proposed here invariably provide a unique notation. To illustrate this, we use the structure of $PbCl_2$, cotunnite. According to the IMA-recommendations, the structural formula reads $Pb^{[3]}Cl_2$, $Pb^{[7]}Cl_2$, or $Pb^{[9]}Cl_2$, according to the somewhat arbitrary choice of the coordination number. In the here proposed modification, the formula reads $\frac{1}{\infty}\,Pb^{[3\,po+4]}\,Cl_2$ or $\frac{1}{\infty}\,Pb^{[3\,po+4+1]}\,Cl_2$ if one stresses the chain character of the atomic arrangement which is indicated by the three shortest PbCl-bonds, otherwise $Pb^{[3+4+1]}Cl_2$ or $Pb^{[7+1]}Cl_2$.

III. The Effective Radii of Atoms and Ions

For many considerations in crystal chemistry it is useful to attribute effective radii to the atoms and ions.

The most convincing system of radii can be derived for the inert gases in the solid state and for the metals; of these only those for metals are of importance in geochemistry. For the majority of metallic elements, the atomic arrangement of at least one modification is the cubic or the hexagonal close packing of equal spheres. In both of them, each atom has a 12-coordination. It seems to be clear that the metallic radius should be defined as half the minimum metal-metal-distance. The "metallic radii" thus obtained can well be used to predict interatomic distances in intermetallic compounds. E.g. the metallic radii for Cu and Mg as derived from the elements are: $r_{Cu} = 1.28$ Å, $r_{Mg} = 1.60$ Å. In $MgCu_2$, one of the intermetallic compounds in the system magnesium — copper, the shortest interatomic distances are: Cu-Cu = 2.49 Å, Mg-Mg = 3.04 Å, Mg-Cu = 2.92 Å. The agreement with the expectations from the metallic radii is quite satisfactory, the maximum deviation being only ca. 5%.

Much more problematic is the derivation of "ionic radii" (see Table 12-8). As an example, let us consider the crystal structure of rocksalt, NaCl (Fig. 2-3), which is built from Na^+- and Cl^--ions. Each Na^+ is surrounded by six Cl^- in the form of an octahedron and vice versa; the smallest distance between Na^+ and Cl^- measures 2.82 Å, the smallest distance between equal kinds of ions 3.99 Å. The only unquestionable conclusion for the effective radii of the ions which can be drawn therefrom is that the "radii" of Na^+ and Cl^- cannot be larger than 2.0 Å. If we make the additional but, from the mutual electrostatic attraction of differently charged ions extremely plausible, assumption that the Na^+ and Cl^- ions "touch" each other, we obtain:

$$r_{Na^+} + r_{Cl^-} = 2.82\ \text{Å},$$

$$0.82\ \text{Å} \leqq r_{Na^+},\ r_{Cl^-} \leqq 2.00\ \text{Å}.$$

This is obviously only the order of magnitude.

A way to obtain much better empirical values for the size of the alkali- and the halogen-ions is, instead of considering only one compound, to gather information from a whole set of alkali halogenides which have the same crystal structure as NaCl; cf. Table 2-1. From this table it is apparent, that the interionic distances increase with atomic number of the cations and anions. LiI is therefore an extreme case, as the smallest cation is combined with the largest anion. From the Li-I distance 3.00 Å follows $r_I \leqq 2.12$, with a high probability that the true effective radius of I^- is close to this value (the I^--ions "touch" each other); as a consequence, $r_{Li} \sim 0.88$ Å. From these two values and the regularities in the change of the lattice constants when going from one alkali halogenide to the next, a fairly self-consistent set of radii for the alkali- and halogen-ions can be derived.

Table 2-1. *Cation-anion-distances in alkalihalogenides of the NaCl-type*

	F^-	Cl^-	Br^-	I^-
Li^+	2.10	2.56	2.75	3.00
Na^+	2.31	2.82	2.99	3.24
K^+	2.67	3.15	3.30	3.53
Rb^+	2.82	3.29	3.43	3.67

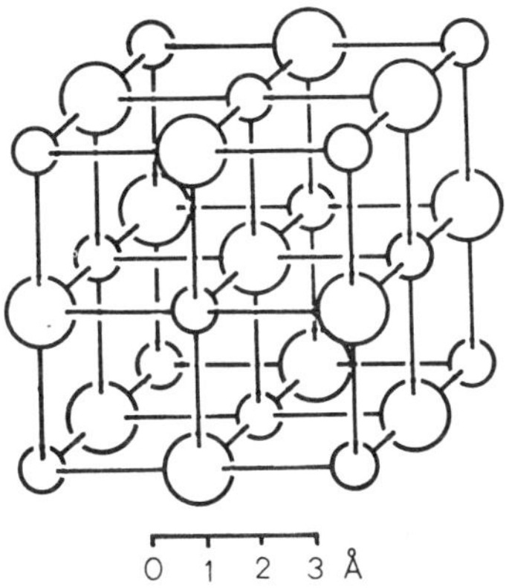

Fig. 2-3. Unit cell of rocksalt. Small spheres: Na^+, large spheres: Cl^-

An analogous set for the alkaline earth- and the chalcogenideions can be derived from the interatomic distances of the alkaline Earth-chalcogenides of the NaCl structure type. It is very important, that the radii thus derived also allow the prediction of the cation-anion distances in the alkali-chalcogenides and the halogenides of the alkaline Earths with rather good approximation.

This way to derive effective ionic radii dates back to LANDÉ (1920). Another way was used by WASASTJERNA (1923). He derived the observed cation-anion distances in the ratio determined by the mole refraction values of the ions. His results are in principle similar to those determined by the method of LANDÉ. WASASTJERNA'S values $r_{O^{2-}} = 1.32$ Å and $r_{F^-} = 1.33$ Å served as the basis for the extensive set of empirical ionic radii of V. M. GOLDSCHMIDT (1926). In 1927 PAULING introduced a new principle for the derivation of ionic radii. He divided the cation-anion distances of crystals whose ions have the same electronic configuration (NaF, KCl, RbBr), in the ratio of the $1/(Z-S)$-values, where Z is the charge of the nucleus of the ion (measured in multiples of the charge of the electron) and S is the "screening constant" (cf. PAULING, 1960). PAULING's results again agree reasonably well with the others (Table 12-8).

All the systems mentioned up to now take the effective radius of O^{2-} between 1.3 and 1.4 Å. There are however also some indications that one should use a system with smaller values for the anions and larger values for the cations. One of them is the fact, that very careful determinations of the electron density in ionic crystals show the minimum of electron density shifted toward the anion compared with the expectations from the usual ionic radii. It has however to be borne in mind that this minimum is extremely flat, and that therefore its exact location is difficult to determine. But another observation also points in the same direction in some structures: the observed anion-anion-distance is considerably shorter than the sum of the usual radii. E.g. in corundum, Al_2O_3, the shortest O-O-distance (edge of the triangle which forms a common face between two AlO_6-"octahedra") measures only 2.52 ± 0.03 Å (NEWNHAM and DE HAAN, 1962), in rutile, the common O-O-edge between two TiO_6-octahedra measures also only 2.52 ± 0.02 Å (BAUR, 1956). An extreme case is provided in andalusite, $Al_2O[SiO_4]$, where two AlO_5-polyhedra share an edge which measures only 2.25 Å (BURNHAM and BUERGER, 1961).

Although the "effective ionic radii" are of great use in large parts of inorganic and mineralogical crystal chemistry, it must be emphasized, that the attribution of rigid radii to the ions is an extremely crude approximation to nature. For a better approximation cf. p. 26.

Also for essentially covalent crystals is it possible to attribute "radii" to the atoms (see Table 12-7), provided that the type of bond in the different crystals is essentially the same (e.g. electron pair bond). It must however be borne in mind that here the term "radius" must not suggest that the bond is non-directional; very often the reverse is true: the covalent bonds point into definite directions. — Standard values for "covalent radii" are derived from the bond lengths between equal atoms. E.g. "tetrahedral radii" can be derived which are of great value in predicting interatomic distances in derivative structures of the wurtzite- and sphalerite-type.

IV. The Chemical Bonds in Crystals

All chemical bonds are fundamentally governed by the laws of quantum theory of the electron shell of the atoms. Nevertheless it is usual to distinguish four main types of bond: (a) the covalent bond, (b) the metallic bond, (c) the ionic bond, and (d) the van der Waals' bond.

a) Covalent Bond

The covalent (or homopolar) bond is characterized by the sharing of electrons between several (as a rule two) atoms. Very often the bond between two neighbouring atoms is effectuated by two electrons which occupy the same orbital with antiparallel spins ("electron pair bond"). There are only few examples of crystals which are held together exclusively by electron pair bonds, the best representative being diamond. Good examples of electron pair bonds within finite groups of atoms (molecules) in mineralogy are orthorhombic sulfur, which contains covalently bonded puckered S_8-rings, and realgar, which contains As_4S_4-molecules (Fig. 2-4). The bonds between the molecules are very weak. In orpiment, As_2S_3, the units held together by electron pair bonds are indefinitely large sheets; between neighbouring sheets the bonds are again very weak. Crystals which are held together by electron pair bonds are transparent and are good insulators. The two usual methods of treating the covalent bond are the valence bond- and the molecular orbital-theory, which start from different approximations; if the approximations are gradually reduced, the results will of course converge.

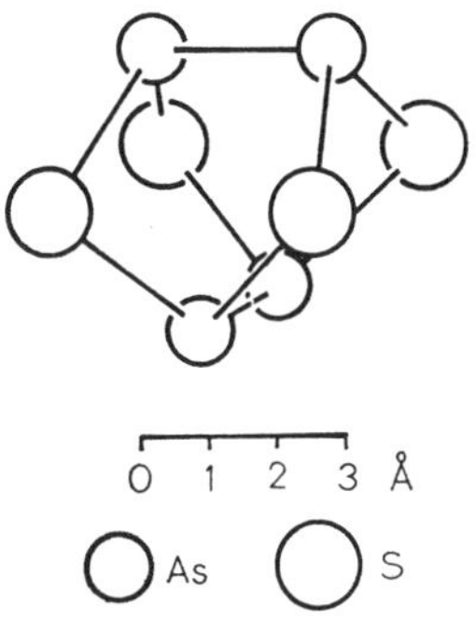

Fig. 2-4. As_4S_4-molecule of realgar. The electron-pair bonds are indicated by lines

b) Metallic Bond

The metallic bond resembles in some respects the covalent bond. There are however not sufficient electrons to effectuate electron pair bonds. The theory most commonly used to deal with metal crystals is the electron band theory. Crystals with metallic bonds are good conductors of electricity; the resistivity increases with increasing temperature. They have very high absorption coefficients for visible light. Good mineralogical examples are native copper, silver, and gold.

c) Ionic Bond

The physical properties of crystals with ionic (also heteropolar or electrostatic) bonds are similar to those of covalent crystals. In typical cases the type of bond is, however, very different in that the crystal is built up from charged ions. Here the laws of quantum chemistry favour the formation of charged particles, and the bond in the crystal is effectuated by the electrostatic attraction of the differently charged ions. Good examples in mineralogy are rocksalt (NaCl), sylvite (KCl), periclase (MgO), and others.

d) Van der Waals' Bond

The van der Waals' bonds are weaker by an order of magnitude compared with the bonds discussed previously. The weak van der Waals' attraction is caused by permanent and induced dipole interaction, and also to a great extent by the interaction due to dynamic polarisation ("dispersion effect" of F. LONDON); the attractive force is proportional to d^{-6}, d being the interatomic distance. Good examples in mineralogy are the forces between the S_8-rings in orthorhombic sulfur, and also the forces between the As_4S_4-molecules in realgar and between the As_2S_3-sheets in orpiment.

The main types of the chemical bonds are not without transitions; the pure types are exceptions rather than the rule.

Of special importance are the transitions between the covalent bond and the ionic bond. With his system of electronegativites (Table 12-10), PAULING has provided a relatively primitive tool for estimating the percentage of ionic bond, but one which is very useful for qualitative and semiquantitative considerations. The system is derived from thermochemical data. The absolute figures are not of great importance as they contain an arbitrary additive constant. The greater the figure for the electronegativity, the greater the tendency to form a negatively charged ion. Most important is the difference between the electronegativities of the two elements under consideration: the greater this difference, the greater is the ionicity of the bond. Fig. 2-5 gives the approximate correlation between electronegativity difference and the partial ionic character of the bond.

The greatest difference in the table of electronegativities is found between Cs and F; the bond between these two elements is therefore the best example of a purely electrostatic bond. But from Table 12-10 and Fig. 2-5 one can see that many bonds

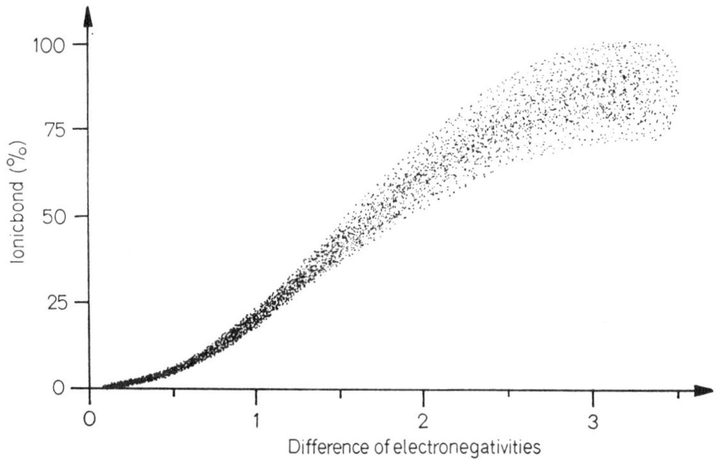

Fig. 2-5. Approximate correlation between electronegativity difference and ionicity of bond

are intermediate between purely ionic and purely covalent. A very important example in mineralogy is the bond between silicon and oxygen, which is approximately 50% ionic — this figure of course gives only the order of magnitude.

When estimating the ionicity of the bond from PAULING's electronegativities it should be borne in mind that the amount of ionic bond does not depend only upon the two ligands, but also upon the rest of the atomic arrangement. E.g. there is no doubt that the bond between Na and Cl is more ionic in rocksalt than in the NaCl-molecule of the vapour. The electronegativity of an atom does of course also depend somewhat on its state of oxydation.

e) Crystal Field- and Ligand Field-Theory

Especially for the explanation of some characteristic features of the stereochemistry of transition metal compounds, the *crystal field-* and the *ligand field-theory* are of great value[1]. Both theories deal mainly with the splitting of the *d*-electron terms of

[1] Sometimes the expression ligand field theory is used as a common term for both closely related theories (cf. SCHLÄFER and GLIEMANN, 1967).

transition elements in the electrostatic field of a coordination polyhedron, the fields of the ligands being approximated by point charges. The former theory neglects a possible partial covalent character of the bonds between the transition metal ions and the ligands, while the latter takes it into account.

While in an isolated cation all five *d* orbitals have the same energy, this is not true any longer if the cation is surrounded by negative charges (ions). For the different coordination polyhedra, the splitting of the energy levels is different. E.g. for regular octahedral coordination the five 3d-electron terms split into a triplet (d_{xy}, d_{yz}, d_{zx}) of lower energy and a doublet ($d_{x^2-y^2}$, d_{z^2}) of higher energy, the "low energy level" being the one which is energetically more favourable. For regular tetrahedral coordination the splitting is weaker and inverse: there is now a high energy triplet (d_{xy}, d_{yz}, d_{zx}) and a low energy dublet ($d_{x^2-y^2}$, d_{z^2}). Coordination polyhedra with high energy levels occupied (with electron lobes pointing into electrostatic unfavourable directions) will be relatively unstable.

Of considerable importance in this connection is the Jahn-Teller theorem (Jahn and Teller, 1937) which can be formulated as follows: In (symmetrical) non-linear molecules a system cannot remain in equilibrium in an orbitally-degenerated state, but will distort in such a way that the degeneracy is lifted. As a consequence of this theorem distortions of coordination octahedra around Cu^{2+}, Mn^{3+}, and Cr^{2+} are to be expected which correspond to those observed in crystal structures.

In connection with the chemical bond in crystals, special consideration is to be given to semiconductors and crystals with hydrogen bridges.

f) Semiconductors

Semiconductors are defined by their electric conductivity, which is intermediate between the very low values of typical metals (spec. resistivity: ca. 10^{-5}—$10^{-6}\ \Omega \cdot cm$) and the extreme high values for good insulators (spec. resistivity up to ca. $10^{18}\ \Omega \cdot cm$). In contrast to metals, the resistivity of semiconductors decreases with increasing temperature, and, again in contrast to metals, may depend greatly on very small amounts of impurities; e.g. the resistivity of silicon of highest grade purity decreases by a factor of 10^3 if only one of every 100,000 Si atoms is replaced by boron. Natural examples for semiconductors are numerous, e.g. a great part of the sulfide minerals, such as galena, belong to this class of compounds.

The chemical bond in semiconductors is often closely related to the covalent bond. In good insulators having this type of bond, such as diamond, the valence bands are completely filled with electrons, and the energy gap to the next higher band is so large that the thermal energy is not sufficient to allow electrons to jump from the filled band to the next higher empty band, the "conduction band". If in the atomic arrangement of an insulator a small part of the atoms is replaced by another kind of atoms with a different number of electrons in the outermost shell, either the outermost band is not completely filled (e.g. B for Si), or the conduction band will be partly filled (e.g. As for Si). If the energy gap between the highest completely filled band and the next higher empty band is only small, it may happen that at higher temperatures some electrons jump into this band, and electric conduction occurs without the introduction of impurities (intrinsic semiconductors).

g) Hydrogen Bond

A bond type which is in some cases clearly related to the electrostatic and to van der Waals' bond, but which also shows very individual features is the hydrogen bond. It is geometrically characterized by a hydrogen atom near the line connecting two strongly electronegative atoms (but in the vast majority of cases not equidistant from those two atoms). In the examples which are of special importance in geochemistry, the two atoms are oxygens, and we will restrict ourselves to this case.

The length of hydrogen bonds between oxygens varies considerably: from ca. 2.4 Å to ca. 3.0 Å. The strength of the bond decreases with increasing length, the most frequently encountered values being 2.65 Å—2.85 Å. Any numerical value for an upper limit for the hydrogen bond is problematic. There is evidently a continuous transition from the strongest, very short hydrogen bonds, where the hydrogen is equidistant from both oxygens ("symmetric hydrogen bond") via the usual hydrogen bond, where the hydrogen atom is clearly bound to one of the two "ends" of the bridge (but with a relatively strong interaction also to the "acceptor oxygen") to cases where the interaction with any second neighbour seems to be negligibly small. With the exception of the symmetric hydrogen bridges, the H-atom is strongly bound to one of the "ends" of the bridge and therefore belongs to a hydroxyl group or a water molecule. The O-H-distances in both groups measure ca. 1.0 Å and the exact value depends only slightly on the distance from the H-atom to the acceptor-oxygen, while the latter distance can vary considerably in length and strength. Breaking a hydrogen bridge means of course breaking the longer part of the bridge. The acceptor oxygen may be bound only to cations (example: diaspore), it may however also be the oxygen of a water molecule (examples: water, and many water-rich hydrates), or a hydroxyl-group.

It is important to realize that the hydrogen atoms of OH-groups and of H_2O-molecules are not necessarily involved in a hydrogen bridge; e.g. in brucite, $Mg(OH)_2$, there are probably no hydrogen bridges at all. In some few cases the O-H-direction points in between two oxygen atoms — such a configuration is called a "bifurcated hydrogen bond".

As the hydroxyl group and the water molecule both have an electrical dipole moment, at least part of the binding energy of the hydrogen bond results from dipole interaction. But especially for short hydrogen bridges (OH-O $<$ 1.6 Å), other bond contributions have also to be taken into account.

V. The Theoretical Treatment of Ionic Crystals

Relatively easily accessible to a theoretical treatment are the ionic crystals. In many of the ions the distribution of the negatively charged electron cloud around the nucleus has spherical symmetry. This is the case for all ions whose electron configuration corresponds to that of the inert gases, and also for Mn^{2+}, Zn^{2+} etc. For such ions the electrostatic interaction at distances which are distinctly greater than the sum of the traditional ionic radii (say, above $1^1/_2$ of this value) can be treated as if the ions were points which carry the net charge of the ion — of course on the assumption that polarization effects can be neglected. Mathematical methods have been developed to compute the electrostatic part of the lattice energy, that is the gain of energy (usually per mole substance) which is set free in building up an indefinitely large periodic pattern of point charges from an initial stage where the points were at infinitely great distances from each other (cf. EWALD, 1921; BERTAUT, 1952).

To prevent the point charge model from a collapse between the positive and negative charges, it is necessary to introduce non-electrostatic repulsion terms between the ions. The most primitive way to do this is to attribute to each ion a "radius" and then to treat it as a rigid sphere.

A much better approximation to reality are models in which the ions are characterized by two constants: one containing more or less the information about the "size", the other about the "compressibility". The most commonly used forms for the non-electrostatic repulsion between two ions are $b \cdot d^{-n}$ and $\lambda \cdot e^{-d/\varrho}$, where d is the interionic distance — b, n, λ and ϱ are constants. If appropriate constants are chosen, both systems give reasonable results and both have been used successfully. For neither of them, however, there exists a system of constants as complete as the tables of ionic radii.

The electrostatic lattice energy and the repulsion term are the most important contributions to the lattice energy of ionic crystals. Several problems of the crystal chemistry of this class of compounds can be solved by their application, but some simple looking problems remain difficult to answer. A typical example is, that the CsCl-type of structure is slightly more favourable from the point of view of electrostatic lattice energy than the NaCl-type of structure but nevertheless it is not found among the alkali fluorides although the ratio of the "sizes" of the atoms sometimes seems to be very favourable.

For the larger anions in particular, the neglect of the polarizability is not a good approximation if they coordinate small two- or higher-valent cations. HUND (1925) showed in the early days of the theory of crystal structures that the transition from the rutile-type to the CdI_2-type of structure can be explained by the polarizability of the larger anions. The change of structure is common when going from the difluorides with $r_{cation} \sim 0.7$ Å, to the dichlorides, dibromides and diiodides of divalent cations of the same size; the latter belong to the CdI_2- or the very similar $MgCl_2$-type of structure.

The effective radius is not spherical for all ions. A typical exception is offered by Cu^{2+}. It shows e.g. in CuF_2, the most ionic Cu(2)-compound, a characteristic distorted octahedral coordination, as if its shape were a rotational ellipsoid.

VI. Polymorphism

Polymorphism is the phenomenon where a chemical substance crystallizes in different crystal structures depending on the conditions of crystal growth. As a rule, each of the polymorphic modifications has a well defined p-T-field of thermodynamic stability. There are however some crystal species which have no known field of stability — an example is vaterite, μ-$Ca[CO_3]$. The answer to the question whether every crystal species has a field of stability seems not to have been settled. There is no doubt, however, that many crystals grow well outside of their field of thermodynamic stability.

The different polymorphic modifications of a chemical substance may have entirely different atomic arrangements: graphite ⇌ diamond, $\alpha-\mathrm{Fe} \rightleftharpoons \gamma-\mathrm{Fe}$, coesite ⇌ stishovite, etc. It is also possible that the coordinations do not radically change: quartz ⇌ (alkali-free) tridymite, sphalerite ⇌ wurtzite, low-quartz ⇌ high-quartz.

A very approximate, but nevertheless important, regularity for the structural relations of polymorphs can be formulated as follows: As a rule, the polymorph which is stable at higher temperatures (lower pressures) has the smaller coordination number[1].

Examples: (a) change of structure at constant temperature

$$\mathrm{Rb}^{[6]}\,\mathrm{Cl}^{[6]} \xrightleftharpoons{20\ \mathrm{kbar}} \mathrm{Rb}^{[8]}\,\mathrm{Cl}^{[8]} \quad \text{room temperature}$$

$$\mathrm{Ca}^{[6]}\,[\mathrm{CO_3}] \xrightleftharpoons{3{,}5\ \mathrm{kbar}} \mathrm{Ca}^{[9]}\,[\mathrm{CO_3}] \quad \text{room temperature}$$

$$\overset{2}{\infty}\,\mathrm{C}^{[3]}\ (\text{graphite}) \xrightleftharpoons{75\ \mathrm{kbar}} \mathrm{C}^{[4]}\ (\text{diamond}) \quad 2{,}000^\circ\ \mathrm{C};$$

(b) change of structure at constant pressure

$$\mathrm{Cs}^{[8]}\,\mathrm{Cl}^{[8]} \xrightleftharpoons{445^\circ\mathrm{C}} \mathrm{Cs}^{[6]}\,\mathrm{Cl}^{[6]} \quad 1\ \text{atm.}$$

$$\mathrm{K}^{[9]}\,[\mathrm{NO_3}] \xrightleftharpoons{126^\circ\mathrm{C}} \mathrm{K}^{[6]}\,[\mathrm{NO_3}] \quad 1\ \text{atm.}$$

The rule implies that the general features of the boundary line between the stability fields of aragonite-calcite (Fig. 2-6) are characteristic also for other cases.

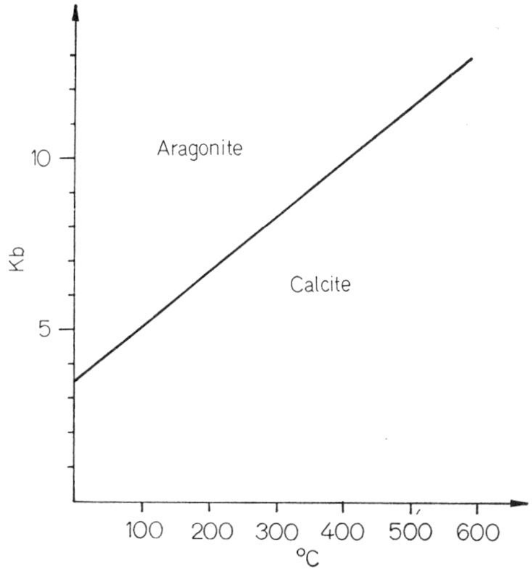

Fig. 2-6. Fields of stability aragonite-calcite (after CLARK, 1957)

In the high-temperature modifications the thermal energy of the atoms is especially large and often leads not only to harmonic oscillations about the equilibrium position, but also to other phenomena of disorder. Statistical distributions with intense change of position, rotations, and oscillations with large amplitudes have been described.

[1] A recent survey article on polymorphic phenomena of geochemical interest at high pressures and temperatures has been given by NEUHAUS (1968).

VII. Isomorphism

Isomorphism is the reciprocal phenomenon to polymorphism: different chemical substances have the same atomic arrangement. Nowadays, the term "isotypism" is often used instead of the older expression isomorphism, and in the definition of isotypism the absolute dimensions of the structures are always neglected, and also no similarity in the chemical bonds is required. Therefore, LiF and PbS, for example, are isotypical, although the lattice constants differ considerably ($a_{LiF} = 4.02$ Å; $a_{PbS} = 5.94$ Å), and the chemical bonds also differ considerably.

Examples for isotypism in mineralogy are numerous; e.g. rocksalt, NaCl — sylvine, KCl — periclase, MgO — galena, PbS; aragonite, $Ca^{[9]}[CO_3]$ — strontianite, $Sr[CO_3]$ — witherite, $Ba[CO_3]$ — cerussite, $Pb[CO_3]$; pyrope, $Mg_3Al_2[SiO_4]_3$ — almandine, $Fe_3Al_2[SiO_4]_3$ — grossularite, $Ca_3Al_2[SiO_4]_3$ — andradite, $Ca_3Fe_2[SiO_4]_3$ — uvarovite, $Ca_3Cr_2[SiO_4]_3$ — cryolithionite, $Na_3Al_2[LiF_4]_3$ — berzeliite, $Ca_2Na(Mg, Mn)_2[AsO_4]_3$.

VIII. Crystal Chemistry of Real Crystals

a) Effects of the Heat Content of Crystals

The heat content of crystals and — to a very small extent — also the "zero point energy" cause the particles to oscillate about their equilibrium positions. In mineral crystals, the amplitudes of the vibrations at room temperature are rather small compared with the interatomic distances. In silicates, for example, the vibrations often have an average oscillation amplitude of the order of 10^{-2} Å, while the interatomic distances range mostly between 1.5 and 3.0 Å. The amplitudes of the thermal oscillations of course increase with temperature. However, it is often found that the "temperature factors" determined at room temperature do not — or in any case do not significantly — decrease, if the diffraction experiment is repeated at much lower temperature, say at ca. 80° K. This proves that the "temperature factor" and the "mean amplitude of oscillation", as determined by standard diffraction methods, may include other kinds of more or less permanent deviations from exact periodicity.

In many cases the oscillation of the atom does not, on average, have spherical symmetry. Good examples are $Ca(OH)_2$ (Busing and Lévy, 1957) and low quartz (Smith and Alexander, 1963; Young and Post, 1962). In the first case, the oscillation of the hydrogen atoms is relatively weak in the direction of the OH-dipoles (mean amplitude: 0.13 Å), while at right angles to this direction the vibrations are much stronger (mean amplitude: 0.26 Å). — In low quartz, the vibration of the oxygens is strongest approximately perpendicular to the plane which is determined by the oxygen in question and its two nearest Si-neighbours[1] (mean amplitude: 0.13 Å); the weakest vibration is approximately parallel to the line connecting the two neighbouring silicons (mean amplitude: 0.08 Å).

The qualitative interpretation of the vibrational anisotropy of hydrogen in $Ca(OH)_2$ seems to be easy. The H-atoms are bound via the O-atoms of the hydroxyl-groups to the 2-dimensional sheets of polymerized calcium-hydroxyl-octahedra.

[1] Note that the angle Si-O-Si is not 180°, but only 144°!

The strong bond to this (indefinitely) large mass causes the mean vibration amplitude of the H-atoms perpendicular to this sheet (= parallel to the direction of the OH-bond) to be small. Only the much weaker bending forces act perpendicular to this direction, and consequently the vibration amplitude is considerably larger.

Similar arguments can serve to explain the anisotropy of the vibration of the O-atoms in low quartz. The vibration perpendicular to the plane defined by an oxygen and its two Si-neighbours mainly causes deformations of bond angles, and the amplitudes are consequently large. It is however worth mentioning that in quartz also the thermal vibration of silicon is definitely anisotropic, the largest amplitudes lying approximately parallel [00.1]. For this atom the arguments just given for oxygens are not valid; to the author's knowledge no explanation for this anisotropy has yet been given.

At higher temperatures (for mineral crystals mostly well above room temperature) the heat content may cause much more severe deviations from the "ideal crystal" than mere oscillation about the equilibrium positions. An important and typical example in mineralogy is Ag_2S. At room temperature, this compound has an ordered monoclinic structure (acanthite; FRUEH, 1958); above 178° C it has a very interesting cubic structure. In this high-temperature modification, only the S-atoms are ordered (they occupy the corners and the center of the cell), while the Ag-atoms move more or lees freely in the interstices.

b) Solid Solutions (Mixed Crystals)

Closely connected with the heat content of a crystal is also the formation of thermodynamically stable solid solutions in which two (or more) kinds of atoms are distributed more or less at random over one point position of a space group. The rules which govern the formation of solid solutions (mixed crystals) are of great importance for the geochemical applications of crystal chemistry.

Simple, typical and well investigated examples are provided by the alkali chlorides. Much can be learnt from the systems KCl-RbCl and NaCl-KCl ,which are used here as examples. The lattice constants are: $a_{NaCl} = 5.64$ Å, $a_{KCl} = 6.29$ Å, $a_{RbCl} = 6.58$ Å. The difference in the lattice constants is 4.8% between KCl and RbCl, and 11.5% between NaCl and KCl (in percentages of the smaller value).

At room temperature (and, of course, also at higher temperatures) KCl and RbCl form mixed crystals (K, Rb)Cl over the whole concentration range. It is, however, important to be aware that even at very slow growth rates (approximating as closely as possible equilibrium conditions) the Rb/K-ratio of the crystal does not correspond to the Rb/K-ratio of the aqueous solution from which the crystal was grown. The value of the partitition coefficient (mol-% Rb in crystal/mol-% Rb in solution) is ca. 0.13 for very small Rb-concentrations (0.1 mol-% in solution,) but ca. 1.0 for high Rb-concentrations in the solution (cf. Fig. 2-7). The partition coefficient varies with concentration, with temperature, and with the concentration of other substances in the solution.

On the other hand, the mutual solubility of NaCl and KCl at room temperature is extremely limited. According to REICHERT (1966) the maximum Na-content in KCl at 40° C (at slow crystallization from aqueous solution) is 1.4×10^{-2} mol-%, for KCl in NaCl (under the same conditions) even only 0.8×10^{-2} mol-%. The mutual solubility increases with temperature (especially from ca. 300° C), and above ~500° C mixed crystals (Na, K)Cl are stable over the whole concentration range.

It should be noted that, in the pair with very similar lattice constants (and at the same time with similar effective radii of the cations), the mutual solubility in the solid state is already complete at room temperature while, in the system NaCl-KCl, complete solid solution occurs only above ca. 500° C. That, in fact, similarity of the effective radii and *not* similarity of the lattice constants is essential for the formation of mixed crystals is proved by the pair NaCl-RbF. The lattice constants of these two isotypical alkali halogenides are almost identical. They do not, however, form solid solutions in any amount worth mentioning: the extreme similarity of the lattice constants merely results from the fact that the differences in effective radii ($r_{Rb} - r_K$) and ($r_{Cl} - r_F$) are almost alike.

Partial mutual substitution of atoms (ions) in crystals is sometimes also possible, if the pure endmembers have different crystal structures. An important example in

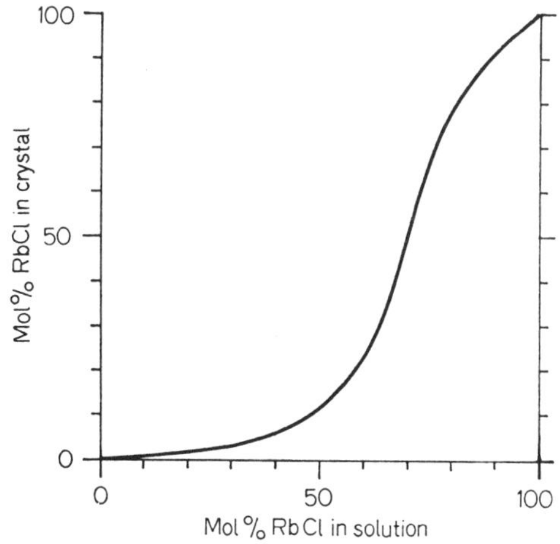

Fig. 2-7. Mol-% RbCl in (K, Rb)Cl mixed crystals in (quasi-)equilibrium with aqueous solutions (room temperature) (after REICHERT, 1966)

mineralogy is sphalerite, cubic ZnS, which often contains several percent Fe, although FeS does not crystallize in the same structure. At ca. 900° C at least 40% of the Zn-atoms can be replaced by Fe in this structure (KULLERUD, 1953); according to the more recent work of BARTON and TOULMIN (1966) the limit of the Fe-substitution for Zn is even higher than 50% at this temperature. On the contrary, the maximal substitution of Zn for Fe in FeS (NiAs-type of structure) is extremely small at equally high temperatures.

The formation of solid solutions is of course not restricted to two components. Much more complicated examples are widespread in mineralogy, e.g. (Mg, Fe^{2+}, Mn^{2+}, $Ni)_2[SiO_4]$ and (Zn, Fe, Mn, Cd)S.

Also of very great importance in earth sciences are cases where a point position of a space group is occupied by ions of different charge (atoms of different valency). In such cases it is important that the effective radii of the particles which replace each other are similar, and that the charge balance is maintained. Such complicated solid solutions occur in many of the rock forming minerals, such as basaltic augite (Ca, Na) ($Mg, Fe^{2+}, Al, Fe^{3+}, \ldots$) [$(Si, Al)_2O_6$], plagioclase (Na, Ca) [$(Si, Al)_4O_8$], etc.

The system Ag-Al demonstrates clearly that, not only the effective radii, but also a sufficient similarity in the nature of the chemical bond is essential for the formation of solid solutions. Both metals crystallize in the densest cubic packing of spheres, the lattice constants being $a_{Ag}=4.09$ Å, $a_{Al}=4.05$ Å; nevertheless, silver is virtually insoluble in solid Al, and the solubility of Al in solid Ag is restricted to a small percentage even at high temperature.

Of the many problems connected with mixed crystals, the distribution of a special kind of atom (often a trace element) between an aqueous solution (or melt) and a mineral grown from it is of special importance in geochemistry.

Two definitions for distribution coefficients are in common use. One form reads:

$$k=C_S/C_L$$

where $C_S=$ concentration of the trace element in the solid phase,
$C_L=$ concentration of the trace element in the liquid phase.

The law of distribution of BERTHELOT and NERNST states that k is a constant for vanishing concentration of the trace element.

Another definition of the distribution coefficient was introduced by HENDERSON and KRACEK (1927), namely:

$$D=(Tr/Cr)_S/(Tr/Cr)_L$$

where $(Tr/Cr)_S=$ ratio tracer/carrier in the solid phase
$(Tr/Cr)_L=$ ratio tracer/carrier in the liquid phase.

The knowledge of distribution coefficients as a function of concentration, temperature and composition of the liquid phase is of great importance in reconstructing the genesis of a mineral association. The most consistent application (up to the present) seems to have been the use of Br-concentrations in chlorides of oceanic salt deposits. The first experimental basis was laid by BOEKE (1908). However, some 30 years had to pass before it was applied to the elucidation of the genesis of salt deposits. Following the reexamination of the experimental basis by BRAITSCH and HERRMANN (1963), a very extensive and critical application of the Br-distribution coefficients to problems of salt-mineralogy was made by BRAITSCH (1962) and BRAITSCH and HERRMANN (1964).

GOLDSCHMIDT (1937) formulated general rules for distribution coefficients. Of special importance is his statement that when a mixed crystal forms from a liquid phase (melt or solution), of two ions of equal charge the one with the smaller ionic radius will be concentrated in the crystal. Although the theoretical basis for this rule is poor (cf. the criticisms of SHAW, 1953, and of WHITAKKER, 1967), it holds good in a considerable number of cases. AHRENS (1953) and RINGWOOD (1955) have tried to develop GOLDSCHMIDT's point of view by paying attention also to the ionization potentials and electronegativities respectively.

Besides the statistical distribution of two or more kinds of atoms (ions) over one equivalent position, another similar, but more drastic deviation from ideal 3-dimensional periodicity is possible, namely that a position is only partly occupied.

A good mineralogical example is yttrofluorite. It is well known that some pegmatitic fluorites contain at least up to 10 weight-% YF_3. By measuring lattice constants and densities of corresponding synthetic products, ZINTL and UDGÅRD (1939) were able to prove that the taking up of YF_3 into the fluorite structure is effected (1) by the partial substitution of Ca^{2+} by Y^{3+} and simultaneously (2) by partial occupancy of the position 1/2 1/2 1/2 by additional F^--ions. YF_3 itself has a different structure (ZALKIN and TEMPLETON, 1953).

Another important mineralogical example is pyrrhotite. In this natural "iron(2)-sulfide" the ratio Fe:S is somewhat less than 1, the empirical composition being often $\sim Fe_7S_8$. At least at relatively high temperatures (and certainly above 300° C) its structure is of the hexagonal NiAs-type with disordered vacancies in the Fe-position. At room temperature, lower symmetry (with partial ordering) has been observed.

Unoccupied positions and (to a lesser degree) occupancy of a position by "wrong" atoms cause distortions in the periodic structure — mainly, of course, in the neighbourhood of the "point defects". Such crystals seem to have high "temperature factors", as determined by standard diffraction methods.

IX. Diffusion and Reactions in the Solid State

The most drastic proofs for deviations from the "static" picture of a crystal are diffusion and reactions in the crystalline state. In earth sciences, diffusion in crystals is most convincingly proved by solid exsolutions in crystals which are, of course, not possible unless there is mobility of at least a part of the atoms (ions). A few of the many well known examples are: perthites, chalcopyrite-exsolutions in sphalerite, ulvite-exsolutions in magnetite, etc. That the observed intergrowth structures are really to be explained by exsolution can often be proved by keeping the „crystal" (which contains exsolutions) for some time at an elevated temperature, and then quenching it. If a homogenous high-temperature solid solution has been formed, and if it can be undercooled, the exsolved phase of the original "crystal" will have disappeared.

Diffusion in a perfectly periodic crystal structure, somewhat resembling a densest packing of spheres, could hardly be understood. The (Schottky- and Frenkel)defects, however, greatly facilitate the "jumping" of the atoms from one equilibrium position to the next (Fig. 2-8). Therefore, diffusion in crystals is most closely bound to the presence of defects in the structure.

There is, of course, an intimate interrelation between diffusion in the solid state and reactions in crystals. A well known example (although it includes a liquid phase) is the reaction:

$$Ag_{\text{solid.}} + S_{\text{liqu.}} = Ag_2S_{\text{solid.}}.$$

This reaction has been investigated above the inversion point acanthite $\rightleftharpoons$ body centred argentite (178° C) by WAGNER (1933). He was able to prove that the reaction takes place at the phase boundary Ag_2S/S, and that Ag-atoms diffuse through solid Ag_2S. In this connection it should be borne in mind that in body centred argentite only the S-atoms have fixed positions (0 0 0 + 1/2 1/2 1/2), whereas in the static picture the Ag-atoms are more or less randomly distributed over the interstices.

Another typical example is provided by the dehydration of diaspore:

$$2\ AlO(OH) \rightarrow Al_2O_3\ (\text{corundum}) + H_2O\ (\text{gas}).$$

DEFLANDRE (1932) had already noticed from X-ray studies that the corundum crystallites which grow by this dehydration have a special orientation to the diaspore crystal, and that both structures are based more or less on a hexagonal densest packing

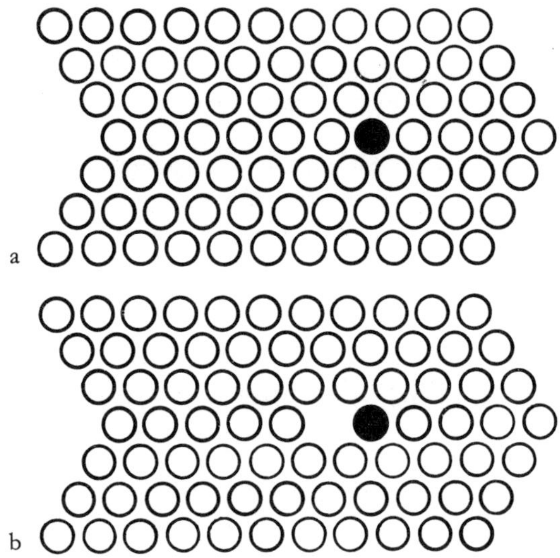

Fig. 2-8. A particular atom (drawn in black) cannot move in a hexagonal closely packed layer of atoms (a), but it can easily move, if there are vacant positions (b). Repetiton of the step indicated make diffusion in the solid state possible

of oxygens, whose positions are maintained during the reaction. Using also additional informations (obtained e.g. from the development of pores during this dehydration and from electron photographs) LIMA-DE-FARIA (1963) was able to show that the dehydration mechanism is best explained by assuming that the Al^{3+}-ions and protons migrate mainly in the [00.1]-direction of the common hexagonal densest packing of oxygens, forming oxide sheets of ca. 40 Å thickness. The protons evidently migrate to the regions between these sheets and combine with oxygen to form H_2O which escapes, leaving elongated pores. The dehydration of diaspore is a typical case of a topotactic reaction.

References

AHRENS, L.: The use of ionization potentials. Part 2. Anion affinity and geochemistry. Geochim. Cosmochim. Acta **3**, 1 (1953).

BARTON JR., P. B., and P. TOULMIN III: Phase relations involving sphalerite in the Fe-Zn-S system. Econ. Geol. **61**, 815 (1966).

BAUR, W.: Über die Verfeinerung der Kristallstrukturbestimmung einiger Vertreter des Rutiltyps: TiO_2, SnO_2, GeO_2 und MgF_2. Acta Cryst. **9**, 515 (1956).

BERTAUT, F.: L'énergie électrostatique de réseaux ioniques. J. Phys. Radium **13**, 499 (1952).
BOEKE, H. E.: Über das Krystallisationsschema der Chloride, Bromide, Jodide, von Kalium, Natrium und Magnesium, etc. Z. Krist. **45**, 346 (1908).
BRAITSCH, O.: Entstehung und Stoffbestand der Salzlagerstätten. (Mineralogie u. Petrographie in Einzeldarstellungen. Bd. 3.) Berlin-Göttingen-Heidelberg: Springer 1962.
—, u. A. G. HERRMANN: Zur Geochemie des Broms in salinaren Sedimenten. Teil I. Experimentelle Bestimmung der Br-Verteilung in verschiedenen natürlichen Salzsystemen. Geochim. Cosmochim. Acta **27**, 361 (1963).
— — Zur Geochemie des Broms in salinaren Sedimenten. Teil II. Die Bildungstemperaturen primärer Sylvin- und Carnallit-Gesteine. Geochim. Cosmochim. Acta **28**, 1081 (1964).
BURNHAM, CH. W., and M. J. BUERGER: Refinement of the crystal structure of andalusite. Z. Krist. **115**, 269 (1961).
BUSING, W. R., and H. A. LÉVY: Neutron diffraction study of calcium hydroxide. J. Chem. Phys. **26**, 563 (1957).
CLARK JR., S. P.: A note on calcite-aragonite equilibrium. Am. Mineralogist **42**, 564 (1957).
DEFLANDRE, M.: La structure cristalline du diaspore. Bull. Soc. Franç. Minéral. Crist. **55**, 140 (1932).
DONNAY, J. D. H., E. HELLNER, and A. NIGGLI: Coordination polyhedra. Z. Krist. **120**, 364 (1964).
DORNBERGER-SCHIFF, K.: Grundzüge einer Theorie der OD-Struk turen aus Schichten. Abhandl. Deut. Akad. Wiss. Berlin, Kl. Chem., Geol. Biol. **1964**, Nr. 3.
EWALD, P. P.: Die Berechnung optischer und elektrostatischer Gitterpotentiale. Ann. Physik [4] **64**, 253 (1921).
FRUEH, A. J.: The crystallography of silver sulfide, Ag_2S. Z. Krist. **110**, 136 (1958).
GOLDSCHMIDT, V. M.: Geochemische Verteilungsgesetze der Elemente. VII. Die Gesetze der Krystallochemie. Skrifter Norske Videnskaps-Akad. Oslo I: Mat.-Naturv. Kl. **1926**, Nr. 2.
— The principles of distribution of chemical elements in minerals and rocks. J. Chem. Soc. **1937**, 655.
HENDERSON, L. M., and F. C. KRACEK: The fractional precipitation of barium and radium chromates. J. Am. Chem. Soc. **49**, 739 (1927).
HUND, F.: Versuch einer Ableitung der Gittertypen aus der Vorstellung des isotropen polarisierbaren Ions. Z. Physik **34**, 833 (1925).
JAGODZINSKI, H.: Kristallstruktur und Fehlordnung des Artinits $Mg_2[CO_3(OH)_2] \cdot 3H_2O$. Tschermaks Mineral. Petrog. Mitt. **10**, 297 (1965).
JAHN, H. A., and E. TELLER: Stability of polyatomic molecules in degenerate electronic states. Proc. Roy. Soc. (London) Ser. A **161**, 220 (1937).
KULLERUD, G.: The FeS-ZnS system, a geological thermometer. Norsk. Geol. Tidsskr. **32**, 61 (1953).
LANDÉ, A.: Über die Größe der Atome. Z. Physik **1**, 191 (1920).
LIMA-DE-FARIA, J.: Dehydration of goethite and diaspore. Z. Krist. **119**, 176 (1964).
MACHATSCHKI, F.: Konstitutionsformeln für den festen Zustand. Monatsh. Chem. **77**, 333 (1947).
NEUHAUS, A.: Über Phasen- und Materiezustände in den tiefen und tiefsten Erdzonen (Ergebnisse der modernen Hochdruck-Hochtemperatur-Forschung zum geochemischen Erdbild). Geol. Rundschau **57**, 972—1001 (1968).
NEWNHAM, R. E., and Y. M. DE HAAN: Refinement of the α-Al_2O_3, Ti_2O_3, V_2O_3 and Cr_2O_3 structures. Z. Krist. **117**, 235 (1962).
PAULING, L.: The sizes of ions and the structure of ionic crystals. J. Am. Chem. Soc. **49**, 763 (1927).
— The nature of the chemical bond, 3rd. edit. New York: Cornell Univ. Press 1960.
PRANDL, W.: Verfeinerung der Kristallstruktur des Grossulars mit Neutronen- und Röntgenstrahlbeugung. Z. Krist. **123**, 81 (1966).
REICHERT, J.: Verteilung anorganischer Fremdionen bei der Kristallisation von Alkalichloriden. Contr. Mineral. and Petrol. **13**, 134 (1966).

RINGWOOD, A. E.: The principles governing trace element distribution during magmatic crystallization. — I. The influence of electronegativity. Geochim. Cosmochim. Acta **7**, 189 (1955).

SAHL, K.: Die Verfeinerung der Kristallstrukturen von $PbCl_2$ (Cotunnit), $BaCl_2$, $PbSO_4$ (Anglesit) und $BaSO_4$ (Baryt). Beitr. Mineral. Petrog. **9**, 111 (1963).

SCHLÄFER, H. L., and G. GLIEMANN: Einführung in die Ligandenfeldtheorie. Frankfurt am Main: Akademische Verlagsgesellschaft 1967.

SHAW, D. N.: The camouflage principle and trace element distribution in magmatic minerals. J. Geol. **61**, 142 (1953).

SMITH, G. S., and L. E. ALEXANDER: Refinement of the atomic parameters in α-quartz. Acta Cryst. **16**, 462 (1963).

SHUBNIKOV, A. V., and N. V. BELOV: Colored symmetry. (Translated by ITZHOFF and GOLLOL.) Oxford: Pergamon Press 1964.

SÜSSE, P.: Verfeinerung der Kristallstruktur des Malachits, $Cu_2(OH)_2CO_3$. Acta Cryst. **22**, 146 (1967).

WAGNER, C.: Beitrag zur Theorie des Anlaufvorganges. Z. Physik. Chem. B **21**, 25 (1933).

WASASTJERNA, J. A.: On the radii of ions. Soc. Sci. Fennica, Commentationes Phys.-Math. I, 38, 1923.

WELLS, A. F.: Malachite: re-examination of crystal structure. Acta Cryst. **4**, 200 (1951).

WHITTAKER, E. J. W.: Factors affecting element ratios in the crystallization of minerals. Geochim. Cosmochim. Acta **31**, 2275 (1967).

YOUNG, R. A., and B. POST: Electron density and thermal effects in alpha quartz. Acta Cryst. **15**, 337 (1962).

ZALKIN, A., and D. H. TEMPLETON: The crystal structures of yttrium fluoride and related compounds. J. Am. Chem. Soc. **75**, 2453 (1953).

ZINTL, E., u. A. UDGÅRD: Über die Mischkristallbildung zwischen einigen salzartigen Fluoriden von verschiedenem Formeltyp. Z. Anorg. Allgem. Chem. **240**, 150 (1939).

CHAPTER 3

K. B. KRAUSKOPF

THERMODYNAMICS USED IN GEOCHEMISTRY

Introduction

Thermodynamics is useful in geochemistry as a means of coordinating data, of predicting reactions, and especially of setting limits to the chemical processes that are possible under various assumed natural conditions. Like any application of mathematical methods to complex natural situations, the use of thermodynamics for geological purposes requires the exercise of judgment. In any particular problem a course must be steered between over-use of thermodynamic equations, which all too easily give numerical results of spurious accuracy, and under-use of the powerful tools that thermodynamics makes available.

Among the limitations of thermodynamic reasoning applied to nature, the following are especially important:

(a) Lack of Information About Reaction Rates. Thermodynamics gives precise information about what reactions are possible from the standpoint of energy, but no information at all about how fast the reactions will take place.

(b) Requirement of Equilibrium. Classical thermodynamics deals almost entirely with closed systems that have reached or are approaching equilibrium, whereas most geologic environments are partly open systems in which complete attainment of equilibrium is uncommon.

(c) Scarcity of Data. Experimental data from which thermodynamic properties can be derived are scarce or lacking for many substances of geologic interest, especially for silicate minerals and for solutions at high temperatures.

(d) Limitation to Materials of Simple Composition. Thermochemical data are available chiefly for pure substances and simple mixtures, whereas in nature most materials contain a variety of impurities. Thermodynamic properties of natural substances are seldom identical with those of the corresponding pure laboratory chemicals.

In view of these limitations, it is perhaps surprising how often thermodynamics can be used to provide insight into geochemical problems.

The following paragraphs are a brief summary of the more important thermodynamic concepts and equations that have found use in geochemistry. Further theoretical details, illustrative examples, and tables of thermochemical properties for materials of geologic interest may be sought from the references in the appended list.

I. Perfect Gases and Perfect Solutions

The perfect (or ideal) gas and the perfect (or ideal) solution play important roles in thermodynamics. They are abstract concepts, corresponding to no real gases or solutions. Real gases approach the behavior of a perfect gas as the pressure is made indefinitely small, and real solutions approach ideal behavior as they become indefinitely dilute. Perfect gases and perfect solutions have the great theoretical advantage that simple and exact thermodynamic formulas can be set up to express their properties; then the properties of real gases and solutions can be conveniently described by noting their deviations from the behavior predicted for their ideal counterparts.

A *perfect gas* is defined as one which (1) obeys the equation of state

$$PV = nRT$$

where P is pressure, V is volume, n is number of moles, R is a universal constant called the *gas-law constant*, and T is absolute temperature; and (2) whose internal energy, E, is a function of temperature only.

A *perfect solution* is defined as one for which the fugacity of any component is equal to its mole fraction, x_1, multiplied by its fugacity in the pure state at the same temperature and pressure:

$$f_1 = x_1 f_1^0.$$

The fugacity will be defined later (p. 41). Approximately, the definition of a perfect solution means that the solution must obey RAOULT's law, i.e. that the vapor pressure of each component must be proportional to its mole fraction.

II. The Equilibrium Constant

Definition. For a reversible chemical reaction,

$$bB + dD + \cdots \rightleftharpoons yY + zZ + \cdots \tag{1}$$

the *equilibrium constant* at a given temperature may be defined as

$$K = \frac{(Y)^y (Z)^z \ldots}{(B)^b (D)^d \ldots} \tag{2}$$

where $B, D, \ldots$ are chemical formulas, $b, d, \ldots$ are coefficients, and parentheses around formulas indicate concentrations. The quotient of concentrations is only approximately constant; it would be strictly constant if all gases behaved as perfect gases and all solutions as perfect solutions. For many approximate calculations in geochemistry the assumption of constancy gives useful order-of-magnitude results, especially in dilute solutions and in gases at low pressures. For accurate calculations activities must be used rather than concentrations (p. 41).

The approximate constancy of K may be regarded as a generalization from experiment, or it may be derived as a necessary consequence of free-energy relationships (p. 57).

Units. The units of concentration commonly employed in equilibrium calculations are:

For gases: *partial pressure* in atmospheres.

For components of solid solutions and non-aqueous liquid solutions: *mole fraction* (symbol x), defined as the ratio of number of moles of a given constituent to the total number of moles of all constituents.

For solutes in aqueous solutions: either moles per liter of solution (*molar* concentrations, symbol M) or moles per kilogram of water (*molal* concentrations, symbol m). In dilute solutions the difference between molarity and molality is so small that for most geochemical calculations they may be used interchangeably. Mole fraction is sometimes used for aqueous solutions, especially for very concentrated solutions and for systems at high temperatures. Molar concentrations are convenient because they are the numbers determined directly by common analytical procedures; molal concentrations and mole fractions have the advantage that they are independent of temperature and pressure.

Conventions. Since equilibrium constants are usually expressed without units, their numerical values have meaning only within a framework of rigid conventions. Values listed in most standard tables are based on the following rules:

(a) Relation to the Equation. An equilibrium constant refers to a specific equation. Reversing the equation or multiplying it by a number does not affect the constancy of K, but changes its numerical value.

(b) Exponents. The concentration of each substance is raised to a power given by the coefficient of its formula in the equation.

(c) Units. Partial pressure in atmospheres for gases, molality for solutes in aqueous solution. When fugacities or activities (see next section) are used in place of concentrations, they are based on standard states which are defined in terms of these same units[1].

(d) Temperature. If no temperature is specified, 25° C is understood.

(e) Pressure. Equilibrium constants do not change with pressure, except as pressure may affect the degree to which gases approximate perfect-gas behavior and solutions perfect-solution behavior. If no pressure is specified, one atmosphere is understood.

(f) Concentrations of Pure Solids and Pure Liquids. Amounts of pure solids and pure liquids present do not affect an equilibrium, and their concentrations are assigned an arbitrary value of 1. Because they do not influence the value of K, they are often omitted from the quotient [Eq. (2)].

(g) Concentration of Water. For reactions that take place in aqueous solution, the concentration of water is generally much larger than that of any other substance, and it does not change appreciably even if H_2O appears in the equation. Because this concentration is practically constant, it is commonly incorporated in the value of K and does not appear in the quotient. (This does not apply, of course, to gas reactions in which water vapor plays a role.)

---

[1] Other units of concentration are sometimes convenient, but tney must be explicitly specified.

Special Kinds of Equilibrium Constants. Equilibrium constants for certain kinds of reactions are given special names:

(a) Solubility Product. For the ionization of a slightly soluble salt,

$$B_n D_m(s) \rightleftharpoons n B^{m+} + m D^{n-} \tag{3}$$

the equilibrium constant, called the *solubility product*, is

$$K = (B^{m+})^n (D^{n-})^m. \tag{4}$$

(b) Dissociation Constant. The ionization of a soluble weak electrolyte may be illustrated by the dissociation reaction for a weak acid:

$$H_n A(\text{aq}) \rightleftharpoons n H^+ + A^{n-}. \tag{5}$$

For this reaction the *dissociation constant* is

$$K = \frac{(H^+)^n (A^{n-})}{(H_n A)} \tag{6}$$

If the ionization reaction is written in the alternative form

$$n\,\text{H}_2\text{O} + H_n A(\text{aq}) \rightleftharpoons n\,\text{H}_3\text{O}^+ + A^{n-} \tag{7}$$

the constant remains the same, in accordance with convention (g) above. Most ionization reactions actually take place in steps, for example:

$$H_n A(\text{aq}) \rightleftharpoons H^+ + H_{n-1} A^- \tag{8}$$

$$H_{n-1} A^- \rightleftharpoons H^+ + H_{n-2} A^= \qquad \text{etc.} \tag{9}$$

For these reactions *stepwise* dissociation constants may be written; the *total* dissociation constant [Eq. (6)] is then the product of the stepwise constants.

(c) Formation Constants for Complexes. The dissociation of a complex ion or molecule in solution may be expressed by an ionization constant analogous to Eq. (6). Often, however, the equation is written in reverse,

$$y Y^{n+} + z Z^{m-} \rightleftharpoons Y_y Z_z^{(ny-mz)} \tag{10}$$

and the stability of the complex is then expressed by a *formation constant*:

$$K = \frac{(Y_y Z_z^{(ny-mz)})}{(Y^{n+})^y (Z^{m-})^z}\,. \tag{11}$$

(d) Partition Coefficient or Distribution Coefficient. When a solute distributes itself between two immiscible liquids, or in general when a substance distributes itself between two phases, an equilibrium is set up for which a constant can be written:

$$K = \frac{\text{concentration of solute in phase 1}}{\text{concentration of solute in phase 2}}\,. \tag{12}$$

This quotient is called a *partition coefficient* or *distribution coefficient*, and the constancy of the quotient is an expression of NERNST's *distribution law*.

III. Fugacity and Activity

Definitions. The K of Eq. (2), only approximately constant for concentrations, becomes strictly constant when fugacities or activities are substituted. For many geochemical purposes these two quantities are used essentially as modifications of analytically determined partial pressures or concentrations, the correction factors being taken from tables or calculated from simple semi-empirical equations. They will be discussed first from this point of view, and then defined later in terms of free energies.

The *fugacity* of a gas may be regarded as its partial pressure corrected for deviation from perfect-gas behavior. Since all gases approach perfect-gas behavior as the pressure is made indefinitely small, the ratio of fugacity and partial pressure approaches one as the pressure approaches zero. At pressures near 1 atm the deviation is generally small enough to be disregarded in most geochemical calculations, but at high pressures it becomes large. Fugacities of most common gases are recorded in standard tables, for a wide range of temperatures and partial pressures, either as fugacities themselves or as *fugacity coefficients*, which are ratios of fugacity to partial pressure.

Fugacity has also a wider meaning, applicable to solids and liquids as well as gases. It is most easily visualized as a vapor pressure, or more precisely the vapor pressure a substance would exert if its vapor behaved as a perfect gas.

More commonly, however, corrected concentrations of substances in liquid and solid solutions are expressed in terms of *activities*, defined as ratios of fugacities to fugacities in a standard state:

$$a = \frac{f}{f^0} \,. \tag{13}$$

The standard state is variously defined for different substances and different problems, and may be arbitrarily changed for convenience in particular situations. In most geochemical applications the following rules hold:

For gases: $a = f$. Thus activity is equal to partial pressure at very low pressures, and to partial pressure corrected for deviation from the perfect-gas law at higher pressures.

For solvents, liquid or solid: $a/x = 1$ when $x = 1$. This means that the activity of a pure substance is equal to 1, and that when the substance acts as a solvent its activity is approximately equal to mole fraction as long as the solution is dilute.

For solutes: either $a \rightarrow m$ when $m \rightarrow 0$, or $a/x = 1$ when $x = 0$. This says that activity is approximately equal to either molality or mole fraction in very dilute solutions.

The difference between activity and concentration for either the solvent or the solute is an expression of how much a solution deviates from perfect-solution behavior.

The ratio of activity to molality for a dissolved substance is called its *activity coefficient*:

$$\gamma = a/m \,. \tag{14}$$

The activity coefficient is a complex function of temperature, pressure, and the concentrations of all substances in the solution. Coefficients rather than activities themselves are commonly measured and listed in tables, and activities are calculated from them.

Units. Fugacity is expressed in the same units as partial pressure, most commonly in atmospheres. Regarding units of activity there is no general agreement. Strictly from its definition as a ratio of fugacities, activity should be dimensionless, and it is often so used. On the other hand activity plays the role of a modified concentration, and for some purposes it may be conveniently expressed in concentration units. If activity is dimensionless, the activity coefficient must have reciprocal dimensions of concentration; if activity is given concentration units, the activity coefficient must be dimensionless. In practice the choice of unit has little importance, since numerical values are the same whether activity is considered a pure number or a modified concentration.

Activity Constants. Activity in algebraic expressions is symbolized either by a or by brackets (or braces) around formulas. Thus for a general chemical equation [Eq. (1)] the equilibrium constant in terms of activities may be written as

$$K_a = \frac{a_Y^y \cdot a_Z^z \cdots}{a_B^b \cdot a_D^d \cdots} \tag{15}$$

or as

$$Ka = \frac{[Y]^y [Z]^z \ldots}{[B]^b [D]^d \ldots} \quad \left(\text{or} \quad \frac{\{Y\}^y \{Z\}^z \ldots}{\{B\}^b \{D\}^d \ldots}\right). \tag{16}$$

(Authors who favor braces for activities generally use square brackets for concentrations.) An equilibrium constant in the form of Eq. (15) or (16) is called an *activity constant* or a *thermodynamic equilibrium constant*, in contrast to the approximate constant (often written K_c) based on analytically determined concentrations. The relation between the two constants may be expressed in terms of activity coefficients:

$$K_a = \frac{(\gamma_Y\, m_Y)^y (\gamma_Z\, m_Z)^z \ldots}{(\gamma_B\, m_B)^b (\gamma_D\, m_D)^d \ldots} = \frac{\gamma_Y^y \cdot \gamma_Z^z \cdots}{\gamma_B^b \cdot \gamma_D^d \cdots} K_c. \tag{17}$$

Sometimes, especially for gas reactions, it is useful to express the equilibrium constant as a quotient of fugacities:

$$K_f = \textit{fugacity constant} = \frac{f_Y^y \cdot f_Z^z \cdots}{f_B^b \cdot f_D^d \cdots}. \tag{18}$$

Activities in Solutions of Strong Electrolytes. When a symmetrical strong electrolyte (e.g. NaCl or $CuSO_4$) is dissolved in water, it dissociates more or less completely into two ions whose activities may be represented by a_+ and a_-. These may be defined so as to approach molal concentrations of the ions at infinite dilution:

$$a_+/m_+ \to 1 \quad \text{and} \quad a_-/m_- \to 1 \quad \text{as} \quad m \to 0. \tag{19}$$

The activity of the electrolyte itself in solution is conveniently *defined* as a product of the ion activities:

$$a = a_+ a_-. \tag{2}$$

According to this definition, the activity of a strong symmetrical electrolyte approaches the square of the molality (rather than the molality itself, as in non-electrolyte solutions) as the solution becomes indefinitely dilute. This is merely a refinement of the familiar observation that a symmetrical electrolyte in dilute solution behaves as if its molal concentration were approximately twice as great as the concentration indicated by its formula.

The *mean activity* of the ions of a symmetrical electrolyte is defined as

$$a_{\pm} = (a_{+} a_{-})^{\frac{1}{2}} = a^{\frac{1}{2}} \tag{21}$$

and the *mean activity coefficient* as

$$\gamma_{\pm} = a_{\pm}/m_{\pm}\,. \tag{22}$$

The quantities $a_{\pm}$ and $\gamma_{\pm}$ are experimentally measurable, but the single-ion activities and activity coefficients are not. As an operational statement, therefore, Eq. (21) serves better than Eq. (20) to define the activity of a symmetrical electrolyte.

For an unsymmetrical electrolyte (e.g. Na_2SO_4 or $AlCl_3$) the relations are more complicated. If dissociation gives a total of n ions (n_+ positive ions and n_- negative ions), then

$$a_{\pm} = a^{1/n} = (m_{+}^{n+} \cdot m_{-}^{n-} \cdot \gamma_{+}^{n+} \cdot \gamma_{-}^{n-})^{1/n} \tag{23}$$

where γ_+ and γ_- are activity coefficients of the individual ions. Also

$$\gamma_{\pm} = (\gamma_{+}^{n+} \cdot \gamma_{-}^{n-})^{1/n} = \frac{a_{\pm}}{(m_{+}^{n+} \cdot m_{-}^{n-})^{1/n}} = a_{\pm}/m_{\pm}\,. \tag{24}$$

Although single-ion activities and activity coefficients ($a_+, a_-, \gamma_+, \gamma_-$) are not directly measurable, values can be assigned to these quantities by assuming a relation between γ_+ and γ_- for some one compound. A common assumption is that the activity coefficients for the two ions of KCl are equal; from this assumption the coefficients for all other ions can be determined. The basis of the assumption is far from secure, but the derived single-ion activity coefficients are often useful in geochemical calculations.

Estimation of Activity Coefficients. Activities differ from concentrations because of complex interactions between solvent and solute molecules. They are not rigorously predictable from megascopic or molecular properties of pure substances, although the Debye-Hückel theory, by making use of a number of simplifying assumptions, gives a basis for calculating approximate values in very dilute aqueous solutions. A variety of experimental methods have been devised for determining activities and activity coefficients in more concentrated solutions, and the results show enough consistency so that some useful empirical rules can be formulated:

In general, activity coefficients in a complex electrolyte solution depend roughly on the *ionic strength* of the solution, defined as

$$I = \tfrac{1}{2} \sum_i c_i z_i^2 \tag{25}$$

where c_i is the concentration of any ion, z_i is its charge, and the summation is taken over all ions in the solution.

For neutral molecules, activity coefficients are generally greater than 1, and are less than 1.02 for ionic strengths under 0.1. Up to $I = 5$ the relation

$$\log \gamma = kI \tag{26}$$

holds approximately; the constant k is about 0.1 for small molecules.

For the solvent water, the activity is measured by the ratio of the fugacity of water in the solution to the fugacity of pure water; since water vapor at usual temperatures behaves approximately as a perfect gas, this is the same as the ratio of vapor pressures. The activity is within 1 per cent of unity at ionic strengths up to 0.5.

For ions, activity coefficients are generally less than 1, although for some kinds of ions in concentrated solutions they may be much greater. Activity coefficients are smaller for multivalent ions than for univalent ions; in solutions with ionic strength 0.1, for example, coefficients for univalent ions are approximately 0.8, for many divalent ions about 0.4, for many trivalent ions about 0.1. This means that activity corrections for multivalent ions are important even in dilute solutions.

The Debye-Hückel theory, by treating ions as point charges surrounded by an "atmosphere" of other ions in a continuous dielectric medium, leads to fairly simple equations for calculating activity coefficients in solutions with ionic strengths less than 0.1. The approximate formulas

$$\log \gamma_{\pm} = -A z_{+} z_{-} I^{\frac{1}{2}} \tag{27}$$

for the mean activity of a binary electrolyte, and

$$\log \gamma = -A z^2 I^{\frac{1}{2}} \tag{28}$$

for individual ion activity coefficients, show good agreement with experiment in solutions of ionic strength up to 0.001. (A is a constant depending on the temperature and dielectric constant of the solvent, equal to 0.51 for water at 25° C, and z is ionic charge.) At higher ionic strengths the formulas from the theory become somewhat more complex, and parameters in addition to A are introduced; the equations and tables of parameters are given in standard references.

Table 3–1. *Activity coefficients calculated from* DAVIES' *equation*

Ionic	Ionic charge		
strength	± 1	± 2	± 3
0.001	0.97	0.87	0.73
0.005	0.93	0.74	0.51
0.01	0.90	0.66	0.40
0.05	0.82	0.45	0.16
0.1	0.78	0.36	0.10
0.2	0.73	0.28	0.06
0.5	0.69	0.23	0.04
0.7	0.69	0.23	0.04

For many geochemical applications a simpler equation suggested by DAVIES, derived empirically but roughly similar in form to the Debye-Hückel equations, gives sufficiently accurate values even at ionic strengths greater than 0.1:

$$\log \gamma = -A z^2 \left(\frac{I^{\frac{1}{2}}}{1 + I^{\frac{1}{2}}} - 0.2\, I \right) \tag{29}$$

Activity coefficients obtained from this equation are shown in Table 3–1. It should be emphasized that these numbers are no more than approximations, generally correct within about 3 per cent at $I = 0.1$ and 8 per cent at $I = 0.5$, but subject to larger errors for some ions and some solutions. More accurate values are obtainable from the Debye-Hückel equations at ionic strengths below 0.1, but for accurate values at higher ionic strengths there is no recourse except to actual measurement.

IV. The Laws of Thermodynamics

a) Internal Energy

The equilibrium constant for a reaction gives information as to whether the substances involved will react when they are mixed, and how far the reaction will go before it reaches a state of equilibrium. The same kind of information can be expressed more concisely by means of *thermodynamic functions* for each of the reacting substances: internal energy, enthalpy, entropy, and free energy. These functions are useful also in formulating equations from which changes of equilibrium constants with temperature, pressure, and other variables may be calculated.

The *internal energy* of a system is the sum of the energies of its constituent particles — molecules, atoms, subatomic particles. When the system undergoes change (for example, when heat is added to it or when an external system does work upon it), the internal energy changes. The absolute value of internal energy is not measurable, but changes in energy may be readily measured. It is only the changes that are important in applications of thermodynamics, but the hypothetical absolute values appear in many thermodynamic equations.

The internal energy is symbolized by E or U, and a change by ΔE or ΔU:

$$\Delta E = E_f \text{ (for final state)} - E_i \text{ (for initial state)}. \tag{30}$$

Specifically, the change in energy during a chemical reaction is equal to the sum of the energies of the products minus the energies of the reactants. If the reaction gives out energy, the change in energy is negative.

b) The First Law

The internal energy of a system may be increased by adding heat to it or by doing work upon it, or by both at once. In symbols,

$$\Delta E = Q + W. \tag{31}$$

This is the *first law of thermodynamics*, which is one way of expressing the law of conservation of energy. In differential form the equation becomes

$$dE = \delta Q + \delta W. \tag{32}$$

The differentials of Q and W are written with a special symbol to indicate that these are not perfect differentials, in other words that a given change in energy dE may be produced by many combinations of changes in Q and W. The change in E, on the other hand, is a perfect differential because it depends only on the initial and final states of the system. The same idea may be expressed by saying that E is a *property* of the system, meaning that E depends only on the thermodynamic state of the system, whereas Q and W are arbitrary quantities which depend on the method by which the change is accomplished. Moreover, E may be described as an *extensive property* of the system, since its value changes with the amount of material present, in contrast to *intensive properties* like temperature and density.

Some authors prefer to state Eqs. (31) and (32) with a negative sign,

$$\Delta E = Q - W \tag{33}$$

$$dE = \delta Q - \delta W. \tag{34}$$

In this formulation a system is visualized as gaining energy by absorbing heat and losing energy by doing work on its surroundings. The two ways of stating the first law are equivalent if one remembers that positive values of Q and W refer to heat and work *added to* a system by its surroundings, while negative values refer to heat and work *lost from* a system and gained by its surroundings.

c) Enthalpy

Definition. The *enthalpy* or *heat content* of a system is defined by the equation

$$H = E + PV. \tag{35}$$

Like E, the enthalpy has no measurable absolute value, but differences in enthalpy are easily measured and play an important role in thermodynamics. For a megascopic change, for example the change resulting from a chemical reaction,

$$\Delta H = \Delta E + \Delta(PV) \tag{36}$$

and for an infinitesimal change

$$dH = dE + d(PV) = dE + PdV + VdP. \tag{37}$$

The enthalpy of a system, again like E, is an extensive property; its changes are determined only by the initial and final states of the system, not by the path by which the changes may be brought about, and the amount of change depends on the amount of material present.

Reactions at Constant Volume and Constant Pressure. Many reactions of chemical and geological interest involve no production of work except the work of expansion against a resisting pressure. For such a reaction the work (done *by* the reaction mixture *on* its surroundings, hence negative) is represented by $-P\Delta V$ if pressure is constant, or by

$$W = -\int PdV \tag{38}$$

if pressure is variable. For a process occurring at constant pressure, Eqs. (31) and (32) become

$$\Delta E = Q - P\Delta V \tag{39}$$

and

$$dE = \delta Q - PdV. \tag{4}$$

Also in a constant-pressure process Eq. (37) becomes

$$dH = dE + PdV \tag{41}$$

and substitution of Eq. (40) gives

$$dH = \delta Q. \tag{42}$$

For a macroscopic change at constant pressure,

$$\Delta H = Q. \tag{43}$$

In other words, when pressure is constant the change in enthalpy of a system is measured by the heat lost or gained.

If a process occurs at constant volume rather than constant pressure, no pressure-volume work can be done, and therefore

$$\Delta E = Q. \tag{44}$$

If both volume and pressure remain constant, a condition that holds approximately for many reactions that do not involve gases,

$$\Delta E = \Delta H = Q. \tag{45}$$

Units. Changes in internal energy and enthalpy are measured in heat units, commonly calories or kilocalories. Work is generally measured in cm^3-atm or liter-atm, and these units may be converted to other energy units by the relations

$$1 \text{ kcal} = 1000 \text{ cal} = 4{,}184.0 \text{ joules} = 41{,}300 \text{ cm}^3\text{-atm} = 41.3 \text{ liter-atm}.$$

Conventions Regarding ΔH. The enthalpy change in a chemical reaction is easily measured by determining the heat of reaction in a constant-pressure calorimeter [Eq. (43)]. If heat is evolved, the heat of reaction is considered positive; but the enthalpy change is the enthalpy of the products minus that of the reactants, and therefore negative. Hence ΔH is numerically equal to the heat of reaction, but the signs are opposite. ΔH is negative for an exothermic reaction, positive for an endothermic reaction. (Note that heat of reaction has a sign opposite to that of Q in Eqs. (39) to (45), since Q is heat *added to* a system while heat of reaction is conventionally regarded as heat *evolved by* a reaction mixture.)

The numerical value of ΔH for a reaction, unless specified otherwise, means the enthalpy change for the number of moles indicated by the equation; if the equation is multiplied by a factor, ΔH is multiplied also. For example, the statement

$$2\,HCl \rightleftharpoons H_2 + Cl_2; \qquad \Delta H_{298,\,1\,\text{atm}} = +44.12 \text{ kcal}$$

means that 44.12 kcal are required to decompose 2 moles of HCl into its elements at 25° C and 1 atm pressure. The gas mixture need not remain at this temperature and pressure while the decomposition is taking place, but the products must be brought back to 25° C and 1 atm before the heat change is measured.

Standard Enthalpies of Formation. The enthalpy of formation of a compound is the enthalpy change when the compound is formed from its elements. The absolute value of enthalpy for neither compound nor elements is known, but for numerical computation it is convenient to *assume* that the enthalpies of all elements in some state selected as standard are zero. The usual *standard state* chosen for a liquid or solid element is its most stable form at a given temperature and 1 atm pressure; the standard state for a gaseous element is the perfect gas state, or the limit for the real gas as pressure approaches zero, which for most gases differs little from the state at 1 atm. Referred to these standard states of the elements, the enthalpy of a compound at any temperature is its *standard enthalpy of formation*, ΔH_f^0. These standard enthalpies are recorded in tables of thermodynamic data, usually for a temperature of 25° C. From them the standard enthalpy change for a reaction, ΔH^0, may be calculated by adding the enthalpies for the products and subtracting the sum of enthalpies for the reactants.

d) Heat Capacity

When heat is supplied to a given mass of a substance, the heat absorbed is proportional to the temperature rise, and the proportionality constant is called the *heat capacity* of the substance. If the mass is one gram, the number is given the special name *specific heat*; more important in thermodynamics is the figure for one mole, the *molal heat capacity*. Since heat capacity changes slightly with temperature, a more rigorous definition is given by the equation

$$C = \frac{\delta Q}{dT} \tag{46}$$

where δQ is an infinitesimal amount of heat added and dT is the corresponding temperature rise. The value of C depends on how the heat is added; if pressure is maintained constant, $\delta Q = dH$ [Eq. (42)], so that

$$C_P = \left(\frac{\partial H}{\partial T}\right)_P \tag{47}$$

while if volume is constant, $\delta Q = dE$, so that

$$C_V = \left(\frac{\partial E}{\partial T}\right)_V . \tag{48}$$

Heat capacities depend on the various ways in which the particles of a substance can take up kinetic and potential energy as the temperature rises. They are not derivable thermodynamically from more fundamental quantities, but for some simple substances may be calculated from atomic and molecular data. Most heat capacities, however, are determined empirically. Generally, measured values of heat capacities can be expressed with sufficient accuracy for chemical purposes by means of power series involving no more than three terms, either

$$C_P = a + bT + cT^2 \tag{49}$$

or

$$C_P = a' + b'T + c'T^{-2}. \tag{50}$$

More terms can be added when greater accuracy is needed. For many geochemical purposes the first two terms give sufficient accuracy, and good estimates can often be made by using the first term alone ($C_P = \text{a constant}$). Standard heat capacities are commonly recorded in tables by listing values of the constants a, b, and c for one of the two equations, the measurements referring to conditions of 1 atm pressure and a specified temperature range.

The change in heat capacity during a reaction may be obtained by summing heat capacities of products and reactants and subtracting:

$$\Delta C_P = \Sigma C_P \text{ (for products)} - \Sigma C_P \text{ (for reactants)} = \Delta a + \Delta b T + \Delta c T^2 \tag{51}$$

[if equation (49) is used for the individual heat capacities].

A few rules about heat capacities, in part empirical and in part derived from theories of atomic and molecular motions, are useful in estimating order-of-magnitude values when experimental data are lacking:

For solids at high temperatures the heat capacity per gram atom is roughly 6 cal/degree; in other words, the number of atoms in the formula multiplied by 6 gives the approximate heat capacity per mole. The temperature above which this rule holds varies from one solid to another; for most common minerals it is valid within 10 per cent above 500° C. *At room temperature* most rock-forming minerals have gram-atomic heat capacities between 3.5 and 5.0 cal/degree. *At still lower temperatures* the heat capacity of all solids falls off, approaching zero in the neighborhood of absolute zero.

For pure liquids with fairly simple formulas, heat capacities are similar to those of the corresponding solids, generally a little lower.

For monatomic gases C_V is approximately 3 cal/degree and C_P approximately 5; *for diatomic gases* C_P is generally between 6 and 8; for gases with more complex molecules no simple generalization is possible.

For solutions no widely applicable rules can be framed. The scarcity of information on heat capacities of ions in electrolyte solutions is a major gap in heat-capacity data.

The difference between C_P and C_V for perfect gases is equal to the gas-law constant, $R = 1.987$ cal/degree-mole. For most solids of geologic interest the difference is very small, generally less than 0.1 cal/degree-mole.

If the enthalpy change in a reaction is known at one temperature, heat capacities provide the necessary data for calculating the change at other temperatures. From Eq. (47) the heat capacity change during a reaction may be expressed as

$$\Delta C_P = \left(\frac{\partial \Delta H}{\partial T}\right)_P . \tag{52}$$

Integration gives

$$\Delta H = \int_0^T \Delta C_P dT = \Delta a T + \frac{\Delta b}{2} T^2 + \frac{\Delta c}{3} T^3 + \Delta H_0 \tag{53}$$

where the heat capacities of reactants and products have been expressed in terms of the empirical constants *a*, *b*, *c* of Eq. (49), and ΔH_0 is the constant of integration. The value of ΔH_0 is found from the known ΔH at one value of T; theoretically it would be the value of ΔH at absolute zero if Eq. (53) could be extrapolated to very low temperatures, which is generally not valid. For most reactions of geochemical interest the change of ΔH with temperature is not large, commonly amounting to only a few kilocalories even for temperature ranges of 1000° and more. The changes are especially small for solid-solid reactions, but may be important for reactions involving gases or aqueous solutions.

e) Entropy

Why Does a Reaction Occur? The answer that first comes to mind is simply, "Because it gives out energy." Seemingly any exothermic reaction (ΔH_0 negative) should take place spontaneously if the reactants and products are mixed in stoichiometric proportions; in other words, the equilibrium constant should be greater than 1, and an equilibrium mixture should contain more of the products than of the reactants. In general this expectation is fulfilled, especially if ΔH^0 has a *large* negative value, say greater than 10 kcal.

Predictions on this basis are sometimes useful in geochemistry, but they cannot be strictly accurate for the simple reason that some spontaneous reactions are endothermic (ΔH^0 positive). Evidently the enthalpy change is not the only factor that determines whether or not a given reaction is possible.

To visualize the other factor, consider a simple change in which ΔH is zero. Let two gases mix, and assume perfect-gas behavior. In effect each gas expands into the space occupied by the other, and for the expansion of a perfect gas there is no enthalpy change. The mixing is spontaneous, but the "driving force" of the process is an increase in randomness rather than a decrease in energy. The gas mixture is a more random, more disordered, more probable state of the system than the two separate pure gases. Evidently a reaction may take place either because it has a negative ΔH, or because it results in an increase of randomness, or both. If the reaction has a positive ΔH, it may still take place spontaneously if the increase in randomness is large enough.

Definition of Entropy. The measure of increase in randomness of a system is a quantity called *entropy*. Any natural spontaneous process involves some increase in randomness, hence an overall increase in entropy. In other words, natural processes always lead to greater disorder, to a progression from a state of low probability to a state of higher probability. Entropy is a quantitative expression of this tendency.

While the entropy of an overall natural process always increases, the process may be dissected into steps in which entropy may increase, decrease, or remain constant. For purposes of quantitative definition, it is convenient to imagine an idealized process, or part-process, in which entropy may be transferred from one substance to another but in which the total amount does not change. Such a process is described as *reversible*, which means that it is conducted in such a manner that at any time a slight change in external conditions would lead to reversal of the process. If, for example, a quantity of gas is supplied with heat so that it expands against an external pressure, the temperature of the gas remaining constant, the experiment can be arranged so that the internal and external pressures are at all times nearly equal; then a very slight increase of external pressure would cause the gas to be compressed and give up heat to its surroundings. Under these conditions the gas is said to increase in entropy while it is absorbing heat, and *the amount of entropy increase is defined as the ratio of heat absorbed to the constant absolute temperature*:

$$dS = \frac{\delta Q}{T} \text{ for infinitesimal absorption of heat} \tag{54}$$

$$\Delta S = \frac{Q}{T} \text{ for megascopic absorption of heat.} \tag{55}$$

The surroundings lose heat, and hence lose entropy, at the same rate, so that the entropy change of the total system (gas + surroundings) does not change. It is characteristic of reversible processes that entropy is conserved, just as energy is conserved.

Reversibility is an ideal, never actually attainable either in nature or in the laboratory. All real processes are to some extent *irreversible*. This is true both for the simple transfer of heat from one system to another and for processes that involve conversion of heat into work. If heat is added to a real gas confined by a movable piston, for example, not all of the heat can go into expansion work against the external pressure, no matter how carefully the experiment is set up. In other words, the entropy of the overall system would increase. To get a quantitative measure of the increase, suppose that the idealized experiment of the last paragraph is repeated, but this time in a completely irreversible manner, by letting the gas expand freely into an evacuated space equal in volume to the space occupied in the preceding experiment. This time the gas will have done no work, and no heat will have been added to it from the surroundings. The change in entropy, however, will be the same as before, because the gas has attained the same state of higher probability or randomness with respect to its original compressed state. In an ideally irreversible process, therefore, entropy may increase with no heat exchange at all; but *the amount of the increase is calculated by reference to the ideally reversible process*, in which heat moves from one part of the system to another.

The increase in entropy of the gas is the same in the above experiments whether the process of expansion is reversible, irreversible, or partly one and partly the other. Its amount depends solely on the initial and final states of the gas. Entropy is therefore a *property* of the gas, just as internal energy and enthalpy are properties. Like them also, entropy is an *extensive* property, since the heat absorbed in the reversible process depends on the amount of gas present.

f) The Second Law

The statement that natural processes in an isolated system always result in an increase of total entropy is one formulation of the *second law of thermodynamics*. The law may be stated in many other ways: natural processes always lead to an overall increase in randomness or disorder; natural changes proceed always from less probable to more probable states; heat flows spontaneously from hot objects to cold, never from cold to hot; a self-acting machine cannot transfer heat from a cold object to a hot object. These are all generalizations from accumulated experience with many kinds of energy exchanges. They are important in that they specify the *direction* in which natural processes may be expected to go. The first law says only that total energy in a natural process remains unchanged, but this requirement is satisfied whether the process goes toward increasing order or increasing disorder. The second law adds the requirement that the direction must be toward disorder.

The mathematical statement of the second law is merely Eq. (54). The equation itself says nothing about randomness or disorder, but it gives a quantitative definition of the function by which randomness and disorder are measured.

A combination of the first and second laws [Eq. (40) and (54)] leads to useful expressions for changes in internal energy and enthalpy:

$$dE = T\,dS - P\,dV \tag{56}$$

$$dH = T\,dS + V\,dP \tag{57}$$

g) The Third Law

A relation between entropy change and heat capacity at constant pressure is obtained by combining Eqs. (46) and (54):

$$dS = C_P \frac{dT}{T} = C_P\, d\ln T. \tag{58}$$

Integration gives

$$S_2 - S_1 = \int_{T_1}^{T_2} C_P \ln T \tag{59}$$

for the increase in entropy of a substance when the temperature is raised from T_1 to T_2. The integral can be evaluated if C_P is expressed in terms of temperature, say by using Eq. (49). At very low temperatures the relation between C_P and T is complicated, and the integration is most easily performed graphically on a plot of C_P against $\log T$. Since heat capacities fall rapidly toward zero as T approaches absolute zero (C_V is approximately proportional to T^3 below about 20° K, according to the Debye law), the graphical integration can be extended to absolute zero with only slight extrapolation, and hence accurate values can be obtained for the increase in entropy when a substance is heated from absolute zero to any desired temperature.

Comparison of such experimentally determined heat-capacity curves for many substances permits the generalization that *if the entropy of each element in some crystalline state is assumed to be zero at absolute zero, then the entropies of other pure crystalline solids are zero at this temperature.* This is the *third law of thermodynamics.* The requirement of pure crystalline solids should be emphasized: a glassy solid, or a crystalline solid showing defects or containing impurities, will in general have a small finite entropy at absolute zero.

The third law is important because it gives a means of obtaining absolute entropies for any substance at any temperature. For a pure crystalline solid that undergoes no phase change between 0° K and a temperature T, one need only measure heat capacities over this temperature range, plot C_P against $\log T$, and measure the area under the curve. If the substance undergoes a phase change, it absorbs heat at a constant temperature and pressure, and its entropy is increased by [Eqs. (43) and (55)]:

$$\Delta S = \frac{Q}{T} = \frac{\Delta H}{T} \tag{60}$$

where ΔH is the enthalpy change of the phase transition (= heat of fusion, heat of vaporization, or heat of a solid-solid transition). Hence the total entropy of a substance at 298° K and 1 atm pressure is given by a summation of terms, some of them integrals like Eq. (59) and some of them phase-transition terms like Eq. (60):

$$\begin{aligned} S^0_{298} = \int_0^{T_1} C_P\, d\ln T + \frac{\Delta H_1}{T_1} + \int_{T_1}^{T_2} C_P\, d\ln T + \frac{\Delta H_2}{T_2} + \cdots \\ + \frac{\Delta H_r}{T_r} + \int_{T_r}^{298} C_P\, d\ln T, \end{aligned} \tag{61}$$

where $T_1, T_2, \ldots T_r$ are temperatures of the various transition points and ΔH_1, $\Delta H_2, \ldots \Delta H_r$ are the corresponding enthalpy changes.

Values of entropies so calculated are recorded as *standard entropies,* S^0, in tables of thermodynamic data. From these the standard entropy change for a reaction, ΔS^0, may be found by adding entropies for the products and subtracting the sum of entropies for the reactants. Customary units are calories/mole-degree, often called *entropy units* and abbreviated eu. Numerical entropy values are handled similarly to enthalpy values, with two notable differences: (1) entropies are absolute values, and are not zero for elements at temperatures above 0° K, whereas enthalpies are relative values based on the assumption of zero enthalpy for elements in their standard states at any temperature; (2) entropies are commonly given in calories, enthalpies usually in kilocalories.

Absolute entropies for many substances with simple molecules can also be calculated from spectroscopic data. In general the agreement with values calculated from heat-capacity measurements is excellent.

V. Free Energy

Entropy as a Criterion of Equilibrium. Any spontaneous natural process involves an overall increase in entropy. Changes that occur in one part of a larger system may seem at times to violate this requirement; for example, the part-system represented by a living organism uses the relatively unorganized energy of food to make highly ordered structures within its body, hence accomplishes a local decrease in entropy. But if all the energy changes in the larger system including the organism — the surrounding air, water, soil — are considered, the increase of entropy in most of these changes will more than counterbalance the local decrease within the organism.

A system in which no overall increase of entropy can occur is a system in which spontaneous change is impossible, hence a system at equilibrium. Entropy thus furnishes a possible criterion for the recognition of equilibrium: if the sum of entropy changes for all possible processes in an isolated system is zero, the system must be at equilibrium. The only proviso is that *all* possible reactions must be considered. The system must be completely isolated, or it must be taken so large that reactions within it can have no appreciable effects on the surroundings.

The requirement that all conceivable changes must be included in the summation of entropies makes the entropy criterion of equilibrium difficult to apply. In most problems attention is focused on a single reaction, which may or may not be part of a larger system. For this particular reaction, under specified conditions of temperature, pressure, and concentration of the reactants, is further change possible or has equilibrium been reached? Such a question cannot be answered easily by the entropy criterion alone, for in general the reaction will take up energy from or add energy to its surroundings. A more convenient criterion of equilibrium for individual reactions, which involves both energy and entropy changes, is provided by the property called free energy.

Definition of Free Energy. Two kinds of free energy are defined by the expressions

$$A = E - TS \tag{62}$$

$$G = H - TS. \tag{63}$$

The first is called *Helmholtz free energy* or *work content,* and the second *Gibbs free energy, Gibbs function,* or *free enthalpy.* The latter is the more important for most chemical and geochemical purposes, and the simple expression "free energy" refers to the Gibbs function unless it is otherwise specified. The use of A and G as symbols for the two kinds of free energy has been established by international agreement[1], but unfortunately a number of other symbols are in common use: A is often re-

[1] International Union of Pure and Applied Chemistry.

presented by F, especially in European literature; G is designated F in much American literature, and Z in many Russian papers.

Since E, H, and S are extensive properties, it follows that A and G are extensive properties also. Likewise, since E and H do not have defined absolute values, A and G also cannot be given absolute values. Macroscopic changes in A and G under isothermal conditions may be formulated

$$\Delta A = \Delta E - T\Delta S \tag{64}$$

$$\Delta G = \Delta H - T\Delta S \tag{65}$$

and infinitesimal changes by

$$dA = dE - TdS \tag{66}$$

$$dG = dH - TdS. \tag{67}$$

Infinitesimal changes without the restriction of constant temperature are expressed by the total differentials

$$dA = dE - TdS - SdT \tag{68}$$

$$dG = dH - TdS - SdT. \tag{69}$$

Substitution of Eqs. (56) and (57) gives

$$dA = -PdV - SdT \tag{70}$$

$$dG = VdP - SdT. \tag{71}$$

The four Eqs. (56), (57), (70), and (71) are useful in deriving other thermodynamic relations, especially the dependence of the thermodynamic functions on pressure, temperature, and volume. The equations may be easily converted to forms applicable to reactions as well as to changes of E, H, A, and G for single substances. For example, if Eq. (71) is written for each substance taking part in a reaction, the overall infinitesimal change in free energy for the reaction is

$$\Delta dG = d\Delta G = \Delta VdP - \Delta S\, dT. \tag{72}$$

Although this equation is derived under the stringent condition of reversibility, it can be shown to hold for irreversible processes as well.

Free-Energy Change in a Chemical Reaction. Free energies for a reaction are handled in the same manner as enthalpies. Since absolute values are undefined, free energies of elements in their standard states may be assigned the arbitrary value zero; the standard states are defined as for enthalpies (p. 47), with the minor difference that the standard state for a gas is chosen as the perfect-gas state at 1 atm rather than at 0 atm. Then the standard free energy of formation of a compound is the change in free energy when a mole of the compound is formed from its elements, and this figure is often called simply "the free energy" of the compound. Such figures are the *standard free energies* (ΔG_f^0 and ΔA_f^0) recorded in tables of thermodynamic data, usually for a temperature of 25° C. The standard free-energy change for a reaction, ΔG^0, is obtained by adding free energies for the products and substracting the sum of free energies for the reactants.

Free-Energy Change as the "driving force" of a Reaction. The change in free energy during a reaction is made up of two energy terms [Eq. (65)]: the change in enthalpy, ΔH, representing chiefly energy changes resulting from the breaking and formation of chemical bonds; and a term involving entropy, $T\Delta S$, which measures the change in randomness or disorder resulting from the reaction. Hence free energy is a better measure than enthalpy of the "driving force", or tendency of a reaction to take place, since it includes not only the heat of reaction but also a quantity expressing change in randomness. A reaction may occur spontaneously *either* because it is exothermic (ΔH negative), *or* because it leads to increased randomness (ΔS positive); the two tendencies may work together to cause a reaction to take place, or they may work in opposite directions so that the effect of one is partly nullified by the other.

Free Energy as a Criterion of Equilibrium. If the $T\Delta S$ term for a reaction exactly balances the ΔH term, the foregoing argument suggests that the reaction would have no tendency to take place, hence should be at equilibrium. This idea can be made more precise by considering the relation of the reaction mixture to its environment.

Suppose that the reaction mixture is a small part of an environment taken large enough so that it may be considered an isolated system; suppose that equilibrium in the mixture has been established, and that temperature and pressure are constant. The general criterion for equilibrium in the large system is that the sum of entropy changes for all possible reactions must be zero (p. 53). The sum of possible entropy changes may expressed as the entropy change within the reaction mixture plus entropy changes in the environment:

$$\Delta S_{\text{large, isolated system}} = \Delta S_{\text{environment}} + \Delta S_{\text{reaction mixture}} = 0. \quad (73)$$

For an equilibrium mixture at constant temperature and pressure, $\Delta S = Q/T = \Delta H/T$ (Eqs. 43 and 55); hence

$$\Delta S_{\text{reac}} = \frac{\Delta H_{\text{reac}}}{T} = -\Delta S_{\text{environment}}. \quad (74)$$

For the reaction mixture, therefore,

$$\Delta S - \frac{\Delta H}{T} = 0 \quad (75)$$

and

$$T\Delta S - \Delta H = -\Delta G = 0. \quad (76)$$

For infinitesimal changes the same steps lead to

$$TdS - dH = -dG = 0. \quad (77)$$

Thus heat may flow from the environment into the reaction mixture or vice versa, and entropy may be exchanged between them, but a necessary and sufficient condition for equilibrium within the reaction mixture is that no process can lead to a free-energy change.

The derivation assumes that temperature and pressure remain constant during any possible change. Actually it is only necessary that the temperature and pressure be the same at the end of a change as at the beginning, for dG and ΔG depend only on the initial and final states of a system.

If volume is kept constant instead of pressure for any infinitesimal change, a similar derivation gives as a criterion of equilibrium

$$dA = 0. \quad (78)$$

For a reaction mixture not at equilibrium, the total entropy of mixture-plus-environment must increase as the reaction takes place:

$$\Delta S_{\text{large system}} = \Delta S_{\text{environment}} + \Delta S_{\text{reaction}} > 0. \quad (79)$$

Heat lost by the mixture must equal heat gained by the environment; and heat absorbed by the environment is a measure of entropy increase in the environment, regardless of whether heat is generated by the reaction mixture reversibly or irreversibly. Hence, for a reaction at constant temperature and pressure,

$$\Delta S_{\text{environment}} = Q_{\text{environment}}/T = -\Delta H_{\text{reaction}}/T \quad (80)$$

and

$$\Delta S_{\text{react.}} - \Delta H_{\text{react.}}/T = -\Delta G/T > 0. \quad (81)$$

Thus a reaction can take place only if $\Delta G < 0$, and the magnitude of ΔG is a measure of how far the reaction mixture is from equilibrium.

The corresponding criterion for a spontaneous reaction at constant volume and constant temperature is the statement that ΔA must be negative. Since reactions of chemical and geochemical interest take place more commonly under conditions of constant pressure than constant volume, ΔG is the more generally useful of the two kinds of free energy.

Free Energy and the Equilibrium Constant. For a change of free energy with pressure at constant temperature ($dT = 0$), Eqs. (71) and (72) give

$$\left(\frac{\delta G}{\delta P}\right)_T = V \qquad \text{and} \qquad \frac{d\Delta G}{dP} = \Delta V. \quad (82)$$

If the second equation is applied to the vaporization of a pure liquid or solid, ΔV is practically equal to the volume of vapor. If the vapor obeys the perfect-gas law, ΔV for one mole of gas may be set equal to RT/P, and Eq. (82) is readily integrable:

$$\Delta G = \int \Delta V dP = \int (RT/P) dP = RT \int d \ln P. \quad (83)$$

Hence

$$\Delta G_2 - \Delta G_1 = RT \int_{P_1}^{P_2} d \ln P = RT \ln (P_2/P_1). \quad (84)$$

If ΔG_1 for $P_1 = 1$ atm is defined as a standard free-energy change, ΔG^0, the equation becomes

$$\Delta G_2 - \Delta G^0 = RT \ln P_2. \quad (85)$$

At some one value of P_2, defined as the vapor pressure at a given temperature, the liquid and vapor are in equilibrium and $\Delta G_2 = 0$. Hence

$$-\Delta G^0 = RT \ln P_{vp}. \quad (86)$$

Since vapor pressure is identical with the equilibrium constant for a reaction liquid $\rightleftharpoons$ vapor,

$$-\Delta G^0 = RT \ln K. \quad (87)$$

To generalize this relationship, fugacity may be substituted for pressure. As defined previously (p. 41), fugacity is an idealized vapor pressure, the vapor pressure a substance would exert if its vapor behaved as a perfect gas. One way of expressing this *definition of fugacity* is by writing Eq. (86) in differential form for a single substance:

$$d \ln f = dG/RT \quad \text{(at constant temperature)} \tag{88}$$

where f and G are the fugacity and molal free energy of the gas. Integration gives

$$G - G' = RT \ln(f/f') \tag{89}$$

where f and f' are fugacities, and G and G' free energies, of a substance in two different states, or at two different pressures (but at the same temperature). If f^0 and G^0 are values of f' and G' for a standard state,

$$G - G^0 = RT \ln(f/f^0) = RT \ln a \tag{90}$$

where a, the activity of the substance at the given temperature, is defined as the ratio f/f^0 (Eq. 13).

Now for a chemical reaction

$$bB + dD \rightleftharpoons yY + zZ \tag{91}$$

Eq. (90) can be set up for each substance, and the change in free energy during the reaction is

$$\begin{aligned} \Delta G &= yG_Y + zG_Z - bG_B - dG_D \\ &= yG_Y^0 + zG_Z^0 - bG_B^0 - dG_D^0 + RT \ln a_Y^y + RT \ln a_Z^z - RT \ln a_B^b - RT \ln a_D^d \\ &= \Delta G^0 + RT \ln \frac{a_Y^y \cdot a_Z^z}{a_B^b \cdot a_D^d} \,. \end{aligned} \tag{92}$$

At equilibrium $\Delta G = 0$, so that for the activities in an equilibrium mixture

$$\Delta G^0 = -RT \ln \frac{a_Y^y \cdot a_Z^z}{a_B^b \cdot a_D^d} = -RT \ln K_a \tag{93}$$

where K_a is the equilibrium constant (p. 42). In the earlier discussion the approximate constancy of K expressed in terms of concentrations was accepted as an *empirical* fact; the above derivation constitutes *thermodynamic proof* that K is constant when expressed in terms of activities, since the constancy follows as a consequence of the condition $\Delta G = 0$ for any mixture at equilibrium.

For numerical calculations of ΔG^0 from K_a or vice versa, Eq. (93) is generally expressed

$$\Delta G^0 = -2.303\, RT \log_{10} K = -4.574\, T \log K \qquad \text{(for } \Delta G^0 \text{ in calories).} \tag{94}$$

At 25° C this simplifies to

$$\begin{aligned} \Delta G^0 &= -1364 \log K \qquad &\text{(for } \Delta G^0 \text{ in calories)} \\ &= -1.364 \log K \qquad &\text{(for } \Delta G^0 \text{ in kilocalories).} \end{aligned} \tag{95}$$

As a rule of thumb, it is worth noting that when ΔG^0 for a reaction is less negative than -13 kcal at ordinary temperatures, the exponent of K is less than 10. This means that for ΔG^0 between approximately $+13$ and -13 kcal, an equilibrium mixture should contain both reactants and products in detectable amounts. If ΔG^0 is less than -13, the reaction at equilibrium is so far displaced toward the products that the reactants would be detectable only with difficulty or not at all; if ΔG^0 is greater than +13, the reaction in the forward direction is practically undetectable. Obviously these rules are generalizations to which there are many exceptions.

Change of ΔG and K with Temperature. Eq. (94), despite the fact that T appears in it, cannot be used to calculate the change of either ΔG^0 of K with temperature, because the two quantities change together as the temperature is raised or lowered. To find an expression for the change with temperature, a convenient starting-point is Eq. (71) or (72). If pressure is held constant ($dP=0$), these equations become

$$\left(\frac{\delta G}{\delta T}\right)_P = -S \quad \text{and} \quad \frac{d\Delta G}{dT} = -\Delta S. \tag{96}$$

Substitution from Eq. (65) gives

$$\frac{d\Delta G}{dT} = \frac{\Delta G - \Delta H}{T} \tag{97}$$

and a simple application of differential calculus leads to

$$\frac{d(\Delta G/T)}{dT} = -\frac{\Delta H}{T^2}. \tag{98}$$

Substitution of Eq. (53) gives

$$-\frac{d(\Delta G/T)}{dT} = \frac{\Delta a}{T} + \frac{\Delta b}{2} + \frac{\Delta cT}{3} + \frac{\Delta H_0}{T^2} \tag{99}$$

from which, by integration and rearrangement,

$$\Delta G = \Delta H_0 - \Delta a T \ln T + IT - \frac{\Delta b}{2} T^2 - \frac{\Delta c}{6} T^3 \tag{100}$$

where I is the constant of integration. This is a general expression for the calculation of ΔG at any temperature. In order to use it, three things must be known: (1) heat capacities of all reactants and products, expressed in terms of the empirical constants a, b, and c in Eq. (49); (2) ΔH at some one temperature, so that the integration constant ΔH_0 of Eq. (53) may be determined; (3) the value of ΔG at some one temperature, so that the constant I may be found. In the most common case, ΔH^0 and ΔG^0 are known at 25° C, and from these the value of ΔG^0 may be calculated over any temperature range for which heat-capacity data are available.

Tables for calculating ΔG at high temperatures often use Eq. (100) in a more general form,

$$\Delta G = A + BT \log T + CT + DT^2 + \cdots. \tag{101}$$

The constants A, B, C, D, calculated from heat capacities or determined experimetally, are listed for each reaction. Often, especially in problems of geochemical interest, the multiplicity of terms in Eqs. (100) and (101) is a needless refinement. For reactions in which ΔC_P may be assumed constant over a temperature range, only the first three terms are needed. If ΔC_P is approximately zero (i.e., if ΔH is nearly constant), the equation reduces to

$$\Delta G = A + CT \tag{102}$$

which has the same form as Eq. (65).

For this last case (ΔH assumed constant), Eq. (65) provides the simplest way to get approximate values of ΔG^0 at temperatures other than 25° C. One need only find ΔH^0_{298} and ΔS^0_{298} for the reaction from the values of ΔH^0_{298} and S^0_{298} for individual substances as tabulated in standard references, and use these as the constants of Eq. (102) to find ΔG^0 at any value of T. This simple procedure serves as a useful check, even when the requirements of a particular problem make it necessary to calculate results of greater accuracy with Eq. (100) or (101).

To find the equilibrium constant a temperatures other than 25° C, an obvious procedure is to calculate ΔG^0 by one of the above methods and then to use Eq. (94). For the special case of constant ΔH, K may be expressed explicitly as a function of T by substituting $-R \ln K$ for $\Delta G/T$ in Eq. (98):

$$\frac{d \ln K}{dT} = \frac{\Delta H}{RT^2}. \tag{103}$$

This relation is often called the *van't Hoff equation*. For vaporization of a liquid or solid, where K is simply the vapor pressure, the equation becomes

$$\frac{d \ln P}{dT} = \frac{\Delta H}{RT^2}. \tag{104}$$

In this form it is called the *Clausius-Clapeyron equation*. Integration of Eq. (103) gives

$$\log K = -\frac{\Delta H}{2.303\ RT} + \text{constant} \tag{105}$$

or

$$\log(K_1/K_2) = \frac{\Delta H}{2.303\ R}\left(\frac{T_2 - T_1}{T_1 T_2}\right) = \frac{\Delta H}{2.303\ R}\left(\frac{1}{T_1} - \frac{1}{T_2}\right). \tag{106}$$

These equations are useful for rough calculations, but it should be emphasized that they depend on the assumption of constant ΔH. Eq. (105) may be modified to allow for some variation in ΔH by assuming that ΔC_P is constant but not equal to zero:

$$\log K = -\frac{\Delta H_0}{2.303\ RT} + \frac{\Delta C_P}{R}\, 2.303 \log T + \text{constant}. \tag{107}$$

Free-Energy Functions. Free energies and equilibrium constants at a variety of temperatures are so frequently needed in chemical and geochemical problems that ways of expediting the calculations are highly desirable. The methods described above have the drawback that the simpler ones give only approximate values and the more accurate ones are cumbersome to use. An obvious solution to the difficulty is to compile tables of free energy at many temperatures over the range of interest. This is often done for special groups of substances, but the compilation of a general table is not feasible because free energies change with temperature in a highly non-linear fashion; hence intervals between successive temperatures must be small, and the tables would be prohibitively bulky. It is possible, however, to calculate derivative functions, called *free-energy functions*, which change only slowly and approximately linearly with temperature, so that a relatively short table may be set up that permits easy and accurate interpolation.

Several free-energy functions are in use, the most common being $(G^0_T - H^0_O)/T$. In this expression G^0_T is the standard molal free energy of a substance at a temperature T, and H^0_O is the standard enthalpy of the substance at absolute zero. Both quantities are hypothetical, but their difference may be found by graphical integration of rather complicated heat-capacity equations or, for many gases, by calculations based on spectrographic data. To find the free-energy change in a reaction, the functions for products and reactants are combined in the usual manner:

$$\frac{\Delta G_T^0 - \Delta H_0^0}{T} = \sum \left(\frac{G_T^0 - H_0^0}{T} \right)_{\text{products}} - \sum \left(\frac{G_T^0 - H_0^0}{T} \right)_{\text{reactants}} \tag{108}$$

whence

$$\Delta G_T^0 = T \sum \frac{G_T^0 - H_0^0}{T} + \Delta H_0^0 . \tag{109}$$

Values of ΔH_0^0 may be calculated from values of ΔH_0^0 of formation for the individual substances, often tabulated with the free-energy function; or, for any given reaction, the value of ΔH_0^0 may be calculated from Eq. (108) if ΔG_T^0 is known for one value of T.

Another common free-energy function is $(G_T^0 - H_{298}^0)/T$. Its use is similar to that of the function based on enthalpy at absolute zero. One function may be calculated from the other by adding or subtracting $(H_{298}^0 - H_0^0)/T$, values for which are commonly tabulated with those of the free-energy functions.

Change of ΔG and K with Pressure. The change of ΔG with pressure is given by Eq. (82):

$$\frac{d\Delta G}{dP} = \Delta V .$$

The usual substitution of $-RT \ln K$ for ΔG^0 gives, for the equilibrium constant,

$$\frac{d \ln K}{dP} = -\frac{\Delta V}{RT} . \tag{110}$$

Integration of these equations requires that ΔV be expressed in terms of pressure. For a reaction giving a gaseous product, where the principal volume change is the volume of gas, Eq. (110) becomes an identity if the gas obeys the perfect-gas laws; in other words, these equations are not applicable to perfect gases, because gas concentrations are expressed as pressures and hence pressure effects are accounted for in the equilibrium constant itself. For a solid-solid or solid-liquid reaction, the change of volume with pressure is expressed by the coefficients of compressibility of the various substances taking part in the reaction.

The Clapeyron Equation. Eq. (72) may be used not only to evaluate the changes of ΔG with temperature and pressure, but also to calculate how temperature and pressure must change if equilibrium in a reaction is to be maintained. For at equilibrium ΔG and $d\Delta G$ must be zero, so that

$$\Delta V dP - \Delta S dT = 0 \tag{111}$$

and

$$\frac{dP}{dT} = \frac{\Delta S}{\Delta V} = \frac{\Delta H}{T \Delta V} . \tag{112}$$

This expression is the Clapeyron equation, derived originally to show how the vapor pressure of a solid or liquid changes with temperature, but applicable to equilibria in general. For the special case of equilibrium between a liquid (or solid) and its vapor, assumption of perfect-gas behavior permits substitution of RT/P for ΔV, and the equation is identical with Eq. (104).

If the pressure necessary to maintain equilibrium in a reaction is plotted against temperature, Eq. (112) gives the slope of a line separating the field of $P-T$ conditions under which the reactants are stable from the field in which the products are stable. For equilibrium between a solid or liquid and its vapor this is the vapor-pressure curve; for equilibrium between a solid and a liquid, it is the freezing-point curve.

Some useful generalizations about $P-T$ curves for geologic reactions may be made from the Clapeyron equation. For a reaction involving only solids (for example, a mineral transformation during metamorphism), ΔS and ΔV are generally small and approximately constant as the temperature changes. This means that equilibrium is represented by a straight line on a $P-T$ plot. For most such reactions (but not all), the line has a positive slope, since the signs of ΔS and ΔV are generally the same; in other words, if heat is absorbed in a reaction, the volume of the products is usually greater than the volume of the reactants. The equilibrium line for a process of melting is generally steeper than the line for a solid-solid reaction, because melting means a considerable increase in disorder and therefore a fairly large increase in entropy. If gas is produced in a reaction, (for example, in dehydration and decarbonation processes), the line is curved with its concave side toward the pressure axis, since the change in volume is large at low pressures but smaller as the gas becomes compressed and behaves more and more like a liquid.

Free Energy and Maximum Work. The two kinds of free energy, A and G, are related in a simple way to the work obtainable from a chemical reaction.

If a reaction takes place at constant temperature and pressure (or, more practically, if the products after the reaction are brought back to the initial temperature and pressure of the reactants), any change in volume must represent production of an amount of mechanical work $P\Delta V$. This may be the only work done, or the reaction may produce varying amounts of additional work, depending on the conditions under which it takes place. The additional work is most easily visualized as electrical work in an oxidation-reduction process, produced when the reaction is set up to function as a galvanic cell, but with suitable arrangements other kinds of work may be produced also.

The relation of free energy to work may be shown by substituting the expression for the first law [Eq. (31)] in the equation for change in the Helmholtz free energy [Eq. (64)] of a reaction carried out isothermally and isobarically:

$$\Delta A = \Delta E - T\Delta S = Q + W - T\Delta S. \tag{113}$$

Rearrangement gives

$$W = \Delta A + (T\Delta S - Q). \tag{114}$$

If the reaction takes place reversibly, the heat absorbed is measured by the entropy change, so that $Q = T\Delta S$ and therefore $W = \Delta A$. If the process is to some extent irreversible, as all actual processes are, some entropy is produced in addition to that resulting from heat absorbed; the quantity $(T\Delta S - Q)$ is a measure of this additional, or irreversible, entropy production. If work is being done *on* the reaction mixture, *more* work must be done to accomplish a given increase in A; in other words, W is greater than ΔA by the amount $(T\Delta S - Q)$. If, on the other hand, the reaction is spontaneous, so that a *decrease* in A is being brought about as the system does work *on* its surroundings, the work done is smaller than $-\Delta A$ by the amount

$(T\Delta S-Q)$. In other words, the maximum work a reaction can perform is equal to the negative value of the Helmholtz free energy, and is obtainable if the reaction is set up so as to take place reversibly; but for any actual process the work done is less than this figure.

The quantity W in Eqs. (113) and (114) includes all the work the reaction can do on its surroundings, pressure-volume work as well as other kinds. The work exclusive of pressure-volume work may be represented by

$$W' = W - P\Delta V. \tag{115}$$

If the reaction is carried out reversibly, both values of W will be maxima:

$$W'_{\max} = W_{\max} - P\Delta V. \tag{116}$$

If $-\Delta A$ is substituted for $W_{\max}$,

$$\begin{aligned} W'_{\max} &= -\Delta A - P\Delta V = -(\Delta E - T\Delta S) - P\Delta V \\ &= -(\Delta H - P\Delta V - T\Delta S) - P\Delta V \\ &= -\Delta H + T\Delta S = -\Delta G. \end{aligned} \tag{117}$$

This means that *the maximum work other than expansion work* obtainable from a reaction carried out isothermally and isobarically is equal to the negative value of the change in Gibbs free energy. It is in this sense that the energy represented by G is "free" energy.

The condition of reversibility, under which the theoretical maximum work of a chemical reaction can be obtained, is often difficult to visualize. For a simple oxidation-reduction process a possible experimental procedure is to use the reaction to power a galvanic cell, the chemical energy being converted to the energy of electrons moving along a wire. Reversibility requires that the current produced by the reaction be opposed by another source of EMF in the circuit, so that the process can take place exceedingly slowly and can be reversed at any time by an infinitesimal increase in the external EMF. The maximum work is thus a hypothetical figure obtainable only if the reaction proceeds infinitely slowly; a reaction going at a finite rate must give less work than the free energy would theoretically allow.

VI. Electrode Potentials

Definitions. For the special case of oxidation-reduction reactions, a third convenient measure of the tendency of a reaction to take place and the position of a given mixture with respect to equilibrium is provided by electrode potentials. These are related to the other common measures, equilibrium constants and free energies, by simple equations.

An oxidation-reduction process involves transfer of electrons from one substance (the *reducing agent*) to another (the *oxidizing agent*). Theoretically any such process may be set up so that the transfer of electrons takes place along a wire, although practical difficulties limit the number of reactions that can actually be used as a source of current. The maximum voltage theoretically obtainable from a reaction is called the *electromotive force* (*EMF*), or *difference of potential,* for the reaction. To measure this voltage requires that the reducing agent be placed in contact with a solution at one end of a wire, and the oxidizing agent in contact with a solution at the other end; the solutions must be in partial contact so that ions can move from one to the other, but should not be allowed to mix freely. The reducing agent supplies electrons to the wire, and the oxidizing agent takes electrons from the wire; the potential difference driving the electrons when the flow of current is made infinitesimally small (by a counter-EMF in the circuit) is the EMF of the cell or of the reaction.

The processes at the two electrodes go on simultaneously, and together they constitute the overall chemical reaction. For many puposes, however, it is convenient to think of the part-reaction at each electrode as a separate process. The two may be expressed in a very general language by the equations

$$\text{reducing agent} \rightarrow \text{oxidized product} + \text{electrons}$$
$$\text{oxidizing agent} + \text{electrons} \rightarrow \text{reduced product.}$$

A simple example is a cell consisting of copper dipping into copper sulfate solution at one side, and iron in iron sulfate solution at the other. The overall reaction is

$$Fe + Cu^{++} \rightarrow Fe^{++} + Cu \tag{118}$$

and the two part-reactions (*electrode reactions* or *half reaction*) are

$$Fe \rightarrow Fe^{++} + 2e^- \tag{119}$$

$$Cu^{++} + 2e^- \rightarrow Cu\,. \tag{120}$$

Here Fe is the reducing agent, Cu^{++} the oxidizing agent, Fe^{++} the oxidized product, Cu the reduced product, and e^- the symbol for an electron.

If the electrode reactions could be studied separately, so that the tendency of a substance to lose or to take up electrons could be directly measured, the measurements would serve as an index of oxidizing and reducing strengths. This program cannot be carried out, but an effective substitute is the measurement of EMF's for cells consisting of various half-reactions combined with some one half-reaction taken as a standard. The standard half-reaction universally employed is

$$H^+ + e^- \rightleftharpoons \tfrac{1}{2} H_2 \tag{121}$$

for which the potential is arbitrarily taken as 0.00 volt when the temperature is 25° C, the pressure is 1 atm, and the activities of H_2 and H^+ are each unity. The electrode potential for any other half-reaction is then the EMF of a cell consisting of the standard hydrogen electrode at one side and the given half-reaction at the other. The half-reaction potential is given the symbol ε; if measured at 25° C, 1 atm pressure, and unit activity of all substances, the potential is called the *standard electrode potential* and is symbolized ε^0.

Because of practical difficulties in working with the hydrogen electrode, actual measurements are usually made with some other electrode that is more convenient for laboratory use. The most widely used *reference electrode* is the calomel electrode, for which the half-reaction is

$$Hg_2Cl_2 + 2e^- \rightleftharpoons 2Hg + 2Cl^-.$$

Careful measurements give a standard potential of 0.2676 volt with respect to the hydrogen electrode.

Electrode potentials may be combined to give potential differences for complete reactions (also symbolized ε). For example, the standard potential for Eq. (118) may be calculated by subtracting potentials (obtained from tables) for the two part-reactions:

$$Cu^{++} + 2e^- \rightleftharpoons Cu \qquad \varepsilon^0 = +0.34 \text{ volt}$$
$$Fe^{++} + 2e^- \rightleftharpoons Fe \qquad \varepsilon^0 = -0.44$$
$$Fe + Cu^{++} \rightleftharpoons Fe^{++} + Cu \qquad \varepsilon^0 = +0.78 \text{ volt}$$

Potentials for both half-reactions and complete reactions are often called *oxidation potentials*, *oxidation-reduction potentials*, or *redox potentials.*

Conventions Regarding Sign. Unfortunately there is no agreement among chemists or geochemists as to the proper sign for electrode potentials. The signs in the example just given follow a convention adopted by the International Union of Pure and Applied Chemistry (Stockholm meeting, 1953): positive for strong oxidizing agents when the half-reaction is written with the oxidized form on the left. This is opposite to the convention used in most geochemical literature, but will probably be the most widely accepted in the future. According to this convention, a complete reaction is spontaneous when the sign of ε^0 is positive, as in the above example. Clearly the numerical quantities are unaffected by the sign convention, but the confusion of signs requires that care be used in comparing ε^0 values from different sources and in translating ε^0 values to corresponding values of ΔG^0 and K.

Electrode Potentials and Free Energy. The number 0.78 volt in the above example expresses the tendency of electrons to move from iron to cupric ion, or the tendency of iron to displace copper from solution. Since free energy gives the same sort of information, it should be possible to express one in terms of the other. A way to do this is suggested by the fact that the maximum work obtainable from a galvanic cell is given by its EMF multiplied by the amount of electricity that moves along the circuit:

$$W'_{\text{max}} = \varepsilon \times \text{amount of charge.} \tag{122}$$

The maximum work is also given by the free-energy change (p. 62) in the cell reaction, so that

$$W'_{\text{max}} = -\Delta G = n\mathscr{F}\varepsilon \tag{123}$$

where $\mathscr{F}$ is the Faraday constant (23.06 kcal/volt-equivalent) and n is the number of electrons transferred in the equation as written. The standard free-energy change for a reaction (either a half-reaction or a complete reaction) is given by

$$\Delta G^0 = -n\mathscr{F}\varepsilon^0 = -23.06\,\text{n}\,\varepsilon^0 \tag{124}$$

Eq. (124) makes possible the calculation of free energies from electrode potentials and vice versa; it is also useful in the experimental determination of free energies, since for some reactions ε^0 can be measured with great accuracy. It is important to note that ε^0 is a *potential*, and thus independent of the amount of substance present, while ΔG^0 is a *quantity of energy* and therefore dependent on the mass of material undergoing reaction. For a given process ε^0 is the same no matter how the equation is written, but ΔG^0 changes if the coefficients in the equation are changed. The factor n in Eq. (124) takes care of this difference, for it also varies with the amount of material represented by the equation.

Combination of Eqs. (123) and (92) gives

$$\varepsilon = \frac{-\Delta G}{n\mathscr{F}} = -\frac{\Delta G^0}{n\mathscr{F}} - \frac{RT}{n\mathscr{F}} \ln \frac{a_Y^y \cdot a_Z^z}{a_B^b \cdot a_D^d} \tag{125}$$

$$= \varepsilon^0 - \frac{2.303\, RT}{n\mathscr{F}} \log \frac{a_Y^y \cdot a_Z^z}{a_B^b \cdot a_D^d}\,. \tag{126}$$

This equation, called the *Nernst equation*, expresses the potential for a reaction at any arbitrary activities of reactants and products. At 25° C it simplifies to

$$\varepsilon = \varepsilon^0 - \frac{0.059}{n} \log \frac{a_Y^y \cdot a_Z^z}{a_B^b \cdot a_D^d}\,. \tag{127}$$

The equilibrium constant for a reaction is related to the standard EMF by the equation

$$\varepsilon^0 = -\frac{\Delta G^0}{n\mathscr{F}} = +\frac{2.303\, RT}{n\mathscr{F}} \log K\,. \tag{128}$$

Change of Potential with Temperature. Any of the expressions for change of free energy with temperature may be used to find the corresponding change for electrode potentials. The simplest expression, useful for approximate calculations when ΔH and ΔS may be considered constant, is

$$\varepsilon = -\frac{\Delta H - T\Delta S}{n\mathscr{F}}\,. \tag{129}$$

Another relation between ε and temperature is given by the *Gibbs-Helmholtz equation*, obtained by combining Eqs. (123) and (97):

$$\frac{d\Delta G}{dT} = -n\mathscr{F}\frac{d\varepsilon}{dT} = \frac{-n\mathscr{F}\varepsilon - \Delta H}{T} \tag{130}$$

or

$$\varepsilon + \frac{\Delta H}{n\mathscr{F}} = T\frac{d\varepsilon}{dT}\,. \tag{131}$$

For some reactions values of ε and its temperature coefficient can be obtained with great accuracy, and use of Eq. (131) then provides a means of finding ΔH with greater accuracy than is possible from calorimetric measurements.

Redox Potentials. If an inert electrode is placed in a large volume of solution or in a natural aqueous medium, and if the electrode is connected to a hydrogen electrode (or other reference electrode) so as to form a cell, the resulting voltage is called the *redox potential* of the medium. It is not the potential of a particular half-reaction, but rather a measure of the ability of the medium to supply or use up electrons. (The term redox potential is sometimes used also as a synonym for electrode potential or oxidation potential, referring to specific reactions as well as to environments.)

The redox potential, commonly symbolized Eh, is somewhat analogous to pH: the latter is a measure of the ability of a medium to supply protons, while Eh measures its ability to furnish electrons. Values of Eh are commonly given in volts or millivolts, and measured values in nature range from about +1.0 to −0.6 volt. Satisfactory measurements of Eh are difficult to make because of slow reactions and polarization of electrodes, so that available data are far fewer than for pH.

Eh-pH Relations. Eh and pH are two important variables in natural aqueous environments that in large measure determine what molecular or ionic species are stable and what reactions are possible. On a plot of Eh against pH the characteristics of an environment can be displayed; fields of stability of minerals and dissolved substances can be outlined, and lines representing equilibrium between different species can be located. Such diagrams are a convenient short-hand method of summarizing the chemistry of various elements in low-temperature aqueous environments.

The conventional Eh-pH diagram shows Eh on the vertical axis and pH on the horizontal axis. Two diagonal lines show limits of Eh in nature: an upper line representing the oxidation of water to O_2, and a lower line showing reduction of water to H_2. The upper line is obtained from the half-reaction

$$\tfrac{1}{2}O_2 + 2H^+ + 2e^- \rightleftharpoons H_2O \qquad \varepsilon^0 = +1.23 \text{ volt} \tag{132}$$

for which the Nernst equation [Eq. (127)] may be written

$$\varepsilon = 1.23 + \frac{0.059}{2} \log [O_2]^{\frac{1}{2}} [H^+]^2 . \tag{133}$$

For equilibrium with the atmosphere the activity of O_2 may be set equal to its partial pressure, 0.2 atm:

$$\text{Eh} = 1.23 + 0.03 \log (0.2)^{\frac{1}{2}} + 0.059 \log [H^+] = 1.22 - 0.059 \text{ pH}. \tag{134}$$

This is the equation of a straight line on the Eh-pH plot. The lower limiting line, derived from the equation for the H_2—H^+ electrode reaction, is

$$\text{Eh} = -0.059 \text{ pH} \tag{135}$$

on the assumption that the extreme limit of $[H_2]$ in nature is 1 atm. Most measured values of Eh in nature fall far within these limiting lines, suggesting that step-reactions in the oxidation and reduction of water may restrict values more than the overall reactions. For pH, most measurements in nature lie within the range 4 to 9, but some unusual environments show values well outside these limits. In general the possible Eh-pH combinations in nature are outlined by a rough parallelogram on the diagram; oxidizing environments appear toward the top of the parallelogram, reducing environments toward the bottom, acid conditions toward the left, and alkaline conditions toward the right.

For any element the important oxidation-reduction reactions can be set up, the Nernst equations calculated, and the corresponding lines plotted on the diagram. Each line separates the oxidized form of an element from the reduced form, or the form stable in acid from the form stable in base. Thus the form of an element stable in various natural situations can be seen at a glance, and stable assemblages of forms of different elements can be visualized by superposing diagrams.

Most Eh-pH diagrams are drawn for a temperature of 25° C. They could be extended to higher temperatures, but generally the state of oxidation of a high-temperature system is indicated more conveniently by the activity of O_2 than by Eh.

Plots of log $[O_2]$ against pH are commonly used for temperatures up to 200 or 300° C; at still higher temperatures other pairs of variables, e.g. log $[O_2]$ against log $[S_2]$, may give more informative diagrams.

VII. Solutions

Partial Molal Quantities. If several substances mix without appreciable energy change, the free energy of the mixture is the sum of the free energies of the constituents. If G_1, G_2, ... are the free energies per mole of the pure substances, and if $x_1, x_2, \ldots$ are the mole fractions in the mixture, the free energy per mole of solution is

$$G = x_1 G_1 + x_2 G_2 + \cdots, \tag{136}$$

The change in free energy resulting from addition of dn_1 moles of the first component, dn_2 moles of the second component, etc., would be

$$dG = G_1 dn_1 + G_2 dn_2 + \cdots. \tag{137}$$

If only a single constituent changes, all other dn's become zero, and at constant temperature and pressure

$$dG = G_1 dn_1 \quad \text{or} \quad G_1 = \left(\frac{\delta G}{\delta n_1}\right)_{T, P, n_2, n_3 \ldots} \tag{138}$$

For the ideal solution in which there is no energy of mixing this equation merely expresses the obvious fact that the molal free energy of a constituent of the mixture is equal to the rate of change of the total free energy per mole of this constituent.

All real solutions, however, involve some energy of mixing. This means that the total free energy cannot be expressed as a simple sum of free energies of the pure constituents, like Eq. (136). The equation can still be given meaning, however, if G_1, G_2 ... are *defined* by Eq. (138). These quantities are called the *partial molal free energies* of the different constituents; each one represents the contribution of one constituent toward the total free energy of the solution. For a nearly ideal solution (low heat of mixing) the partial molal free energies are nearly equal to molal free energies of the pure constituents; for mixtures with a large heat of solution the two quantities may be very different. Partial molal free energies are represented by a bar over the symbol:

$$\bar{G}_1 = \left(\frac{\delta G}{\delta n_1}\right)_{T, P, n_2, n_3 \ldots} \tag{139}$$

Partial molal quantities for other extensive properties of substances in solution may be defined similarly. For example:

$$\bar{V}_1 = \left(\frac{\delta V}{\delta n_1}\right) \quad \bar{S}_1 = \left(\frac{\delta S}{d n_1}\right) \quad \bar{H}_1 = \left(\frac{\delta H}{\delta n_1}\right).$$

Each of these expressions means that the partial molal quantity is equal to the rate of change of the property of the solution (volume, entropy, enthalpy) per mole of substance being added to an infinite amount of solution. In general the partial molal quantities are different from the corresponding molal quantities for pure substances, and they change in a complex manner as the composition of the solution changes. They are not readily predictable, but must be determined experimentally.

Partial molal quantities may be used in place of the corresponding properties of pure substances in most of the equations derived previously. The total value of any extensive property of a solution, per mole of the solution, may be expressed in terms of partial molal properties by an equation analogous to Eq. (136):

$$J = x_1 \bar{J}_1 + x_2 \bar{J}_2 + \cdots = x_1 \left(\frac{\delta J}{\delta n_1}\right) + x_2 \left(\frac{\delta J}{d n_2}\right) + \cdots \tag{140}$$

where J stands for an extensive property and the x's are mole fractions.

Chemical Potential. A quantity of work or energy, in general, may be expressed as a product of a potential and a capacity factor. Potential energy in a gravitational field, for example, is equal to a potential, gh (acceleration of gravity $\times$ height), multiplied by the capacity factor m (mass). By analogy, the free energy of a chemical reaction may be regarded as the product of the molal free energy change (potential) multiplied by the number of moles (capacity). Or for a single pure substance the (hypothetical) value of its free energy per mole may be labeled its *chemical potential*, and the free energy in any given quantity is then the chemical potential multiplied by the number of moles.

This manner of speaking is not particularly useful for pure substances, but is often employed in the thermodynamic treatment of solutions. Here the *chemical potential of any constituent is its partial molal free energy*,

$$\mu_i = \bar{G}_i = \left(\frac{\delta G}{\delta n_i}\right)_{T,\,P,\,\text{other } n\text{'s}} . \tag{141}$$

The symbol μ may be substituted for molal free energy of pure substances or partial molal free energy of dissolved substances in any of the formulas derived previously. If the free energy of a solution is regarded as a function of temperature, pressure, and amounts of its various constituents, $G = G(T, P, n_1, n_2 \ldots)$, the total differential of G is

$$dG = \left(\frac{\delta G}{\delta T}\right) dT + \left(\frac{\delta G}{\delta P}\right) dP + \sum_i \left(\frac{\delta G}{\delta n_i}\right) dn_i . \tag{142}$$

Substitution of values for the partial derivatives from Eqs. (82), (96), and (141) gives

$$dG = -S\,dT + V\,dP + \sum_i \mu_i\,dn_i \tag{143}$$

where the summation is to be taken over all the constituents of the solution. This equation is a generalization of Eq. (71) for application to solutions. The corresponding generalized equation for HELMHOLTZ free energy is

$$dA = -S\,dT - P\,dV + \sum_i \mu_i\,dn_i . \tag{144}$$

Here μ_i is equal to $(\delta A/\delta n_i)_{T,V}$, which is another expression for chemical potential:

$$\mu_i = \left(\frac{\delta G}{\delta n_i}\right)_{T,\,P,\,n_j} = \left(\frac{\delta A}{\delta n_i}\right)_{T,\,V,\,n_j} . \tag{145}$$

Equilibrium in Heterogeneous Systems. The general criterion for equilibrium in a system at constant temperature and pressure (Eq. 76) is that the free-energy change shall be zero for any infinitesimal change in the system. In particular, if the system consists of several phases, movement of material from one phase to another must be accompanied by no change of free energy; in other words, the molal free energy of each substance must be the same in all phases. If some of the phases are solutions, this criterion requires that movement of material between phases produce no change in partial molal free energy. The general requirement can be stated succinctly in terms of chemical potential: *for equilibrium to exist in a heterogeneous system, the chemical potential of each substance must be the same in all phases.*

Another way to specify equilibrium conditions in a heterogeneous system is to think of each substance as tending to escape from one phase into another; then equilibrium requires that the *escaping tendency* of any substance must be the same in all phases. Thus if an aqueous solution is in equilibrium with water vapor, the tendency for water to escape from the solution into the vapor must be balanced by its tendency to move from the vapor into the solution. The chemical potential may then be visualized as a measure of escaping tendency: as long as the chemical potential of a substance is the same in all phases, it will have no tendency to move from one to another.

Fugacity. For a liquid or solid in contact with vapor, an obvious rough measure of escaping tendency is its vapor pressure. This measure can be refined, as noted earlier, by replacing vapor pressure with fugacity, defined according to Eq. (88). For the *fugacity of a constituent of a solution,* this defining equation can be expressed in terms of partial molal free energy:

$$d \ln f = d\overline{G}/RT. \tag{146}$$

Alternatively,

$$\left(\frac{\delta \ln f}{\delta G}\right)_T = \frac{1}{RT} \tag{147}$$

Integration gives

$$\overline{G}_1 = RT \ln f_1 + k \tag{148}$$

or

$$\mu_1 = \mu_0 + RT \ln f_1 \tag{149}$$

where the integration constant (k or μ_0) is constant for a given substance at the temperature T.

Although most easily visualized as an idealized vapor pressure, fugacity is a general property of any substance, whether pure or a constituent of a solution. Like the chemical potential, it is a measure of escaping tendency; and the criterion of equilibrium in heterogeneous systems can be expressed by saying that the fugacity of every substance must be constant throughout the system, just as the chemical potential must be constant. For some purposes fugacity is a more convenient measure of escaping tendency, because the chemical potential has the awkward mathematical property of approaching minus infinity as the concentration of a substance approaches zero.

Laws of the Dilute Solution. The experimental fact that the vapor pressure of a volatile solute is proportional to its concentration in very dilute solutions is called

Henry's law. Generalized for all kinds of solutes the law may be expressed

$$f = kx \tag{150}$$

where f is fugacity, x is mole fraction, and k is the proportionality constant. A solution for which this law holds at all concentrations would be ideal, or perfect, in the sense that its constituents would mix without a change in energy. Real solutions never obey the law rigorously, but it is a good approximation in dilute solutions.

Insofar as HENRY's law holds, simple expressions may be derived for the change in solubility of a solid with temperature and pressure in a binary solution. The equation for change with pressure, derived from Eqs. (82) and (146), is

$$\left(\frac{\delta \ln x}{\delta P}\right)_T = \frac{V_s - \overline{V}}{RT} \tag{151}$$

V_s is the molal volume of the pure solute, and $\overline{V}$ the partial molal volume of the solute in the saturated solution; for an ideal solution the latter quantity can be assumed equal to the molal volume of the pure liquid solute, corrected by extrapolation to the temperature of the solution. The equation for change of solubility with temperature, derived from Eqs. (96) and (146), is

$$\left(\frac{\delta \ln x}{\delta T}\right)_P = \frac{\Delta \overline{H}}{RT^2} \,. \tag{152}$$

The ΔH here is strictly the differential heat of solution, but for an ideal solution it may be equated with the heat of fusion of the solute at the temperature of the solution.

Raoult's law expresses the experimental fact that the vapor pressure of the solvent in a very dilute solution is proportional to its concentration. The general expression in terms of fugacities and mole fractions is

$$f_1 = f_1^0 x_1 = f_1^0 (1 - x_2) \tag{153}$$

where f_1 is fugacity of the solvent, x_1 and x_2 are mole fractions of solvent and solute, and f_1^0 is the proportionality constant, here equal to the fugacity of the pure solvent. Like Henry's law, Raoult's law is not strictly valid for any real solution, but it is a useful approximation as long as solutions are dilute.

If a solute distributes itself between two immiscible solvents and if Henry's law is assumed to hold for each, Eq. (150) leads to the expression

$$\frac{f'}{f} = \frac{x'}{x} = \text{constant} \tag{154}$$

where f and f' are fugacities of the solute in the two solvents, and x and x' are the corresponding mole fractions. Eq. (154), expressing *Nernst's distribution law*, is a slightly more sophisticated version of Eq. (13).

Activity. The definition of activity, like that of fugacity, can be generalized by expressing it in terms of chemical potential.

If the concentration of a solute changes, with the temperature being kept constant, its change in partial molal free energy may be expressed as the definite integral of Eq. (146):

$$\overline{G} - \overline{G}' = \mu - \mu' = RT \ln (f/f') \,. \tag{155}$$

Then if f^0, $\bar{G}^0$, and μ^0 are taken as values of f', $\bar{G}'$, and μ' in a state selected as standard (p. 54), the ratio f/f^0 is defined as the activity of the solute at the given temperature. In the standard state the activity is 1, and in any other state it is given by the expression

$$RT \ln a = \bar{G} - \bar{G}^0 = \mu - \mu^0. \tag{156}$$

Since activity at a given temperature is proportional to fugacity,

$$da = df/f^0 \quad \text{and} \quad d \ln a = d \ln f. \tag{157}$$

By using these expressions many of the previously derived equations can be recast in terms of activities. For example, the pressure coefficient of activity is given by an equation similar to (151):

$$\left(\frac{\delta \ln a}{\delta P}\right)_T = \frac{\bar{V}}{RT} \tag{158}$$

and the temperature coefficient by an equation similar to (152):

$$\left(\frac{\delta \ln a}{\delta T}\right)_P = \frac{\Delta \bar{H}}{RT^2}. \tag{159}$$

VIII. The Phase Rule

Definitions. For heterogeneous systems so complex that free-energy calculations are difficult or pointless, useful descriptive information about equilibrium assemblages can often be obtained from the phase rule. This is a statement about the number of quantities that can be varied without changing the state of a system, in other words the number of *degrees of freedom* possessed by the system. For a system consisting of a single phase of a pure substance, the state (whether gaseous, liquid, or one of several possible crystal forms) is ordinarily fixed by the two variables temperature and pressure, and the system is said to have two degrees of freedom; this means that temperature and pressure can both vary, within limits, without causing the appearance of another phase. If the system has two pure phases in equilibrium, the number of degrees of freedom is reduced to one; for now either temperature or pressure may be varied, but not both, since an arbitrary change in one requires a specific change in the other if both phases are to persist. Three phases of the same pure substance can exist only at a single temperature and pressure (a triple point), so that the number of degrees of freedom is zero. If a second pure substance is added to the first, and if it dissolves so that its chemical potential is the same in all phases present, an additional degree of freedom results; for the state of the system is now dependent on the concentration of the second substance as well as on temperature and pressure. Each substance added to the system means an additional degree of freedom (its concentration) for any given number of phases. Hence the number of degrees of freedom in any system is equal to 2 more than the difference between the number of substances (components) and the number of phases present:

$$f = c - p + 2. \tag{160}$$

The meaning of f, the number of degrees of freedom, and of p, the number of phases, is generally obvious; but the quantity c, the number of components, in some systems gives trouble. It is defined as *the minimum number of chemical formula units needed to specify the composition of all phases present*. It may be a different number for the same combination of elements in different systems; for example, oxygen and hydrogen may constitute two components of a system if the proportion of the two elements is not fixed, or a single component if the elements occur only in the combination H_2O. The number of degrees of freedom, therefore, is dependent on how the number of components is chosen — in other words, on just what chemical ratios are considered to vary significantly under the conditions in which a given system is being studied. The difficulty in defining the number of components becomes especially troublesome when some of the substances involved are solid solutions of limited miscibility.

Because of this difficulty of definition, the phase rule is restricted in its applicability to natural systems. It cannot be used to predict the number of phases to be expected in a complex mineral assemblage, for the possible number of phases depends on how the components are chosen. What the phase rule does is to specify, for an *observed* number of phases and a *probable* number of components (estimated from the chemical composition), the *probable* number of degrees of freedom represented by the mineral assemblage. If this number is less than zero (many phases and few components), the system was definitely *not* at equilibrium when the minerals were formed. If the number is zero or 1 the assemblage may possibly have been at equilibrium, but very probably was not, since assemblages occurring in rocks were almost certainly stable through a range of both temperature and pressure. If the estimated number of degrees of freedom is 2 or more, then the system *may have been* at equilibrium when it was formed. Thus the phase rule serves as a criterion of *lack of equilibrium* in a natural assemblage, but as a criterion of the presence of equilibrium it is permissive only.

Variants of the Phase Rule. Modifications of the phase rule are convenient for special situations. If other variables than temperature and pressure are important in determining the state of a system, for example electric fields and surface forces, the constant in Eq. (160) would be greater than 2. In environments where pressure may be considered essentially constant, for example in systems at or near the Earth's surface, a useful variant of the phase rule is

$$f = c - p + 1. \tag{161}$$

Since equilibrium mineral assemblages in nature must in general be stable over a range of the two variables pressure and temperature, f for such assemblages must have a minimum value of 2; hence the number of phases in a rock should be no greater than the number of components if equilibrium was maintained during its formation:

$$p_{\max} = c. \tag{162}$$

This statement is called the *mineralogical phase rule*.

A variant of the mineralogical phase rule is occasionally useful for systems in which some of the components are fixed and some are free to move in and out. The latter, called *mobile* components, have chemical potentials determined by the larger environment outside the system in question; the most common examples would be volatile substances like water and carbon dioxide. Since the mobile components are simply part of the general environment, they play no role in determining the maximum number of phases at equilibrium. Hence the phase rule may be written

$$p_{max} = c - c_m \tag{163}$$

where c_m is the number of mobile components. This is not a thermodynamic equation in the classical sense, since it assumes knowledge about details of the process by which the mineral assemblage originated, but it is helpful in emphasizing differences in the behavior of various components.

Phase-Rule Diagrams. A system with one degree of freedom is called *univariant*, with two degrees of freedom *divariant*, with three degrees of freedom *trivariant*, and so on. These possibilities may be displayed on various kinds of diagrams, using the important state-determining variables (temperature, pressure, composition) as axes of coordinates. Most commonly a two-dimensional plot is used, on which univariant systems are shown by lines, divariant systems by fields between lines, and invariant systems by points at the intersections of field boundaries. If composition is assumed constant, pressure is plotted against temperature, giving the familiar "*P—T*" diagrams. If pressure (or temperature) is held constant, temperature (or pressure) may be plotted against one of the composition variables (a "T—X" or "P—X" diagram). Two compositional variables are conveniently shown by a triangular plot, on which temperature or pressure may be added as contours. Three-dimensional plots are sometimes employed; here a univariant system is represented by a line, a divariant system by a curved surface, and a trivariant system by a space bounded by curved surfaces. More commonly, however, the mutual variation of three quantities is displayed by sections or projections of a three-dimensional figure.

On a P—T diagram each univariant line shows equilibrium between substances that are stable in the fields on either side. For a one-component system a line shows equilibrium between two phases:

$$p_1 \rightleftharpoons p_2 .$$

For a three-component system a line shows equilibrium in a reaction

$$B + D \rightleftharpoons Y + Z,$$

with all four substances present (and only three present in the adjacent divariant fields). If ΔS, ΔV, and ΔH represent entropy change, volume change, and enthalpy change in such reactions, the slope of a univariant line is given by the Clapeyron equation (Eq. 110):

$$\frac{dP}{dT} = \frac{\Delta S}{\Delta V} = \frac{\Delta H}{T \Delta V} .$$

Useful generalizations about slopes of such lines for reactions of geologic interest are mentioned on p. 61.

A common kind of temperature-composition diagram represents a system including a melt and two or more pure solids (*eutectic diagram*). For the simple case of a binary system, univariant equilibrium is indicated by lines showing how the solubility of each component in the melt changes with temperature (or alternatively, how the melting point of each component changes as the other component is added). The change of solubility of each component may be expressed by an integrated form of Eq. (152), on the assumption that the melt behaves as a perfect solution:

$$\log x = \log a = \frac{\Delta H}{2.303\, R}\left(\frac{1}{T_m} - \frac{1}{T}\right). \tag{164}$$

Here ΔH is the heat of solution of either component in the melt, assumed to be constant and to be equal to the heat of fusion of the pure component at temperature T; T_m is the melting point of the pure component, a temperature at which $a=1$ and $\log a = 0$. Since T_m is a constant, this equation has the form

$$\log x = k - k'(1/T) \tag{165}$$

where k and k' are constants. This equation gives a straight line for $\log x$ plotted against $1/T$, and a curved line with an increasingly negative slope for x plotted against T, agreeing qualitatively with most experimentally determined eutectic diagrams. For most systems of geochemical interest, however, the assumption of ideality of the solution is so far from accurate that the calculated curve fits the experimental points only in the immediate vicinity of a pure component.

For binary pressure-composition diagrams an equation for univariant lines can be derived similarly from Eq. (151). But again ideality of solutions is so far from realized in geologic situations that the theoretical equation is seldom useful.

Thus analytical expressions for the lines and surfaces in phase-rule diagrams can be formulated, but their value is limited because most solutions deviate widely from ideal behavior. Activity corrections are of little help, because activity coefficients in concentrated solutions change so rapidly with concentration and temperature. But the diagrams, despite their theoretical shortcomings, are widely useful as simple summaries of empirical data and as a means of checking the data against requirements of the phase rule.

Symbols

a	Activity; empirical constant in equations expressing heat capacity as a function of temperature.
$a_+, a_-, a_\pm$	Activity of a cation, activity of an anion, mean activity of the ions of a dissolved electrolyte.
(aq)	Dissolved in water (used after a formula in a chemical equation; omitted after formulas of dissolved ions).
A	Helmholtz free energy or work content (some authors use F); general symbol for the anion of an acid; parameter in equations of the Debye-Hückel theory; constant in equation for free energy as a function of temperature.
b	Coefficient in a generalized chemical equation; empirical constant in equations expressing heat capacity as a function of temperature.
B	Chemical formula or symbol in generalized equations; constant in equation for free energy as a function of temperature.
c	Concentration in general, no units specified; number of components; empirical constant in equations expressing heat capacity as a function of temperature.
C	Heat capacity; constant in equation for free energy as a function of temperature.
C_P, C_V	Heat capacity at constant pressure, heat capacity at constant volume.
°C	Degrees Celsius (or centigrade).
d	Coefficient in a generalized chemical equation.
dx	Exact differential of x.

D	Chemical formula or symbol in generalized equations; constant in equation for free energy as a function of temperature.
e^-	Electron (in electrode reactions).
E	Internal energy (some authors use U).
ε	Electrode potential; oxidation potential; electromotive force.
ε^0	Standard electrode potential.
Eh	Redox potential (= oxidation potential for an environment; also used as a synonym for ε for a reaction).
f	Fugacity; number of degrees of freedom.
f^0	Fugacity in a standard state.
$\mathcal{F}$	Faraday constant (= 23,062.3 cal/volt equiv = 96,493.5 coulombs/equiv).
(g)	Gas (used after a formula in a chemical equation).
G	Gibbs free energy, Gibbs function, or free enthalpy (some authors use F, some use Z).
G^0, ΔG_f^0	Standard free energy, standard free energy of formation of a compound from its elements.
H	Enthalpy or heat content.
ΔH_0	Constant of integration in equation expressing change of enthalpy with temperature.
H^0, ΔH_f^0	Standard enthalpy, standard enthalpy of formation of a compound from its elements.
H_0^0	Standard enthalpy at absolute zero.
I	Ionic strength (some authors use μ); integration constant in equation for free-energy change as a function of temperature.
i	Subscript indicating one of a number of components.
J	Any extensive property.
k	Constant, in general.
K	Equilibrium constant.
K_c, K_a, K_f	Concentration constant, activity constant, fugacity constant.
°K	Degrees Kelvin (or absolute).
(l)	Liquid (used after a formula in a chemical equation).
$\ln x$	Natural logarithm of x (= $2.3026 \log x$).
$\log x$	Decimal logarithm of x.
m	Molality (= moles of solute per kilogram of water); generalized subscript in a chemical formula; generalized valence of an ion.
M	Molarity (= moles of solute per liter of solution).
n	Number of moles; number of ions formed by dissociation of a formula-unit of an electrolyte; number of electrons transferred in an oxidation-reduction reaction; generalized subscript in a chemical formula; generalized valence of an ion.
p	Number of phases.
P	Pressure.
pH	Negative logarithm of the hydrogen-ion activity.
Q	Heat.
R	Gas-law constant (= 1.987 cal/degree mole).
(s)	Solid (used after a formula in a chemical equation).
S	Entropy.
S^0	Standard entropy.
T	Absolute temperature.
V	Volume.
W	Work.
x	Mole fraction (some authors use N).
y	Coefficient or subscript in generalized chemical equations.
Y	Chemical formula or symbol in generalized equations.
z	Coefficient or subscript in generalized chemical equations; charge on an ion.
Z	Chemical formula or symbol in generalized equations.
γ	Activity coefficient.

$\gamma_+, \gamma_-, \gamma_\pm$ Activity coefficient of a cation, activity coefficient of an anion, mean activity coefficient of a dissolved electrolyte.
δx Inexact differential of x.
Δ Delta (meaning "change in").
μ Chemical potential.
μ^0 Chemical potential in a standard state.
( ) Parentheses around chemical formulas in algebraic equations indicate concentrations, with no specification of units (some authors use square brackets; others use c with a subscript).
[ ] Square brackets around chemical formulas in algebraic equations indicate activities (some authors use braces; others use a with a subscript).
$\bar{G}$, $\bar{S}$, etc. Bars over symbols indicate partial molal quantities.

References

General Works

A great number of books on chemical thermodynamics are available, in any of which the topics covered in this summary article may be pursued in greater depth. The two standard texts on which much of this article is based are:

KLOTZ, I.: Chemical thermodynamics, revised ed. New York: W.A. Benjamin 1964.
LEWIS, G. N., and M. RANDALL: Thermodynamics, 2nd ed., revised by K. S. PITZER and L. BREWER. New York: McGraw-Hill Book Co. 1961.

Books on Specific Applications of Thermodynamics to Geologic Problems

BARTH, T. F. W.: Theoretical petrology, 2nd ed. New York: John Wiley Sons 1962.
EITEL, W.: Thermochemical methods in silicate investigation. New Brunswick: Rutgers University Press 1952.
— The physical chemistry of the silicates. Chicago: Chicago University Press 1954.
FYFE, W. S., F. J. TURNER, and J. VERHOOGEN: Metamorphic reactions and metamorphic facies. Geological Society of America 1958, Memoir 73.
GARRELS, R. M., and C. L. CHRIST: Solutions, minerals, and equilibria. New York: Harper & Row 1965.
KORZHINSKII, D.: Fiziko-khimicheskie osnovy analiza paragenezisov mineralov. Moscow: Akademiya Nauk SSSR 1957. English translation by Consultants Bureau, New York 1959.
SMITH, F. G.: Physical geochemistry. Reading, Mass.: Addison-Wesley 1963.
TURNER, F. J., and J. VERHOOGEN: Igneous and metamorphic petrology, 2nd ed. New York: McGraw-Hill Book Co. 1960.
WINKLER, H. G. F.: Die Genese der metamorphen Gesteine. Berlin-Heidelberg-New York: Springer 1965.

Articles on General Aspects of Thermodynamics

FYFE, W. S.: Hydrothermal synthesis and the determination of equilibrium between minerals in the subliquidus region. J. Geol. **68**, 553—566 (1960).
MACWOOD, G. E., and F. H. VERHOEK: How can you tell whether a reaction will occur? J. Chem. Educ. **38**, 334—337 (1961).
MCINTIRE, W. L.: Trace element partition coefficients — a review of theory and applications to geology. Geochim. Cosmochim. Acta **27**, 1209—1264 (1963).
THOMPSON, J. B.: The thermodynamic basis for the mineral facies concept. Am. J. Sci. **253**, 65—103 (1955).
WEILL, D. F., and W. S. FYFE: A discussion of the KORZHINSKII and THOMPSON treatment of thermodynamic equilibrium in open systems. Geochim. Cosmochim. Acta **28**, 565—576 (1964).

Compilations of Thermodynamic Data

GARRELS, R. M., and C. L. CHRIST: Solutions, minerals, and equilibria. New York: Harper & Row 1965. The appendix includes tables of ΔH^0, ΔG^0, and S^0 at for 25° C for minerals, compounds, and ions of interest in geochemistry.

CLARK, S. P. (ed.): Handbook of physical constants, revised ed. Geological Society of America 1966, Memoir 97. Includes tables of ionization constants and of thermodynamic properties of minerals.

KELLEY, K. K.: Contributions to the data on theoretical metallurgy: XIII, High-temperature heat-content, heat-capacity, and entropy data for the elements and inorganic compounds. U.S. Bur. Mines, Bull. No. 584 (1960).

— Contributions to the data on theoretical metallurgy: XIV, Entropies of the elements and inorganic compounds (with E. G. KING). U.S. Bur. Mines, Bull. No. 592 (1961). Critical review and summary of data on S^0_{298}, $H^0_T - H^0_{298}$, and $S^0_T - S^0_{298}$ at 100° intervals.

KUBASCHEWSKI, O., and E. L. EVANS: Metallurgical thermochemistry, 3rd ed. London: Pergamon 1958. Comprehensive tables of thermodynamic data for substances and reactions of interest in metallurgy; also summaries of experimental methods, methods of solving problems, and methods of estimating data.

LATIMER, W. M.: Oxidation potentials, 2nd ed. Englewood Cliffs, New Jersey: Prentice-Hall 1952. Somewhat out-of-date and marred by many errors, but still useful as a compilation of data on electrode potentials and on ΔH^0, ΔG^0, and S^0 at 25° C, especially for substances in aqueous solution.

ROSSINI, F. D., D. D. WAGMAN, W. H. EVANS, S. LEVINE, and I. JAFFE: Selected values of physical and thermodynamic properties of hydrocarbons and related compounds. Pittsburgh: Carnegie Press 1953.

WAGMAN, D. D., W. H. EVANS, S. LEVINE, and I. JAFFE: Selected values of chemical thermodynamic properties. U.S. Nat. Bur. Std. Circular **500** (1952). Data on ΔH^0, ΔG^0, and S^0 at 25° C for elements and compounds; some values for ΔH at 0° K; and data for phase transitions. Supplements are published in the J. Res. Natl. Bur. Std.

SILLÉN, L. G.: Stability constants of metal-ion complexes. Section I: Inorganic ligands. Chem. Soc. (London) Spec. Publ. No. 17 (1964). A comprehensive tabulation of ionization constants, acid-dissociation constants, and formation constants of complexes, mostly at 25° C but including some data at higher and lower temperatures. Also a table of standard electrode potentials.

Chapter 4

K. Keil

METEORITE COMPOSITION

I. Introduction

Meteorites are solid objects which reach the Earth from space. The science devoted to their study is referred to as meteoritics. Meteorites have been known to man for thousands of years. The oldest recorded meteorite fall, from which material is still preserved, is the stone which fell near Ensisheim (Alsace) on November 16, 1492. The true nature of meteorites was, however, recognized only about 170 years ago by Chladni (1794), whose views were at first sharply criticized by many of his contemporaries but became generally accepted after the well-documented and thoroughly investigated fall of the stone meteorite shower of L'Aigle (France) in 1803. Once the extraterrestrial nature of meteorites was recognized they became objects of intensive study for several generations of mineralogists and chemists. During this classical period of meteorite research, which began in 1794 and can be said to have ended around 1900 with Cohen's (1894, 1903, 1905) monumental "Meteoritenkunde", some of the most eminent mineralogists and chemists of that time studied these objects. Between the two world wars, interest in meteorites diminished somewhat and only a few scientists were actively engaged in meteorite research. Due to increasing interest in space research in the years since World War II and availability of new analytical techniques such as mass spectrometry, neutron activation analysis, and electron microprobe analysis, meteorites once again became most attractive objects of scientific study. Meteoritics now is a rather unique scientific discipline inasmuch as it attracts students from a great many different fields, such as geochemistry, mineralogy, petrology, inorganic and organic chemistry, nuclear chemistry and nuclear physics, metallurgy, astronomy and biology.

Geochemists have, for a long time, been particularly enthusiastic students of meteorites. Some of the unique qualities of meteorites which attract the attention of geochemists, are the following:

a) Meteorites represent the only truly extraterrestrial material available at the present time in reasonable quantities for study in the laboratory. This in itself is reason enough to study in detail their composition, structure, texture, and physical properties.

b) Most meteorites are very old; reliable age values cluster around 4.7 billion years. In contrast, the oldest known terrestrial rocks are only about 3.6 billion years of age. Meteorites, therefore, represent extraterrestrial matter in a very early stage of development.

c) Certain meteorite types, in spite of their old age, are remarkably unaltered; i.e., after initial formation 4.5 billion years ago they were apparently little affected by geological processes, at least in comparison to the highly differentiated and metamorphosed rocks of the crust of the Earth. Study of meteorites may therefore well provide important clues to the composition and early history of solid matter in the solar system. It is intriguing to consider that among the meteorites there may be some unaltered objects representing the average composition of the essentially unfractionated nonvolatile solar material from which the planets formed. Such absolutely unaltered[1] material, however, has not been found, even though certain carbonaceous chondrites seem to come close to it.

d) Meteorites have been in space for millions of years as small, decimeter-to meter-sized bodies. During this time they were bombarded by cosmic rays and are therefore most suitable objects to study the processes of interaction of cosmic rays with matter.

e) The recently-established cooling rates of iron meteorites suggest that these objects were formed in depths of between about 10—120 km in their parent body (WOOD, 1964; GOLDSTEIN and OGILVIE, 1965; SHORT and ANDERSEN, 1965).

f) Study of the distribution of chemical elements between the three major phases of meteorites, namely, metallic nickel-iron, troilite, and silicate, provides valuable information as to the geochemical behavior of elements (siderophile, chalcophile, and lithophile elements).

Meteorites are named after their place of fall or find (closest village, post office, railroad station, landmark, etc.). They vary considerably in size: the largest single meteorite, a nickel-iron body weighing about 60 tons, was found in 1920 near Hoba, Southwest Africa, and is still at the place of fall; the smallest meteorites are objects of often 0.1 grams or less in weight which still exhibit the characteristics of typical meteorites. Such small objects have frequently been collected in the vicinity of large meteorite falls such as at the place of fall of the gigantic 1947 iron meteorite shower of Sikhote-Alin, Siberia (smallest meteorite weighs 0.0003 gram). There is no doubt that even smaller masses reach the Earth. These objects usually go unnoticed unless special efforts for their identification and recovery are made.

The number of meteorites reaching the Earth per year is small. BROWN (1961) estimates an influx of about 500 meteorites per year[2]. Averaging over the time period from 1800 to 1960, however, only about 4 observed meteorite falls were actually recovered annually (for more details on fall statistics compare MILLARD, 1963; and MILLARD and BROWN, 1963). If meteorite finds (i.e., meteorites which are found on the surface of the Earth without record of their exact time of fall[3]) are included in the statistics, the number of new meteorites recovered per year increases

[1] The term "primitive" often used in this connection has frequently been the cause of misunderstanding and is therefore avoided, for even the most "primitive" meteorites often reveal, after careful study, signs of chemical, structural, and/or textural alterations.

[2] A single fall may, however, bring many thousands of individual meteorites due to breakup of the original body upon entry into the atmosphere (e.g., the meteorite shower which fell near Pultusk, Poland in 1868 is estimated to have brought on the order of 100,000 pieces).

[3] Because the terrestrial age of meteorite finds is often considerable (e.g., stone meteorite Potter $\geqq$21,000 years; iron meteorite Sardis $\geqq$16,000 years; GOEL and KOHMAN, 1962), the composition of these objects has often been altered appreciably by terrestrial processes, and students of meteorites are well-advised to use falls and avoid finds whenever feasible, especially if minor or trace elements are of interest.

to 7 or 8 (according to HEY, 1966; the total number of meteorite falls and finds preserved in collections is about 2,000).

Table 4-1 lists recovered meteorite falls for the time period of 1492 to 1961 (KEIL, 1960, 1962a). Only meteorites from which samples are preserved in the meteorite

Table 4-1. *Frequency and weight shares of observed meteorite falls from 1492—1961, from which material is preserved.* (After KEIL, 1962a)

Class	Number	Frequency (%)	Weight (kg)	Weight (%)
Achondrites	55	7.9	1,625	4.4
Carbonaceous chondrites	17	2.4	104	0.3
Chondrites	575	82.4	10,777	29.0
Total stones	647	92.7	12,506	33.7
Siderolites	9	1.3	575	1.6
Lithosiderites	3	0.4	46	0.1
Irons	39	5.6	24,000	64.7
Total meteorites	698		37,127	

collections of the world are included in the table. It is apparent that, in terms of frequency, stone meteorite falls make up more than 92% of the total falls. In terms of weight, on the other hand, iron meteorites dominate over stone meteorites. It is significant, however, that from the 39 recovered iron meteorite falls, one individual fall, namely, the one of Sikhote-Alin, Siberia, yielded in terms of weight 23 times as much material as all the other 38 falls combined. It is apparent that even the nearly 500 years of meteorite statistics are too short a period of time for representative sampling. The significance of presently available meteorite statistics is further subject to doubt because of possible preferential loss of certain friable meteorites of low physical strength due to collisions in space, space erosion, and breakup during entry into the Earth's atmosphere. The carbonaceous chondrites, for example, which are of considerable importance for geochemical and cosmochemical considerations of the origin of meteorites and the solar system, and which make up only 2.4% of the recovered meteorite falls, may in fact have been more abundant in the meteorite parent body(s) than is apparent from current statistics.

In addition to large meteorites, fine-grained solid extraterrestrial material might be expected to accumulate on the Earth's surface. This material has commonly been called "dust" and can be of various modes of origin (SCHMIDT, 1965); a) *cosmic dust*— micron- and submicron-sized particles small enough to enter the atmosphere without undergoing melting; b) *meteoric dust* — particles resulting from the disintegration of meteors; and c) *meteoritic dust*— micron-sized particles resulting from ablation and disintegration of meteorites during their entry into the Earth's atmosphere. Commonly quoted values for the accumulation of this dust range from 30 tons (based on the abundance of spherules in deep-sea sediments) to 10^6 tons (based on satellite measurements) per year on the total Earth (e.g., SCHMIDT, 1965). Over the 4.5 billion years of history of the Earth this would amount to an increase in weight of the Earth of about 2.3×10^{-11}% and 2.7×10^{-4}%, respectively[1]. Recent studies indicate,

[1] This statement implies that the influx rate has not varied appreciably with time.

however, that satellite data are probably too high by orders of magnitude (e.g., NILSSON, 1966), and this would significantly lower the maximum estimate of the total material accumulated on the Earth.

Because of the intensive geological forces constantly shaping and modeling the surface of the Earth, extraterrestrial material has little chance, under normal circumstances, to accumulate in identifiable form to significant deposits. Only under very special conditions (e.g., in extremely slowly-accumulating sediments which remain essentially undisturbed for long periods of time) can one expect to find noticeable amounts of extraterrestrial material. Study of deep-sea sediments for instance, has indeed resulted in the discovery of magnetite spherules (MURRAY and RENARD, 1884) which, mainly on the basis of their nickel-iron-cobalt contents, are interpreted as being ablation droplets of meteorites (e.g., CASTAING and FREDRIKSSON, 1958; SCHMIDT and KEIL, 1966). Furthermore, small amounts of excess ^{36}Ar and ^{38}Ar were found in magnetic material separated from deep-sea sediments, this excess being attributed to the presence of cosmic dust in the sediment (MERRIHUE, 1964). Recently, LAL and VENKATAVARADAN (1966), and WASSON et al. (1967) have shown that samples of sediment from the Pacific have ^{26}Al activities which can only be explained by the presence of cosmic dust in the sediment, the dust having been exposed to solar-flare particles before being deposited in the sediment. However, the bulk chemical compositions of the few such sediments studied have apparently not been altered significantly by extraterrestrial material, for TUREKIAN (1958), WISEMANN (1964), and others point out that the bulk nickel content of deep-sea sediments can largely be accounted for by attributing the nickel to terrestrial sources. Hence, for most practical purposes, the influence of accumulating extraterrestrial material on the chemistry of the Earth's crust can be neglected.

The present article is mainly concerned with the composition of meteorites. For details on other important aspects of meteoritics the reader is referred to the books by COHEN (1894, 1903, 1905), FARRINGTON (1915), PERRY (1944), BROWN (1953), KRINOV (1960), MASON (1962a), HEIDE (1964), HEY (1966), and the review articles by ANDERS (1963, 1964), WOOD (1963), and RINGWOOD (1966).

II. Meteorite Classification

Meteorites have been divided into three main groups: stones, stony-irons and irons, a useful classification which was established in the middle of the last century and has since been accepted by most workers in the field. Subdivision of these groups was attempted at an early stage and over the years has resulted in a large number of meteorite classification systems, many of which, however, are now obsolete. The most practical system was originally proposed by PRIOR (1920) and later modified by MASON (1962a). In this system meteorites are classified according to their mineralogical and chemical compositions.

In Table 4-2 a modified version of the Prior-Mason system is presented, taking into account results of recent mineralogical and chemical investigations. An attempt was made to choose class names which are consistent with the chemical and mineralo-

Table 4-2. *Classification of meteorites*

Group		Class	Class symbol	Other names	Number of falls and finds[a]
Stones	Ca-poor achondrites	Enstatite achondrites	Ae	Aubrites, bustites	9
		Bronzite achondrites	Ab	Diogenites, hypersthene achondrites	8
		Olivine achondrite	Ao	Chassignite	1
		Olivine-pigeonite achondrites	Aop	Ureilites	5
	Ca-rich achondrites	Augite achondrite	Aa	Angrite	1
		Diopside-olivine achondrites	Ado	Nakhlites	2
		Polymict orthopyroxene-pigeonite-plagioclase achondrites	Aor	Howardites	13
		Monomict pigeonite-plagioclase achondrites	Ap	Eucrites	30
	Chondrites	Enstatite chondrites	Type I: Ce_1	Daniel's Kuil type chondrites, Group 1 chondrites	5
			Intermediate Type: Ce_i		2
			Type II: Ce_2	Daniel's Kuil type chondrites, Group 1 chondrites	8
		High iron (H)-group chondrites	CH	Bronzite-olivine chondrites, Cronstad-type chondrites Group 2 chondrites	about 400[b]
		Low iron (L)-group chondrites	CL	Hypersthene-olivine chondrites, Baroti-type chondrites, Group 3 chondrites	about 500[b]
		Low iron-low metal (LL)-group chondrites	CLL	Sometimes classified with hypersthene-olivine and Soko-Banja-group chondrites. Group 4 chondrites, amphoterites	at least 51, probably more[c]

(Table 4-2: continued)

Group		Class	Class symbol	Other names	Number of falls and finds [a]
Stones	Chondrites	High iron-low metal (HL)-group chondrites	CHL	Pigeonite-olivine chondrites, ornansites, Group 5a chondrites, type 3 carbonaceous chondrites, type 2 carbonaceous chondrites. Earlier classified with Soko-Banja-group and hypersthene-olivine chondrites	10
		Carbonaceous chondrites 1	Cc1	Group 5c chondrites, earlier classified with Soko-Banja group and hypersthene-olivine chondrites	4
		Carbonaceous chondrites 2	Cc2	Group 5b chondrites, earlier classified with Soko-Banja group and hypersthene-olivine chondrites. Type 1 carbonaceous chondrites	16
Stony-Irons		Pallasites	P		43
		Siderophyre	Si		1
		Lodranite	Lo		1
		Mesosiderites	M		25
Irons		Hexahedrites [d]	H		81
	Octa-hedrites	Coarsest octahedrites	Ogg		24
		Coarse octahedrites	Og		96
		Medium octahedrites	Om		215
		Fine octahedrites	Of		67
		Finest octahedrites	Off		22
		Nickel-rich ataxites	D		42

[a] Data mostly after HEY (1966).

[b] The number of meteorites in L- and H-groups was calculated from the combined total of 900 chondrites given by MASON (1962a) assuming a frequency ratio of 100:76 for L:H-group chondrites (KEIL, 1962b).

[c] Some L-group chondrites may actually belong to the LL-group.

[d] Following MASON (1962a), the nickel-poor ataxites have been included in the hexahedrite group.

gical compositions of the respective meteorites, but even though this classification is quite practical and represents the most straightforward approach to the problem, it may have to be modified somewhat in the future when new data become available.

Many of the old meteorite class names which are often based on less important properties such as color and veining of the meteorites, are still retained and in frequent use, a situation which has resulted in an unfortunate complication of meteorite classification terminology. The mineralogical and chemical compositions of certain rare meteorites such as some achondrites have, in addition, not been thoroughly studied by modern mineralogical and chemical methods, and they are often classified on the basis of incomplete and sometimes doubtful data. Furthermore, many meteorite classes are known under a number of names and this also adds to the complication of terminology. The high iron-low metal (HL) chondrites (KEIL and FREDRIKSSON, 1964), are, for example, also known as pigeonite-olivine chondrites (MASON, 1962a), ornansites (BREZINA, 1904), group 5a chondrites (YAVNEL, 1963), type 3 carbonaceous chondrites (WIIK, 1956), and type 2 carbonaceous chondrites (BOATO, 1954), and have also previously been included in the Soko-Banja group (PRIOR, 1916), and in the hypersthene-olivine chondrite group (PRIOR, 1920). Further confusion arises from the improper use of mineral names to designate meteorite classes. Several stone meteorite classes are, for example, named after the specific orthopyroxene (isomorphous series $MgSiO_3$—$FeSiO_3$) which they contain. Orthopyroxene nomenclature used by petrologists is as follows: enstatite 0—12, bronzite 12—30, hypersthene 30—50, ferrohypersthene 50—70, eulite 70—88, and orthoferrosilite 88—100 mole % $FeSiO_3$. As this nomenclature is generally accepted for terrestrial orthopyroxenes (e.g., DEER et al., 1963) it should as well be adopted to name meteoritic orthopyroxenes if confusion is to be avoided. Several meteorite class names are, however, not in accordance with the true compositions of their respective orthopyroxenes. The "bronzite"-olivine chondrites and "hypersthene"-olivine chondrites (PRIOR, 1920; MASON, 1962a, b), for example, both contain orthopyroxene of bronzite composition (average mole % $FeSiO_3$ is 16.2 and 20.5, respectively, KEIL and FREDRIKSSON, 1964; also compare Table 3). In the meteorite classification presented in the present paper (Table 4-2) these classes are instead referred to as high iron (H) and low iron (L) group chondrites (UREY and CRAIG, 1953). Furthermore, some Russian scientists refer to bronzite as enstatite (e.g., the "enstatite"-olivine chondrites Kunashak and Svonkovoje of YUDIN, 1952, and RODIONOV, 1959, respectively, are actually bronzite-olivine chondrites), which can lead to confusion with true enstatite chondrites containing enstatite nearly free of FeO. The "hypersthene" achondrites (PRIOR, 1920; MASON, 1962a, b, 1963) contain orthorhombic pyroxene having between 25—27 mole % $FeSiO_3$ (MASON, 1963). Hence, they are better referred to as bronzite achondrites (Table 4-2). The "hypersthene"-plagioclase achondrites of MASON (1962a, p. 107) which contain orthopyroxene ranging in composition from enstatite to ferrohypersthene, and in addition, pigeonite, are better referred to as polymict orthopyroxene-pigeonite-plagioclase achondrites. Finally, the pigeonite-olivine chondrites (MASON, 1962a, b) not only contain pigeonite (which, incidentally, occurs in some other chondrites as well), but a whole suite of pyroxenes. For reasons discussed below, this group is therefore referred to as HL-group (high iron—low metal) chondrites (KEIL and FREDRIKSSON, 1964).

a) Classification of Stone Meteorites

Stone meteorites are divided into two groups, achondrites and chondrites. Distinction of these two groups from each other is, in most cases, readily possible on the basis of the presence of chondrules in chondrites and their absence from achondrites. Chondrules are spherical or elliptical silicate bodies, usually a few tenths of a millimeter to a few millimeters in diameter (Fig. 4-1), containing essentially the same minerals as the matrix of the chondrite in which they are embedded. The quantities of certain minerals in chondrules and matrix can, however, be different. KEIDEL (1965) has shown that the matrix of the Borkut chondrite is richer by a factor of two

Fig. 4-1. Chondritic structure of the H group chondrite Pantar. *p* orthopyroxene chondrule; *o* olivine chondrules (one of which is slightly deformed). *Dark* nickel-iron, troilite, and fine-grained interstitial material. Light material outside the chondrules is mostly olivine. Thin section; transmitted plane polarized light

in metallic nickel-iron and troilite than its chondrules, a finding which, on the basis of qualitative microscopical observation, holds for at least some other chondrites as well. Chondrules most likely represent rapidly-cooled silicate droplets which apparently solidified largely as individual bodies, even though indentation and inclusion of one chondrule in another has been observed. Some stones, such as the Type II enstatite chondrites, have lost their chondritic structure (KEIL, 1968b). Nevertheless, their chemical and mineralogical similarities to the chondritic Type I enstatite chondrites leave no doubt as to their classification with the chondrites. The Type I carbonaceous chondrites do not contain chondrules at all, but because of their close chemical and mineralogical relationships to chondrites they are properly classified with them. Furthermore, certain polymict breccias consist of fragments of both chondritic and achondritic material, and these meteorites are usually classified on the basis of the meteorite type dominating in the breccia.

1. Classification of Achondrites

Achondrites are subdivided into Ca-poor enstatite, bronzite, olivine, and olivine-pigeonite achondrites, and into Ca-rich augite, diopside-olivine, polymict orthopyroxene-pigeonite-plagioclase, and monomict pigeonite-plagioclase achondrites. As can be seen from Fig. 4-2 the various achondrite sub-groups are quite distinct with

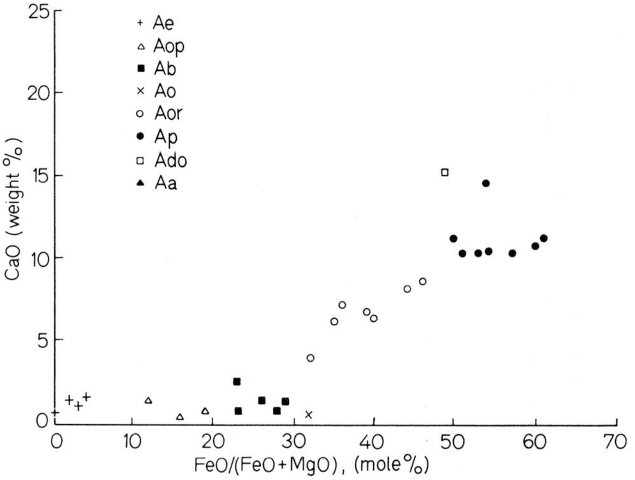

Fig. 4-2. Classification of achondrites on the basis of bulk CaO content (in weight percent) and ratio FeO/(FeO+MgO) (modified version of MASON's, 1962a, diagram). The four Ca-poor and four Ca-rich achondrite groups are quite distinct. *Ae* enstatite achondrites; *Aop* olivine-pigeonite achondrites; *Ab* bronzite achondrites; *Ao* olivine achondrite; *Aor* polymict orthopyroxene-pigeonite-plagioclase achondrites; *Ap* monomict-pigeonite-plagioclase achondrites; *Ado* diopside-olivine achondrites; *Aa* augite achondrite

regard to their bulk CaO and FeO/FeO + MgO values. Achondrites resemble terrestrial rocks more closely than any other meteorite group, for in contrast to the metal and sulfide-rich chondrites, they contain only small amounts of metallic nickel-iron and troilite.

Enstatite Achondrites: Aubres, Bishopville, Bustee, Cumberland Falls, Khor Temiki, Norton County, Pena Blanca Spring, Pesyanoe, and Shallowater. Major minerals[1]: Enstatite (< 0.1% FeO)[2], forsterite (< 0.1% FeO), plagioclase (~25 mole % An). Minor minerals[3]:

[1] The minerals listed do not necessarily occur in every one of the individual meteorites of the respective class.

[2] All numbers in weight percent unless stated otherwise.

[3] Less than about 3 weight percent.

Kamacite (remarkably low Ni content of 3.6% in Norton County), taenite (in Pesyanoe), ferromagnesian and ferroan alabandite (the mineral MgS described from Pesyanoe by Du Fresne and Anders, 1962a, is actually ferroan alabandite (Fe, Mn)S, Yudin and Smishlaev, 1964; Keil and Snetsinger, 1967), schreibersite, rhabdite, titanium-bearing troilite (0.4—4.1% Ti in Norton County; up to about 10% in Khor Temiki, Keil, 1968a), daubreelite, graphite, metallic copper, osbornite, djerfisherite (in Pena Blanca Spring, Ramdohr, 1963a; Grögler und Liener, 1968), chromite (Yudin and Smishlaev, 1964, has to be verified). Structure: Coarse, often brecciated. Other important references: Michel (1912); Beck and LaPaz (1951); mineral compositions after Keil and Fredriksson (1963).

Bronzite Achondrites: Ellemeet, Garland, Ibbenbühren, Johnstown, Manegaon, Roda, Shalka, and Tatahouine. Major minerals: Bronzite (25—27 mole % $FeSiO_3$; bronzite from Johnstown has 0.81% Cr_2O_3 and <0.001% Ni), clinobronzite. Minor minerals: Plagioclase (85 mole % An), olivine ($\sim$28 mole % Fe_2SiO_4 in Ellemeet), chromite, troilite (in Tatahouine with pentlandite exsolution lamellae, Ramdohr, 1963a), kamacite (<3% Ni) (Mason, 1963). Structure: Coarse, brecciated (except Tatahouine).

Olivine Achondrite: Chassigny (sole example). Major minerals: $\sim$90% olivine ($\sim$33 mole % Fe_2SiO_4, Mason, 1962a), $\sim$3.7—4.8% chromite (Kvasha, 1958). Minor minerals: Pyroxene, plagioclase (probably oligoclase, Mason, 1962a), nickel-iron (Ni-rich, Kvasha, 1958), glass (Tschermak, 1883). Structure: Allotriomorphic-granular, resembling terrestrial dunite.

Olivine-pigeonite Achondrites: Dingo Pup Donga, Dyalpur, Goalpara, North Haig, and Novo-Urei. Major minerals: Olivine ($\sim$21 mole % Fe_2SiO_4, Mason, 1962a), pigeonite, kamacite ($\sim$2—4% Ni). Minor minerals: Troilite, graphite, diamond (identified by X-ray diffraction in Goalpara by Urey et al., 1957, in Novo-Urei by Ringwood, 1960, and in Dyalpur by Lipschutz, 1962). Description of Dingo Pup Donga and North Haig by McCall and Cleverly (1968). Structure: Grains of olivine and pigeonite embedded in fine-grained carbonaceous matrix.

Augite Achondrite: Angra dos Reis (sole example). Major minerals: $\sim$93% titanian augite (Al-rich member of the diopside-hedenbergite series, Ludwig and Tschermak, 1887; Wahl, 1907), olivine, troilite. Structure: Not brecciated.

Diopside-olivine Achondrites: Lafayette and Nakhla. Major minerals in Nakhla: $\sim$75% diopside [$(Ca_{0.39}Mg_{0.37}Fe_{0.24})SiO_3$, Prior, 1912], $\sim$15% olivine ($\sim$65 mole % Fe_2SiO_4, Prior, 1912), plagioclase ($\sim$35 mole % An, Michel, 1912). Minor mineral: Magnetite. Structure: Not brecciated.

Polymict Orthopyroxene-Pigeonite-Plagioclase Achondrites: Bholgati, Binda, Bununu, Chaves, Frankfort, Jodzie, Kapoeta, Le Teilleul, Pampa del Infierno, Pavlovka, Washougal, Yurtuk, Zmenj (Mason, 1962a; Duke and Silver, 1967). Major minerals: Orthopyroxene of bronzite to ferrohypersthene composition (most grains have <40 mole % $FeSiO_3$), dominating over pigeonite (usually >45 mole % $FeSiO_3$) (Duke and Silver, 1967; also compare analyses by Fredriksson and Keil, 1963). Minor minerals: Olivine ($\sim$8.5—37.7 mole % Fe_2SiO_4), feldspar ($\sim$75—97 mole % An, Michel, 1912; in Kapoeta $\sim$80 mole % An), augite [$\sim(Fe_{0.25}Mg_{0.32}Ca_{0.43})SiO_3$], kamacite (3.8% Ni), taenite (38.2% Ni), troilite (all data for Kapoeta after Fredriksson and Keil, 1963). Further minor constituents (Wood, 1963): Glass, chromite, magnetite, tridymite. According to Duke (personal communication, 1966): quartz, ilmenite. Structure: Highly polymict brecciated, consisting of crystal and rock fragments of various types of materials formed under different conditions. Exception: Binda, which contains orthopyroxene with $\sim$30 mole % $FeSiO_3$ but is a monomict breccia (Duke and Silver, 1967).

Monomict Pigeonite-Plagioclase Achondrites: Adalia, Bereba, Bialystok, Brient, Cachari, Chervony Kut, Emmaville, Haraiya, Ibitira, Jonzac, Juvinas, Kirbyville, Lakangaon, Luotolax, Macibini, Mässing, Medanitos, Moore County, Nagaria, Nobleborough, Nuevo Laredo, Padvarninkai, Pasamonte, Peramiho, Petersburg, Serra de Mage, Shergotty, Sioux County, Stannern, Zagami, (Mason, 1962a; Duke and Silver, 1967). Major minerals: Pigeonite [$(Ca_{0.17}Mg_{0.33}Fe_{0.50})SiO_3$, Wahl, 1907; $(Ca_{0.08}Mg_{0.25}Fe_{0.67})SiO_3$ in Nuevo Laredo, Duke and Silver, 1962; $(Ca_{0.10}Mg_{0.35}Fe_{0.55})SiO_3$ in Stannern, v. Engelhardt, 1963], orthopyroxene ($\sim$55 mole % $FeSiO_3$, v. Engelhardt, 1963), augite [$(Ca_{0.32}Mg_{0.25}Fe_{0.43})(SiAl)O_3$, Duke and Silver, 1962], plagioclase ($\sim$84 to 92 mole % An, Michel, 1912;

v. ENGELHARDT, 1963). Minor minerals: Magnetite, ilmenite, chromite, quartz, cristobalite, tridymite, maskelynite, whitlockite (FUCHS, 1962a), kamacite, troilite. Structure: Ophitic, brecciated. Exceptions: Moore County, Shergotty, and Serra de Mage are not brecciated; Bialystok, Luotolax, Mässing, Nobleborough, and Petersburg are polymict.

2. Classification of Chondrites

Chondrites are divided into enstatite chondrites of Type I, Intermediate Type, and Type II, high iron (H), low iron (L), low iron — low metal (LL), high iron — low metal (HL) groups, and carbonaceous chondrites of class I and class II. The

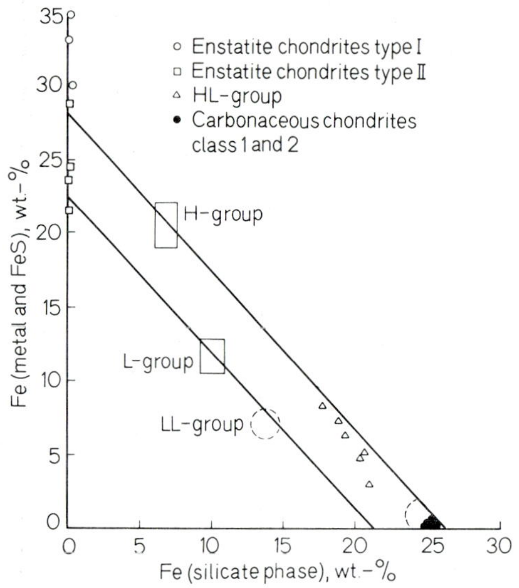

Fig. 4-3. Modified version of CRAIG's (1964) latest Urey-Craig diagram of chondrites. Classification of chondrites into enstatite chondrites of Type I and Type II, H, L, LL, and HL group chondrites, and carbonaceous chondrites of class 1 and class 2 on the basis of oxidized iron in silicates (expressed as weight % Fe) and combined iron in metal and sulfide phases (also expressed as weight % Fe). The analyses of the enstatite chondrites Pillistfer and St. Mark's, which in CRAIG's diagram plot around 2 and 6% oxidized Fe, respectively, are almost certainly wrong and have been deleted. H and L groups: Ranges in oxidized iron were calculated from olivine and pyroxene compositions; ranges in combined iron of metal and sulfide phases represent twice the average deviation of total iron contents of groups. LL group: Plotted with total iron of group distributed between metal and silicate as in the superior Soko-Banja analysis. Carbonaceous chondrites calculated on water and carbon free basis (points are raised somewhat from abscissa for clarity). The three HL chondrites with highest oxidized iron contents (Kaba, Ornans, Karoonda) are probably not reliable analyses. Diagonal lines of constant iron are corrected for change in oxygen content of meteorite

division of chondrites into classes on chemical (and mineralogical) grounds is best illustrated when the analyses are plotted in a diagram with the weight percent of iron present in the silicate as FeO on the abscissa and the combined weight percent of iron present in the metal and sulfide phases on the ordinate (UREY and CRAIG, 1953). In Fig. 4-3 a modified version of CRAIG's (1964) most recent diagram is presented. It is apparent from this diagram that chondrites form quite distinct classes.

Enstatite chondrites, which contain less than 0.5% FeO, plot close to the ordinate and are characterized by highly variable metallic and sulfide iron contents. Subdivision of enstatite chondrites into two types was first suggested by YAVNEL (1963), ANDERS (1964), and MASON (1966), and was recently confirmed by KEIL (1968b). According to KEIL (1968b) there seems to be a relation between the chemical and mineralogical compositions of the enstatite chondrites and their degree of recrystallization. Type I enstatite chondrites are only slightly recrystallized, have high total contents of Fe, S, and volatile trace elements, and contain predominantly clinoenstatite (Mn-bearing), kamacite of high Si-content, (Fe, Mg, Mn)S poor in Mn, often free SiO_2, etc. Type II enstatite chondrites are, on the other hand, highly recrystallized, have low total contents of Fe, S, and volatile trace elements, and contain predominantly ortho-enstatite (Mn-free; Mn $<$ 100 ppm), kamacite of low Si content, (Fe,Mg,Mn)S rich in Mn, Si_2N_2O instead of free SiO_2, etc. It is remarkable that the boundaries between the enstatite chondrite types are not sharp, but diffuse.

The high iron (H), low iron (L), and low iron — low metal (LL) chondrites, which are identical to MASON's (1962a, b) bronzite-olivine and hypersthene-olivine chondrites, respectively, appear as distinct groups in the diagram. The H-group chondrites are named in accordance with their high total iron, the L-group chondrites in accordance with their low total iron contents, following the original suggestion of UREY and CRAIG (1953) as modified by CRAIG (1964) and KEIL and FREDRIKSSON (1964). The LL-group chondrites also have low total iron, but in addition, less iron in the metal phase (i.e., more oxidized iron). Accordingly, they are named low iron — low metal (LL)-group chondrites. The H, L, and LL group chondrites are further distinguished on the basis of metallic nickel-iron contents, Fe/Ni ratios in bulk metallic nickel-iron, ratio Fe/(Fe+Mg) in olivine and orthopyroxene, chromite composition (SNETSINGER *et al.*, 1967), and specific gravity (Table 4-3). MASON (1965a) and RINGWOOD (1966) have recently argued against a separate LL-group and suggest including these meteorites into the L-group ("hypersthene"-olivine chondrites). More recent data on the Fe/(Fe+Mg) ratios in olivines and rhombic pyroxenes (REID and FREDRIKSSON, 1967; VAN SCHMUS and WOOD, 1967), however, confirm the existence of an independent LL-group separated by a compositional hiatus from the L-group as was previously suggested by PRIOR[1] (1916), CRAIG (1964), and KEIL and FREDRIKSSON (1964).

DODD and VAN SCHMUS (1965) recently suggested that some of the so-called "unequilibrated" chondrites containing zoned olivines and pyroxenes (a typical example is the Chainpur chondrite; KEIL et al., 1964) are the parent materials of the so-called "equilibrated" H, L, and LL group chondrites[2] which make up the vast majority of the ordinary chondrites and which contain homogeneous olivines and pyroxenes (e.g., KEIL and FREDRIKSSON, 1964). They suggested that the "equilibrated" chon-

[1] PRIOR (1916) originally classified in his "Soko-Banja" group nine chondrites which were shown to be typical L-group chondrites (KEIL and FREDRIKSSON, 1964): Albareto, Alfianello, Bachmut, Bluff, Eli Elwah, Ergheo, Knyahinya, Mocs, and Nerft. Furthermore, he also included the carbonaceous chondrites and HL-group chondrites ("ornansites") in this group, which, however, definitely constitute separate groups.

[2] The term "equilibrated" is misleading inasmuch as even the most highly recrystallized chondrites contain nickel-iron with steep concentration gradients across kamacite-taenite interfaces, suggesting that the nickel-iron alloys in the "equilibrated" chondrites are not in fact in equilibrium.

drites are metamorphosed "unequilibrated" chondrites. VAN SCHMUS and WOOD (1967) subdivided the major chondrite groups on the basis of varying petrologic (implied metamorphic) properties into thirty possible types (e.g., L_1 to L_6; H_1 to

Table 4-3. *Classification of chondrites into LL, L, and H groups on the basis of chemical-mineralogical parameters.* (From KEIL and FREDRIKSSON, 1964)

	LL Group A, Average B, Range in composition	L Group A, Average B, Range in composition	H Group A, Average B, Range in composition	References
Bulk iron, on the basis of 'superior' bulk chemical analyses (weight %)	A. 20.87 (7) [a] B. 19.85—22.55	A. 21.82 (30) B. 20.55—23.33	A. 27.52 (27) B. 25.34—30.88	CRAIG (1964)
Weight % ratio Fe/Ni in metallic nickel-iron, on the basis of 'superior' chemical analyses	A. 3.57 (7)	A. 6.87 (30)	A. 10.90 (27)	CRAIG (1964)
Metallic nickel-iron, on the basis of planimetric integration analyses of polished sections (weight %)	A. 2.64 (3) B. 1.72—3.23	A. 6.85 (38) B. 4.40—11.65	A. 16.72 (28) B. 14.17—19.81	KEIL (1962a, b)
Ratio Fe/(Fe + Mg) in olivine (mole %)	A. 27.3 (5) B. 26.3—29.0	A. 23.6 (44) B. 21.6—24.6	A. 17.9 (36) B. 16.1—19.4	KEIL and FREDRIKSSON (1964)
Ratio Fe/(Fe + Mg) in rhombic pyroxene (mole %)	A. 23.1 (5) B. 22.2—24.6	A. 20.5 (44) B. 17.9—21.7	A. 16.2 (36) B. 14.7—17.2	KEIL and FREDRIKSSON (1964)
Specific gravity	A. 3.48 (10) B. 3.44—3.54	A. 3.52 (87) B. 3.26—3.75	A. 3.69 (62) B. 3.50—3.90	Based on POKRZYWNICKI (1959) and KEIL (1962b)

[a] The number of chondrites from which the averages were calculated are given in parentheses.

H_6, etc.), twenty of which are known. This subdivision of meteorite classes seems valid for the ordinary and enstatite chondrites and has proven valuable as a basis for detailed comparative chemical and mineralogical studies of chondrites (e.g. BUNCH *et al.*, 1967).

The pigeonite-olivine chondrites of MASON (1962a, b), which have total iron contents similar to the H group chondrites but low metal contents, are named high iron — low metal (HL)-group chondrites in accordance with the terminology of H, L, and LL group chondrites (KEIL and FREDRIKSSON, 1964).

The carbonaceous chondrites of classes 1 and 2 have little or no metal and iron sulfide, but high total iron similar to H and HL group chondrites (i.e., nearly all the iron is present in oxidized form). When the analyses are recalculated on a water- and carbon-free basis, both classes of carbonaceous chondrites plot very close to each other in the Urey-Craig diagram (Fig. 4-3). They are, however, quite distinct with regard to their water and carbon contents (Table 4-5) and their mineralogy and, hence, are justly divided into two classes.

Enstatite Chondrites Type I: Indarch, South Oman, Kota-Kota, Bethune, Adhi-Kot, Abee. Major minerals: Clinoenstatite (0.28—0.65% Fe, 0.04—0.23% Ca, up to 0.39% Mn, < 0.03% Al), kamacite (3.3—8.2% Ni, 2.6—3.5% Si), troilite (0.29—0.57% Ti, 0.60—3.2% Cr). Minor minerals: Taenite (9.8% Ni in Adhi-Kot), graphite, oldhamite, daubreelite (none in Adhi-Kot, Abee), zincian daubreelite ($\sim$5% Zn) (in Kota-Kota), niningerite [(Fe,Mg,Mn)S, ranges in composition from 37.1% Fe, 10.1% Mg, 4.02% Mn, 0.03% Ca to 15.6% Fe, 23.5% Mg, 11.6% Mn, 0.39% Ca, KEIL and SNETSINGER, 1967)], sphalerite (30.6% Fe in Adhi-Kot), djerfisherite (Kota-Kota, $\sim K_3(Na, Cu), (Fe, Ni)_{12}S_{14}$,), enstatite, plagioclase ($\sim$2 mole % An; 5 mole % Or), olivine (in Kota-Kota, South Oman, and Indarch,; BINNS, 1967), roedderite [in Indarch $(Na_{1.30}K_{0.69})(Mg_{4.86}Fe_{0.27})(Si_{11.88}Al_{0.07})O_{30}$, FUCHS et al., 1966; also in Kota-Kota, FUCHS, 1966], tridymite (in Indarch, MASON, 1966), cristobalite, schreibersite, cohenite (in Indarch and Abee, MASON, 1966), lawrencite (in Indarch), perryite. All data are after KEIL (1968b), unless stated otherwise. Structure: chondritic, more or less pronounced.

Enstatite Chondrites Intermediate Type: St. Mark's and Saint Sauveur (for a description see KEIL, 1968b).

Enstatite Chondrites Type II: Daniel's Kuil, Hvittis, Atlanta, Blithfield, Jajh deh Kot Lalu, Khairpur, Pillistfer, Ufana. Major minerals: Enstatite (0.02—0.18% Fe, 0.43—0.58% Ca, < 0.01% Mn, < 0.03% Al, kamacite (6.1—6.8% Ni, 1.1—1.6% Si), troilite (0.55—0.77% Ti, 0.65—1.35% Cr), plagioclase ($\sim$15 mole % An, 4 mole % Or). Minor minerals: Taenite (9.1% Ni in Hvittis), native copper (in Hvittis, RAMDOHR, 1963a), graphite, cohenite (Hvittis, MASON, 1966), oldhamite (none in Atlanta and Blithfield), daubreelite, schreibersite, ferroan alabandite (average 15.4% Fe, 3.55% Mg, 41.3% Mn, 0.34% Ca, 0.31% Cr, 38.8% S), sphalerite (Hvittis, Atlanta, Khairpur, Pillistfer), quartz (Hvittis; MASON, 1966), tridymite (ANDERSEN et al., 1964), clinoenstatite, sinoite (ANDERSEN et al., 1964; KEIL and ANDERSEN, 1964, 1965a, b). All data after KEIL (1968b), unless stated otherwise. Structure: Highly recrystallized, chondrules are extremely rare or absent.

High Iron (H)-Group Chondrites: About 400 meteorites. Major minerals: 25—40% olivine (MASON, 1962a) (16.1—19.4 mole % Fe_2SiO_4, KEIL and FREDRIKSSON, 1964), 20—35% bronzite (MASON, 1962a) (14.7—17.2 mole % $FeSiO_3$, KEIL and FREDRIKSSON, 1964), clinopyroxene (MASON, 1965a), 5—10% plagioclase or maskelynite ($\sim$20 mole % An, MASON, 1962a), kamacite (5.5 ± 1.5% Ni, SHORT and KEIL, 1968), taenite (8—55% Ni, SHORT and KEIL, 1968), total metal content is 14.2—19.8% (KEIL, 1962a, b) (kamacite dominates over taenite), 4.9[1] (3.6—7.2%) troilite (KEIL, 1962a, b). Minor minerals: 0.23[1] (0.04—0.61%) chromite (KEIL, 1962a, b, analyses by SNETSINGER *et al.*, 1967; BUNCH *et al.*, 1967), spinel (in Forest Vale, RAMDOHR, 1963b), ilmenite, rutile (in Allegan, BUSECK and KEIL, 1966), cristobalite, (in Nadiabondi, CHRISTOPHE-MICHEL-LEVY et al., 1965), magnetite (in Plainview, RAMDOHR, 1963a), chlorapatite, whitlockite (FUCHS, 1962b), chalcopyrrhotite (RAMDOHR, 1963a, b), valleriite (perhaps identical to mackinawite, RAMDOHR, 1963b), native copper (YUDIN, 1952; RAMDOHR, 1963a, b). Structure: More or less chondritic.

Low Iron (L)-Group Chondrites: About 500 meteorites. Major minerals: 35—60% olivine (MASON, 1962a) (21.6—24.6 mole % Fe_2SiO_4, KEIL and FREDRIKSSON, 1964), 25—35% bronzite (MASON, 1962a) (17.9—21.7 mole % $FeSiO_3$, KEIL and FREDRIKSSON, 1964),

[1] This average was calculated from the original values of KEIL (1962a, b) excluding enstatite chondrites.

clinopyroxene (Mason, 1965a; Kurat and Kurzweil, 1965), 5—10% plagioclase or maskelynite (~20 mole % An, Mason, 1962a), kamacite (5.5 ± 1.5% Ni, Short and Keil 1968), taenite (8—55% Ni, Short and Keil, 1968), total metal content is 4.4—11.7% (Keil, 1962a, b) (kamacite dominates over taenite), 5.6[1] (3.8—6.9%) troilite (Keil, 1962a,b). Minor minerals: diopside (in Lanzenkirchen; Kurat and Kurzweil, 1965), 0.27 (0.05—0.54%) chromite (Keil, 1962a, b; analyses by Snetsinger *et al.*, 1967; Bunch *et al.*, 1967), ilmenite, rutile (in Farmington, Ramdohr, 1963a; Buseck and Keil, 1966), titanomagnetite (in Alfianello, Farmington and Holbrook, Ramdohr, 1963a), chlorapatite, whitlockite (Fuchs, 1962b), native copper (Yudin, 1958; Ramdohr, 1963a, b), chalcopyrrhotite (Ramdohr, 1963b), valleriite (perhaps identical to mackinawite, Ramdohr, 1963b), merrihueite (Dodd et al., 1965). Structure: More or less chondritic.

Low Iron — Low Metal (LL)-Group Chondrites: Aldsworth, Appley Bridge, Arcadia, Athens, Bandong, Benares, Benton, Bhola, Boelus, Borgo San Donino, Caratash, Cherokee Springs, Chico, Chicora, Dhurmsala, Douar Mghila, Ensisheim, Galim, Guidder, Hamlet, Holman Island, Isoulane-n-Amahar, Jelica, Karakol, Kelly, Khanpur, Krähenberg, Lake Labyrinth, Mainz, Manbhoom, Mangwendi, Mellenbye, Meru, Oberlin, Okniny, Olivenza, Ottawa, Oubari, Perth, Pevensey, Savtschenskoje, Sevilla, Sharps, Soko-Banja, St. Mesmin, Uden, Umbala, Varvik, Vavilovka, Witsand Farm, Yukan (mostly after Mason and Wiik, 1964). Major minerals: 50—60% olivine (Mason and Wiik, 1964) (26.3—29.0 mole % Fe_2SiO_4, Keil and Fredriksson, 1964), 20—25% bronzite (Mason and Wiik, 1964) (22.2—24.6 mole % $FeSiO_3$, Keil and Fredriksson, 1964), 5—10% plagioclase (~10 mole % An, Mason and Wiik, 1964), kamacite (5.7—6.8% Ni for Cherokee Springs and Dhurmsala, Short and Keil, 1968), taenite (25—55% Ni for Cherokee Springs and Dhurmsala, Short and Keil, 1968), total metal content is 1.7—3.2% (for Dhurmsala, Ensisheim, and Mainz, Keil, 1962a, b), taenite dominates over kamacite, 4.0—6.1% troilite (for Dhurmsala, Ensisheim, and Mainz, Keil, 1962a, b). Minor minerals: clinopyroxene (Preston et al., 1941; Jeremine and Lelubre, 1952), 0.22—0.32% chromite (in Dhurmsala, Ensisheim, and Mainz, Keil, 1962a, b; analyses by Snetsinger *et al.*, 1967; Bunch *et al.*, 1967), apatite (or merrillite, Jeremine et al., 1956), magnetite and pentlandite (in Isoulane-n-Amahar, Jeremine et al., 1956). Structure: Chondritic, often poorly developed.

High Iron — Low Metal (HL)-Group Chondrites: Felix, Grosnaja, Kaba, Karoonda, Lancé, Leoville, Mokoia, Ornans, Vigarano, Warrenton. Major minerals (mineral contents after Mason, 1962a): ~70% olivine, ~5% clinopyroxene and pigeonite, orthopyroxene, ~10% plagioclase (in Kaba 68—78 mole % An, Sztrokay et al., 1961), primary glass, ~5—6% troilite and/or pentlandite, ~0—5% kamacite and taenite. Minor minerals: Magnetite, sometimes Ni-magnetite (Sztrokay et al., 1961), chalcopyrite and pyrite (in Karoonda, Ramdohr, 1963a, b), carbon, hydrocarbons, spinel (Sztrokay et al., 1961). Structure: Chondritic.

Carbonaceous Chondrites I: Alais, Ivuna, Orgueil, Tonk. Major minerals: ~63% layer lattice silicate [probably chlorite, $(Mg,Fe^{2+},Ni,Al)_{5.93}OH_{7.86}O_{0.14}Si_4O_{10}$ or $(Mg,Fe^{3+},Ni,Al)_{5.93}OH_{5.75}O_{2.25}Si_4O_{10}$], ~6.7% magnesium sulfate with nH_2O (epsomite), ~6% magnetite (irregular grains and spherules, the latter are nearly pure Fe_3O_4 with < 100 ppm Ni and ~400 ppm Mn), ~4.6% troilite (1.3—1.6% Ni). Minor minerals: ~1.7% elementary sulfur, ~2.9% gypsum (Nagy and Andersen, 1964), sodium sulphate (closely associated with $MgSO_4 \cdot nH_2O$), astrakhanite, ~2.8% breunnerite (~12% Fe, Nagy and Andersen, 1964), dolomite (Du Fresne and Anders, 1962b), brucite or periclase (one 10μ crystal having about 50% Mg, no X-ray data available), ~0.5% limonite (Nagy et al., 1963; probably of extraterrestrial origin), ~0.8% merrillite (Nagy and Andersen, 1964) (or whitlockite, Bostrom and Fredriksson, 1966), olivine (very small amounts, Kerridge, 1964, 1968; Bostrom and Fredriksson, 1966), graphite (in Ivuna, Vdovykin, 1964), organic materials (fatty acids, porphyrins, nucleic acid bases, amino acids, free radicals, hydrocarbons; reviews by Briggs and Mamikunian, 1962—63; Urey, 1966), "organized elements" [microscopic-sized particles of uncertain nature interpreted by some investigators (e.g., Claus and Nagy, 1961) as of biological origin; some of these particles have been

[1] This average was calculated from the original values of Keil (1962a,b) excluding LL-group chondrites.

shown to be altered hexagonal troilite plates (e.g., MUELLER, 1963), still others are possibly terrestrial contaminants (e.g., FITCH and ANDERS, 1963)]. All mineralogical and chemical data are after BOSTROM and FREDRIKSSON (1966) for the Orgueil meteorite, unless stated otherwise. Structure: Like bituminous clay with clastic texture, sometimes with angular inclusions, but no chondrules.

Carbonaceous Chondrites 2: Al Rais, Bells, Boriskino, Cold Bokkeveld, Crescent, Erakot, Essebi, Haripura, Mighei, Murray, Nawapali, Nogoya, Pollen, Renazzo, Revelstoke, Santa Cruz. Major minerals: ~70% hydroxyl containing layer lattice silicate (~22% Fe, ~9—10% Mg, ~0.5—1% Al, ~1—4% Ni), olivine (0.3—91.2 mole % Fe_2SiO_4; remarkably high Ca content of 0.20% in comparison to olivine of other meteorites), pyroxene (probably ortho- and clinopyroxene, 0.4—45.4 mole % $FeSiO_3$), magnetite. Minor minerals: Troilite and pentlandite (0.9—21.5% Ni), metallic nickel-iron, magnesium sulfate with nH_2O (epsomite), calcite and breunnerite (KVASHA, 1948), glass (DU FRESNE and ANDERS, 1961), amorphous carbon, complex hydrocarbons. All chemical data after FREDRIKSSON and KEIL (1964) for MURRAY. Recent study of olivine and pyroxene compositions by WOOD (1967). Structure: Chondritic, olivine-pyroxene fragments and chondrules embedded in layer lattice silicate matrix.

b) Classification of Stony-Iron Meteorites

The stony-iron meteorites are rather arbitrarily divided into pallasites, siderophyre, lodranite, and mesosiderites. Members of this group are characterized by appreciable nickel-iron contents.

Pallasites: 39 finds, 4 falls. Major minerals: ~25—65% kamacite and taenite, ~35—75% olivine (10—20 mole % Fe_2SiO_4). Minor minerals: Native copper (in Molong, RAMDOHR, 1963b), troilite, schreibersite, farringtonite (in Springwater, DU FRESNE and ROY, 1961), ilmenite (in Mt. Dyrring, RAMDOHR, 1963b), magnetite (RAMDOHR, 1965), chromite, graphite, lawrencite (WOOD, 1963). Structure: Coarse grained; well-developed olivine crystals; nickel-iron has Widmanstätten structure.

Siderophyre: Steinbach (sole example). Major minerals: Kamacite (6.73% Ni, 0.54% Co, 0.08% P), taenite (16.5—39.7% Ni, 0.2—0.5% Co, 0.02—0.05% P) (total nickel-iron content about 50%), ~35% bronzite (~12 mole % $FeSiO_3$, 0.34% Mn), ~15% tridymite. Minor minerals: troilite (0.61% Cr), schreibersite (46.3—50.0% Ni), chromite (1.36% Mn, 1.21% Mg), (data on mineral contents are from STORY-MASKELYNE, 1871; compositional data are from DÖRFLER et al., 1965). Structure: Coarse grained; nickel-iron has Widmanstätten structure.

Lodranite: Lodran (sole example). Major minerals: Kamacite and taenite (~32% total metal), ~29% olivine (~14 mole % Fe_2SiO_4), ~31% bronzite (~17 mole % $FeSiO_3$), ~7% troilite. Minor minerals: chromite, anorthite (all data after TSCHERMAK, 1870). Structure: Pallasitic, but crystal diameter < 1 mm. Widmanstätten structure.

Mesosiderites: 7 falls, 18 finds. Major minerals: Kamacite and taenite (~50% total metal), orthopyroxene (often bronzite of ~20—30 mole % $FeSiO_3$; in Vaca Muerta, three phases of 30, 25, and 19 mole % $FeSiO_3$, respectively, were found, MARVIN and KLEIN, 1964; 12 mole % $FeSiO_3$ in Estherville, FUCHS, pers. comm.), anorthite (~87—97 mole % An, MICHEL, 1912), olivine (~9—14 mole % Fe_2SiO_4), troilite. Minor minerals: schreibersite, chromite, ilmenite, tridymite, zircon, rutile (MARVIN and KLEIN, 1964; RAMDOHR, 1965; BUSECK and KEIL, 1966), graphite (and cliftonite), whitlockite, chlorapatite, clinopyroxene (MARVIN and KLEIN, 1964). After RAMDOHR (1965): Native copper, chalcopyrrhotite, sphalerite, magnetite (questionable), lawrencite (questionable), mackinawite. Structure: often brecciated. Nickel-iron fragments sometimes exhibit Widmanstätten structure, but the metal usually does not form a continuous network.

c) Classification of Iron Meteorites

Iron meteorites are classified on the basis of their structure and, to some extent, their nickel contents, into hexahedrites, octahedrites, and nickel-rich ataxites.

The metal phase in hexahedrites is entirely kamacite (~5.5% Ni) forming either large single crystals or coarse aggregates. The name hexahedrite is derived from the large cubic kamacite crystals (hexahedrons) which make up these meteorites.

The octahedrites are the most common iron meteorites ranging in nickel content from about 6.5—18%. When heated slightly or etched they exhibit a characteristic structure named after their discoverer "Widmanstätten" (Fig. 4-4). This structure

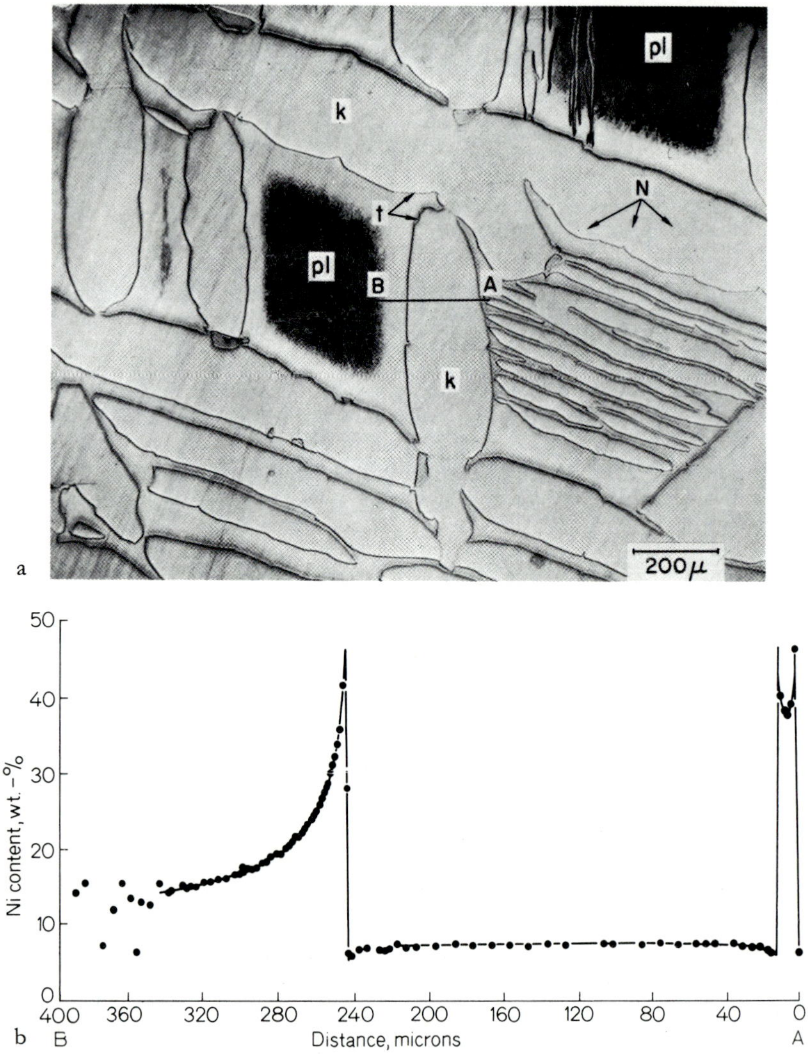

Fig. 4-4. Widmanstätten structure of fine octahedrite Edmonton, Kentucky. Polished section, etched. Reflected light, bright field. Kamacite plates (*k*) are intergrown parallel to the faces of an octahedron. Kamacite (α-nickel-iron) is rimmed by the nickel-rich γ-nickel-iron taenite (*t*). Steep diffusion gradients across kamacite-taenite interface along track *AB* were measured with the electron microprobe and correspond to a cooling rate during formation of the Widmanstätten structure (~600—300° C) of $1.9 \pm 0.3°$ C/10^6 years. *pl* plessite; *N* Neumann lines [mechanical twin lamellae after (211)]. (Courtesy of J. M. Short.)

consists of four systems of kamacite plates (α nickel-iron) which parallel the faces of an octahedron, hence the name of the class "octahedrites". The kamacite plates are rimmed by taenite, the nickel-rich γ nickel-iron. Angular spaces between the kamacite band systems are often filled with a fine intergrowth of kamacite and taenite commonly referred to as plessite (Fig. 4-4).

Octahedrites are sub-divided rather arbitrarily according to the width of their kamacite lamellae, a quite practical and convenient system which, however, has the following shortcomings. First, the width of the kamacite lamellae may vary from place to place in a given iron meteorite. Second, this width is not only a function of the bulk nickel content of the meteorite but also a function of its cooling rate. In systems with identical nickel contents slower cooling results in wider kamacite plates, thus complicating proper classification of iron meteorites. An example of

Table 4-4. *Influence of cooling rate on kamacite band width in octahedrites. Subdivision of octahedrites on the basis of width of kamacite plates and bulk nickel contents is complicated due to differences in cooling rates during formation of Widmanstätten structure. In systems with identical Ni contents, slower cooling results in wider kamacite lamellae.* (After GOLDSTEIN and SHORT, 1967)

Name of meteorite	Classification according to width of kamacite lamellae	Bulk Ni content, in weight percent	Cooling rate, in °C/10^6 years
Toluca	Om	8.4	1.6 ± 0.5
Bristol	Of	8.3	20 ± 3

the magnitude of this effect is given in Table 4-4. Studies to subdivide octahedrites on the basis of their gallium and germanium contents, first attempted by LOVERING et al. (1957), are now in progress (e.g., WASSON, 1967).

The nickel-rich ataxites usually consist of a fine-grained plessite-like intergrowth of kamacite and taenite. Their bulk nickel content is larger than 12% and sometimes exceeds 25%.

Hexahedrites: 74 finds, 7 falls. Included in this group are the chemically and mineralogically similar nickel-poor ataxites, which are probably thermally metamorphosed hexahedrites. Major minerals: ~92% kamacite (~5.5% Ni). Minor minerals: troilite, schreibersite (and rhabdite), daubreelite, graphite, ureyite (in Coahuila and Hex River Montains, FRONDEL and KLEIN, 1965); chromite (in Coahuila; BRETT and HENDERSEN, personal communication). Structure: Frequently large single kamacite crystals, sometimes granular aggregates. Neumann lines (probably mechanically introduced twin lamellae along (211) of kamacite) in kamacite are common.

Octahedrites: At least 424 (mostly finds, only about 43 observed falls). They are subdivided on the basis of the width of their kamacite lamellae (in mm) into coarsest (Ogg, >2.5 mm, 6—7% Ni), coarse (Og, 1.2—2.5 mm, 6.5—7.5% Ni), medium (O_m, 0.5—1.2 mm, 6.5—10% Ni), fine (O_f, 0.2—0.5 mm, 7.5—10.5% Ni), and finest (O_{ff}, <0.2 mm, 8—15% Ni) octahedrites (WOOD, 1963). Medium octahedrites are the most abundant. Major minerals: Kamacite, taenite, troilite, schreibersite (the latter two minerals are sometimes only present in minor amounts). Minor minerals: Native copper (in Odessa, EL GORESY, 1965), graphite, diamond (in Cañon Diablo, FOOTE, 1891), cohenite, daubreelite, sphalerite (EL GORESY, 1965; PARK et al., 1966), ferroan alabandite (EL GORESY, 1965), chalcopyrite (in Delegate, RAMDOHR, 1963b), chalcopyrrhotite, (EL GORESY, 1965), valleriite (perhaps identical to mackinawite, RAMDOHR, 1963b), mackinawite (EL GORESY, 1965), cirstobalite (in Carbo, MARVIN, 1962), rutile (in Odessa, EL GORESY, 1965), ilmenite (in Dalgaranga, EL GORESY,

1965), chromite, lawrencite, chlorapatite (~4.4% Cl), Ca-free sarcopside and Ca-free graftonite (OLSEN and FREDRIKSSON, 1966), brianite and panethite (in Dayton, FUCHS et al., 1967), whitlockite (2.55 $CaO \cdot 0.16\ Na_2O \cdot 0.28\ MgO \cdot 0.01\ FeO \cdot P_2O_5$, FUCHS, personal communication), forsterite (2.8 mole % F_2SiO_4), enstatite (5.4 mole % $FeSiO_3$), chromian endiopside (1.01% Cr, 1.0% Fe, 14.4% Ca, < 0.1% Al), ureyite (in Toluca, FRONDEL and KLEIN, 1965), albite (0.76% Ca), krinovite (in Cañon Diablo, Wichita County, and Youndegin; OLSEN and FUCHS, 1968), richterite (in Cañon Diablo and Wichita County; OLSON, 1967), zircon (in Toluca, MARVIN and KLEIN, 1964), moissanite (doubtful; MASON, 1962a). All mineral compositions are for the Odessa iron meteorite unless stated otherwise (MARSHALL and KEIL, 1965). The amounts of accessory minerals in nodules of one and the same octahedrite vary drastically (KEIL, 1965). Structure: Widmanstätten, usually pronounced; kamacite sometimes has Neumann lines.

Nickel-Rich Ataxites: 42 finds. Major minerals: Taenite, kamacite. Minor minerals: schreibersite, troilite, graphite, silicates (WOOD, 1963).

III. Chemical Composition of Meteorites

In this chapter the chemical composition of meteorites in terms of major and minor elements is discussed. Conventionally, the elements present in the silicate portions of meteorites are reported as oxides. In addition to the elements occurring in the metal phase (Fe, Ni, Co), several elements of particular importance in meteorites are also listed. These are Fe and S combined as FeS, NiO (in carbonaceous chondrites), and C and Cu (in iron meteorites). Trace element abundances are not discussed here since they are reported in chapter 5 and in Volume II of this "Handbook".

a) Chemical Composition of Stone Meteorites

Bulk chemical analysis of major and minor elements in stone meteorites is considered by many to present one of the most difficult problems of rock analysis. The determination of oxidized iron is particularly troublesome owing to the fact that in a particular meteorite, iron occurs in various oxidation states and chemical combinations with other elements, namely as ferrous iron in the major silicate minerals olivine and pyroxene, as ferrous iron combined with sulfur in troilite, pentlandite (in carbonaceous chondrites), ferroan alabandite, niningerite and daubreelite (in enstatite chondrites and enstatite achondrites), and in the neutral state in kamacite and taenite. The portion of the total ferrous (and ferric) iron combined in chromite ($FeCr_2O_4$), ilmenite ($FeTiO_3$), and magnetite (Fe_3O_4) is usually small, but the magnetite content of carbonaceous chondrites can be appreciable. Other iron-bearing phases such as schreibersite, $(Fe, Ni)_3P$, occur in only minor amounts in stone meteorites. Oxidized iron is, in fact, computed after three independent determinations: the iron in metallic nickel-iron is determined by analyzing mechanically or chemically separated metal phase; then iron in the sulfide phase is calculated from the independently-determined total sulfur content of the meteorite; a third determination gives the total iron content, and oxidized iron is then derived in the following manner:

$$Fe_{oxidized} = Fe_{total} - [Fe_{sulfide} + Fe_{metal}]. \quad (1)$$

This procedure has several inherent sources of error. The total iron value is probably the best number, even though it may be severely affected by sampling errors when too small samples are used for analysis of heterogeneous stone meteorites. The value for iron in the sulfide phase tends to be too high since it is calculated by assuming the total sulfur to be present as troilite (FeS), a fact which applies to ordinary LL, L and H group chondrites but not to enstatite chondrites, which contain daubreelite, oldhamite, ferroan alabandite, and niningerite, nor to carbonaceous and HL group chondrites, which contain elementary sulfur (Class I carbonaceous chondrites) and pentlandite. The value for metallic nickel-iron is probably the most uncertain of the three for the following reasons. Metallic nickel-iron is often heterogeneously distributed in stone meteorites and sampling errors can therefore be severe (KEIL, 1962a, b). Furthermore, some of the metallic nickel-iron included within silicate

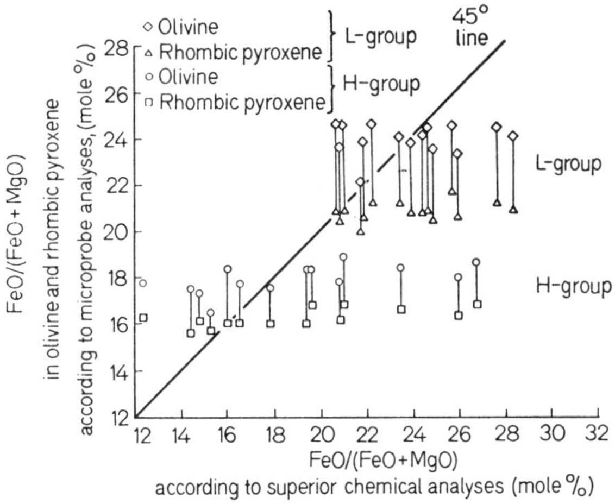

Fig. 4-5. Mole percent ratios FeO/(FeO+MgO) in olivine and rhombic pyroxene of H and L group chondrites according to microprobe analyses are plotted vs. FeO/(FeO+MgO) ratios calculated from FeO and MgO of "superior" bulk chemical analyses. Since olivine and pyroxene are the only major FeO-bearing minerals in ordinary chondrites and since the FeO/(FeO+MgO) ratios are slightly higher in olivine than in coexisting pyroxene the wet chemically obtained bulk FeO/(FeO+MgO) values should lay somewhere between these two extremes, i.e., the lines connecting coexisting olivine-pyroxene pairs should intersect a 45° line. This is true for only a few of the "superior" analyses. Most others plot to the right of the 45° line indicating too high FeO values in the wet chemical analyses. (After KEIL and FREDRIKSSON, 1964)

grains may not show in the analysis. Although its amount can be estimated by measuring the Ni-content of the essentially NiO-free olivine-pyroxene portion[1] errors may be introduced due to inhomogeneous nickel distribution in the metal phase (kamacite vs taenite). Finally, broken surfaces of even fresh meteorite falls show iron oxide halos after only a few days, and this introduces errors into the determination of iron present in the metallic phase.

The magnitude of the errors thus introduced into wet chemically determined bulk FeO values of stone meteorites is illustrated in Fig. 4-5 (after KEIL and FREDRIKSSON,

[1] Olivines and pyroxenes of ordinary LL, L, and H, and of enstatite chondrites contain <100 ppm nickel as NiO (KEIL and FREDRIKSSON, 1964), but carbonaceous chondrites may contain appreciable amounts of NiO in their layer lattice silicate mineral matrix (~1—4% Ni in Murray; FREDRIKSSON and KEIL, 1964).

1964). In this figure the FeO/(FeO+MgO) ratios of olivines and pyroxenes of L and H group chondrites as obtained by electron microprobe analysis are plotted vs the FeO/(FeO+MgO) ratios of "superior" bulk wet chemical analyses of the same meteorites. Most of the "superior" chemical analyses were taken from UREY and CRAIG (1953); some more recent analyses have, however, also been included. Since olivine and pyroxene are the only major FeO-bearing minerals in ordinary chondrites, and since the FeO/(FeO+MgO) ratios are slightly higher in olivine than in coexisting pyroxene, the wet chemically obtained bulk FeO/(FeO+MgO) ratios should lay somewhere between these two extremes; i.e., the lines connecting coexisting olivine-pyroxene pairs should intersect a 45° line (note that the precision of the microprobe data is better than $\pm 1\%$ of the amounts present). As can be seen from Fig. 4-5 this is the case for only some of the "superior" chemical analyses. Most others plot to the right of the 45° line, indicating too high FeO values in the wet chemical analyses (KEIL and FREDRIKSSON, 1964). It is also worth noting that the spread of the FeO/(FeO+MgO) values as obtained by wet chemical analyses is considerably larger than the spread of the microprobe data, indicating analytical errors in the former analyses.

On the basis of bulk chemical analyses PRIOR (1916) observed an apparent relationship between the amounts of metallic and oxidized iron in chondrites. He expressed this apparent relationship in what has become known as PRIOR's rule (second part) which states that "in meteoritic stones generally, the poorer they are in nickel-iron, . . . the richer in iron are the magnesium silicates". He thought this to be the result of changes in the reduction-oxidation state of the meteorites without changes in their bulk composition and concluded that there exists a continuous reduction-oxidation sequence from the most highly oxidized carbonaceous chondrites to the most highly reduced enstatite chondrites. This second part of PRIOR's rule in its original sense has been supported on the basis of bulk chemical analyses (which are, as was shown above, liable to be erroneous with regard to their FeO values) by MASON (1962a) and RINGWOOD (1961, 1966), but was considered a direct result of erroneous chemical analyses by UREY and CRAIG (1953), KEIL and FREDRIKSSON (1964), and CRAIG (1964). Assuming the Fe_{total} to be a reasonably well determined parameter in most wet chemical chondrite analyses, it is apparent from Eq. (1) that a too low $[Fe_{sulfide}+Fe_{metal}]$ value results in a too high $Fe_{oxidized}$, and a too high $[Fe_{sulfide}+Fe_{metal}]$ in a too low $Fe_{oxidized}$. Thus, errors in the determination of the iron in the metal phase artificially create a "PRIOR's rule"-type relationship.

KEIL and FREDRIKSSON (1964), and CRAIG (1964) point out that although the second part of PRIOR's rule holds qualitatively when considering different chondrite groups, it does not hold quantitatively (since the total iron content changes), and does not hold within individual groups. This is illustrated in Fig. 4-6, where the most reliable values of metallic nickel-iron contents of chondrites as obtained by planimetric integration of large polished sections (KEIL, 1962a, b) and by wet chemical analysis are plotted vs oxidized iron values (expressed as Fe) in olivine and rhombic pyroxene of L and H group chondrites as obtained by electron microprobe analysis. The dashed lines were calculated from the group averages assuming straight oxidation-reduction relationships. As can be seen from Fig. 4-6 the analyses do not plot along the dashed lines as would be expected if the second part of PRIOR's rule would hold within chondrite groups. Furthermore, it can readily be seen from the large and definitely real spread in the metallic nickel-iron contents of chondrites of a given group (e.g., H group: 4.40—11.65%) as compared to the small spread in the iron content of olivine and rhombic pyroxene in the same group (e.g., H group: 16.3—17.9% for olivine and 9.0—10.0% for rhombic pyroxene), that a quantitative relationship between metallic and oxidized iron, in the strict sense of the second part of PRIOR's rule, does not exist.

ANDERS (1964) and RINGWOOD (1966) have recently presented data which indicate that the first part of PRIOR's rule ("The less the amount of nickel-iron in chondritic

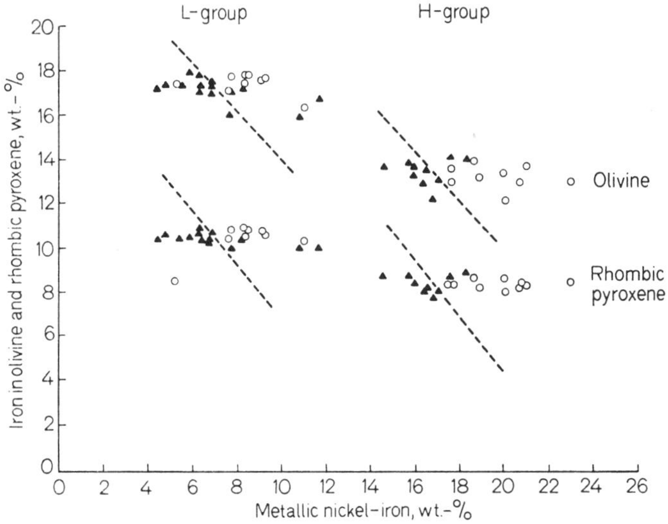

Fig. 4-6. Relation between the amount of metallic nickel-iron and of oxidized iron (expressed as Fe) in olivine and rhombic pyroxene of H and L group chondrites. Triangles (▲) are metallic nickel-iron values obtained by planimetric integration of polished sections; circles (○) are metallic nickel-iron values from superior bulk chemical analyses. Iron contents of olivines and pyroxenes were obtained by electron microprobe techniques. Dashed lines were calculated from the group average assuming a simple oxidation-reduction relationship between metallic and oxidized iron (second part of PRIOR's rule). Analyses do not plot along dashed lines; hence, within groups the second part of PRIOR's rule is not strictly valid. From group to group PRIOR's rule holds only qualitatively (i.e., in addition to changes in the red-ox state of the iron the total iron content changes from group to group). (After KEIL and FREDRIKSSON, 1964.)

stones the richer it is in nickel . . .") may hold true. It appears, however, that more accurate nickel determinations of chondrite samples large enough to exclude sampling errors are necessary to decide on the validity of this portion of PRIOR's rule.

In Table 4-5 the average compositions of stone meteorites are listed.

b) Chemical Composition of Stony-Iron Meteorites

Stony-iron meteorites are characterized by their comparatively high metallic nickel-iron contents and by the coarseness of their mineralogical constituents. Representative sampling for bulk chemical analysis is therefore exceedingly difficult,

Table 4-5. *Average chemical composition of stone*

Class	Silicate portion											
	SiO_2	MgO	FeO	Fe_2O_3[b]	Al_2O_3	CaO	Na_2O[c]	K_2O[c]	Cr_2O_3	MnO	TiO_2	P_2O_5
Enstatite achondrites	54.01	35.92	0.97	—	0.67	0.91	1.32	0.10	0.06	0.14	0.06	0.22
Bronzite achondrites	52.11	25.85	16.05	—	1.18	1.41	0.004	0.001	0.80	0.32	0.19	0.01
Olivine achondrites	37.12	32.05	26.82	—	1.26	0.56	0.19	0.09	0.88	0.49	0.16	0.10
Olivine-pigeonite achondrites	40.83	37.43	12.16	—	0.54	0.87	0.11	0.04	0.85	0.40	0.15	0.08
Augite achondrite	44.58	10.50	8.50	1.81	8.86	24.51	(0.26)	(0.19)	—	—	—	—
Diopside-olivine achondrites	48.96	12.01	19.63	1.29	1.74	15.17	(0.41)	(0.14)	0.33	0.09	0.38	—
Polymict ortho-pyroxene-pigeonite-plagioclase achondrites	49.75	16.10	13.26	2.62[f]	8.71	6.53	(0.95)	(0.28)	0.42	0.78	0.11	0.07
Monomict pigeonite-plagioclase achondrites	48.17	7.10	15.99	1.17	13.91	10.94	(0.67)	(0.13)	0.39	0.46	0.51	0.11
Enstatite chondrites Type I	35.81	17.90	—	—	1.95	1.43	0.93	0.13	Cr=0.24	Mn=0.18	Ti=0.07	—
Enstatite chondrites Type II	41.15	22.89	—	—	1.98	1.24	0.85	0.14	Cr=0.29	Mn=0.10	Ti=0.05	—
High iron (H)-group chondrites	36.52	23.48	8.87	—	2.43	1.82	(0.85)	(0.14)	0.36	0.25	0.13	0.23
Low iron (L)-group chondrites	39.88	24.98	13.12	—	2.31	1.90	(0.88	(0.14)	0.44	0.27	0.15	0.26
Low iron-low metal (LL)-group chondrites	39.38	25.40	17.70	—	2.21	1.96	(0.87)	(0.21)	0.30	0.32	0.19	0.22
High iron-low metal (HL)-group chondrites	33.88	23.68	24.00	—	2.67	2.25	0.59	0.049	0.50	0.20	0.13	0.30
Carbonaceous chondrites 1	23.08	15.56	10.32	—	1.77	1.51	0.76	0.07	0.28	0.19	0.08	0.27
Carbonaceous condrites 2	27.31	19.00	20.06	—	2.31	2.03	0.54	0.05	0.39	0.17	0.10	0.27
Pallasites	17.05	19.83	6.65	—	0.38	0.28	(0.07)	(0.03)	0.68	0.08	0.00	—
Siderophyre	34.61	10.08	4.40	—	—	—	—	—	—	—	—	—
Lodranite	28.94	23.33	7.71	—	0.19	0.18	—	—	0.17	—	—	—
Mesosiderites	19.51	6.36	5.73	1.95	4.10	2.89	(0.17)	—	0.36	0.22	—	—

[a] This table in principle follows Table 10 of Wood (1963).

[b] Most analysts report all oxidized iron as FeO. In case of meteorites containing clinopyroxene, a mineral which commonly has ferric iron, the actually reported Fe_2O_3 values are usually given (following a suggestion of Wood, 1963).

[c] Superior Na_2O and K_2O analyses by Edwards and Urey (1955) and Wiik (1966): Older and probably less reliable values are given in parentheses.

[d] Considering the highly reduced state of enstatite achondrites as is indicated by the presence of nearly FeO-free $MgSiO_3$, CaS, $FeCr_2S_4$, TiN, etc., the reported NiO-values (e.g., 0.78% NiO in Bustee; 0.54% in Bishopville) are exceedingly doubtful.

[e] CaS according to the analyses. These analyses are incorrect inasmuch as enstatite achondrites contain minor amounts of Fe and S in $FeCr_2S_4$; Fe, Mg, Mn, and S in ferroan alabandite and ferromagnesian alabandite; and most of the Ti in troilite combined with sulfur and in osbornite combined with nitrogen.

Silicate portion			Metal portion					FeS	C	Others	Number of mete-orites averaged	Reference
H_2O	NiO	total silicate portion	Fe	Ni	Co	P	total metal portion					
1.14	0.26[d]	95.78	2.29	0.17	—	—	2.46	1.25	—	0.51[e]	4	Urey and Craig (1953) and Wiik (1956), weighed equally
0.14	—	98.06	0.79	0.03	—	—	0.82	1.12	—	—	4	Urey and Craig (1953)
0.24	—	99.96	—	0.06	—	—	—	0.60	—	—	1	Dyakonova and Charitonova (1960)
—	—	93.46	3.94	0.11	0.05	—	4.10	1.66	2.23	—	3	Wiik (1966)
—	—	98.76	—	—	—	—	—	1.26	—	—	1	Ludwig and Tschermak (1887)
0.24	—	99.84	—	—	—	—	—	0.06 (S)	—	—	1	Prior (1912)
0.25	—	99.83	0.33	0.11	—	—	0.44	0.73	—	—	6[g]	Urey and Craig (1953)
0.44	—	99.99	0.80	—	—	—	—	0.41	—	—	13	Urey and Craig (1953), von Engelhardt (1963)
—	—	—	total Fe=32.85	1.77	0.09	0.17	—	S=5.77	0.43[h]	—	5	i
—	—	—	total Fe=24.88	1.53	0.11	0.12	—	S=3.31	0.34[h]	—	6	i
0.33	—	75.42	17.23	1.58	0.085	—	18.90	5.35	0.10[h]	—	27	Craig (1964)
0.34	—	84.67	7.70	1.12	0.059	—	8.88	6.17	0.09[h]	—	30	Craig (1964)
0.37	—	89.58	3.39	0.95	0.052	—	4.39	5.87	—	—	7	Craig (1964)
0.53	—	88.78	3.64	1.43	0.077	—	5.15	5.90	0.33	—	4	Craig (1964)
20.54	1.17	75.60	0.11	0.02	0.00	—	0.13	16.88[j]	3.62	3.77[k]	3	Wiik (1956)
13.23	1.56	87.02	0.00	0.16	0.00	—	0.16	8.58[j]	2.44	1.80[k]	8	Wiik (1956)
—	0.29	45.34	48.98	4.66	0.30	0.11	54.05	0.53	0.08	—	10	Wood (1963)
—	—	49.09	46.02	4.74	0.15	—	50.91	—	—	—	1	Story-Maskelyne (1871)
—	—	60.52	27.77	4.05	—	—	31.82	7.40	—	0.26[l]	1	Tschermak (1870)
0.69	0.40	42.38	45.95	4.39	0.28	0.11	50.73	2.83	—	4.06[l]	4	Wood (1963)

[f] This average was calculated excluding the extraordinarily high and probably doubtful value for Pavlovka.

[g] Meteorites averaged are Binda, Frankfort, LeTeilleul, Pavlovka, Yurtuk, and Zmenj.

[h] After Moore and Lewis (1965, 1966).

[i] Averages were calculated from analyses listed by Mason (1966). Type I includes Saint Sauveur, Adhi Kot, Indarch, St. Mark's, and Abee; Type II includes Daniel's Kuil, Atlanta, Hvittis, Khairpur, Jajh deh Kot Lalu and Blithfield. (For simplicity, Intermediate Type chondrites were included into Type I). The data for Cr, Mn, Ti, and P are presented in metal rather than oxide form since this accounts better for their actual mineralogical occurrence.

[j] Contains little FeS, but most of the sulfur as elementary S, water-soluble sulfates, and pentlandite.

[k] Ignition loss considered to be approximately equal to content of complex organic matter (Wiik, 1956).

[l] Insoluble.

and the average compositions listed in Table 4-5, based mostly on analyses made before 1900, are probably not very accurate.

c) Chemical Composition of Iron Meteorites

Iron meteorites contain nodules up to several centimeters in diameter (e.g., MARSHALL and KEIL, 1965; EL GORESY, 1965) which consist of the minerals troilite, graphite, schreibersite, cohenite, sphalerite, pyroxene, olivine, chromian endiopside, and others. Practically all bulk chemical analyses of iron meteorites were, however, performed on samples carefully cleaned of inclusions. The data reported in Table 4-6 are therefore not representative of the respective meteorite classes, but only give the average compositions of their metallic phases. The published abundances of S, P, C, Si, Ca, Mg, Zn, Al, Cr, Na, K, and others in iron meteorites are probably orders of magnitude too low since these elements form minerals which are enriched in the nodules. The real sulfur contents of iron meteorites, for example, can be estimated from their troilite contents obtained by point counting of polished meteorite sections and are one to two orders of magnitude higher than is apparent from the bulk chemical meteorite analyses (Table 4-7). Much more work, particularly planimetric integration analyses of large polished meteorite sections, is necessary before meaningful values for the average composition of iron meteorites can be given.

Table 4-6. *Average composition of iron meteorites on the basis of bulk chemical analyses* (after BUDDHUE, 1946) *(not accounting for inclusions of troilite, graphite, schreibersite, cohenite, etc.)*

Class	Weight percent									Number of meteorites averaged
	Fe	Ni	Co	Cu	P	S	C	misc.	total	
Hexahedrites [a]	92.62	6.07	0.61	0.29	0.25	0.06	0.15	0.18	100.23	78
Octahedrites:										
coarsest	92.33	6.54	0.50	0.01	0.16	0.02	0.23	0.27	100.06	20
coarse	91.22	7.39	0.54	0.18	0.18	0.08	0.21	0.54	100.34	40
medium	90.67	8.22	0.59	0.03	0.18	0.09	0.08	0.30	100.16	126
fine	90.53	9.00	0.57	0.05	0.17	0.08	0.61	0.18	101.19	43
finest	86.75	11.65	0.61	0.11	0.24	0.63	0.01	0.45	100.45	13
Nickel-rich ataxites	79.63	18.85	1.01	0.05	0.12	0.08	0.10	0.19	100.03	38

[a] Following MASON (1962a) the nickel-poor ataxites have been included in the hexahedrite group.

Table 4-7. *Troilite (FeS) content of octahedrites as obtained by planimetric integration of large sawed and polished meteorite surfaces.* (Data compiled by WOOD, 1963)

Name	Classification	Weight percent troilite
Cañon Diablo	Ogg	5.95
Osseo	Ogg	3.42
Coolac	Og	4.76
Odessa	Og	3.43
Wichita Co.	Og	2.72
Kingston	Om	1.57
Toubil River	Om	1.23
Cape York	Om	0.49
Augustinovka	Om-Of	3.89
Gibeon	Of	1.00
St. Genevieve Co.	Of	0.30

Table 4-8. *Minerals occurring in meteorites. Minerals marked by asterisk are not known from terrestrial rocks*

Mineral class	Mineral	Chemical composition and crystal structure	Occurrence		Comments	Reference
			major constituent of	minor constituent of		
Elements	Copper	Cu Cubic, Fm3m		Ae; Ce_2; CH; CL; P, M; Og		Yudin (1952); Ramdohr (1963a, b); Keil and Fredriksson (1963); Duke and Brett (1965)
	Gold	Au Cubic, Fm3m		D; Ce_2 (?)	Questionable; in Atlanta (Ce_2) probably confused with TiN (Ramdohr pers. comm.)	Edwards (1953); Ramdohr (1963a)
	Kamacite	(Fe, Ni) (Ni $<7.5\%$) Cubic, Im3m	Aop; Ce_1; Ce_2; CH; CL; CLL; CHL; P; Si; Lo; M; H; O	Ae; Ab; Aor; Ap; Cc_2		Goldstein and Ogilvie (1965); Short and Andersen (1965); Wood (1967); Short and Keil (1968)
	Taenite	(Fe, Ni) (Ni ~ 8—55%) Cubic, Fm3m or Im3m (depending on Ni content and temperature)	CH; CL; CLL; CHL; P; Si; Lo; M; O; D	Ae; Ao; Aor; Ce_1; Ce_2		
	Graphite	α-C Hexagonal, $P6_3/mmc$		Ae; Aop; Ce_1; Ce_2; CHL; Cc_1; P; M; H; O; D	Cliftonites are graphite cubes, but probably not pseudomorphs after diamond	El Goresy (1965); Marshall and Keil (1965); Brett and Higgins (1967)

Table 4-8. (continued)

Mineral class	Mineral	Chemical composition and crystal structure	Occurrence		Comments	Reference
			major constituent of	minor constituent of		
	Diamond	β-C Cubic, Fd3m		Aop; Og	Recently suggested origin by shock transformation of graphite	Foote (1891); Urey et al. (1957); Ringwood (1960); Lipschutz (1962
	Sulfur	α-S Rhombic, Fddd		Cc_1		Du Fresne and Anders (1962b)
Carbides, nitrides, phosphides, silicides	Cohenite	Fe_3C Rhombic, Pbnm		Ce_1; M; O		Brett (1966)
	Moissanite	SiC Hexagonal, Pb_3mc		O	Doubtful	Mason (1962a)
	Osbornite*	TiN Cubic, rock-salt structure		Ae		Bannister (1941)
	Sinoite*	Si_2N_2O Orthorhombic, $Cmc2_1$		Ce_2		Andersen et al. (1964); Keil and Andersen (1964, 1965a, b)
	Schreibersite*	$(Fe, Ni, Co)_3P$ Tetragonal, $I\bar{4}$	O	Ae; Ce_1; Ce_2; P; Si; M; H; O; D	Rhabdite is needle-and spindle-shaped schreibersite	Reed (1965)
	Perryite*	$(Ni, Fe)_x(Si, P)_y$ (Ni ~81%, Fe ~3%, Si ~12%, P ~5%)		Ce_1; O	In Horse Creek, Col. iron meteorite	Fredriksson and Henderson (1965)
Sulfides	Pentlandite	$(Fe, Ni)_9S_8$ Cubic, $4/m\ \bar{3}2/m$		Ab; CLL; CHL; Cc_2	Of terrestrial origin in many iron meteorite nodules	Jeremine et al. (1956); Fredriksson and Keil (1964)

Sulfides	Sphalerite	α-ZnS Cubic, F $\bar{4}$3m		Ce_1; Ce_2; M; O		El Goresy (1965); Keil (1968b)
	Chalcopyrite	$CuFeS_2$ Tetragonal, I$\bar{4}$2d		CHL; Og	In O of terrestrial origin	Ramdohr (1963a, b)
	Chalco-pyrrhotite	$CuFe_2S_3$ Orthorhombic, Pcmn		CL; CH; M; Og		Ramdohr (1963a, b)
	Troilite	FeS Hexagonal, $P6_3/mmc$	Aa; Ce_1; Ce_2; CH; CL CLL; CHL; Lo; M; O	Ae; Ab; Aop; Aor; Ap; Cc_2; P; Si; H; D	Titanium-bearing in Ae and Ce	Keil and Fredriksson (1963); Keil (1968a, b)
	Oldhamite*	CaS Cubic, Fm3m		Ce_1; Ce_2		Keil (1968b)
	Ferroan alabandite	$(Fe_{0.23}Mg_{0.12}Mn_{0.63}Ca_{0.007})S$				
	Ferromag-nesian alabandite	$(Fe_{0.2}Mg_{0.3}Mn_{0.5})S$ Cubic, Fm3m		Ae; Ce_2; M; O		Keil and Fredriksson (1963); El Goresy (1965); Keil (1968b)
	Niningerite*	$(Fe_{0.19}Mg_{0.66}Mn_{0.14}Ca_{0.007})S$ to $(Fe_{0.52}Mg_{0.33}Mn_{0.06}Ca_{0.03})S$ Cubic, rock-salt structure		Ce_1		Keil (1968b); Keil and Snetsinger (1967)
	Valleriite	$CuFeS_2$ Hexagonal		CH; CL; O	Perhaps confused with mackinawite	Ramdohr (1963a, b)
	Mackinawite	FeS Tetragonal, P4/nmm		CH; CL; M; O	Tetragonal	Ramdohr (1963a, b)
	Daubreelite*	$FeCr_2S_4$ Cubic, Fd3m		Ae; Ce_1; Ce_2; H; O		Keil and Fredriksson

(Table 4-8. continued)

Mineral class	Mineral	Chemical composition and crystal structure	Occurrence		Comments	Reference
			major constituent of	minor constituent of		
						(1963); El Goresy (1965); Keil (1968b) Marshall and Keil (1965)
	Pyrite	FeS_2 Cubic, Pa3		CHL		Ramdohr (1963a, b)
	Djerfisherite*	$K_3(Na, Cu)(Fe, Ni)_{12}S_{14}$ Cubic		Ce_1; Ae		Fuchs (1966) Grögler and Liener (1968)
Halides	Lawrencite*	$FeCl_2$		Ce_1; P; M; O		
Oxides and hydroxides	Spinel	$\sim MgAl_2O_4$ Cubic, Fd3m		CH; CHL		Sztrokay et al. (1961); Ramdohr (1963b)
	Magnetite	Fe_3O_4 Cubic, Fd3m	Cc_2	Ado; Aor; Ap; CH; CLL; CHL; Cc_1; P; M	Sometimes has Ni	Bostrom and Fredriksson (1966)
	Chromite	$FeCr_2O_4$ Cubic, Fd3m	Ao	Ae(?); Ab; Aor; Ap; CH; CL; CLL; P; Si; Lo; M; O	Sometimes has considerable Mg, Mn, Al	Snetsinger et al., (1967); Bunch et al. (1967)
	Ilmenite	$FeTiO_3$ Hexagonal, $R\bar{3}$		Aor; Ap; CH; CL; P; M; O		Buseck and Keil (1966)
	Quartz	SiO_2 Hexagonal		Aor; Ap Ce_1; Ce_2		Mason (1966)

Oxides and hydroxides	Tridymite	SiO_2 Rhombic (low temp.), hexagonal (high temp.)	Si	Aor; Ap; Ce_1; Ce_2; M; H		Götz (1962)
	Cristobalite	SiO_2 Tetragonal (low temp.), cubic (high temp.)		Ap; Ce_1; CH; Om		Marvin (1962)
	Rutile	TiO_2 Tetragonal, P 4_2/mnm		CH; CL; M; Og		Buseck and Keil (1966)
	Limonite	$FeOOH \cdot nH_2O$ Rhombic		Cc_1	Claimed to be of extraterrestrial origin	Nagy et al. (1963)
Carbonates and sulfates	Breunnerite	$(Mg, Fe)CO_3$ Hexagonal, $R\bar{3}c$		Cc_1; Cc_2		Kvasha (1948); Nagy and Andersen (1964)
	Calcite	$CaCO_3$ Hexagonal, $R\bar{3}c$		Cc_2		Kvasha (1948)
	Dolomite	$CaMg(CO_3)_2$ Hexagonal, $R\bar{3}$		Cc_1		Du Fresne and Anders (1962b)-
	Epsomite	$MgSO_4 \cdot 7H_2O$ Rhombic, $P2_12_12_1$	Cc_1	Cc_2	Origin of water uncertain	Du Fresne and Anders (1962b)
	Astrakhanite	$Na_2Mg(SO_4)_2 \cdot 4H_2O$ Monoclinic, $P2_1/a$		Cc_1	Origin of water uncertain	Bostrom and Fredriksson (1966)
	Gypsum	$CaSO_4 \cdot 2H_2O$ Monoclinic, A 2/n		Cc_1		Nagy and Andersen (1964)
Phosphates	Farringtonite*	$Mg_3(PO_4)_2$ Monoclinic		P	Has some Fe and Mn	Du Fresne and Roy (1961)
	Graftonite	$(Fe, Mn)_3(PO_4)_2$ Monoclinic, $P2_1/c$		Of	Ca-free	Olsen and Fredriksson (1966)
	Sarcopside	$(Fe, Mn)_3(PO_4)_2$ Monoclinic		Of	Ca-free	Olsen and Fredriksson(1966)

Table 4-8. (continued)

Mineral class	Mineral	Chemical composition and crystal structure	Occurrence		Comments	Reference
			major constituent of	minor constituent of		
	Whitlockite	β-$Ca_3(PO_4)_2$ Hexagonal, R$\bar{3}$m		Ap; CH; CL; Cc_1(?); M; Off	With some Na, Mg, and Fe^{2+} (Fuchs, personal communication)	Fuchs (1962b)
	Chlorapatite	$Ca_5(PO_4)_3Cl$ Hexagonal, $P6_3/m$		CH;CL; CLL; M; O		Marshall and Keil (1965)
	Merrillite*	$Na_2Ca_3(PO_4)_2O$		CLL; Cc_1	Fuchs (1962b) doubts the existence of meteoritic merrillite	
	Brianite*	$Na_2CaMg(PO_4)_2$ Probably orthorhombic		Off		Fuchs et al. (1967)
	Panethite*	$Na_{1.42}K_{0.06}Mg_{1.70}Ca_{0.28}Fe_{0.21}Mn_{0.07}(PO_4)_2$ Monoclinic, $P2_1/n$		Off		
	Stanfieldite*	$Mg_{0.98}Ca_{1.35}Fe_{0.59}Mn_{0.06}(PO_4)_2$ Monoclinic, Pc or P2/c		M		Fuchs (1967)
Silicates	Forsterite-fayalite	Mg_2SiO_4—Fe_2SiO_4 Rhombic, Pmcn	Ae; Ao; Aop; Aa; Ado; CH; CL; CLL; CHL; Cc_2; P; Lo; M	Ab; Aor; Ce_1; Cc_1; O	Often ~0.5% Mn	Keil and Fredriksson (1964) Keil et al. (1964)
	Zircon	$ZrSiO_4$ Tetragonal, I 4_1/amd		M; Om		Marvin and Klein (1964)
	Merrihueite*	$(K,Na)_2(Fe, Mg)_5Si_{12}O_{30}$ Hexagonal		CL	K dominates over Na	Dodd et al. (1965)
	Roedderite*	$(Na, K)_2(Fe, Mg)_5Si_{12}O_{30}$ Hexagonal		Ce_1	Na dominates over K	Fuchs et al. (1966)

	Yagiite*	$(Na_{1.2}K_{0.3})Mg_{2.0}(Mg_{0.6}Fe_{0.3}Ti_{0.1}Al_{2.0})(Si_{10.2}Al_{1.8})O_{30}$ Hexagonal		Om		BUNCH and FUCHS (1968)
	Clinoenstatite-clinoferrosilite (including pigeonite)	$Mg_2Si_2O_6$—$Fe_2Si_2O_6$—$Ca_2Si_2O_6$ Monoclinic, $P2_1/c$	Ab; Aop; Aor; Ap Ce_1; CH; CL; CHL; Cc_2	Ce_2; CLL; M	Including pigeonite	MASON (1966)
Silicates	Diopside	$CaMgSi_2O_6$ Monoclinic, C 2/c	Ado	Ae; CL; O	Including Cr-bearing endiopside	MARSHALL and KEIL (1965); KURAT and KURZWEIL (1965)
	Ureyite*	$NaCrSi_2O_6$ Monoclinic, C 2/c		H; Om	Cosmochlore	FRONDEL and KLEIN (1965); NEUHAUS (1967)
	Augite	$(Ca, Mg, Fe^{2+}, Al)_2(Si, Al)_2O_6$ Monoclinic, C 2/c	Aa	Aor; Ap	Including titanian augite	LUDWIG and TSCHERMAK (1887)
	Enstatite-orthoferro-silite	$Mg_2Si_2O_6$—$Fe_2Si_2O_6$ Rhombic, Pbca.	Ae; Ab; Aor; Ap; Ce_2; CH; CL; CLL; CHL; Cc_2; Si; Lo; M	Ao; Ce_1; O	Ca $\leqq 0.5\%$ often $\sim 0.5\%$ Mn	KEIL and FREDRIKSSON (1964); MARSHALL and KEIL (1965)
	Richterite	$Na_2Ca(Mg, Fe)_5Si_8O_{22}(OH,F)_2$ Monoclinic		Og		OLSEN (1967)
	Layer lattice silicate (chlorite-like)	$\sim(Mg, Fe^{2+}, Ni, Al)_{5.93}OH_{7.86}O_{0.14}Si_4O_{10}$	Cc_1; Cc_2			BOSTROM and FREDRIKSSON (1966)
	Plagioclase	$NaAlSi_3O_8$—$CaAl_2Si_2O_8$ Triclinic	Ae; Ado; Ap; Ce_2; CH; CL; CLL; CHL; M	Ab; Ao; Aor; Ce_1; O	Little K. Maskelynite is shock transformed plagioclase glass	MASON (1965b)
	Krinovite*	$NaMg_2CrSi_3O_{10}$ Monoclinic		Og		OLSEN and FUCHS (1968)

IV. Mineralogical Composition of Meteorites

In general, the degree of reduction of meteorites is higher than that of most rocks of the Earth's crust and, as a result, meteoritic mineral assemblages are often unlike terrestrial assemblages. Nineteen of the 66 minerals presently recognized from meteorites are not known from terrestrial rocks. Due to the extensive application of the electron microprobe to analysis of meteoritic minerals, a great wealth of new data has become available and a number of new minerals have been discovered (e.g., sinoite, perryite, merrihueite, roedderite, djerfisherite, niningerite, brianite, panethite, stanfieldite, krinovite, yagiite). A brief summary of some vital data on the compositions and occurrences of minerals found in meteorites are given in Table 4-8. Some further data are presented in the sections describing the various meteorite classes. For more detailed information the reader is referred to the literature (e.g. Mason, 1962a, 1967).

Most of the minerals listed in Table 4-8 are well-established as meteoritic constituents. A number of only tentatively identified compounds (e.g., Fe-C-S-mineral, minerals A, B, and D through L of Ramdohr, 1963a; Mg_2TiO_4, Ramdohr, 1963b; CrS, El Goresy, 1965; brucite or periclase, Bostrom and Fredriksson, 1966) are not included in the table. Furthermore, igneous acidic glass (e.g., Fredriksson and Reid, 1965) and organic compounds (e.g., Urey, 1966), which occur in certain meteorites, are not listed. Secondary minerals, which originated by terrestrial weathering of primary meteorite constituents and, hence, contribute little to knowledge of history and origin of meteorites, were also deleted. A brief discussion of these phases is given by Mason (1967).

Acknowledgement

I would like to thank Drs. R. Brett, M. M. Duke, A. El Goresy, L. H. Fuchs, G. G. Goles, B. Mason, P. Ramdohr, R. Van Schmus, J. A. Wood, and J. M. Short for stimulating discussions and for making their unpublished data available to me, and Drs. K. G. Snetsinger, and T. E. Bunch, for critically reading the manuscript. Dr. J. M. Short kindly provided Fig. 4-4. Miss E. L. Craig deserves credit for preparing the microphotographs.

References

Anders, E.: Meteorite ages. In: The solar system, vol. 4, p. 402 (ed. Middlehurst and Kuiper). Chicago: Chicago University Press 1963.

— Origin, age, and composition of meteorites. Space Sci. Rev. **3**, 583 (1964).

Andersen, C. A., K. Keil, and B. Mason: Silicon oxynitride: a meteoritic mineral. Science **146**, 256 (1964).

Bannister, F. A.: Osbornite, meteoritic titanium nitride. Mineral. Mag. **26**, 36 (1941).

Beck, C. W., and L. La Paz: The nortonite fall and its mineralogy. Am. Mineralogist **36** 45 (1951).

Binns, R. A.: Olivine in enstatite chondrites. Am. Mineralogist **52**, 1549 (1967).

Boato, G.: The isotopic composition of hydrogen and carbon in the carbonaceous chondrites. Geochim. et Cosmochim. Acta **6**, 209 (1954).

Bostrom, K., and K. Fredriksson: Surface conditions of the Orgueil meteorite parent body as indicated by mineral associations. Smiths, Misc. Coll. **151**, No. 3, 39 p. (1966).

Brett, R.: Cohenite in meteorites: A proposed origin. Science **153**, 60 (1966).
—, and G. T. Higgins: Cliftonite in meteorites: a proposed origin. Science **156**, 819 (1967).
Brezina, A.: The arrangement of collections of meteorites. Proc. Am. Phil. Soc. **43**, 211 (1904).
Briggs, M. H., and G. Mamikunian: Organic constituents of the carbonaceous chondrites. Space Sci. Rev. **1**, 647 (1962—63).
Brown, H.: A bibliography on meteorites. Chicago: Chicago University Press 1953.
— The density and mass distribution of meteoritic bodies in the neighborhood of the Earth's orbit. J. Geophys. Research **66**, 1316 (1961).
Buddhue, J. D.: The average composition of meteoritic iron. Pop. Astr. **54**, 149 (1964).
Bunch, T. E., K. Keil, and K. G. Snetsinger: Chromite composition in relation to chemistry and texture of ordinary chondrites. Geochim. Cosmochim. Acta **31**, 1569 (1967).
—, and L. H. Fuchs: Yagiite, the sodium-magnesium analogue of osumilite. Am. Mineralogist, in press (1968).
Buseck, P. R., and K. Keil: Meteoritic rutile. Am. Mineralogist **51**, 1506 (1966).
Castaing, R., and K. Fredriksson: Analysis of cosmic spherules with an X-ray microanalyzer. Geochim. et Cosmochim. Acta **14**, 114 (1958).
Chladni, E. F. F.: Über den Ursprung der von Pallas gefundenen und anderer ihr ähnlicher Eisenmassen und über einige damit in Verbindung stehende Naturerscheinungen. Riga: J. F. Hartknoch 1794.
Christophe-Michel-Levy, M., H. Curien, et J. Goni: Etude a la microsonde electronique d'un chondre d'olivine et d'un fragment riche en cristobalite de la meteorite de Nadiabondi. Bull. soc. franç. minéral. et crist. **88**, 122 (1965)
Claus, G., and B. Nagy: A microbiological examination of some carbonaceous chondrites. Nature **192**, 594 (1961).
Cohen, E.: Meteoritenkunde, H. 1, 2, and 3. Stuttgart: Schweizerbart 1894, 1903, and 1905.
Craig, H.: Petrological and compositional variations in meteorites. In: Isotopic and cosmic chemistry, p. 401. Amsterdam: North-Holland-Publ. Co. 1964.
Deer, W. A., R. A. Howie, and J. Zussman: Rock-forming minerals: Chain silicates, vol. 2. New York: John Wiley & Sons 1963.
Dodd, R. T., and W. R. van Schmus: Significance of the unequilibrated ordinary chondrites. J. Geophys. Research **70**, 3801 (1965).
—, W. R. van Schmus, and U. B. Marvin: Merrihueite, a new alkali-ferromagnesian silicate from the Mezö-Madaras chondrite. Science **149**, 972 (1965).
Dörfler, G., F. Hecht u. E. Plöckinger: Elektronenstrahl-Mikroanalyse des Meteoriten von Steinbach. Tschermak's mineral. u. petrogr. Mitt. **10**, 413 (1965).
Du Fresne, E. R., and E. Anders: The record in the meteorites — V. A thermometer mineral in the Mighei carbonaceous chondrite. Geochim. et Cosmochim. Acta **23**, 200 (1961).
— — On the retention of primordial noble gases in the Pesyanoe meteorite. Geochim. et Cosmochim. Acta **26**, 251 (1962a).
— — On the chemical evolution of the carbonaceous chondrites. Geochim. et Cosmochim. Acta **26**, 1085 (1962b).
—, and S. K. Roy: A new phosphate mineral from the Springwater pallasite. Geochim. et Cosmochim. Acta **24**, 198 (1961).
Duke, M. B., and R. Brett: Metallic copper in stony meteorites. U.S. Geol. Survey Profess. Papers No. 525-B, B 101 (1965).
—, and L. T. Silver: Paper presented at 43rd Ann. Meet., Am. Geophys. Un., Washington, D.C. 1962.
— — Petrology of eucrites, howardites and mesosiderites. Geochim. Cosmochim. Acta **31**, 1637 (1967).
Dyakonova, M. I., and V. Y. Charitonova: Results of chemical analyses of some stone and iron meteorites from the collection of the Academy of Sciences of the USSR. Meteoritika **18**, 48 (1960) [in Russian].
Edwards, A. B.: The Wedderburn meteoritic iron. Proc. Roy. Soc. Victoria **64**, 73 (1953).

EDWARDS, G., and H. C. UREY: Determination of alkali metals in meteorites by a distillation process. Geochim. et Cosmochim. Acta **7**, 154 (1955).

EL GORESY, A.: Mineralbestand und Strukturen der Graphit- und Sulfideinschlüsse in Eisenmeteoriten. Geochim. et Cosmochim. Acta **29**, 1131 (1965).

ENGELHARDT, W. v.: Der Eukrit von Stannern. Beitr. Mineral. u. Petrog. **9**, 65 (1963).

FARRINGTON, O. C.: Meteorites. Chicago: Publ. by the author 1915.

FITCH, F. W., and E. ANDERS, Observations on the nature of the "organized elements" in carbonaceous chondrites. Ann. N.Y. Acad. Sci. **108**, 495 (1963).

FOOTE, A. E.: A new locality for meteoritic iron with a preliminary notice of the discovery of diamonds in the iron. Proc. Am. Ass. Advancement Sci. **40**, 279 (1891).

FREDRIKSSON, K., and E. P. HENDERSON: The Horse Creek, Baca County, Colorado, iron meteorite. Trans. Am. Geophys. Union. **46**, 121 (1965).

—, and K. KEIL: The light-dark structure in the Pantar and Kapoeta stone meteorites. Geochim. et Cosmochim. Acta **27**, 717 (1963).

— — The iron, magnesium, calcium, and nickel distribution in the Murray carbonaceous chondrite. Meteoritics **2**, 201 (1964).

—, and A. M. REID: A chondrule in the Chainpur meteorite. Science **149**, 856 (1965).

FRONDEL, C., and C. KLEIN: Ureyite, $NaCrSi_2O_6$: A new meteoritic pyroxene. Science **149**, 742 (1965).

FUCHS, L. H.: Identification of phosphate minerals in stony meteorites. Am. Geophys. Un., 43 Ann. Meet., Washington, D. C. 1962a.

— Occurrence of whitlockite in chondritic meteorites. Science **137**, 425 (1962b).

— Djerfisherite, alkali copper-iron sulfide, a new mineral from the Kota-Kota and St. Mark's enstatite chondrites. Science **153**, 166 (1966).

— Stanfieldite: a new phosphate mineral from stony-iron meteorites. Science **158**, 910 (1967).

—, C. FRONDEL, and C. KLEIN: Roedderite, a new mineral from the Indarch meteorite. Am. Mineralogist **51**, 949 (1966).

—, E. OLSEN, and E. P. HENDERSON: On the occurrence of brianite and panethite, two new phosphate minerals from the Dayton meteorite. Geochim. Cosmochim. Acta **31**, 1711 (1967).

GOEL, P. S., and T. P. KOHMAN: Cosmogenic carbon-14 in meteorites and terrestrial ages of "finds" and craters. Science **136**, 875 (1962).

GÖTZ, W.: Untersuchungen am Tridymit des Sideophyrs von Grimma in Sachsen. Chem. Erde **22**, 167 (1962).

GOLDSTEIN, J. I., and R. E. OGILVIE: The growth of the Widmanstätten pattern in metallic meteorites. Geochim. et Cosmochim. Acta **29**, 893 (1965).

—, and J. M. SHORT: Cooling rates of 27 iron and stony-iron meteorites. Geochim. Cosmochim. Acta **31**, 1001 (1967).

GRÖGLER, N., and A. LIENER: Cathodoluminescence and thermoluminescence observations of aubrites. Spoleto Conf. Appl. Thermoluminescence to Geol. Problems, Academic Press, in press (1968).

HEIDE, F.: Meteorites. Chicago and London: Chicago University Press (1964) (English Transl.).

HEY, M. H.: Catalogue of meteorites. British Museum (Natural History), London (1966).

JEREMINE, E., et M. LELUBRE: Sur la meteorite d'Oubari. Geochim. et Cosmochim. Acta **2**, 217 (1952).

— —, et A. SANDREA: La meteorite d'Isoulane-n-Amahar (Nord-Nord-Est de Fort Polignac, confins Algero-Fezzanais). Compt. rend. **242**, 2369 (1956).

KEIDEL, W.: Untersuchungen am Meteoriten von Borkut und anderen Chondriten über Form, Aufbau und Entstehung der Chondren. Beitr. Mineral. u. Petrog. **11**, 487 (1965).

KEIL, K.: Fortschritte in der Meteoritenkunde. Fortschr. Mineral. **38**, 202, (1960).

— On the phase composition of meteorites. J. Geophys. Research. **67**, 4055 (1962a).

— Quantitativ-erzmikroskopische Integrationsanalyse der Chondrite (Zur Frage des mittleren Verhältnisses von Nickeleisen-: Troilit-: Chromit-: Silikatanteil in den Chondriten). Chem. Erde **22**, 281 (1962b).

KEIL, K.: Mineralogical modal analysis with the electron microprobe X-ray analyzer. Amer. Mineralogist **50**, 2089 (1965).
— Chalcophile tendencies of titanium in meteorites. Earth and Planetary Sci. Lett. (in press) (1968a).
— Mineralogical and chemical relationships among enstatite chondrites. J. Geophys. Research (in press) (1968b).
— Zincian daubreelite from the Kota-Kota and St. Mark's enstatite chondrites. Am. Mineralogist **53**, 491 (1968c).
—, and C. A. ANDERSEN, Electron microprobe study of the Jajh deh Kot Lalu enstatite chondrite. Trans. Am. Geophys. Union **45**, 86 (1964).
— — Electron microprobe study of the Jajh deh Kot Lalu enstatite chondrite. Geochim. et Cosmochim. Acta **29**, 621 (1965a).
— — Occurrences of sinoite, Si_2N_2O, in meteorites. Nature **207**, 745 (1965b).
—, and K. FREDRIKSSON: Electron microprobe analysis of some rare minerals in the Norton County achondrite. Geochim. et Cosmochim. Acta **27**, 939 (1963).
— — The iron, magnesium, and calcium distribution in coexisting olivines and rhombic pyroxenes of chondrites. J. Geophys. Research **69**, 3487 (1964).
—, B. MASON, H. B. WIIK, and K. FREDRIKSSON: The Chainpur meteorite. Am. Museum Novitates No. 2173, 28 p. (1964).
—, and K. G. SNETSINGER: Niningerite: a new meteoritic sulfide. Science **155**, **451** (1967).
KERRIDGE, J. F.: Low temperature minerals from the fine-grained matrix of some carbonaceous meteorites. Ann. N.Y. Acad. Sci. **119**, 41 (1964).
— Occurence of olivine in a Type I carbonaceous meteorite. Nature **217**, 729 (1968).
KRINOV, E. L.: Principles of meteoritics. New York: Pergamon Press 1960 (English transl.).
KURAT, G., and H. KURZWEIL: Der Meteorit von Lanzenkirchen. Ann. naturhist. Museum Wien **68**, 9 (1965).
KVASHA, L. G.: Investigation of the stony meteorite Staroye Boriskino. Meteoritika **4**, 83 (1948) [in Russian].
— Über einige Typen von Steinmeteoriten. Chem. Erde **19**, 249 (1958).
LAL, D., and V. S. VENKATAVARADAN: Low-energy protons. Average flux in interplanetary space during the last 100,000 years. Science **151**, 1381 (1966).
LIPSCHUTZ, M. E.: Diamonds in the Dyalpur meteorite. Science **138**, 1266 (1962).
LOVERING, J. F., W. NICHIPORUK, A. CHODOS, and H. BROWN: The distribution of gallium, germanium, cobalt, chromium, and copper in iron and stony-iron meteorites in relation to nickel content and structure. Geochim. et Cosmochim. Acta **11**, 263 (1957).
LUDWIG, E., u. G. TSCHERMAK: Der Meteorit von Angra dos Reis. Mineral. u. petrog. Mitt. **8**, 341 (1887).
MARSHALL, R. R., and K. KEIL: Polymineralic inclusions in the Odessa iron meteorite. Icarus **4**, 461 (1965).
MARVIN, U. B.: Cristobalite in the Carbo iron meteorite. Nature **196**, 634 (1962).
—, and C. KLEIN: Meteoritic zircon. Science **146**, 919 (1964).
MASON, B.: Meteorites. New York and London: John Wiley & Sons 1962a.
— The classification of chondritic meteorites. Am. Museum Novitates No. 2085, 20 p. (1962b).
— The hypersthene achondrites. Am. Museum Novitates No. 2155, 13 p. (1963).
— The chemical composition of olivine-bronzite and olivine-hypersthene chondrites. Am. Museum Novitates No. 2223, 38 p. (1965a).
— Feldspar in chondrites. Science **148**, 943 (1965b).
— The enstatite chondrites. Geochim. et Cosmochim. Acta **30**, 23 (1966).
— Extraterrestrial mineralogy. Am. Mineralogist **52**, 307 (1967).
—, and H. B. WIIK: The amphoterites and meteorites of similar composition. Geochim. et Cosmochim. Acta **28**, 533 (1964).
MCCALL, G. J. H., and W. H. CLEVERLY: New stony meteorite finds including two ureilites from the Nullarbor Plain, Western Australia. Mineral. Mag. **36**, 691 (1968).
MERRIHUE, C. M.: Rare gas evidence for cosmic dust in modern Pacific red clay. Ann. N.Y. Acad. Sci. **119**, 351 (1964).
MICHEL, H.: Die Feldspäte der Meteoriten. Mineral. u. petrog. Mitt. **31**, 563 (1912).

Millard, H. T.: The rate of arrival of meteorites at the surface of the Earth. J. Geophys. Research **68**, 4297 (1963).
—, and H. Brown: Meteoritic time-of-fall patterns. Icarus **2**, 137 (1963).
Moore, C. B., and C. F. Lewis: Carbon abundances in chondritic meteorites. Science **149**, 317 (1965).
— — The distribution of total carbon content in enstatite chondrites. Earth and Planetary Sci. Letters **1**, 376 (1966).
Mueller, G.: Interpretation of micro-structures in carbonaceous meteorites. In: Advances Org. Geochem. vol. 1. Oxford-London-New York-Paris: Pergamon Press 1963.
Murray, J., and A. F. Renard: On the microscopic characters of volcanic ashes and cosmic dust and their distribution in deep-sea deposits. Proc. Roy. Soc. Edinburgh **12**, 474 (1884).
Nagy, B., and C. A. Andersen, Electron probe microanalysis of some carbonate, sulfate, and phosphate minerals in the Orgueil meteorite. Am. Mineralogist **49**, 1730 (1964).
—, K. Fredriksson, H. C. Urey, G. Claus, C. A. Andersen, and J. Percy: Electron probe microanalysis of organized elements in the Orgueil meteorite. Nature **198**, 121 (1963).
Neuhaus, A.: Über Kosmochlor. Naturwissenschaften **54**, 440 (1967).
Nilsson, C. S.: Some doubts about the Earth dust cloud. Science **153**, 1242 (1966).
Olsen, E.: Amphibole: first occurence in a meteorite. Science **156**, 61 (1967).
—, and K. Fredriksson: Phosphates in iron and pallasite meteorites. Geochim. et Cosmochim. Acta **30**, 459 (1966).
—, and L. H. Fuchs: Krinovite: $NaMg_2CrSi_3O_{10}$, a new meteorite mineral. Science, in press (1968).
Park, F. R., T. E. Bunch, and T. B. Massalski: A study of the silicate inclusions and other phases in the Campo del Cielo meteorite. Geochim. et Cosmochim. Acta **30**, 399 (1966).
Perry, S. H.: The metallography of meteoric iron. U.S. Natl. Museum Bull. **184**, 206 p. (1944).
Pokrzywnicki, J.: The specific gravity of meteorites. Acta Geophys. Polon. **6**, 127 (1959).
Preston, F. W., E. P. Henderson, and J. R. Randolph: The Chicora (Butler Co., Pa.) meteorite. Proc. U.S. Natl. Museum **90**, 387 (1941).
Prior, G. T.: The meteoric stone of El Nakhla el Baharia, Egypt. Mineral. Mag. **16**, 274 (1912).
— On the genetic relationship and classification of meteorites. Mineral. Mag. **18**, 26 (1916).
— The classification of meteorites. Mineral. Mag. **19**, 51 (1920).
Ramdohr, P.: The opaque minerals in stony meteorites. J. Geophys. Research **68**, 2011 (1963a).
— Beobachtungen am Opakerzbestand einiger Meteoriten besonders von New South Wales. Chem. Erde **23**, 119 (1963b).
— Über Mineralbestand von Pallasiten und Mesosideriten und einige genetische Überlegungen. Monatsber. Deut. Akad. Wiss. **7**, 923 (1965).
Reed, S. J. B.: Electron probe microanalysis of schreibersite and rhabdite in iron meteorites. Geochim. et Cosmochim. Acta **29**, 513 (1965).
Reid, A. M., and K. Fredriksson: Chondrules and chondrites. In: Researches in Geochemistry, vol. 2, p. 170, ed. P. H. Abelson. New York-London-Sidney: John Wiley & Sons 1967.
Ringwood, A. E.: The Novo-Urei meteorite. Geochim. et Cosmochim. Acta **20**, 1 (1960).
— Chemical and genetic relationships among meteorites. Geochim. et Cosmochim. Acta **24**, 159 (1961).
— Genesis of chondritic meteorites. Rev. Geophys. **4**, 113 (1966).
Rodionov, S. P.: Mineralogical-petrographical study of the stone meteorite Svonkovoje. Meteoritika **17**, 47 (1959) [in Russian].
Schmidt, R. A.: A survey of data on microscopic extraterrestrial particles. NASA TN D-2719, 132 p. (1965).
—, and K. Keil: Electron microprobe study of spherules from Atlantic Ocean sediments. Geochim. et Cosmochim. Acta **30**, 471 (1966).

SCHMUS, W. R. VAN, and J. A. WOOD: A chemical—petrologic classification for the chondritic meteorites. Geochim. et Cosmochim. Acta **31**, 747 (1967).
SHORT, J. M., and C. A. ANDERSEN: Electron microprobe analyses of the Widmanstätten structure of nine iron meteorites. J. Geophys. Research **70**, 3745 (1965).
—, and K. KEIL: Composition of metallic phases in 33 chondrites. Geochim. et Cosmochim. Acta (in press) (1968).
SNETSINGER, K. G., K. KEIL, and T. E. BUNCH: Chromite from "equilibrated" chondrites. Am. Mineralogist **52**, 1322 (1967).
STORY-MASKELYNE, N. H. M:. On the mineral constituents of meteorites. Phil. Trans. Roy. Soc. London, **161**, 359 (1871).
SZTROKAY, K. I., V. TOLNAY, and M. FOLDVARI-VOGL: Mineralogical and chemical properties of the carbonaceous meteorite from Kaba, Hungary. Acta Geol. Acad. Sci. Hung. **7**, 51 (1961).
TSCHERMAK, G.: Der Meteorit von Lodran. Sitzber. Akad. Wiss. Wien, Math.-nat. Kl., Abt. II, **61**, 465 (1870).
— Beitrag zur Klassifikation der Meteoriten. Sitzber. Akad. Wiss. Wien, Math.-nat. Kl. **88**, 347 (1883).
TUREKIAN, K. K.: Rate of accumulation of nickel in Atlantic equatorial deep-sea sediments and its bearing on possible extra-terrestrial sources. Nature **182**, 1728 (1958).
UREY, H. C.: Biological material in meteorites: A review. Science **151**, 157 (1966).
—, and H. CRAIG: The composition of the stone meteorites and the origin of the meteorites. Geochim. et Cosmochim. Acta **4**, 36 (1953).
—, A. MELE, and T. MAYEDA: Diamonds in stone meteorites. Geochim. et Cosmochim. Acta **13**, 1 (1957).
VDOVYKIN, G. P.: Carbonaceous matter of meteorites in connection with their origin. Geokhimiya **4**, 299 (1964) [in Russian].
WAHL, W.: Die Enstatitaugite. Monokline Pyroxene mit kleinem Winkel der optischen Achsen und niedrigem Kalkgehalt. Mineral. u. petrog. Mitt. **26**, 1 (1907).
WASSON, J. T.: The chemical classification of iron meteorites: I. A study of iron meteorites with low concentrations of gallium and germanium. Geochim. Cosmochim. Acta **31**, 161 (1967).
—, B. ALDER, and H. OESCHGER: Aluminum — 26 in Pacific sediment: Implications. Science **155**, 446 (1967).
WIIK, H. B.: The chemical composition of some stony meteorites. Geochim. et Cosmochim. Acta **9**, 279 (1956).
— On the genetic relationship between meteorites. Preprint No. 18 (1966).
WISEMAN, J. D. H.: Rates of sedimentation of nickel, cobalt, copper and iron on the equatorial mid-Atlantic floor and its bearing on the nature of cosmic dust. Nature **202**, 1286 (1964).
WOOD, J. A.: Physics and chemistry of meteorites. In: The solar system, vol. 4, p. 337 (ed. MIDDLEHURST and KUIPER). Chicago: Chicago University Press 1963.
— The cooling rates and parent planets of several iron meteorites. Icarus **3**, 429 (1964).
— Chondrites: Their metallic minerals, thermal histories, and parent planets. Icarus **6**, 1 (1967).
— Olivine and pyroxene compositions in Type II carbonaceous chondrites. Geochim. Cosmochim. Acta **31**, 2095 (1967).
YAVNEL, A. A.: On genetic relationships in chemical composition of chondrites. Meteoritika **23**, 36 (1963) [in Russian].
YUDIN, I. A.: Mineralogical and chemical study of the stone meteorite Kunashak. Meteoritika **10**, 42 (1952) [in Russian].
— Preliminary investigation of the mineralogical composition of the stone meteorite Vengerovo. Doklady Akad. Nauk S.S.S.R. **84**, 123 (1952) [in Russian].
— Opaque minerals of stone meteorites. Meteoritika **16**, 78 (1958) [in Russian].
—, and S. I. SMISHLAEV: Chemical-mineralogical investigation of opaque minerals in the Norton County and Staroe Pesyanoe achondrites. Meteoritika **25**, 96 (1964) [in Russian].

Chapter 5

G. G. Goles

COSMIC ABUNDANCES

I. Introduction

Estimates of the bulk composition of the Earth are required in order to construct and test models of the mechanisms which produced its present structure. Unfortunately, neither geochemistry nor geophysics has as yet developed to the point where it is possible to estimate the bulk composition of the Earth without relying on information derived from extra-terrestrial sources. Hence, data on the composition of extra-terrestrial objects and their roles in the task of defining a "cosmic" abundance curve are the principal concerns of this chapter. The table of "cosmic" abundances which is presented here was compiled from such sources using two arbitrary assumptions. The arbitrariness of these assumptions is unavoidable in our present state of knowledge.

The results thus obtained may be used either to look inward (what is the bulk composition of the Earth?) or outward (how did the elements originate?). The first of these questions is central to the logical structure of geochemistry. Accordingly, even though the topic is a controversial and speculative one, it is considered at some length. The second question leads to problems of nucleosynthesis and of time scales for the formation and evolution of stars. These topics, no less in a state of ferment than that of the composition of the Earth, are peripheral to most geochemical discussions and hence are mentioned only very briefly.

II. Cosmic Abundances

We cannot discuss the elemental abundances of the "cosmos" in either of the senses of that word, since we can do no more than speculate about the composition of the universe as a whole, while our knowledge of abundances in our immediate environment is not an orderly system. It is possible to make a rather detailed examination of elemental abundances in our solar system, a less detailed examination of abundances in stars of our Galaxy, but only a series of sketchy outlines of abundances in other galaxies, especially the distant ones. Even upon limiting our considerations to our immediate astronomical environment, unsolved problems and divergent interpretations introduce a very real aspect of (partial) chaos to this topic. The convention first introduced by Noddack and Noddack (1930), of using the adjective "cosmic" to characterize abundances of elements in our solar system and in stars of our Galaxy, will be followed here but the reader should note that this convention could be misleading.

The classic paper on "cosmic" abundances is that by SUESS and UREY (1956). (References to earlier sources of data and of critical ideas foreshadowing their approach are given by SUESS and UREY.) Their fundamental contributions arose from the simultaneous application of two ideas: 1) there should be "some kind of a correlation of nuclear properties with the distribution of the nuclear abundances" (ibid, p. 53), so that isotopic relative abundances "are meaningful and not the result of 'chance' variations" (ibid, p. 54); 2) "the moon and also the chondrites would best represent the average composition of the non-volatile part of solar matter" (ibid, p. 56). The first idea, that nuclear properties are reflected in systematic ways in the abundances of the isotopes, gave SUESS and UREY a recipe of sorts to follow in adjusting and interpreting observed abundances. The second concept had the effect of directing the attention of many geochemists to the analysis of chondritic meteorites. Thus, for several years the terms "cosmic abundances" and "chondritic abundances" were thought to be almost synonymous for most chemical elements.

There are compelling reasons for emphasizing chondritic abundance data in estimating the composition of the solar system. Meteorites in general are the only ponderable samples of undeniably extra-terrestrial matter which we can now take into a geochemical laboratory for intensive examination. Chondrites in particular are thought to be close approximations to representative samples of the parent bodies of meteorites, either because they are mixtures of materials from various portions of differentiated parent bodies or because they are more primitive (and thus less differentiated) than any other class of meteorites. Furthermore, for many elements chondritic and solar abundances agree within the errors assigned to the data, an observation which strengthens one's confidence in the usefulness of chondritic abundances for which such comparisons are not meaningful. (The composition of the Sun *is*, of course, that of the solar system to within about one part in 10^5, owing to its great mass. Unfortunately, solar abundances of many elements have not been determined precisely enough to delineate by themselves an acceptable "cosmic" abundance curve.) Finally, chondrites comprise about 80% to 90% by number of observed meteoritic falls, and about 50% by mass. Accordingly, it is reasonable to assign special significance to elemental abundances observed in such common interplanetary materials.

Granted that these arguments are valid, there still remains a fundamental uncertainty in the choice of chondritic data to use in estimating solar system abundances. As discussed by KEIL in Chapter 4 of this Handbook, there are at least eight distinguishable classes of chondritic meteorites. Although these classes might be defined on the basis of mineralogical criteria alone, it has become apparent during the last decade that there are very significant chemical differences as well among the various classes. Thus, we are faced with two problems: 1) Which among the classes of chondrites (if any) best approximates the composition of the non-volatile fraction of primitive solar matter? 2) Is it possible to unravel the geochemical history which led to the formation of the members of these classes sufficiently well so that one might infer, with reasonable confidence, the even more primitive parent composition? Before seeking answers to these questions it is necessary to examine the principal remaining source of data on extraterrestrial abundances, solar and stellar spectra.

ALLER's monograph, "The Abundances of the Elements" (1961), gives an excellent review of the older work on determinations of solar abundances, as well as illustrations of the principles involved and problems encountered in this task. Recent work on solar abundances has been compiled, with brief critical comments, by UREY (1967). In Table 5-1 are presented solar abundance data from GOLDBERG, MÜLLER and ALLER (1960), UREY's (1967) choices of the best values (based largely on ALLER, 1965), the recent results of O'MARA (1967) on abundances of light elements and ROSS, ALLER and MOHLER (1968) on the Pb abundance, and the values adopted for use here. Note that all data in Table 5-1 are given according to the astronomical convention, as logarithms to the base 10 of atomic abundances normalized to lg A(H) ≡ 12.00. This convention arises because hydrogen is overwhelmingly the most abundant element in the outer envelope of the Sun, all other elements (except He) being mere traces, so that solar abundances are always determined as atomic ratios versus H. (Abundances in the outer envelope do not necessarily represent those of the Sun as a whole [ALLER, 1965], although they probably exemplify best the primitive abundance distribution, unmodified by nuclear processes such as hydrogen-burning. Exceptions must be made for deuterium, Li, Be and B, which are readily destroyed by nuclear reactions, even under the comparatively mild conditions within convection zones underlying stellar envelopes.) In less than a decade there have been large changes in published estimates of a few solar abundances (Table 5-1), reflecting the intensive efforts which have been directed toward improving the solar models and oscillator strength values upon which these estimates depend. There is of course no guarantee that we now have the "right" values, an uncertainty which is of special concern in comparing solar and meteoritic abundances.

In principle, stellar spectroscopic abundance data might be used to confirm the solar data. In practice, stellar spectra are much more useful in extending the range of elements for which abundances can be estimated than in improving the precision of abundances determined in the Sun. Stellar spectroscopists tend, for obvious reasons, to apply simplified forms of solar models in their calculations, while the same oscillator strength values are used in both types of determinations. Only rarely is it feasible to carry out detailed computations of absorption line shapes for stellar spectra, as is being done for investigations of solar abundances. Accordingly, stars whose surface temperatures are similar to that of the Sun generally yield no really new abundance information, while those whose temperatures differ markedly from that of the Sun are principally useful in making relatively imprecise estimates of abundances of elements which cannot readily be observed in the Sun.

In order to compare solar and meteoritic abundances, we must agree on two conventions: 1) which non-volatile element to use to normalize the two abundance scales; 2) which class (or classes) of meteorites, presumably chondrites, may be assumed to be most primitive. (Note the essentially arbitrary nature of these choices. If we understood the geochemical differentiations which have affected the various classes of meteorites well enough to predict either which element was least fractionated or which class is most primitive, a rational choice could be made. This is not yet the case, and it is necessary to choose conventions on the basis of precedent or of speculative models of meteorite origins.) Customarily, Si is chosen as the non-volatile element to bring the solar and meteoritic abundance scales to a common normalization. This is not a good choice, both because Si is not very reliably determined in

Table 5-1. *Solar abundances, normalized to lg $A(H) \equiv 12.00$*

Element	Logarithms of the abundances			
	GOLDBERG, MÜLLER, and ALLER (1960)	UREY (1967)	O'MARA (1967)	Adopted values
Li	0.96	1.54		1.54
Be	2.36	2.34		2.34
C			8.56 ± 0.13	8.56 ± 0.13
N			7.98 ± 0.12	7.98 ± 0.12
O			8.96 ± 0.23	8.96 ± 0.23
Na	6.30	6.27 ± 0.1	6.28 ± 0.08	6.28 ± 0.08
Mg	7.40	$7.40 \pm < 0.1$	7.19 ± 0.25	7.19 ± 0.25
Al	6.20	$6.22 \pm < 0.15$	6.16 ± 0.08	6.16 ± 0.08
Si	7.50	7.50 ± 0.3	7.31 ± 0.25 (Si I) 7.33 ± 0.03 (Si II)	7.32 ± 0.15
P	5.34	5.40	5.59 ± 0.03	5.59 ± 0.03
S	7.30	$7.35 \pm > 0.3$	7.22 ± 0.15	7.22 ± 0.15
K	4.70	4.82	4.67 ± 0.19	4.67 ± 0.19
Ca	6.15	6.04 ± 0.1	6.10 ± 0.15(?)	6.10 ± 0.15(?)
Sc	2.82	2.80 ± 0.3		2.8 ± 0.3
Ti	4.68	4.58 ± 0.1		4.58 ± 0.10
V	3.70	4.12 ± 0.1		4.12 ± 0.10
Cr	5.36	4.90 ± 0.15		4.90 ± 0.15
Mn	4.90	4.80 ± 0.15		4.80 ± 0.15
Fe	6.57	6.71 ± 0.1		6.71 ± 0.10
Co	4.64	4.70 ± 0.1		4.70 ± 0.10
Ni	5.91	5.69 ± 0.15		5.69 ± 0.15
Cu	5.04	3.50 ± 0.5		3.5 ± 0.5
Zn	4.40	3.66 ± 0.4		3.7 ± 0.4
Ga	2.36	2.63 ± 0.4		2.6 ± 0.4
Ge	3.29	2.49		2.5
Rb	2.48	2.35		2.3
Sr	2.60	2.70 ± 0.3		2.7 ± 0.3
Y	2.25	3.20 ± 0.5		3.2 ± 0.5(?)
Zr	2.23	2.65 ± 0.3		2.6 ± 0.3
Nb	1.95	2.30		2.3
Mo	1.90	2.30		2.3
Ru	1.43	1.82		1.8
Rh	0.78	1.37		1.4
Pd	0.78	1.27 ± 0.4		1.3 ± 0.4
Ag	0.14	0.92 ± 0.4		0.9 ± 0.4
Cd	1.46	1.66		1.7
In	1.16	1.36 ± 0.5		1.3 ± 0.5
Sn	1.54	2.15		2.1(?)
Sb	1.94	0.42		0.4(?)
Ba	2.10	2.50 ± 0.4		2.5 ± 0.4
Yb	1.53	2.28		2.2(?)
Pb	1.33	1.64 ± 0.3	1.80[a]	1.9 ± 0.2

[a] ROSS, ALLER, and MOHLER (1968).

the Sun (O'MARA, 1967) owing to uncertainties in the proper oscillator strengths to use, and because Si may have been fractionated in chondrites. Variations in Mg/Si ratios among chondrite classes have been recognized for some time (UREY, 1961;

AHRENS, 1964), and somewhat similar variations in Al/Si ratios have been pointed out by LOVELAND, SCHMITT and FISHER (1968). These ratios, as well as Mg/Al ratios, are given for chondrites and some of the achondrites in Table 5-2, where the first-order fractionations in these elements among chondrites as a whole and the two groups of achondrites are clearly exhibited. We are concerned here especially with the second-order fractionations among the chondrite classes themselves. Mg/Si ratios are low in enstatite chondrites compared with both "ordinary" (CH, CL, and CLL)[1] chondrites and carbonaceous chondrites. Probably there is a meaningful

Table 5-2. *Mg/Si, Al/Si and Mg/Al atomic ratios in stone meteorites and the Sun*[a]

Meteorite class	Mg/Si	(Al × 100)/Si	(Mg × 0.1)/Al
Chondrites:			
Cc_1	1.07 ± 0.05	8.5 ± 0.2	1.26 ± 0.06
Cc_2	1.04 ± 0.13	8.4 ± 0.6	1.24 ± 0.18
CHL	1.07 ± 0.05	10 ± 2	1.1 ± 0.2
CH	0.96 ± 0.03	6.1 ± 0.4	1.57 ± 0.11
CL	0.94 ± 0.03	6.1 ± 0.7	1.54 ± 0.18
CLL	0.94 ± 0.03	6.2 ± 0.3	1.52 ± 0.09
Ce_1	0.73 ± 0.05	5.0 ± 0.5	1.46 ± 0.18
Ce_2	0.81 ± 0.09	4.7 ± 0.5	1.7 ± 0.3
Achondrites:			
Ca-rich (Aor and Ap)	0.30	28	0.11
Ca-poor (Ae)	0.86	2.4	3.6
Solar:	$0.7^{+0.8}_{-0.4}$	7^{+3}_{-2}	$1.1^{+0.9}_{-0.5}$

[a] Data tabulated by SCHMITT, GOLES and SMITH (1968) from various sources; Al abundances from LOVELAND, SCHMITT and FISHER (1968). Dispersions given for Mg/Si and Al/Si ratios are estimates of population standard deviations (not errors in the abundance determinations themselves), assuming that there are no sampling errors for Si contents. Population standard deviations for Mg/Al ratios were computed assuming that abundances of these elements are neither correlated nor anti-correlated. A significant anti-correlation, which is more likely than correlation owing to the mineralogy of Mg and Al in meteorites, would tend to increase the true population dispersion over the estimates given above.

distinction in Mg/Si ratios between the ordinary and carbonaceous chondrites as well, if CHL are grouped with the latter as is traditional. On the other hand, Al/Si and Mg/Al ratios serve to distinguish carbonaceous from ordinary and enstatite chondrites, perhaps with some fine structure in the Al/Si distribution. These data indicate that both Si and Al have been fractionated in chondrites; in particular, carbonaceous chondrites may have lost Si and enstatite chondrites gained Si.

GREENLAND and LOVERING (1965) suggested the use of Ti as a normalization element, rather than Si, and LARIMER and ANDERS (1967) pointed out some advantages of using Mg for this purpose. Both elements are well determined in the Sun, but there are relatively few reliable determinations of Ti in meteorites. I personally favor Mg as the choice of normalization element, but the arguments for that choice do not out-

[1] For abbreviations see Chapter 4 of this Handbook.

weigh the disadvantages of the confusion inherent in uprooting so firmly established a convention. Furthermore, the Mg/Si ratios even for the carbonaceous chondrites agree within O'MARA's (1967) error estimates with the solar ratio, so that there is no compelling argument from the solar data for normalizing to Mg.

CAMERON (1967) normalizes the solar and meteoritic abundances by averaging the abundance ratios (as logarithms) of ten non-volatile elements. Some of these elements, however, appear to be fractionated in meteorites (Na, Si, S and K), so that this approach introduces complexities of interpretation far more severe than those of single-element normalization. UREY (1967) in effect also normalizes his two sets of abundances by averaging ratios of fifteen elements, but that is intended specifically to test whether there is a meaningful abundance anomaly for Fe and only secondarily to define an improved abundance curve. In this chapter, I shall uphold tradition, unsatisfactory as it may be in some ways, and normalize to $Si \equiv 10^6$ atoms.

In Table 5-3 are given the adopted solar abundances of Table 5-1, expressed in c.a.u. (c.a.u., cosmic abundance units, are atoms per 10^6 Si atoms), observed abundances in carbonaceous chondrites of classes Cc_1 and Cc_2, and for comparison the SUESS-UREY (1956) and CAMERON (1967) abundances for the same elements. Attention has been focussed on carbonaceous chondrites since it was shown (REED, KIGOSHI and TURKEVICH, 1960) that their abundances of Tl, Pb and Bi corresponded much more closely to values predicted by CAMERON (1959) than did abundances of these elements in ordinary chondrites. The solar abundances are of course independent of the observed meteoritic abundances. The SUESS-UREY abundances are essentially independent of the meteoritic abundances given; most of the meteoritic data represented in the table were published after 1956. CAMERON's abundances, however, are strongly dependent on abundances observed in Type I carbonaceous chondrites, Cc_1. Error estimates for the solar abundances, though not given in Table 5-3, may readily be derived from Table 5-1.

Let us compare the solar and chondritic abundances in a general way, ignoring for the moment differences between the two classes of carbonaceous chondrites. Among relatively well-determined elements, Mg, Al, Ca, Sc, Ti, Co and Ni (probably) are present in the Sun and in both classes of chondrites in the same abundance, within reasonable estimates of errors in the solar data. With the exception of Co and Ni, these are strongly lithophilic elements, forming refractory oxides under almost any conceivable cosmochemical conditions. Other elements are depleted, to varying degrees, in meteorites relative to the Sun. These include H and the noble gases (which are not listed in Table 5-3 but are well known to be depleted by factors of 10^4 or greater in meteorites and the Earth — see BROWN, 1949, SUESS, 1949, and SIGNER and SUESS, 1963), and C, N and O, depleted by factors of about ten to a hundred. It is generally agreed that a large-scale fractionation based upon differences in volatility was responsible for this pattern of abundances in meteorites (UREY, 1952a, 1952b, 1954, 1957; SUESS ,1965).

Iron is conspicuously enriched in chondrites, compared to the Sun, even though Co and probably Ni are not fractionated. Careful and repeated studies have been made of the solar abundance of Fe, but the difference in Fe/Si ratios in the Sun and the meteorites persists. The implications of this discrepancy for histories of the evolution of the solar system and the meteorites have been discussed by UREY (1964, 1967 and previous publications) and by ANDERS (1964), among others. Since Fe seems to have

Table 5-3. *Comparisons of solar, meteoritic and "cosmic" abundances*

	Solar, lg *A* [lg *A* (Si) = 6.00]	Solar, c.a.u.[a]	Cc_1, c.a.u.	Cc_2, c.a.u.	SUESS and UREY (1956), c.a.u.	CAMERON (1967),[b] c.a.u.
H	10.68	4.8×10^{10}	5.5×10^{6}	3.0×10^{6}	4.00×10^{10}	2.6×10^{10}
Li	0.22	1.7	50	16	100	45
Be	1.02	11	—	0.81	20	0.69
C	7.24	1.7×10^{7}	8.2×10^{5}	4.5×10^{5}	3.5×10^{6}	1.35×10^{7}
N	6.66	4.6×10^{6}	4.9×10^{4}	2.6×10^{4}	6.6×10^{6}	2.44×10^{6}
O	7.64	4.4×10^{7}	7.7×10^{6}	5.5×10^{6}	2.15×10^{7}	2.36×10^{7}
Na	4.96	9.1×10^{4}	6.0×10^{4}	3.5×10^{4}	4.38×10^{4}	6.32×10^{4}
Mg	5.87	7.4×10^{5}	1.07×10^{6}	1.04×10^{6}	9.12×10^{5}	1.050×10^{6}
Al	4.84	6.9×10^{4}	8.5×10^{4}	8.4×10^{4}	9.48×10^{4}	8.51×10^{4}
Si	≡6.00	$\equiv 1.0 \times 10^{6}$	$\equiv 1.0 \times 10^{6}$	$\equiv 1.0 \times 10^{6}$	$\equiv 1.0 \times 10^{6}$	$\equiv 1.0 \times 10^{6}$
P	4.27	1.9×10^{4}	1.27×10^{4}	8,100	1.00×10^{4}	1.27×10^{4}
S	5.90	8.0×10^{5}	5.1×10^{5}	2.3×10^{5}	3.75×10^{5}	5.06×10^{5}
K	3.35	2,200	3.200	2,100	3,160	3,240
Ca	4.78	6.0×10^{4}	7.2×10^{4}	7.2×10^{4}	4.90×10^{4}	7.36×10^{4}
Sc	1.5	30	31	35	28	33
Ti	3.25	1,800	2,300	2,400	2,240	2,300
V	2.80	630	298	590	220	900
Cr	3.58	3,800	1.27×10^{4}	1.24×10^{4}	7,800	1.24×10^{4}
Mn	3.48	3,000	9,300	6,200	6,850	8,800
Fe	5.39	2.5×10^{5}	9.0×10^{5}	8.3×10^{5}	6.00×10^{5}	8.90×10^{5}
Co	3.38	2,400	2,200	1,900	1,800	2,300
Ni	4.37	2.3×10^{4}	4.9×10^{4}	4.5×10^{4}	2.74×10^{4}	4.57×10^{4}
Cu	2.2	160	590	420	212	919
Zn	2.4	250	1,500	630	486	1,500
Ga	1.3	20	46	28	11.4	45.5
Ge	1.2	16	130	76	50.4	126
Rb	1.0	10	6.0	4.1	6.5	5.95
Sr	1.4	25	24	25	18.9	58.4
Y	1.9(?)	80(?)	4.6	4.7	8.9	4.6
Zr	1.3	20	32	23	54.5	30
Nb	1.0	10	—	—	1.00	1.15
Mo	1.0	10	—	—	2.42	2.52
Ru	0.5	3	1.85	1.83	1.49	1.6
Rh	0.1	1	—	—	0.214	0.33
Pd	0.0	1	1.28	1.33	0.675	1.5
Ag	−0.4	0.4	0.95	0.33	0.26	0.5
Cd	0.4	3	2.1	1.2	0.89	2.12
In	0.0	1	0.22	0.10	0.11	0.217
Sn	0.8(?)	6(?)	4.2	1.7	1.33	4.22
Sb	−0.9(?)	0.1(?)	0.40	0.20	0.246	0.381
Ba	1.2	16	4.7	5.0	3.66	4.7
Yb	0.9(?)	8(?)	0.21	0.22	0.220	0.21
Pb	0.6	4	2.9	1.3	0.47	2.90

[a] Cosmic abundance units (c.a.u.) are atoms per 10^6 Si atoms.
[b] Cameron relies heavily on Cc_1 data for non-volatile elements.

been fractionated not only from Si, Mg, Al, etc. but also from Co and perhaps Ni, it is unlikely that a simple metal-silicate fractionation model can be devised to account for the observations. TAYLOR (1965) has proposed an ingenious solution to this problem, involving the preferential accumulation of magnetite from the primitive solar nebula.

For obviously volatile elements, and where the solar data seem to be reliable, it is best to take the solar values as estimates of "cosmic" abundances. Similarily, for Fe whose abundances apparently reflect a special fractionation process in the primitive solar nebula, the solar abundance should be used. In estimating "cosmic" abundances of non-volatile elements, use of chondritic values has many advantages as discussed above, but which class of chondrites shall we choose? The chondritic abundances vary both in their reliability from an analytical standpoint and in their dispersions about the mean values for the chondrite classes (MASON, 1963; GREENLAND and LOVERING, 1965; SCHMITT, GOLES and SMITH, 1968), yet even after taking these effects into account many elements in Table 5-3 are significantly fractionated between Cc_1 and Cc_2 classes. These elements include H, C, Na, S, K, Mn, Cu, Zn, Ga, Ge, Rb, Ag, Cd, In, Sn, Sb, and Pb. LARIMER and ANDERS (1967), in their comprehensive treatment of this topic, identify fourteen additional elements as being significantly fractionated. In most cases, abundances of fractionated elements differ between the Cc_1 and Cc_2 classes by a factor of two or three, so that the problem is not a trivial one.

A very common choice as the most primitive meteorites has been the Cc_1 class, although there are many problems raised by this assumption (see UREY, 1967, pp. 40—42). SCHMITT, SMITH and GOLES (1966) argued that Cc_2 chondrites might a better choice for least-differentiated material than are Cc_1 chondrites, but LARIMER and ANDERS (1967) imply that Cc_1 chondrites are more nearly representative of the fraction of the solar nebula condensable at $\leqq 315°$ K than are Cc_2 chondrites. With few exceptions, Cc_1 chondrites exhibit the highest known abundances among meteorites of the fractionated elements. Thus, if they are indeed the most primitive of the meteorites, there must be a missing fraction which is a complement to the materials represented by the other classes of meteorites. This missing fraction may well be the nebular gases, which in this view would have carried off not only the H, noble gases, C, N, and O in which the meteorites are depleted, but also some part of the fractionated elements. Alternatively, it may be that meteorites as a whole are an incomplete sampling of their parental material (ANDERS and GOLES, 1961), or that Cc_1 chondrites have been enriched in fractionated elements relative to the composition of truly primitive non-volatile matter.

In selecting values of solar system abundances for this chapter (Table 5-4), I have assumed that Cc_2 chondrites are the least differentiated of the meteorites. This is an arbitrary choice, based largely on personal judgement as to which are the most primitive chondrites (and, for that matter, what is meant by "primitive" — see definition given by SCHMITT, SMITH and GOLES, 1966, note *10*). This choice has been influenced by the description of Cc_1 chondrites as hydrothermally-altered tuffs (NAGY, 1966); such alteration would afford many opportunities for modification of the chemical composition of this class. If one prefers instead to utilize solar system abundances chosen under the assumption that Cc_1 chondrites are more closely related to undifferentiated solar system material than are Cc_2 condrites, the compliation by CAMERON (1967) is well suited for that purpose. It seems necessary to choose observed abundances from one or the other of these classes, rather than attempting to make corrections to approximate the composition of their parental materials. The

Table 5-4. *Selected solar system abundances*

Element	Abundance (c.a.u.)	Notes and references
H	4.8×10^{10}	From solar abundances (Table 5-1 and 5-3)
He	3.9×10^{9}	Normalized to O abundance following CAMERON (1967)
Li	16	PINSON (1954)
Be	0.81	SILL and WILLIS (1962)
B	6.2	HARDER (1961); SHIMA (1962)
C	1.7×10^{7}	From solar abundances
N	4.6×10^{6}	From solar abundances
O	4.4×10^{7}	From solar abundances
F	2,500	FISHER (1963)
Ne	4.4×10^{6}	Normalized as for He
Na	3.5×10^{4}	SCHMITT, GOLES and SMITH (1968)
Mg	1.04×10^{6}	MASON (1963)
Al	8.4×10^{4}	LOVELAND, SCHMITT and FISHER (1968)
Si	$\equiv 1.0 \times 10^{6}$	By definition
P	8,100	MASON (1963)
S	8.0×10^{5}	From solar abundances
Cl	2,100	REED and ALLEN (1966)
Ar	3.4×10^{5}	Interpolated after CAMERON (1967)
K	2,100	MASON (1963)
Ca	7.2×10^{4}	MASON (1963)
Sc	35	SCHMITT, GOLES and SMITH (1968)
Ti	2,400	MASON (1963)
V	590	MASON (1963)
Cr	1.24×10^{4}	SCHMITT, GOLES and SMITH (1968)
Mn	6,200	SCHMITT, GOLES and SMITH (1968)
Fe	2.5×10^{5}	From solar abundances
Co	1,900	SCHMITT, GOLES and SMITH (1968)
Ni	4.5×10^{4}	MASON (1963)
Cu	420	SCHMITT, GOLES and SMITH (1968)
Zn	630	LARIMER and ANDERS (1967)
Ga	28	LARIMER and ANDERS (1967)
Ge	76	LARIMER and ANDERS (1967)
As	3.8	Interpolated after CAMERON (1967)
Se	27	AKAIWA (1966); GREENLAND (1967)
Br	5.4	Normalized to Cl abundance following GOLES, GREENLAND and JÉROME (1967)
Kr	25	Interpolated after CAMERON (1967)
Rb	4.1	SMALES *et al.* (1964); MURTHY and COMPSTON (1965)
Sr	25	MURTHY and COMPSTON (1965)
Y	4.7	HASKIN, FREY, SCHMITT and SMITH (1966)
Zr	23	SCHMITT, BINGHAM and CHODOS (1964)
Nb	0.90	Interpolated after CAMERON (1967)
Mo	2.5	KURODA and SANDELL (1954)
Ru	1.83	CROCKETT, KEAYS and HSIEH (1967)
RH	0.33	SCHINDEWOLF and WAHLGREN (1960)
Pd	1.33	CROCKETT, KEAYS and HSIEH (1967)
Ag	0.33	GREENLAND (1967)
Cd	1.2	LARIMER and ANDERS (1967)
In	0.10	AKAIWA (1966); SCHMITT, SMITH and GOLES (1966)
Sn	1.7	LARIMER and ANDERS (1967)
Sb	0.20	EHMANN and TANNER (1966)
Te	3.1	GOLES and ANDERS (1962); AKAIWA (1966); REED and ALLEN (1966)

Table 5-4 (Continued)

Element	Abundance (c. a. u.)	Notes and references
I	0.41	Normalized as for Br
Xe	3.0	Extrapolated after Cameron (1967)
Cs	0.21	Smales *et al.* (1964)
Ba	5.0	Reed, Kigoshi and Turkevich (1960); Moore and Brown (1963)
La	0.47	Haskin, Frey, Schmitt and Smith (1966)
Ce	1.38	Haskin, Frey, Schmitt and Smith (1966)
Pr	0.19	Haskin, Frey, Schmitt and Smith (1966)
Nd	0.88	Haskin, Frey, Schmitt and Smith (1966)
Sm	0.28	Haskin, Frey, Schmitt and Smith (1966)
Eu	0.10	Haskin, Frey, Schmitt and Smith (1966)
Gd	0.43	Haskin, Frey, Schmitt and Smith (1966)
Tb	0.061	Haskin, Frey, Schmitt and Smith (1966)
Dy	0.45	Haskin, Frey, Schmitt and Smith (1966)
Ho	0.093	Haskin, Frey, Schmitt and Smith (1966)
Er	0.28	Haskin, Frey, Schmitt and Smith (1966)
Tm	0.041	Haskin, Frey, Schmitt and Smith (1966)
Yb	0.22	Haskin, Frey, Schmitt and Smith (1966)
Lu	0.036	Haskin, Frey, Schmitt and Smith (1966)
Hf	0.31	Setser and Ehmann (1964)
Ta	0.019	Ehmann (1965)
W	0.16	Amiruddin and Ehmann (1962)
Re	0.059	Morgan and Lovering (1967)
Os	0.86	Crockett, Keays and Hsieh (1967)
Ir	0.96	Crockett, Keays and Hsieh (1967)
Pt	1.4	Crockett, Keays and Hsieh (1967)
Au	0.18	Crockett, Keays and Hsieh (1967)
Hg	0.60	Interpolated after Cameron (1967)
Tl	0.13	Anders and Stevens (1960)
Pb	1.3	Reed, Kigoshi and Turkevich (1960); Marshall (1962); (primordial Pb component only)
Bi	0.19	Reed, Kigoshi and Turkevich (1960)
Th	0.04 (0.05)	Morgan and Lovering (1967); value for 4.5 AE ago in parentheses
U	0.01 (0.025)	Morgan and Lovering (1967); value for 4.5 AE ago in parentheses

latter undertaking would require more knowledge of the chemical history of these meteorites than we presently have.

Values given in Table 5-4 are based on abundances either in the Sun or in Cc_2 chondrites, wherever possible. For B, Mo and Rh, where there is no evidence of fractionation among chondrites in general and for which abundances in Cc_2 chondrites are not available (as of November, 1967), data on ordinary chondrites were used. The values for Zn, Ga, Ge, and Cd were taken from Larimer and Anders (1967) rather than from the original sources of data (in part relatively inaccessible) which they averaged. Only six elements (Ar, As, Kr, Nb, Xe, and Hg) are represented by interpolated abundances. These interpolations are equivalent to assuming a condition of continuity for the abundances of neighboring isotopes whose mass numbers have the same parity. They were carried out after the procedures of Cameron (1967), who found it necessary to interpolate eight elements. Abundances of the heavy noble gases must be determined in the Sun if those interpolations are to be avoided, while adequate data on As and Nb in chondrites are lacking. Hg has been affected by far-reaching fractionations in chondrites (EHMANN and LOVERING, 1967; REED and JOVANOVIČ, 1967), so that its solar system abundance cannot at present be estimated from observed data.

III. The Bulk Composition of the Earth

The composition of the Earth is treated in detail by SCHMUCKER in the following chapter of this Handbook. Accordingly, here we shall merely consider two closely related problems for which the preceding discussion of solar system abundances is particularly useful. These are the controversy on the nature of the Earth's core and the origin of the density variations among the inner planets, and the question of which meteoritic type, if any, best approximates the composition of Earth's mantle plus crust.

The Earth and other inner planets resemble the meteorites in being volatile-poor. The analogy may be carried further, however: variations in abundance ratios between the Sun and chondritic meteorites and among chondrites themselves may be interpreted as evidence for fractionation by volatilization to degrees intermediate between nearly total loss and total retention. Surely, the Earth was subjected to similar fractionations during its early history, but there is no consensus on which elements were partially lost, nor on the details of the mechanisms by which they were fractionated.

The abundances of two elements, Fe and S, are of particular interest in connection with discussions on the composition of Earth's core and on the varying densities of the inner planets. UREY (1951, 1952a, 1960, 1967), who first called attention to this

Table 5-5. *Masses, radii, densities, and zero-pressure densities of bodies in the inner solar system*

	Mass	Radius	Density	Density $P=0$ $T=25°$ C
Mercury	0.0543	0.377	5.59	5.2
Venus	0.8137	0.957	5.12	~4.0
Earth	≡1.0	≡1.0	5.515	~4.0
Moon	0.0123	0.2728	3.34	3.41
Mars	0.1077	{0.520	4.22	3.8
		{0.530	3.99	3.6

Masses and radii are given in units of the Earth's mass and radius. Densities are given in g/cm³. An uncertainty in the radius of Mars is reflected in the table. Adapted after UREY (1966).

interesting problem, has consistently maintained that a metal-silicate fractionation occurred in the primitive solar system and that the decrease in planetary density (corrected for compressional effects) with distance from the Sun of the inner planets is a reflection of this fractionation (Table 5-5). UREY's hypothesis implies that the overall oxidation-reduction state for the Earth is not markedly different from that observed in meteorites, and that the core is composed of Fe-Ni alloy with a significant admixture of some lower Z component, perhaps S. RINGWOOD (1966 and previous publications) interprets both the density variations among the inner planets and the high average density of the Earth relative to chondrites or to the Moon as reflections of differing degrees of reduction. RINGWOOD suggests that Earth's core is an Fe-Ni-Si alloy, about 11 weight percent Si being present owing to the more highly reduced state of the Earth relative to that of meteorites, and that the overall chemical composition of the Earth is closely related to that of Cc_1 chondrites except that many semivolatile elements have been lost. Thus, the fundamental differences between these hypotheses lie in what the respective authors choose to conserve. UREY conserves the oxidation-reduction state of materials in the inner solar system, but not the Fe/Si ratio; RINGWOOD conserves the Fe/Si ratio, but not the oxidation-reduction state. Both agree that the Earth has lost volatile elements, but RINGWOOD would extend this class to include alkali metals, Zn, Cd, Ge, Hg, Tl, Pb, and Cl.

It is not possible at present to test either hypothesis in a conclusive way. The most definitive tests, of course, would be to determine the abundances of Fe, S and elemental Si in the Earth, but geophysical information about the composition of the core does not yet lead to a sufficiently precise model to undertake this task. Evidence from meteorites can be used, arguing by analogy, to support either of these hypotheses. Meteorites display both variations in oxidation-reduction state (see discussion by KEIL in the preceding chapter) and in Fe/Si ratios (UREY and CRAIG, 1953; CRAIG, 1964; SUESS, 1964). In neither case are these variations as large as those called for by the respective hypotheses, which may merely indicate that meteorites are derived from a limited region within the inner solar system, across which such compositional variations were relatively minor.

A more illuminating argument arises from the difference in Fe/Si ratios in the Sun and chondrites which was discussed above. This has been interpreted by UREY as evidence for large-scale Fe/Si fractionations, while RINGWOOD has claimed in effect that the solar Fe/Si ratio is in error, so that the problem does not exist. From Tables 5-1 and 5-3, comparing the adopted solar abundances with the Fe/Si ratio in Cc_2 chondrites, one may infer that for RINGWOOD's suggestion to be valid, the error in the solar ratio must have been under-estimated somehow by almost a factor of *five*. Using the larger error estimate given by O'MARA (1967) for his Si(I) abundance, the factor by which this error (or that for the Fe determination) must have been underestimated remains *2.6*. Thus, it seems that there is strong evidence for large-scale Fe/Si fractionations in the inner solar system. As noted above, these may well be due to differentiation of magnetite from non-ferromagnetic oxides and silicates as suggested by TAYLOR (1965), rather than to a metal-silicate differentiation as proposed by UREY.

RINGWOOD has also argued that sulphur cannot be the low-Z component required in the Earth's core by geophysical models because this element is not present in his estimated mantle composition in high enough abundance. However, he postulates that the core and mantle of the Earth are not in chemical equilibrium with each other and never were so equilibrated. If this is the case, one cannot conclude what the S content of the core might be from that of mantle-derived materials, let alone from that of a hypothetical mantle composition. The geophysical models would require that about 15% by mass of the core should be S, so that the S/Si ratio for the Earth as a whole would be almost precisely the same as that in Cc_2 chondrites. It does not seem implausible that the Earth could have retained such a fraction of the initially-available S.

A related problem is the question of whether the composition of Earth's mantle plus crust more closely resembles that of chondrites or of Ca-rich achondrites. (An excellent review of earlier arguments bearing upon this question was given by Harris and Rowell, 1960.) Ratios of K/Rb, Rb/Sr (as evidenced by $^{87}Sr/^{86}Sr$ ratios), and K/U in crustal and mantle derived materials have been compared with those of chondrites and achondrites as a means of dealing with this question (Gast, 1960, 1965; Taylor, 1964; Wasserburg, MacDonald, Hoyle and Fowler, 1964; Murthy and Stueber, 1967). Hurley (1968) has presented an ingenious set of limiting arguments on K, Rb and Sr abundances in the Earth, based on the assumptions that ^{40}Ar is more highly enriched in the crust (plus atmosphere) than is Rb, which in turn is more enriched in the crust than is K, all relative to their abundances in the total Earth. The conclusions of these authors may be summarized as follows: 1) The Rb/Sr ratios in the source regions of most if not all of the terrestrial samples available to us were lower than those in chondrites during much of the histories of both types of materials. The Rb/Sr ratio by mass for the crust plus mantle is about 0.032, which may be compared to ratios of 0.16 for Cc_2 chondrites and about 0.003 for Ca-rich achondrites. 2) K/Rb ratios by mass in the source regions for magmas which give rise to sub-alkaline abyssal basalts may be greater than 1,500, while for the Earth as a whole the ratio is about 350. In Ca-rich achondrites, K/Rb ratios average about 1,650, while in ordinary chondrites the average is about 300. From Table 5-4, this ratio would be about 230 in solar material, assuming that Cc_2 chondrites contain unfractionated abundances of both elements. 3) Wasserburg *et al.* (1964) present evidence that the average K/U ratio by mass in surficial materials (crust and upper mantle) on the Earth is about 1×10^4. For chondrites, they give values for this ratio of 6.7×10^4 and 7.7×10^4, while in Ca-rich achondrites the average ratio is about 0.5×10^4. However, if one assumes that ordinary chondrites are not reliable samples for K (Larimer and Anders, 1967) and perhaps U as well, and replaces their abundances by those for Cc_2 chondrites, the chondritic K/U ratio becomes about 1.7×10^4 (Table 5-4). Note, however, that this latter estimate is based only on six U analyses of two meteorites, Mighei and Murray (Mason, 1963).

These comparisons imply that the Earth is in several characteristics intermediate between chondrites and Ca-rich achondrites, although there seem to be regions in the upper mantle which resemble Ca-rich achondrites closely. K/Rb ratios appear to be sensitive indices of large-scale differentiation (presumably by a volatilization mechanism), and the Earth, while not as depleted in semi-volatile elements as are achondrites, has lost an appreciable fraction of at least the heavy alkali metals among this group. A fundamental assumption of all of the arguments leading to these conclusions, however, is that the compositions of the lower mantle and the core hold no surprises for us. We are restricted to dealing with surficial materials, and those only as represented by biased or geochemically differentiated samples. If the lower mantle or core differs from the upper mantle in unexpected ways, our conclusions might be in error. For instance, the principal difficulty with postulating an early differentiation of the Earth which buried significant fractions of K and Rb beyond our geochemical reach is that these elements are consistently associated with phases (felsic minerals or low-melting magmas) which are concentrated in the crust, and even more so in the most highly differentiated portions of the crust. There are no purely geophysical constraints on putting K and Rb into the lower mantle, since models of the thermal history of the Earth are sufficiently elastic to accomodate heat production from an appreciable fraction of the Earth's ^{40}K in that region (MacDonald, 1959).

In this context, the discovery by Fuchs (1966) of djerfisherite, K_3 (Na, Cu) $(Fe, Ni)_{12}S_{14}$, in two enstatite chondrites and an enstatite achondrite is of great interest. As Fuchs points out, the stability field for this mineral is apparently narrowly restricted, but that it should exist at all must be due to the relatively reduced state of these meteorites. The calculated density, 3.9 g/cm³, indicates that the djerfisherite stability field relative to other potassium-bearing minerals would probably be extended at high pressures. Fuchs reports some indications, from observations on the St. Marks chondrite, that troilite is soluble in djerfisherite at elevated temperatures; complete miscibility at quite high temperatures would not be surprising. If such miscibility were to occur, it would of course greatly reduce the vapor pressure of K above the sulfide (or metal-sulfide?) melt. Accordingly, it is possible that chalcophilic properties of K and presumably Rb as well become significant under high-pressure, reducing conditions such as may have existed in the primitive mantle of the Earth, and that rather than having been lost by volatilization, these elements are present in appreciable amounts in the lower mantle and even perhaps in the core. Note that it is the existence of djerfisherite which lends some plausibility to this suggestion.

To sum up, there is strong evidence for large-scale Fe/Si fractionations in the inner solar system. The materials which now make up the Earth's interior have been reduced, but there is no agreement on whether the degree of reduction was roughly comparable to that observed in meteorites, or more severe. Volatile elements have been lost from the Earth, but there are marked differences of opinion on which elements were in effect volatile.

IV. Other Applications of Cosmic Abundances

Models for the origin of the elements depend in a crucial way upon the details of the "cosmic" abundance curve. These nucleosynthetic models focus on abundances of individual nuclides, rather than those of the elements, following the principle utilized by Suess and Urey (1956). The discussion of nucleosynthesis in stars by Burbidge, Burbidge, Fowler and Hoyle (1957) has been employed as the starting point for most of the recent investigations of this field, but it is clear that important modifications must be made to their model (see Tayler, 1966, and Fowler, 1967, for critical reviews of current ideas of nucleosynthesis and their interrelationships to cosmological models and astronomical observations). Among other topics, there is now much interest in possible contributions of a "big bang" (primeval cosmologic fireball, which may have initiated the present expanding phase of the known universe) to the synthesis of nuclides. Despite their close links to the subject of this chapter, it is not feasible here to discuss ideas in such an active state of development.

Another application of "cosmic" abundances is to the question of when nucleosynthesis began. Evidence bearing on this question is found in the isotopic ratios of U, Pb, and Os, and the abundance ratio Th/U (FOWLER, 1961; CLAYTON, 1964). The decay of ^{187}Re to ^{187}Os, the differential decays of ^{235}U, ^{238}U, and ^{232}Th relative to each other, and the decay of these three radionuclides to Pb isotopes, provide clocks running at varied rates. The clocks may be read with the aid of a nucleosynthetic model for predicting the ratios in which these nuclides are produced. Even though the nucleosynthetic models are subject to uncertainties, the outlines of a concordant solution for the group of cosmochronological clocks are clear, and it may be tentatively stated that nucleosynthesis began in the Galaxy at least 5×10^9 years before the formation of the solar system. Since a number of independent methods of determining the age of the solar system converge on a value of about 4.6×10^9 years, one may conclude that some of the elements comprising our solar system were formed at least 10×10^9 years ago. It seems unlikely that this conclusion can be seriously modified by any foreseeable changes in nucleosynthetic models.

References

AHRENS, L. H.: Si-Mg fractionation in chondrites. Geochim. Cosmochim. Acta **28**, 411—423 (1964).

AKAIWA, H.: Abundances of selenium, tellurium, and indium in meteorites. J. Geophys. Res. **71**, 1919—1923 (1966).

ALLER, L. H.: The abundances of the elements. New York: Interscience 1961.

— The abundance of elements in the solar atmosphere. Advan. Astron. Astrophys. **3**, 1—25 (1965).

AMIRUDDIN, A., and W. D. EHMANN: Tungsten abundances in meteoritic and terrestrial materials. Geochim. Cosmochim. Acta **26**, 1011—1022 (1962).

ANDERS, E.: Origin, age, and composition of meteorites. Space Sci. Rev. **3**, 583—714 (1964).

—, and G. G. GOLES: Theories on the origin of meteorites. J. Chem. Educ. **38**, 58—66 (1961).

—, and C. M. STEVENS: Search for extinct lead 205 in meteorites. J. Geophys. Res. **65**, 3043—3047 (1960).

BROWN, H.: Rare gases and the formation of the earth's atmosphere. The atmospheres of earth and planets (ed. by G. P. KUIPER), p. 260—268. Chicago: University of Chicago Press 1949.

BURBIDGE, E. M., G. R. BURBIDGE, W. A. FOWLER, and F. HOYLE: Synthesis of the elements in stars. Rev. Mod. Phys. **29**, 547—650 (1957).

CAMERON, A. G. W.: A revised table of abundances of the elements. Astrophys. J. **129**, 676—699 (1959).

— A new table of abundances of the elements in the solar system. In: L. H. AHRENS (Ed.), Origin and Distribution of the Elements (1st Meeting, International Association of Geochemistry and Cosmochemistry, Paris, 8—11 May, 1967). Oxford: Pergamon Press 1968.

CLAYTON, D. D.: Cosmoradiogenic chronologies of nucleosynthesis. Astrophys. J. **139**, 637—663 (1964).

CRAIG, H.: Petrological and compositional relationships in meteorites. Isotopic and cosmic chemistry (ed. by H. CRAIG, S. L. MILLER, and G. J. WASSERBURG), p. 401—451. Amsterdam: North-Holland 1964.

CROCKETT, J. H., R. R. KEAYS, and S. HSIEH: Precious metal abundances in some carbonaceous and enstatite chondrites. Geochim. Cosmochim. Acta **31**, 1615—1623 (1967).

EHMANN, W. D.: On some tantalum abundances in meteorites and tektites. Geochim. Cosmochim. Acta **29**, 43—48 (1965).

—, and J. F. LOVERING: The abundance of mercury in meteorites and rocks by neutron activation analysis. Geochim. Cosmochim. Acta **31**, 357—376 (1967).

—, and J. T. TANNER: The abundance of antimony in meteorites. Earth Plan. Sci. Letters **1**, 276—279 (1966).

FISHER, D. E.: The fluorine content of some chondritic meteorites. J. Geophys. Res. **68**, 6331—6335 (1963).

FOWLER, W. A.: Rutherford and nuclear cosmochronology. Proc. Rutherford Jubilee International Conference (ed. by J. B. BIRKS), p. 640—676. New York: Academic Press 1961.

— The empirical foundations of nucleosynthesis. In: L. H. AHRENS (Ed.), Origin and Distribution of the Elements (1st Meeting, International Association of Geochemistry and Cosmochemistry, Paris, 8—11 May, 1967). Oxford: Pergamon Press 1968.

FUCHS, L. H.: Djerfisherite, alkali copper-iron sulfide: A new mineral from enstatite chondrites. Science **153**, 166—167 (1966).

GAST, P. W.: Limitations on the composition of the upper mantle. J. Geophys. Res. **65** 1287—1297 (1960).

— Terrestrial ratio of potassium to rubidium and the composition of earth's mantle. Science **147**, 858—860 (1965).

GOLDBERG, L., E. A. MÜLLER, and L. H. ALLER: Abundances of the elements in the solar atmosphere. Astrophys. J., Suppl. Ser. **5**, 1—137 (1960).

GOLES, G. G., and E. Anders: Abundances of iodine tellurium and uranium in meteorites. Geochim. Cosmochim. Acta **26**, 723—737 (1962).

—, L. P. GREENLAND, and D. Y. JÉROME: Abundances of chlorine, bromine and iodine in meteorites. Geochim. Cosmochim. Acta **31**, 1771—1787 (1967).

GREENLAND, L.: The abundances of selenium, tellurium, silver, palladium, cadmium, and zinc in chondritic meteorites. Geochim. Cosmochim. Acta **31**, 849—860 (1967).

—, and J. F. LOVERING: Minor and trace element abundances in chondritic meteorites. Geochim. Cosmochim. Acta **29**, 821—858 (1965).

HARDER, H.: Beitrag zur Geochemie des Bors. Teil III. Bor in metamorphen Gesteinen und im geochemischen Kreislauf. Nachr. Akad. Wiss. Goettingen, II. Math.-Physik. Kl. **1**, 1—26 (1961).

HARRIS, P. G., and J. A. ROWELL: Some geochemical aspects of the Mohorovičic discontinuity. J. Geophys. Res. **65**, 2443—2459 (1960).

HASKIN, L. A., F. A. FREY, R. A. SCHMITT, and R. H. SMITH: Meteoritic, solar and terrestrial rare-earth distributions. Phys. Chem. Earth **7**, 167—321 (1966).

HURLEY, P. M.: Absolute abundance and distribution of Rb, K and Sr in the Earth. Geochim. Cosmochim. Acta **32**, 273—283 (1968).

KURODA, P. K., and E. B. SANDELL: Geochemistry of molybdenum. Geochim. Cosmochim. Acta **6**, 35—63 (1964).

LARIMER, J. W., and E. ANDERS: Chemical fractionations in meteorites — II. Abundance patterns and their interpretation. Geochim. Cosmochim. Acta **31**, 1239—1270 (1967).

LOVELAND, W., R. A. SCHMITT, and D. E. FISHER: Aluminium abundances in stony meteorites. Submitted to Geochim. Cosmochim. Acta (1968).

MACDONALD, G. J. F.: Calculations on the thermal history of the earth. J. Geophys. Res. **64**, 1967—2000 (1959).

MARSHALL, R. R.: Mass spectrometric study of the lead in carbonaceous chondrites. J. Geophys. Res. **67**, 2005—2015 (1962).

MASON, B. H.: The carbonaceous chondrites. Space Sci. Rev. **1**, 621—646 (1963).

MOORE, C. B., and H. BROWN: Barium in stony meteorites. J. Geophys. Res. **68**, 4293—4296 (1963).

MORGAN, J. W., and J. F. LOVERING: Rhenium and osmium abundances in chondritic meteorites. Geochim. Cosmochim. Acta **31**, 1893—1909 (1967).

— — Uranium and thorium abundances in carbonaceous chondrites. Nature **213**, 873—875 (1967).

MURTHY, V. R., and W. COMPSTON: Rb-Sr ages of chondrules and carbonaceous chondrites. J. Geophys. Res. **70**, 5297—5307 (1965).

—, and A. M. STUEBER: Potassium-rubidium ratios in mantle derived rocks. Ultramafic and related rocks (ed. by P. J. WYLLIE), p. 376—380. New York: John Wiley & Sons 1967.

NAGY, B.: Investigations of the Orgueil carbonaceous meteorite. Geol. Foren. Stockholm Forh. **88**, 235—272 (1966).

NODDACK, I., u. W. NODDACK: Die Häufigkeit der Chemischen Elemente. Naturwissenschaften **18**, 757—764, especially p. 759 (1930).

O'MARA, B. J.: The solar abundance of light elements. Ph.D. Dissertation, University of California, Los Angeles 1967.

PINSON, JR., W. H.: Trace-element composition of a carbonaceous chondrite. Trans. Am. Geophys. Union **35**, 380 (1954).

REED, JR., G. W., and R. O. ALLEN JR.: Halogens in chondrites. Geochim. Cosmochim. Acta **30**, 779—800 (1966).

—, and S. JOVANOVIČ: Mercury in chondrites. J. Geophys. Res. **72**, 2219—2228 (1967).

—, K. KIGOSHI, and A. TURKEVICH: Determinations of concentrations of heavy elements in meteorites by activation analysis. Geochim. Cosmochim. Acta **20**, 122—140 (1960).

RINGWOOD, A. E.: The chemical composition and origin of the earth. Advances in earth science (ed. by P. M. HURLEY), p. 287—356. Cambridge, Mass.: M.I.T. Press 1966.

ROSS, J., L. H. ALLER, and O. C. MOHLER: The abundance of lead in the sun. Proc. Nat. Acad. Sci. U.S. **59**, 1—6 (1968).

SCHINDEWOLF, U., and M. WAHLGREN: The rhodium, silver and indium content of some chondritic meteorites. Geochim. Cosmochim. Acta **18**, 36—41 (1960).

SCHMITT, R. A., E. BINGHAM, and A. A. CHODOS: Zirconium abundances in meteorites and implications to nucleosynthesis. Geochim. Cosmochim. Acta **28**, 1961—1979 (1964).

—, G. G. GOLES, and R. H. SMITH: Elemental abundances in stone meteorites. To be submitted to Geochim. Cosmochim Acta (1968).

—, R. H. SMITH, and G. G. GOLES: Chainpur-like chondrites: Primitive precursors of ordinary chondrites? Science **153**, 644—647 (1966).

SETSER, J. L., and W. D. EHMANN: Zirconium and hafnium abundances in meteorites, tektites and terrestrial materials. Geochim. Cosmochim. Acta **28**, 769—782 (1964).

SHIMA, M.: Boron in meteorites. J. Geophys. Res. **67**, 4521—4523 (1962).

SIGNER, P., and H. E. SUESS: Rare gases in the sun, in the atmosphere, and in meteorites. Earth science and meteoritics (ed. by J. GEISS and E. D. GOLDBERG), p. 241—272. Amsterdam: North-Holland 1963.

SILL, C. W., and C. P. WILLIS: The beryllium content of some meteorites. Geochim. Cosmochim. Acta **26**, 1209—1214 (1962).

SMALES, A. A., T. C. HUGHES, D. MAPPER, C. A. J. MCINNES, and R. K. WEBSTER: The determination of rubidium and caesium in stony meteorites by neutron activation analysis and by mass spectrometry. Geochim. Cosmochim. Acta **28**, 209—233 (1964).

SUESS, H. E.: Die Häufigkeit der Edelgase auf der Erde und im Kosmos. J. Geol. **57**, 600—607 (1949).

— The Urey-Craig groups of chondrites and their states of oxidation. Isotopic and cosmic chemistry (ed. by H. CRAIG, S. L. MILLER, and G. J. WASSERBURG), p. 385—400. Amsterdam: North-Holland 1964.

— Chemical evidence bearing on the origin of the solar system. Ann. Rev. Astron. Astrophys. **3**, 217—234 (1965).

—, and H. C. UREY: Abundances of the elements. Rev. Mod. Phys. **28**, 53—74 (1956).

TAYLER, R. J.: The origin of the elements. Rept. Progr. Phys. **29**, 489—538 (1966).

TAYLOR, S. R.: Trace element abundances and the chondritic earth model. Geochim. Cosmochim. Acta **28**, 1989—1998 (1964).

— Enrichment of iron during accretion in the solar nebula. Nature **208**, 886—887 (1965).

UREY, H. C.: The origin and development of the earth and other terrestrial planets. Geochim. Cosmochim. Acta **1**, 209—277 (1951).

— The Planets. New Haven: Yale University Press 1952a.

— Chemical fractionation in the meteorites and the abundance of the elements. Geochim. Cosmochim. Acta **2**, 269—282 (1952b).

UREY, H. C.: On the dissipation of gas and volatilized elements from protoplanets. Astrophys. J., Suppl. **1**, 147—173 (1954).
— Boundary conditions for theories of the origin of the solar system. Progr. Phys. Chem. Earth **2**, 46—76 (1957).
— On the chemical evolution and densities of the planets. Geochim. Cosmochim. Acta **18**, 151—153 (1960).
— Criticism of Dr. B. Mason's paper on "The origin of meteorites." J. Geophys. Res. **66**, 1988—1991 (1961).
— A review of atomic abundances in chondrites and the origin of meteorites. Rev. Geophys. **2**, 1—34 (1964).
— Chemical evidence relative to the origin of the solar system. Monthly Notices Roy. Astron. Soc. **131**, 199—223 (1966).
— The abundance of the elements with special reference to the problem of the iron abundance. Quart. J. Roy. Astron. Soc. **8**, 23—47 (1967).
—, and H. CRAIG: The composition of the stone meteorites and the origin of the meteorites. Geochim. Cosmochim. Acta **4**, 36—82 (1953).
WASSERBURG, G. J., G. J. F. MACDONALD, F. HOYLE, and W. A. FOWLER: Relative contributions of uranium, thorium, and potassium to heat production in the earth. Science **143**, 465—467 (1964).

Chapter 6

U. Schmucker

GEOPHYSICAL ASPECTS OF STRUCTURE AND COMPOSITION OF THE EARTH

Introduction

Our knowledge about the physical and chemical state of the Earth's interior comes from three sources: (i) from geophysical observations on and above the Earth's surface, (ii) from experiments with rocks and minerals in the laboratory, and (iii) from cosmochemical and geochemical considerations.

Transient natural phenomena such as earthquakes, earth tides or magnetic storms constitute "geophysical experiments" on a grand scale within the Earth. The surface response of the Earth, when recorded with appropriate instruments, allows conclusions about pertinent physical properties of the interior. The depth distribution of some properties can be derived in a straightforward way. The propagation velocity of seismic waves is one of them (Sec. I). Other properties are obtained from indirect "trial-and-error" fits between observed and calculated data or on the basis of some physical concept, relating the properties among themselves and to thermodynamical conditions. This applies to the determination of the internal density (Sec. II).

Additional information about the Earth's interior comes also from certain non-transient geophysical phenomena. Among them are the Earth's gravity field, the geomagnetic main field, and the terrestrial heat flow. The interpretation of the pertinent surface data is plagued, however, by an inherent ambiguity which arises ultimately from their stationary or quasi-stationary nature (Sec. III, IV, V).

Extensive efforts have been made to measure the physical properties of various rocks and minerals in the laboratory under extreme temperature-pressure conditions, comparable to those at some particular depth within the Earth. They can be compared with those from geophysical investigations and permit us to specify potential constituents for the various layers of the Earth's interior. A comparison of this kind is of course purely conjectural and has to be supplemented by geochemical arguments concerning the probable abundances of chemical elements (Sec. VI).

Principal Subdivision of the Earth's Interior

Seismological evidence and the dynamics of the Earth-Moon system establish that the Earth has a high-density core beginning at 2,900 km depth. Hence, its radius measures 3,470 km or $\frac{11}{20}$ of the Earth's radius. It probably consists of molten iron which is in turbulent motion with flow velocities of the order of km/yr. This is indicated by the speed of geomagnetic secular variations (Sec. IV, c). In BULLEN's alphabetical nomenclature for the Earth's interior (Fig. 6-6) the core embraces the concentric regions E (*outer core*), F (*transition layer*), and G (*inner core*).

The core is surrounded by a solid mantle of iron-magnesium silicates and oxides. It contains the regions B (*upper mantle*), C (*transition layer* of the upper mantle), and D (*lower mantle*), the latter extending from 1,000 to 2,900 km depth. There are indications that the mantle is not in perfect hydrostatic equilibrium and that its upper portion is in slow convective motion, involving flow velocities of the order of cm/yr. (Sec. III, b).

The outer shell on top of the mantle is a skinlike *crust* (region A), covered by a thin veil of geological rock formations and the oceans. Composition and structure of this uppermost region reflect a variable and highly diversified tectonic-magmatic history since early Precambrian times. Mid-ocean ridges and island arcs intersect deep ocean basins, stable shields stand in contrast to mobile mountain belts and rift zones.

The ultimate cause for these strikingly different structural surface features is as yet unknown. There is mounting evidence that they extend downward through the crust into a mantle of considerable lateral complexity, connecting in this way upper mantle imbalances with geological surface phenomena, such as mountain building and volcanism.

Bibliography

GASKELL, T. F. (editor): The earth's mantle. New York: Academic Press 1967.

GUTENBERG, B.: Physics of the earth's interior. New York: Academic Press 1959.

JACOBS, J. A.: The earth's core and geomagnetism. Oxford: Pergamon Press 1963.

JEFFREYS, H.: The earth, 4th ed. London: Cambridge University Press 1959.

STEINHART, J. S., and T. J. SMITH: The earth beneath the continents. Washington D.C.: American Geophysical Union (Geophysical Monograph 10) 1966.

TAKEUCHI, H.: Theory of the earth's interior. Waltham, Mass.: Blaisdell 1966.

Handbooks

FLÜGGE, S. (editor, J. Bartels group editor): Handbuch der Physik, Bd. XLVII (Geophysics I). Berlin-Göttingen-Heidelberg: Springer 1956.

KUIPER, G. P. (editor): The earth as a planet. Chicago: The University of Chicago Press 1954.

Tables

BARTELS, J., and P. TEN BRUGGENCATE (editors): Landolt-Börnstein III (Zahlenwerte und Funktionen aus Astronomie und Geophysik). Berlin-Göttingen-Heidelberg: Springer 1952.

CLARK, S. P. (editor): Handbook of physical constants. New York: The Geological Society of America, Memoir 97, 1966.

Section I and II

BULLEN, K. E.: Introduction to the theory of seismology, 3rd ed. London: Cambridge University Press 1963.

MELCHIOR, P.: The earth tides. Oxford: Pergamon 1966.

RICHTER, C. F.: Elementary seismology. San Francisco: Freeman 1958.

Section III

CAPUTO, M.: The gravity field of the earth. New York: Academic Press 1967.
GARLAND, G. D.: The earth's shape and gravity. Oxford: Pergamon Press 1965.
HEISKANEN, W. A., and F. A. VENING-MEINESZ: The earth and its gravity field. New York: MacGraw-Hill 1958.

Section IV

CHAPMAN, S., and J. BARTELS: Geomagnetism, 3rd impression. London: Oxford University Press 1962.
FANSELAU, G. (editor): Geomagnetismus und Aeronomie, Bd. III (Über das aus dem Erdinneren stammende Magnetfeld). Berlin: VEB Deutscher Verlag der Wissenschaften 1959.
IRVING, E.: Paleomagnetism and its application to geological and geophysical problems. New York and London: John Wiley 1964.
NAGATA, T.: Rock magnetism, revised ed. Tokyo: Maruzen 1961.
RIKITAKE, T.: Electromagnetism and the earth's interior. Amsterdam: Elsevier 1966.

Section V

CARSLAW, H. S., and J. C. JAEGER: Conduction of heat in solids, 2nd ed. London: Oxford University Press 1959.
INGERSOLL, L. R., O. J. ZOBEL, and A. C. INGERSOLL: Heat conduction with engineering and geological applications. New York: Mc Graw-Hill 1948.
LEE, W. H. K. (editor): Terrestrial heat flow. Washington D. C.: American Geophysical Union (Geophysical Monograph 8) 1965.

I. Inferences from Earthquakes and Earth Tides

a) Principles of Seismic Wave Propagation

From a sufficiently large distance we may regard an earthquake as a point source (=*focus*) from which strain energy is emitted in all directions. Its propagation through the Earth's interior in form of *body waves* and along the Earth's surface in form of *surface waves* is well described by the theory of perfect elasticity, involving linear stress-strain relations (generalized HOOKE's law). This was the fortuitous and by no means self-evident result of early research in seismology by OLDHAM and WIECHERT around 1900.

Assuming that the stress-strain relations are isotropic, two elastic constants besides the density are sufficient to express the elastic response of the transmitting medium. A convenient choice is the pair *rigidity* (= *shear modulus*) μ_S and *incompressibility* (= *bulk modulus*)

$$K_S = \varrho \left(\frac{\partial p}{\partial \varrho}\right)_S, \tag{1}$$

both defined for adiabatic shear and compression (entropy S constant). The reciprocal K_S^{-1} gives the adiabatic increase in density under increasing hydrostatic pressure p, defined as the arithmetic mean of the three principal stresses exerted upon a volume element of the medium. The reciprocal μ_S^{-1} measures similarly the change in shape which is inflicted upon a volume element by deviatory or shear stress components [Eq. (31)]. The rigidity of fluids is zero.

Body waves traverse the Earth's interior in two modes, as compressional *P-waves* (particle motion parallel to the direction of propagation) and as shear or *S-waves* (particle motion transverse to the direction of propagation). Their respective velocities,

$$v_P = \sqrt{(K_S + \tfrac{4}{3}\mu_S)/\varrho}$$
$$v_S = \sqrt{\mu_S/\varrho}, \tag{2}$$

amount to several kilometers per second. The lead of the faster travelling P-waves (undae *p*rimae) over the S-waves (undae *s*ecundae), written in the form

$$v_B = \sqrt{v_P^2 - \tfrac{4}{3} v_S^2} = \sqrt{K_S/\varrho}, \tag{3}$$

gives the *sound* or *bulk velocity* of the medium. Further useful relations are

$$\left(\frac{v_P}{v_S}\right)^2 = \frac{1-\sigma}{\frac{1}{2}-\sigma} \tag{4}$$

$$E = \frac{3K}{K/\mu + \frac{1}{3}} \tag{5}$$

involving Poisson's ratio σ and YOUNG's elastic modulus E of the medium.

The path and travel time of body waves from the source to a receiving station (seismograph) is determined by refraction and reflexion phenomena. In fact, travel times of P- and S-waves have been the prime source of information about the Earth's interior and will be discussed on the basis of simple ray theory in the following sections.

The particle motion of *surface waves* is elliptic and bound to the free surface of the Earth, dying away with increasing depth. The motion of long-period surface waves (period $\tau \approx 100$ sec) extends downward into the mantle and provides seismologists with the unique opportunity to study mean crustal and upper mantle properties, averaged over some well defined distance along the surface (see below). Surface waves arrive later than the more directly travelling body waves and produce a sequence of oscillatory, long-period ground motions.

They are generated in two principal modes: as *Love-waves* (particle motion in horizontal planes) and *Rayleigh-waves* (particle motion in vertical planes parallel to the direction of propagation). Their velocities are not very different from the S-wave velocity but for a given period τ complicated functions of compressibility, rigidity, and density are involved.

During their propagation seismic waves undergo dispersion, absorption, scattering on small inhomogeneities, and diffraction at obstacles. Body waves show no dispersion, but surface waves do. It is indeed the most important aspect of their analysis.

Dispersion implies that the phase velocity c of a harmonic wave component is a function of its period. If the signal contains a narrow band of frequencies, centered at the frequency τ^{-1}, it will travel with the group velocity

$$U = c\Big/\left(1 + \frac{\tau}{c}\,\frac{dc}{d\tau}\right) \tag{6}$$

from the source to the receiver. In the case of seismic surface waves the dispersion is normal, i.e. c increases with increasing period and the signal travels slower than its harmonic components.

The travel-time T of a surface wave signal of a certain mean period τ can be inferred directly from the seismograms, yielding with $U = \Delta/T$ the group-velocity of the signal for a given path Δ. The determination of the phase velocity c requires a spectral cross-correlation between signals received at adjacent sites. The resulting dispersion curves $U(\tau)$, or $c(\tau)$, reflect the mean internal properties (K, μ, ϱ) either between the two recording sites (c) or between source and receiver (U).

Scattering and *absorption* reduce the transmitted amplitude of the wave according to $\exp(-kx)$; x denotes the distance from the source and k the absorption coefficient of the medium for a given period of the seismic vibrations. A typical value for P-waves ($\tau \approx 10$ sec) is $k = 1/5{,}000\ \text{km}^{-1}$, implying that the amplitude of the wave is reduced by energy dissipation to about one third after a travel distance of 5,000 km.

Considering a single sinusoidal wave component of the propagation velocity U, the expression

$$Q^{-1} = k\, U\, \tau/\pi \tag{7}$$

represents a characteristic dimensionless measure for the degree of anelasticity of the absorbing medium. The thus defined *specific dissipation constant* Q^{-1} is of particular interest when the possibility of viscous motion within seismically solid regions is considered (Sec. II, f). *Diffracted* waves appear in the shadow zone of low-velocity layers. They are difficult to recognize and have played only a minor role in seismological research.

b) The Jeffreys-Bullen Curves

We ignore the specific wave character of body waves and assume that their energy propagates in the form of sharp pulses along *seismic rays*. Suppose such a hypothetical pulse in the P- or S-mode leaves the focus F of an earthquake and is recorded after a time T by seismograph S at the epicentral distance Δ. It travels from F to S along the "fastest" possible path (= minimum condition for T according to FERMAT's principle). The ray path and the travel time of the pulse are readily derived from SNELL's generalized law of refraction for any given velocity distribution (Fig. 6-1).

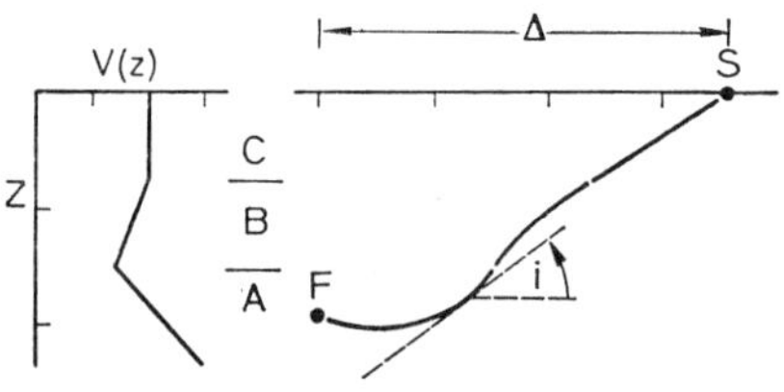

Fig. 6-1. Seismic ray in a plane elastic medium of variable seismic velocity $v(z)$, indicating the path of energy transmission from the earthquake focus F to the receiving seismograph S (P or S-mode of body waves); $p = (\cos i)/v$ is a constant parameter for each ray according to the generalized SNELL's law of refraction. Hence, the rays are concave (when seen from the surface) in region A ($dv/dz > 0$), convex in region B ($dv/dz < 0$), and straight in region C ($dv/dz = 0$). The reciprocal of p is the velocity at the deepest point of penetration where $i = 0°$

If the pulse meets a velocity discontinuity, it is partially reflected at the angle of incidence and partially transmitted at the proper angle of refraction. At the same time P-pulses can be converted into S-pulses and vice versa. Plots of T versus Δ

for one particular mode (or combination of modes in the presence of discontinuities) are called *travel time curves* (Fig. 6-2). Their slope defines the *apparent velocity*

$$v_A = \frac{d\Delta}{dT} \tag{8}$$

of the pulse at the surface. Its change with epicentral distance Δ is a direct measure for the change of true seismic velocity with depth. As is readily inferred from Fig. 6-1, the difference in travel time dT of signals, arriving at adjacent sites, is $p \cdot d\Delta$ with $p = (\cos i)/v$ as constant ray parameter (SNELL's law); $d\Delta$ denotes the small difference in epicentral distance of the two sites. Hence, the true velocity at the deepest point ($i = 0°$) of a ray is identical with the apparent velocity where this ray reaches the surface.

Empirical $T(\Delta)$-curves can be obtained by observing either a single earthquake with a network of seismic stations or by observing at a fixed site many earthquakes of changing epicentral distance. Both methods yield identical travel-time curves in the absence of lateral non-uniformities. This is indeed more or less the case, at least for large epicentral distances, which proves that the deeper interior is basically of spherical symmetry.

After minor regional irregularities had been smoothed out, a set of standard $T(\Delta)$-curves emerged from seismological research during the last decades from which the mean internal velocity distribution has been deduced with an accuracy of 2% or better (Fig. 6-6). These curves, known as Jeffreys-Bullen curves, serve now as reference standards for local traveltime studies to contour regional deviations from the above defined internal mean state (Sec. I, f).

c) The Interpretation of Travel-Time Curves

The interpretation of empirical travel-time curves requires in essence that the observed apparent velocity as a function of epicentral distance, $v_A(\Delta)$, be inverted into the true internal velocity as a function of depth, $v(z)$. This is in theory a single-solution problem when v is either constant or increasing with depth. (The correct formulation of this condition is $dv/dz > -v/(a-z)$ with a as Earth's radius.) In practice, complications arise from all zones of rapidly changing velocity.

Four examples may illustrate the principal features of travel-time curves (Fig. 6-2). The chosen velocity models consist of a plane top-layer with a constant positive velocity gradient, a variable transition zone and a uniform substratum. We consider three different rays, the third one being tangential to the surface of the substratum. Their points of re-emergence are labeled S_1, S_2, S_3.

The first model A is without a transition layer and the three rays are circular arcs. The resulting $T(\Delta)$-curve between F and S_3 is slightly curved, reflecting the gradual velocity increase in the top-layer. Part of the seismic energy, contained in ray 3, is guided as "head wave" along the surface of the substratum, from where it is radiated in the form of *refracted signals* to the surface. These signals produce the straight segment of the $T(\Delta)$-curve beyond S_3. (A correct explanation of the guided wave requires that the wave character of energy transmission is taken into account. It may suffice here to assume that the substratum has a minute downward velocity gradient which bends rays beyond ray 3 back to the surface.)

Model A, B

Model C, D

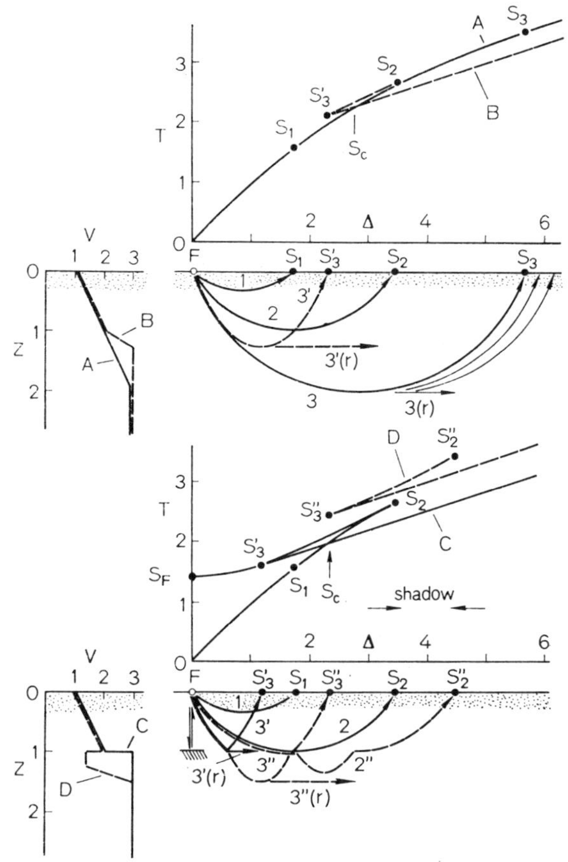

Fig. 6-2. Four seismic velocity models $v(z)$, their travel-time curves $T(\Delta)$, and the pattern of selected rays. *Model A:* Constant velocity gradient between $z=0$ and $z=2$ and uniform velocity below this depth; *model B:* discontinuous increase of the velocity gradient at $z=1$ (*2nd order discontinuity*); *model C:* discontinuous increase of the velocity at $z=1$ (*1st order discontinuity*); *model D:* discontinuous reduction of the velocity at $z=1$ (*low-velocity layer*), followed by a steep velocity increase. The travel-time curve characteristics are: retrograde branches ("loops") between S_2 and S_3' (B and C), near-vertical reflections $S_F - S_3'$ from first-order discontinuity (C), and offset of the $T(\Delta)$ curve between S_2 and S_2'' as *shadow* of the low-velocity layer (D). Guided waves along the surface of the substratum are denoted by the letter "r" in parenthesis. The ray parameters (cf. legend of Fig. 6-1) are 3/4 (ray 1), 1/2 (ray 2), and 1/3 (ray 3). Arbitrary time and distance units

Model B has a high-gradient transition zone between the top-layer and the substratum. As a consequence, ray 3 is sharply bent upwards and emerges now as ray 3′ closer to F than ray 2. Because the same happens to all rays between ray 2 and 3, we obtain between S_2 and S_3' a *retrograde* branch of the travel-time curve before the straight segment of refraction signals sets in. Thus the high-gradient zone produces a loop or *cusp* in the $T(\Delta)$-curve and a seismograph between S_3' and S_2 would record a sequence of three distinct pulses, all belonging to the same mode. Notice the focal point of later arrivals at S_3' and the discontinuous change in the apparent velocity of first arrivals at a critical distance S_c.

The focal point disappears in model C, when the transition from the top to the substratum is a first-order discontinuity. We now receive *reflected signals* from the substratum at all distances between F and S_2, so-called *near-vertical* reflections between F and S_3' and *wideangle* reflections between S_3' and S_2. Reflections near S_3' are extra strong and are termed *critical* reflections.

Empirical travel-time curves of the type A, B, or C are uniquely interpretable, when the loops in the curves B and C are well documented by later arrivals. The insertion of a low-velocity layer in model D, however, introduces an irrevocable ambiguity. It cuts the $T(\Delta)$-curve into two separate branches, the first ending with ray 2 at S_2. Because this ray grazes the top of the low-velocity layer, an adjacent ray 2'' will be deflected into it and bent upwards in the underlying high-gradient zone to emerge at S_2''. Here begins the second branch with a retrograde segment which is terminated by ray 3 at S_3''. The gap between S_2 and S_2'' represents the *shadow zone* of the low-velocity layer, while the time gap between the first and second branch reflects the lag in pulses which traverse this layer.

These models show that detailed information on high-gradient zones, first-order discontinuities and low-velocity layers is hardly possible without proper evaluation of later arrivals. Their identification in seismograms is, however, not an easy task; in particular their correct correlation from site to site may impose formidable difficulties.

Supporting evidence for one or the other interpretation of travel-time data may come from amplitude considerations and from dispersion studies of surface waves. The amplitude of body-wave signals is determined, at least partially, by the geometry of the ray pattern i.e. by the focussing or de-focussing of seismic rays at certain epicentral distances. However, such geometric effect is often obscured by amplitude modulations which arise from a variable absorption along the different rays and from local geological peculiarities at the recording site.

No direct inversion of empirical dispersion curves $c(\tau)$ or $U(\tau)$ into the desired depth-distributions of K, μ and ϱ is possible. Their evaluation is indirect by matching the dispersion curves of layered model distributions with empirical curves in the best possible way.

A comparison of P- and S-wave velocities provides an ultimate test for the compatibility of seismic interpretations. The ratio v_p/v_s is solely a function of Poisson's number σ, which should not differ greatly from $^1/_4$ for solid material. Thus we may expect v_p/v_s to be close to $\sqrt{3}$ in crust and mantle [Eq. (4)].

d) Seismic Velocities in the Crust

Early evidence for a shallow first-order discontinuity of seismic velocities evolved some fifty years ago from earthquake studies. It has been named the MOHOROVIČIĆ-*discontinuity* in honour of its discoverer. Subsequent intensive research with the aid of artificial explosions established it as a world-wide phenomenon.

The *M. discontinuity* now defines the lower boundary of the Earth's crust which is distinguished by a mean P-wave velocity of 6 km/sec from the underlying mantle of Earth with velocities around 8 km/sec in its uppermost portion. A working definition places the crust-mantle interface at that depth below the surface where a steep rise of P-wave velocity levels off to a nearly constant value in excess of 7.8 km/sec. This depth, as well as the crustal and subcrustal velocities, is subject to considerable regional variations (Fig. 6-4, 6-5). Hence, it would be desirable to define the threshold velocity in dependence of the prevailing pressure-temperature conditions.

The results of explosion seismology in continental areas can be summarized as follows: P-wave signals traveling on a path which is completely contained in the crust appear as first arrivals in the seismogram up to focal distances between 120 and 180 km. This *P_1-phase* has an almost constant apparent surface velocity v_A of 6 km/sec and forms the first, nearly straight, segment of a typical explosion travel time curve (Fig. 6-3).

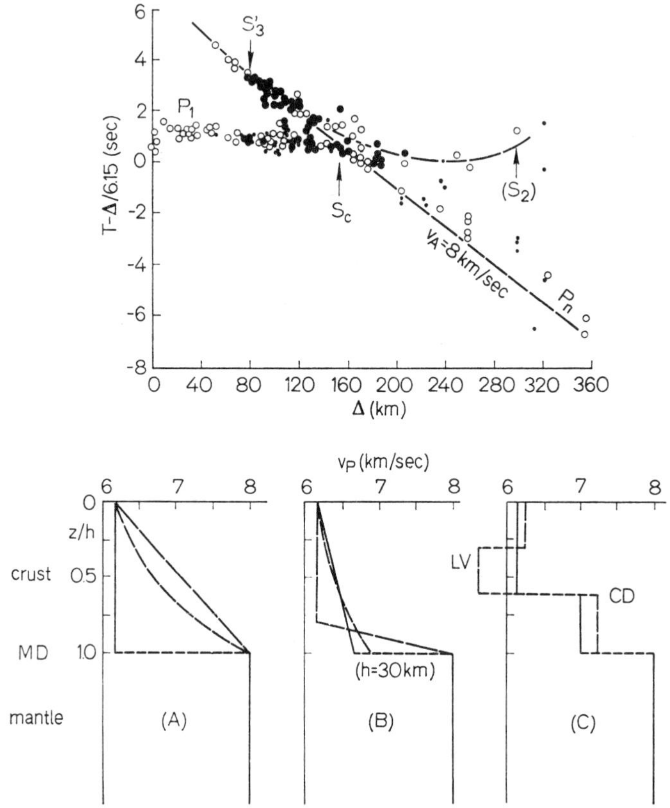

Fig. 6-3. Upper diagram: Time-distance plot of reduced travel-times, for seismic explosion studies in the eastern United States (Chesapeake Bay region, Maryland). (For a direct comparison with the unreduced $T(\Delta)$-curves of Fig. 6-2 rotate the upper diagram by 45° anti-clockwise.) It comprises the results of numerous individual shots at different locations. The crustal P_1-phase (apparent velocity ≈ 6.15 km/sec) is overtaken at about 160 km distance from the explosion site by the faster subcrustal P_n-phase (cf. Fig. 6-2 for notations). The cluster of extra strong second arrivals (full circles) between 80 and 120 km is interpreted by *critical reflections* from the M. discontinuity (MD) or an equivalent high-gradient zone at the bottom of the crust. Lower diagrams: Series of considered velocity distributions (*h:* depth of the M. discontinuity). All models explain equally well the travel-times of first arrivals, but only the models (B) and (C) also explain those of later arrivals. Hence, the models (A) are rejected. The models (C) involve intra-crustal first-order discontinuities (Conrad discontinuity CD and low-velocity layer LV) and require various separate loops in the $T(\Delta)$-curve which are not discernible in the upper diagram. Hence, the models (C) represent an over-interpretation of the available data. The models (B) are adequate. Thus, when the M. discontinuity is a first-order discontinuity, there has to be some gradual velocity increase in the overlying crust. (Adapted from TUVE, TATEL, and HART, 1954)

The apparent P_1-velocity may be somewhat reduced near the explosion site, yielding a slightly curved $T(\Delta)$-curve for $\Delta < 50$ km. This implies that the velocity increases with depth in the uppermost crust (model A, Fig. 6-2).

The crustal P_1-phase is overtaken at some critical distance S_c by the faster subcrustal P_n-*phase* ($v_A \approx 8$ km/sec), a refracted signal from the guided wave following the M. discontinuity. The P_n-phase prevails as first arrival over many hundred kilometers with a strikingly constant apparent velocity. Hence, the subcrustal velocity gradient appears to be exceedingly small.

Travel times of these usually well documented first arrivals are not sufficient to reveal the crustal velocity structure in any detail. They indicate merely (i) that the P-wave velocity increases from a subsurface value of 4 or 5 km/sec to 6 km sec within the upper 5 to 10 km of the crust, (ii) that the velocity is then either constant or increases slightly down to a depth of 20 km (the deepest point of penetration of the P_1-phase when it is a first arrival), and (iii) that the steep ascent to a subcrustal velocity of 8 km/sec takes place at a depth h, given by the critical distance S_c ($h \approx 30$km for $S_c = 160$ km).

There is uncertainty about the lower part of the crust, because P_1-signals which have their deepest point between 20 and 30 km are overtaken by the subcrustal P_n-phase before they reach the surface. Hence, they appear as one of the later arrivals in the seismogram.

In some areas prominent second arrivals almost certainly represent critical reflections from the M. discontinuity. Their combined analysis with first arrivals proves that the crustal velocity in these areas cannot be uniform but that it must increase gradually with depth at some intermediate level (Fig. 6-3).

Occasionally it seems justified to assume between P_1 and P_n an intermediate P^*-*phase* with a distinctly different apparent velocity of about 7 km/sec. It may appear as first arrival within a limited distance range, but it is most commonly found among the later arrivals.

The P^*-phase is usually explained by refractions from an intra-crustal CONRAD *discontinuity*, subdividing the crust into an upper *granite layer* ($v_P = 6.1$ km/sec) and a lower *gabbro* or *basalt layer* (v_P between 6.4 and 7 km/sec). This concept of a layered crust can be disputed on the ground that the complicated picture conveyed by later arrivals rarely allows an unambiguous identification of the required intermediate refraction phase. The Conrad discontinuity, moreover, is not a clearly discernible world-wide characteristic of the crust.

Some investigators see in otherwise unexplained later arrivals indications for an *intra-crustal low-velocity layer*. These arrivals show a slight lag with respect to the P_1-phase but not enough to be explainable as critical reflections from the M. discontinuity (branch $S_3'' - S_2''$ in model D, Fig. 6-2). MÜLLER and LANDISMAN (1966) re-interpreted in this way the travel-time curves of various large explosions and postulate a world-wide low-speed zone of 3 km thickness ($v_p = 5.5$ km/sec) at 10 km depth. GIESE *et al.* (1967) found indications of a similar velocity inversion below the western Alps.

Of particular interest among later arrivals are S-wave signals and, close to the explosion site, near-vertical reflections from the M. discontinuity or from intra-crustal discontinuities (branch $S_F - S_3'$, model C, Fig. 6-2). The correct identification of near-vertical reflections of truly deep origin is complicated by the possibility of multiple reflections within the uppermost geological strata. Nevertheless, they have been shown to occur in certain areas. The depth range around 15 km appears here to be remarkably void of reflecting discontinuities or steep gradient zones, indicating a fair petrological homogeneity (Liebscher, 1962; Dohr and Fuchs, 1967).

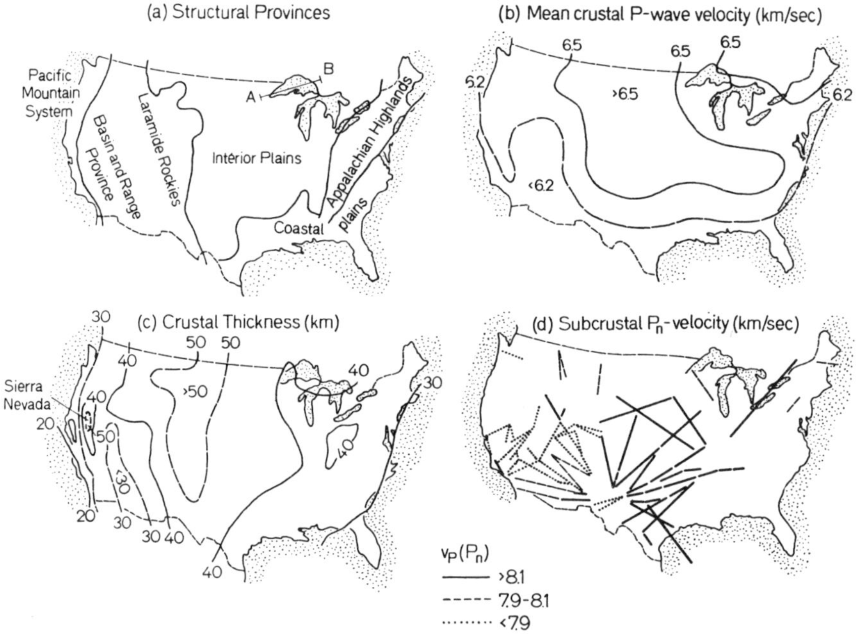

Fig. 6-4a–d. Regional variations of crustal and sub-crustal conditions in the continental United States as inferred from explosion seismology. The mean crustal velocity (b) does not show a visible correlation to structural surface features (a), but the subcrustal velocity (d) does. (Lines in map *d* indicate the position of reversed refraction profiles and the magnitude of the sub-crustal velocity measured on them.) Subcrustal velocities are consistently smaller in the western region of young mountain building than in the central region of long tectonic stability. The changing depth of the M. discontinuity (c) shows that the Sierra Nevada and the eastern front of the Rocky Mountains are supported by a thicker than average crust (Airy's concept of isostasy, Sec. III, d). The thin crust beneath the southwestern Basin and Range province coincides with a low in crustal and sub-crustal velocities. Taking the latter as indicative for lower than average density [Eq. (21)], the surface relief of this province appears to be compensated by a low-density (and perhaps "hot") crust and uppermost mantle (Pratt's concept of isostasy). [(b) and (c) from Pakiser and Steinhart, 1964; (d) from Herrin and Taggart, 1962]

The evaluation of the *S*-mode travel-time curves gives 3.5 km/sec as the representative shear-wave velocity for the crust and 4.7 km/sec for the uppermost mantle.

Deviations from this standard crust will be discussed in terms of the three comprehensive parameters: (i) depth of the M. discontinuity, (ii) mean crustal velocity, and (iii) subcrustal velocity (Fig. 6-4, 6-5). Crustal material is presumably lighter than the underlying mantle (Sec. II, b). Hence, Airy's concept of an isostatic "floating" crust leads us to exspect a thin crust beneath ocean basins, a crust of

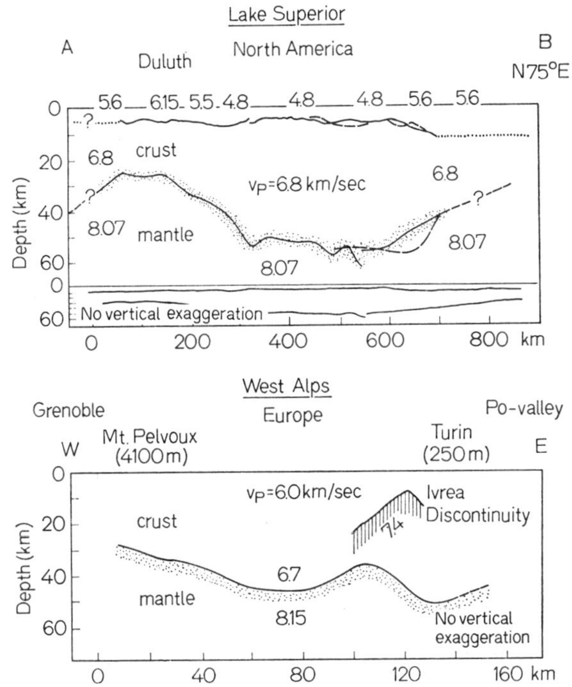

Fig. 6-5. Crustal structures along selected profiles of detailed seismic explosion studies, crossing areas of long-lasting tectonic stability (Lake Superior) and recent mountain building (West Alps). Cf. Fig. 6-4, map *a*, for the exact location of the upper profile. Both cross-sections show a crust of remarkable complexity and reveal that the M. discontinuity can have considerable local relief, even under stable shields. The unusually high mean crustal velocity beneath Lake Superior (6.8 km/sec) indicates that the crust is here composed of more mafic material than elsewhere. The "Ivreabody" at the eastern slope of the arc of the West Alps is an intra-crustal high-velocity lense of mafic to ultramafic material (cf. Fig. 6-19). [From SMITH *et al.* (1966) and FUCHS *et al.* (1963)]

moderate thickness beneath continental lowlands, and a thick crust beneath high plateaus or mountain belts (Sec. III, d).

World-wide explosion studies on land and at sea confirm the general validity of this concept, at least on a grand scale. The standard continental crust of 30 km thickness stands in clear contrast to an average oceanic crust of 5—10 km. Many young mountain belts, among them the European Alps, the Sierra Nevada of North America and the Andes of South America, have indeed compensating "roots" of light crustal material, so that the M. discontinuity sinks here to a depth of 50 km or more, if it is discernible at all. However, a comparable root is missing for instance under most of the Rocky Mountain system of North America, while the adjacent Great Plains, a stable region of moderate elevation, is underlain by an unusually thick crust (Fig. 6-4).

In areas where the lateral variations in mean crustal velocity are well known, we do not find any marked correlation, neither to variations in crustal thickness nor to geological or topographic surface features. The opposite is true of regional differences in subcrustal velocity which are clearly bound to structural provinces. Lower than normal values ($v_p \approx 7.8$ km/sec) are predominant in regions of recent tectonic and magmatic activity, as in Japan (JEFFREYS, 1962) and in the western United States (HERRIN and TAGGART, 1962; PAKISER and ZIETZ, 1965). High subcrustal velocites (8.2 km/sec) are characteristic of stable shield areas.

e) Seismic Velocities in Mantle and Core

The subcrustal P_n-phase of compressional waves remains first arrival up to epicentral distances near 15° (measuring Δ henceforth in degrees on a great circle, $1° \triangleq 111$ km). There is usually a strikingly rapid decrease in the P_n-amplitude which means that this phase is hardly discernible beyond 8°. When strong signals set in again as the *P-phase* near 15°, first as second arrival, we observe a marked bending of their travel-time curve toward higher apparent velocities.

The Jeffreys Bullen curves put a distinct break into the slope of the P-phase travel-time curve at 20°. Accordingly, JEFFREYS' original interpretation, as shown in Fig. 6-6, assumes for the upper mantle a smooth uniform velocity increase down to 410 km depth. Here the velocity gradient changes abruptly to a higher value corresponding to the presumed discontinuous change of apparent velocity of the P-phase.

GUTENBERG offered an alternative interpretation as long ago as 1926. He regards the extremely small amplitude of P_n-arrivals between 8 and 15° as evidence for a "shadow zone", resulting from the geometric spreading of seismic rays by an *upper mantle low-velocity layer*. The P-wave velocity is assumed to decrease from 8.1 km/sec beneath the M. discontinuity to a minimum of 7.9 km/sec between 100 and 200 km depth. Here a steep rise sets in, so that at greater depth GUTENBERG's solution merges into JEFFREYS's solution (Fig. 6-6).

In this way the strong P-phase, beginning at 15°, can be explained by delayed signals from a high-gradient zone (or discontinuity) beneath the low-velocity layer (model D, Fig. 6-2). This would imply, however, that the P_n-phase is cut off at a certain epicentral distance and that the P-phase sets in on a disconnected branch in the $T(\Delta)$-plot. Yet neither the cut-off nor the existence of separate P_n and P travel-time branches are clearly established facts. This does not necessarily disprove GUTENBERG's interpretation, since the cut-off could occur when the P_n-phase is no longer first arrival and therefore hardly observable anyway.

The travel time of shear waves, on the other hand, provides unambiguous evidence for at least one S-wave velocity minimum within the uppermost 200 km of the mantle. The lag of the deeply penetrating S-phase relative to subcrustal S_n-arrivals is clearly established for many regions, amounting occasionally to more than 10 seconds (LEHMANN, 1961; IBRAHIM and NUTTLI, 1967).

Additional support for GUTENBERG's interpretation comes from the dispersion of long-period surface waves. Their phase velocity for a given period is sufficiently reduced by an upper mantle low-velocity layer to enable a distinction to be made between GUTENBERG's and JEFFREYS's velocity curves (Fig. 6-7). A steady velocity increase in the upper mantle leads to acceptable dispersion curves for neither Love nor Rayleigh waves, while the inclusion of a velocity minimum does do this. Depth

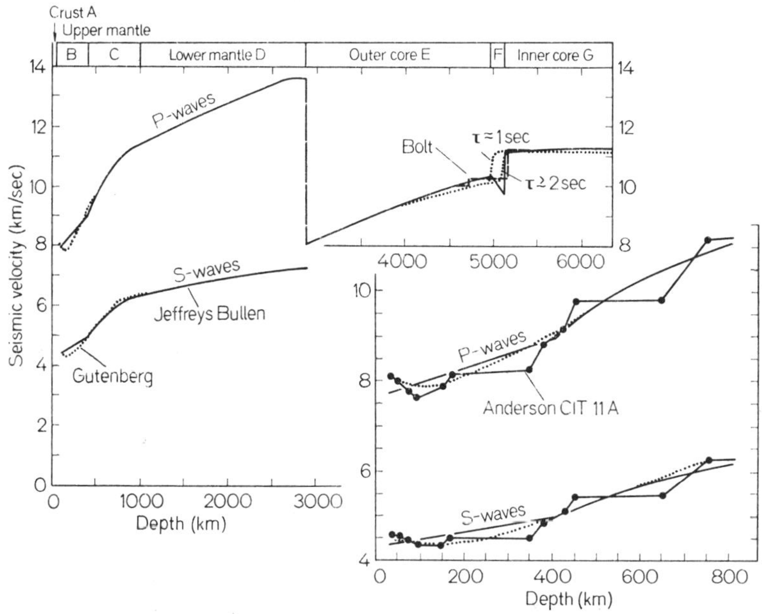

Fig. 6-6. BULLEN's subdivisions of the Earth's interior together with the standard velocity curves for *P*- and *S*-waves. The various curves differ most strongly in the upper mantle (see insert to the right) and in the transition zone *F* of the core (τ: period of seismic signal). Problematic is also the lowest portion of region *D* which is shown here with a zero velocity gradient. Particular upper mantle features are GUTENBERG's velocity minimum near 100 km depth and the 2nd order discontinuities near 400 and 700 km depth in ANDERSON's velocity distribution. The latter evolved from the study of surface wave dispersion (cf. Fig. 6-7), the Jeffreys-Bullen and Gutenberg distributions from body wave data. [From BULLEN (1963, p. 238), GUTENBERG (1953, 1958), BOLT (1964), and ANDERSON (1967, p. 363)]

and velocity within the low-speed layer could be fixed within reasonable limits by trial and error fitting of theoretical and empirical dispersion curves (ANDERSON's model CIT 11 A, Fig. 6-6).

Dispersed surface waves also give comprehensive information about regional variations in the upper mantle structure. There are significant and reproducible discrepancies in dispersion when surface waves arrive at a certain site from different directions. This cannot be explained by crustal differences alone, e.g. by the contrast between oceanic and continental crust, and implies that sub-crustal peculiarities exist along the path which the particular signal takes from the source to the receiver.

ANDERSON (1967) was able to distinguish in this way three structural units, all including a low-velocity layer for *S*-waves. He postulates beneath oceans a pronounced v_S-minimum of 4.4 km/sec (sub-crustal value 4.7 km/sec) at 120 km depth, beneath stable shields a hardly noticeable minimum of 4.6 km/sec at 250 km depth. Continental regions of recent tectonic activity lie somewhere between the oceanic and the stable shield substructure.

Improved instrumental techniques and the availability of nuclear underground explosions as sources have led in recent years to greatly refined travel-time curves

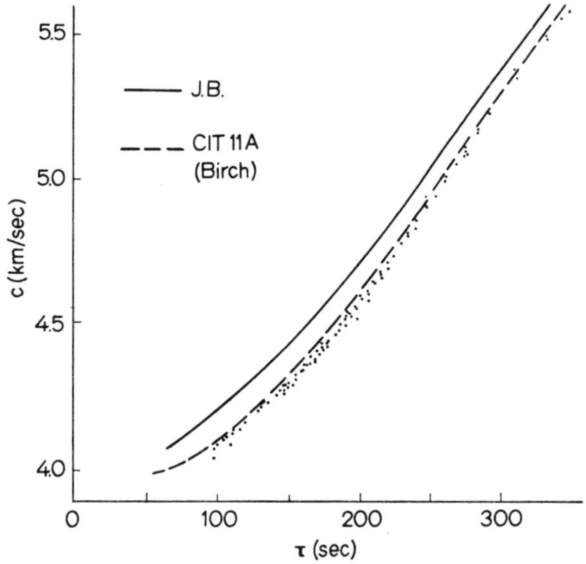

Fig. 6-7. Phase velocity c of Rayleigh waves as function of their period τ. The empirical velocities (dots) fall systematically below the calculated dispersion curve $c(\tau)$, when the calculations are based on the Jeffreys-Bullen velocity distribution. The distribution CIT 11 of Fig. 6-6 which incorporates an upper mantle low-velocity layer leads to a satisfactory fit between observed and calculated dispersion, using a density distribution according to BIRCH's linear velocity-density relation [Eq. (21)]. (From ANDERSON, 1967)

for epicentral distances of up to 25°. Observation with a closely spaced array of seismometers at a single site also made it possible to determine by cross-correlation the direction and apparent velocity of the incoming signals with greater accuracy than ever before (cf. NUTTLI, 1963).

JEFFREYS' original assumption of a single 2nd order discontinuity at 20° now appears to be an oversimplification of a complicated and regionally variable behaviour of the P and S-phase between 15° and 25°. In some areas, e.g. the central and western United States, there is evidence of at least two breaks or even loops in the $T(\Delta)$-curve. JEFFREYS' concept of a steeply increasing velocity below 400 km depth remains in essence valid, but it dissolves into a number of (second-order) discontinuities between 200 and 600 km (model CIT 11 A, Fig. 6-6) which may or may not be traceable from continent to continent. They establish BULLEN's region C following the low-velocity layer as a world-wide transition zone where the P- and S-wave velocities rise more rapidly than anywhere else in the mantle.

This increase levels off near 1,000 km depth where the *lower mantle* begins. After a smooth rise, maximum velocities of 13.7 and 7.3 km/sec respectively are reached 100 to 200 km above the mantle-core interface. In the remaining innermost portion of the mantle the velocities either remain constant or they decrease slightly with depth until the P-wave velocity drops abruptly at 2,900 km depth to 8 km/sec inside the core.

This velocity decrease causes the P-phase to disappear at 103° epicentral distance, where the shadow zone of the core begins. The first P-waves traversing the core re-appear at 143°. Shear waves are not transmitted through the core. The exact depth of the mantle-core boundary is known to within ten kilometers, the greatest uncertainty arising from the peculiar behaviour of mantle velocities just above this boundary. PRESS (1968) concluded, on the basis of statistical trial-and-error experiments to find the best fitting Earth model for all available seismic and free oscillation data, that the core radius is 15—20 km greater than so far assumed.

In 1936 Mme. LEHMANN called attention to very weak P-signals within the shadow of the core, more or less in continuation of the mantle P-phase. They appear in comparable amplitude over the full width of the shadow and cannot therefore be explained by diffracted waves penetrating into the core's geometric shadow. Hence, they have to be "bent-up" signals from some high-gradient zone within the core itself, which now defines the boundary between the *outer* and *inner core* (cf. branch $S_3''-S_2''$ in model D, Fig. 6-2).

The standard velocity distributions of Fig. 6-6 show the resulting rapid increase of P-wave velocities at 5,100 km depth, i.e. at 1,300 km distance from the Earth's center. Arguments for the solidity of the inner core will be discussed in Sec. II, d. It should be noted here that there is no seismic proof that shear waves could traverse it.

f) Travel Time and Amplitude Anomalies

Even minor lateral variations of crustal or subcrustal velocity complicate the ray pattern and travel time of body waves to such an extent that the inversion of their $T(\Delta)$-curves cannot yield more than an average picture of the true velocity distribution. Hence, investigations in areas of obvious lateral complexity are supplemented effectively by a regional comparison of the travel times of those waves which are known to traverse the crust and uppermost mantle in a nearly vertical direction. This applies to near-vertical reflections from first-order discontinuities and to arrivals from deep-focus earthquakes observed either close to or far away from the epicenter.

The travel times of near-vertical reflections from the M. discontinuity show considerable statistical scatter from site to site and cannot give information about sub-crustal conditions. Hence, current interest focuses on the travel-time residuals ΔT of arrivals from deep-focus earthquakes, defined as the remainder of the observed minus the calculated travel time. The latter is deduced from a given epicentral distance of the recording site on the basis of the standard Jeffreys-Bullen curves (Sec. I, b).

Let $v(z)=\bar{v}(z)-\Delta v(z)$ be the velocity-depth profile benath the recording site, Δv denoting local deviations from the global mean $\bar{v}$ as function of depth. If z_0 is the maximum depth to which such deviations occur, the delay of a vertical ray is

$$\Delta T=\int_0^{z_0}\frac{\Delta v}{\bar{v}^2}\,dz \tag{9}$$

for $\Delta v\ll\bar{v}$.

Fig. 6-8 shows a map of travel time residuals for southern Peru. It exhibits a clear pattern of positive and negative values. HERRIN and TAGGART's residual map for the United States reveals a similarly consistent behaviour of the P_n-phase of artificial explosions (Fig. 6-9). In both cases we find positive residuals in zones of

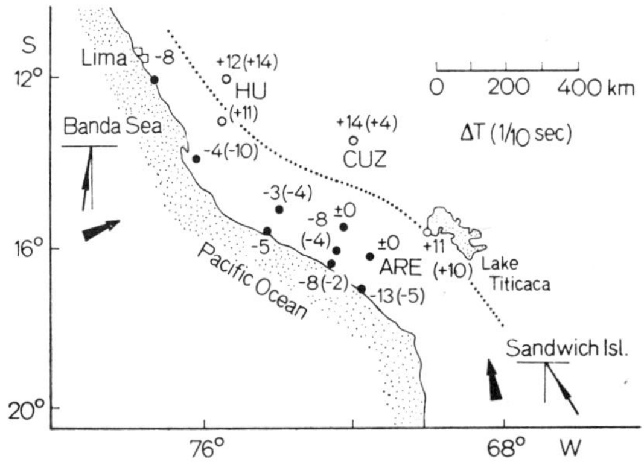

Fig. 6-8. Seismic travel-time anomaly in southern Peru (*ARE* Arequipa, *CUZ* Cuzco, *HU* Huancayo). The numbers refer to the travel-time residuals (cf. text) for first arrivals from two distant deep-focus earth-quakes as recorded at the various sites. The first numbers are the residuals (minus 4.1 sec) of the Banda Sea quake (focal depth $H = 326$ km), the numbers in parenthesis those of the Sandwich Islands quake ($H = 120$ km), both given in tenth of a second. The arrows indicate the direction of wave propagation and their angle of incidence. The mountain belt of the Andes is characterized by positive residuals, i.e. the incoming waves are slowed down by a thicker than normal crust, lower than average internal velocities or a combination of both. Dotted line indicates general trend of the Andes. (From SACKS *et al.*, 1966)

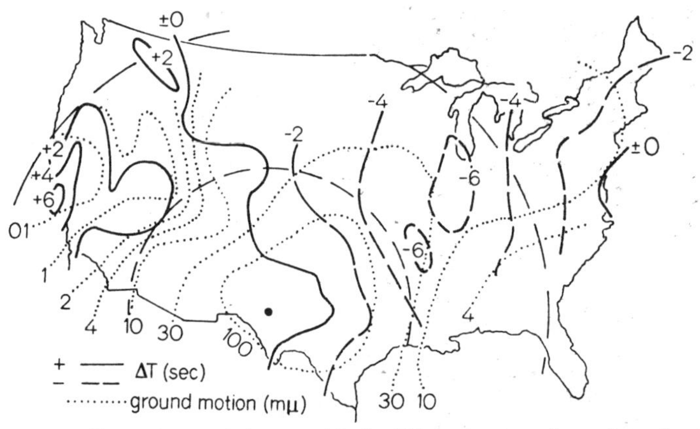

Fig. 6-9. Contours of equal travel-time residual ΔT and ground motion (peak-to-peak deflection) for first arrivals from the "Gnome" experiment, an underground explosion near the eastern front of the Rocky Mountains. Circles around the explosion site (large dot) give the 1,000 and 2,000 km epicentral distance. — Positive residuals and small first ground motion prevail westward of the explosion site, negative residuals and strong ground motion to the east. Hence, the substructure of the mobile western regions of young mountain building shows high seismic absorption and low *P*-wave velocity. This reflects either a general reduction of strength of the upper mantle or the presence of numerous interspersed "magma pockets" filled with partially molten material. [From HERRIN and TAGGART (1962)]

young mountain building. This delay implies a thicker than normal crust, a lower than normal sub-crustal velocity, or a combination of both (cf. concluding remark of Sec. III, d).

These regions are also distinguished by unexpectedly low amplitudes of seismic signals. The Carnegie Andean expedition of 1957 discovered a striking attenuation of seismic waves which propagate across the mountain range in comparison with those propagating parallel to its trend. A similar, although less dramatic directional dependence of the amplitude reduction exists in the continental United States (HERRIN and TAGGART, 1962; JORDAN *et al.*, 1965). Lines of equal amplitude — which would be concentric circles around the epicenter under ordinary conditions — are clearly deformed and reflect an overall seismic attenuation which is higher in the mountainous western states than in the midwestern plains (Fig. 6-9). This difference can not be explained in terms of ray seismology alone. It implies that the dissipation of seismic energy is unusually great under the Peruvian Andes as well as under the Rocky Mountain system. It establishes their substructures as zones of unusually large non-elastic deformability (Sec. II, g).

g) Inferences from free Oscillations and Earth Tides

After major earthquakes the frequency spectrum of seismic energy shows a sequence of spectral peaks at discrete frequencies. They have been identified as *eigen-frequencies* of the solid Earth. Hence, large earthquakes excite the Earth's body as a whole to vibrate in free oscillations like a bell struck by the clapper.

There are two principal classes of vibratory motion: *Torsional* (= *toroidal*) *T-modes* vibrate without radial displacement or compression. *Spheroidal S-modes* combine radial with tangential motion and produce an alternating compression and dilatation of matter. The resulting up-and-down movement of the ground makes *S*-modes detectable with recording gravimeters or tiltmeters of adequate stability and sensitivity. The accelerations associated with the *S*-modes are of the order of 10^{-4} cm/sec^2 (0.1 milligal), the variations of the vertical in the order of 0.1 seconds of arc.

We may regard free oscillations as "stationary" seismic surface waves, each mode having its characteristic pattern of *nodes* of zero motion. Their order n refers to the number of nodal lines which subdivide the surface motion. A second index m gives the number of internal nodal surfaces. Each fundamental S or T-mode ($m=0$) of the order n has an infinite number of more rapidly oscillating overtones ($m=1, 2, \ldots$).

The observed spectrum of eigen-frequencies f_n^m is an overall measure of the internal distribution of density and elastic properties. The anelastic or viscous properties of the Earth's interior produce a dampening of the free oscillations which occurs exponentially according to $\exp(-f_n^m\, t/Q)$, Q^{-1} being the specific dissipation constant of strain energy [Eq. (35)].

The fundamental S-mode of the solid Earth ($n=2$, $m=0$) oscillates at the rate of one cycle in 54 minutes between an oblate and a prolate ellipsoid (Fig. 6-10). The highest resolvable modes extend to one cycle in 3 minutes and are thus well within the continuous spectrum of propagating surface waves.

Free oscillations escaped observation until appropriate instruments and high speed computers for data reduction and interpretation became available. The first earthquake to be successfully analysed was the great Chilean earthquake of May 1960 which yielded eigenfrequencies of some seventy modes with an average precision of 1 part in 500. They were in excellent agreement with theoretical predictions made some years earlier by PEKERIS *et al.* (1958) on the basis of the Jeffreys-Bullen Earth model, an impressive proof for the basic soundness of seismological deductions about the internal constitution of the Earth. The Alaska Good Friday earthquake of 1964 excited the same modes of free oscillations once again. Inferences from these and other quakes are as follows:

Fundamental *S*-modes of low order ($n \leq 16$) affect more or less the entire mantle and core. Their frequencies favor BULLEN's density *model A* over his alternative *model B*, when both are used in conjunction with either JEFFREYS' or GUTENBERG's velocity distribution (Fig. 6-10).

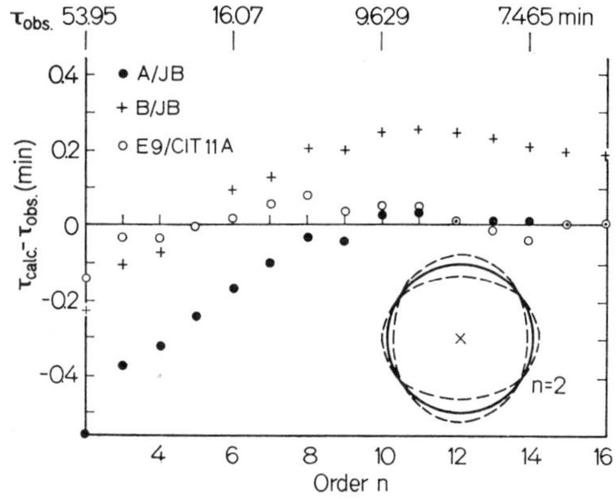

Fig. 6-10. Interpretation of the spheroidal modes of the Earth's free oscillations up to the 16th order. The diagram shows the difference between observed and calculated eigenperiods τ for three different Earth models. The combination of the Jeffreys-Bullen velocity distribution with the Bullen A density distribution (Fig. 6-6 and Fig. 6-12) gives negligible residuals for the high-order modes but considerable discrepancies at low orders. The use of BULLEN's density model B reverses the situation. ANDERSON's density model *E* 9 (Fig. 6-12) has been designed especially to yield minimum residuals when used in combination with the velocity distribution *CIT* 11*A*, Fig. 6-6. [From PRESS (1966) and ANDERSON (1967).] The second spheroidal mode oscillates between the indicated figures of an oblate and prolongated spheroid

T-modes, higher order *S*-modes, and all overtones are sensitive to mantle properties because of their limited downward extent. (The shear motion of the *T*-modes naturally ends at the mantle-core boundary.) Their frequencies supplement the conclusions drawn from the dispersion of surface waves and confirm the existence of an *S*-wave velocity minimum near 150 km depth.

Amplitude and phase of free oscillations, including their regional variability, are usually not considered. Hence, the conclusions drawn from the frequency spectrum of free oscillations refer to mean internal properties of spherical symmetry. Contrary viewpoints guide the investigations of *Earth tides*. These are forced oscillations which the Earth undergoes because of the tidal pull of Moon and Sun. Tidal periods are considerably longer than the gravest mode of free oscillations, thus eliminating resonance problems which plague the study of oceanic and atmospheric tides.

The response of the Earth to these external forces leads to measurable deformations of its surface which can be recorded as tidal-periodic variations of gravity

($\varkappa$ 20% of the primary tidal pull) and tilt at a fixed site. Their amplitude is again quite small, amounting to less than 1 milligal and to fractions of a second of arc, respectively. It was already noted by Lord KELVIN that the tidal deformation of the Earth body as a whole is comparable to that of a sphere made of steel (Fig. 6-14).

A regional intercomparison of the various tidal harmonics reveals significant variations from place to place in amplitude and phase. It is thought that the tidal deformation of the upper layers (including crust and uppermost mantle) occurs in rigid blocks, separated by "soft zones" which act as movable hinges. It would be premature, however, to correlate these zones with structural features of known tectonic mobility, such as faults and rift systems, since there is a host of secondary effects. Prominent among them is the tidal movement of ocean water which leads to oscillatory load deformations near coastlines. This effect may be noticeable even at some distance from the coast (cf. MELCHIOR, 1966; KUO and EWING, 1966).

II. Density, Elasticity, Viscosity

a) Theoretical Density-Seismic Velocity Relations

The task of finding the internal density distribution is by no means a straightforward matter and numerous uncertainties persist up to this date. In the following, only changes of density with depth, $\varrho(z)$, are considered, leaving a discussion of lateral imbalances to Sec. III on gravity and isostasy.

The most immediate effect of the internal masses is the surface gravity g_0 which allows a ready calculation of the total mass M of the Earth (Sec. III, a). The overall distribution of mass is contained in the mean moment of inertia C_0 [Eq. (37,41)]. The exact knowledge of M and C_0 provides important integral constraints upon acceptable density distributions as obtained from seismic body wave data.

As seen from Eqs. (2) and (3) the P and S-wave velocities yield immediately the ratios K_S/ϱ and μ_S/ϱ as a function of depth. The elimination of ϱ itself, however, requires the use of an additional functional dependence of ϱ from one of the elastic constants or velocities. Surface wave dispersion and the periods of free oscillations provide an ultimate test to select from a family of possible density distributions the most likely one.

The differential downward increase of hydrostatic pressure p [cf. comment on Eq. (1)] in gravitating homogeneous matter is

$$\frac{dp}{dz} = \varrho g \tag{10}$$

with g denoting the graviational acceleration at the considered level z beneath the surface. In the case of a spherical Earth of radius a, NEWTON's law of attraction yields

$$g(r) = \frac{4\pi G}{r^2} \int_0^r \varrho(\hat{r})\, \hat{r}^2\, d\hat{r}$$

with G as constant of gravity and $r=(a-z)$ as distance from the Earth's center. The downward increase in gravity readily follows by differentiation as

$$\frac{dg}{dz}=-4\pi G\varrho+\frac{2g}{a-z}. \tag{11}$$

If ϱ is constant, dg/dz is likewise constant, given by $-\frac{4}{3}\pi\varrho G$.

The third gradient to be considered is $d\varrho/dz$. In homogeneous matter the downward change in density is determined by the balance between self-compression under its own weight and thermal expansion due to the general downward increase of temperature T. The total change in density is accordingly

$$d\varrho=\left(\frac{\partial\varrho}{\partial p}\right)_T dp+\left(\frac{\partial\varrho}{\partial T}\right)_p dT=\varrho\left[\frac{dp}{K_T}-\alpha dT\right]; \tag{12}$$

$$K_T=\varrho\left(\frac{\partial p}{\partial\varrho}\right)_T \tag{12a}$$

denotes the isothermal bulk modulus (incompressibility) of the material and α its thermal expansion coefficient at constant pressure as defined in Eq. (105).

Isothermal and adiabatic incompressibility are connected by the thermodynamical relation

$$K_S=K_T(1+\alpha\gamma T) \tag{13}$$

(T: absolute temperature)

with γ as thermal Grüneisen ratio [Eq. (106)]. This ratio, a dimensionless number, is for solids near 2. Because the thermal expansion of silicates is small, K_S will exceed K_T by just a few percent within the Earth (Table 6-4). Eq. (13) reduces Eq. (12) to

$$\frac{d\varrho}{dp}=\varrho/K_S+\alpha\varrho\left(\frac{\gamma T}{K_S}-\frac{dT}{dp}\right). \tag{14}$$

Self-compression of homogeneous matter is termed *adiabatic* when the expression in parenthesis is zero. This defines in combination with (10)

$$\tau_0=\gamma g T\varrho/K_S \tag{15}$$

as *adiabatic temperature gradient* and

$$\tau=\frac{dT}{dz}-\tau_0 \tag{15a}$$

as *super-adiabatic* temperature gradient.

Inserting τ along with the bulk velocity v_B from Eq. (3) and dp/dz from Eq. (10) into (14) yields the generalized Williamson-Adams (W.-A.) differential equation

$$\frac{1}{\varrho}\frac{d\varrho}{dz}=\frac{g}{v_B^2}-\alpha\tau \tag{16}$$

which establishes the desired connection between density and seismic properties in the case of homogeneous gravitating material.

A combined downward integration of the W.-A. equation and Eq. (11) enables us to find $\varrho(z)$ for any given velocity distribution $v_B(z)$, using proper estimates for the usually small superadiabatic term ($\alpha\tau$). The gravity at the upper boundary of the depth range considered must be known. As arbitrary integration constant we may use the density at some specified depth, the total mass or the moment of inertia for the depth range of integration.

The general application of this method to find $\varrho(z)$ is limited by the problematic condition of homogeneity. This deficiency led BIRCH (1952) to introduce a *homogeneity test*, based on the internal pressure derivative of the isothermal bulk modulus,

$$\beta_T = \left(\frac{\partial K_T}{\partial p}\right)_T, \tag{17}$$

a dimensionless number of important theoretical implications. It stands in the following relation to density and bulk velocity:

Differentiate $K_s = \varrho\, v_B^2$ [Eq. (3)] with respect to p and replace $v_B^2\, d\varrho/dp$ by $1 - \alpha\tau v_B^2/g$ [Eqs. (10) and (16)]. This gives

$$\frac{dK_S}{dp} + \frac{\alpha\tau}{\varrho g} K_S = 1 + 2\varrho v_B \frac{dv_B}{dp}. \tag{18}$$

The total change of K_s as a function of pressure and temperature is

$$dK_S = \left[\left(\frac{\partial K_S}{\partial p}\right)_T + \left(\frac{\partial K_S}{\partial T}\right)_p \frac{dT}{d_p}\right] dp \tag{19}$$

with

$$\left(\frac{\partial K_S}{\partial p}\right)_T = \beta_T + \gamma T \left(\frac{\partial [\alpha K_T]}{\partial p}\right)_T + \alpha K_T T \left(\frac{\partial \gamma}{\partial p}\right)_T$$

$$\left(\frac{\partial K_S}{\partial T}\right)_p = \left(\frac{\partial K_T}{\partial T}\right)_p \frac{K_S}{K_T} + K_T \left[\alpha\gamma + T\left(\frac{\partial [\alpha\gamma]}{\partial T}\right)_p\right]$$

in virtue of Eq. (13); $dT/dp = (\tau_0 + \tau)/(\varrho g)$ as seen from Eq. (10) and (15a).

The temperature derivative of K_T at constant pressure is written in the form

$$\left(\frac{\partial K_T}{\partial T}\right)_p = -\,\delta_T\, \alpha\, K_T$$

with δ_T as dimensionless Anderson-Grüneisen parameter (cf. ANDERSEN, 1967). Straightforward simplifications become possible by adopting the approximations

$$\left(\frac{\partial [\alpha K_T]}{\partial p}\right)_T \approx 0, \quad \left(\frac{\partial \gamma}{\partial T}\right)_p \approx 0, \quad \left(\frac{\partial \alpha}{\partial T}\right)_p \approx \alpha^2 (\delta_T - 1)$$

in accordance with high-pressure research on solids (ANDERSON, 1967; BIRCH, 1968). They yield when inserted above

$$\left(\frac{\partial K_S}{\partial p}\right)_T = \beta_T + \alpha K_T\, T \left(\frac{\partial \gamma}{\partial p}\right)_T$$

$$\left(\frac{\partial K_S}{\partial T}\right)_p = \alpha K_S (\varkappa - 1)$$

with

$$\varkappa = \frac{1 + \gamma - \delta_T}{1 + \alpha\gamma T},$$

whence in combination with (18) and (19)

$$\beta_T = 1 + \frac{2}{g}\frac{dv_B}{dz} - \alpha\gamma T \left[\varkappa\left(1 + \frac{\tau}{\tau_0}\right) - 1 + \frac{\varrho}{\gamma}\left(\frac{\partial \gamma}{\partial \varrho}\right)_T\right]. \tag{20}$$

For a first approximation the last product with $(\alpha\gamma T)$ may be omitted, since the value of the parenthesis should be not too far from unity and $(\alpha T) \ll 1$ (Table 6-4). Thus,

$$\beta_T \approx 1 + \frac{2}{g}\frac{dv_B}{dz} \tag{20a}$$

yields in β_T the internal pressure derivative of the isothermal bulk modulus K_T when the internal gravity and bulk velocity gradient are known.

There are theoretical arguments (Sec. II, e) that β_T is either constant (MURNAGHAN) or smoothly decreasing from a value of 4 near the surface to 3.2 at the mantle-core boundary (BIRCH), when the material of the Earth's mantle is homogeneous and self-compressed.

If the application of Eq. (20) to the empirical $v_B(z)$ distribution yields for some intermediate depth strongly divergent values for β_T, the assumption of homogeneity cannot be correct. It implies that the velocity gradient dv_B/dz at this depth is partially due either to a change in chemical composition or to a phase transition within chemically uniform matter (Fig. 6-11).

The transition from a low-density to a high-density phase produces a reduction in deformability and compressibility of the material. The resulting increase of K_S and μ_S outweighs, as a rule, the increase of ϱ. Hence, high-density phases have also higher seismic velocities than their low-density counterparts of the same *mean atomic weight* m (molecular weight divided by the number of constituents in the molecular formula).

BIRCH (1961) proposed on the basis of extensive experimental evidence a linear density-velocity relation for silicatic rocks and minerals,

$$\varrho = a + b\, v_P; \tag{21}$$

a and b are constant parameters for materials of comparable mean atomic weight and thus invariant against phase transitions. Representative values for the constituents of the upper mantle ($m \approx 21$) are $b = 0.38$ (gr./cm³)/(km/sec) and $a = 0.25$ gr./cm³ (BIRCH's 2nd solution, 1964).

b) Density Models for the Earth's Interior

Density values for crustal and sub-crustal matter are chosen on the basis of combined petrological-seismological considerations. Granitic igneous and metamorphic rocks, the probable constituents of the upper crust, have densities between 2.6 and 2.8 gr./cm³. The gradual or stepwise transition to more mafic material in the lower crust (gabbro, amphibolite) is accompanied by a density increase of 0.1 to 0.2 gr./cm³.

Standard mean values are 2.67 gr./cm³ for the upper "granitic" crust, a value often used for the isostatic correction of gravity data (Sec. III, d), 2.75 gr./cm³ for the continental crust and 2.9 gr./cm³ for the oceanic crust. Estimates for subcrustal densities depend on the specific interpretation of the M. discontinuity which may or may not represent a phase boundary. They range from 3.25 (peridotite) to 3.5 (eclogite). Since eclogite cannot be the main constituent of the upper mantle, a proper sub-crustal density value is 3.25 or 3.3 gr./cm³.

WIECHERT's classical subdivision of the Earth's interior into mantle and core attributes to them uniform densities of 4.3 and 12.2 gr./cm³, respectively. He obtained these values from the total mass M, the axial moment of inertia C_0, and the radius of the core as determined by seismology in 1910.

In the light of the subsequent refined analysis of travel-time curves it became clear that the density must increase from the top to the bottom of the mantle and that this increase cannot be due to self-compression of homogeneous matter alone.

This was shown by BULLEN in 1936. He made the reasonable assumption that the density within the mobile core is either constant or increases with depth to ensure gravitational stability. Hence, the axial moment of inertia of the core, C_c, cannot exceed that of a uniform sphere or $0.4\, M_c\, r_c^2$, when M_c and r_c denote its mass and radius. If M_m and C_m denote similarly the mass and the axial moment of inertia of mantle plus crust, $C_c = C_0 - C_m$ and $M_c = M - M_m$. This establishes $C_0 - 0.4\,(M - M_m)\, r_c^2$ as the lower bound for C_m.

Using these constraints for acceptable values of C_m and M_m, BULLEN integrated the W.-A. Eq. (16) with JEFFREYS' velocity distribution, assuming adiabatic self-compression ($\tau = 0$). He obtained 3.8 gr./cm³ as lowest possible density at the top of the mantle. This value is too high from a petrological point of view. BULLEN concluded that the density of the mantle increases at some intermediate level more rapidly than can be explained by adiabatic self-compression of homogeneous matter. The inclusion of any super-adiabatic correction would only tend to strengthen BULLEN's arguments, as is evident from Eq. (16).

The application of the BIRCH-homogeneity test [Eq. (20)] to seismic mantle velocities reveals that the strongest deviation from homogeneity occurs within region C, while region D, the lower mantle, appears as a zone of fair homogeneity

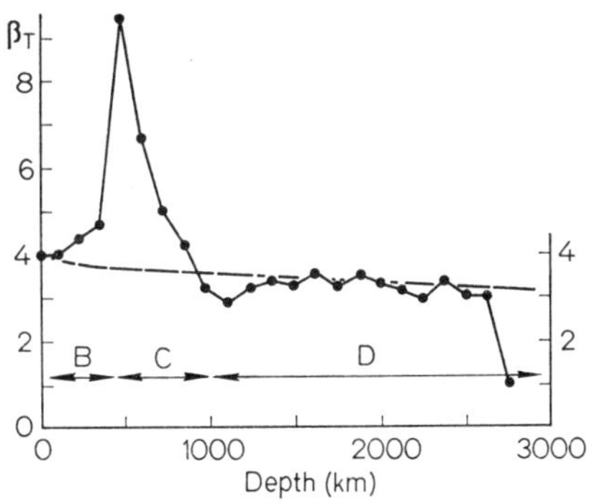

Fig. 6-11. BIRCH's homogeneity test for the Earth's mantle, provided by the pressure derivative β_T of the isothermal bulk modulus K_T. Empirical estimates (dots) have been obtained for the Jeffreys-Bullen velocity distribution (Fig. 6-6) and BULLEN's density model A (Fig. 6-12) according to Eq. (20a). They are compared with a theoretical β_T-curve (dashed line) which has been calculated from BIRCH's equation of state [Eq. (27)] under the assumption of homogeneity. The calculations employed the pressure and the adiabatic bulk modulus (in place of the isothermal modulus) of BULLEN's model A (Fig. 6-13 and 6-14). The transition zone of the upper mantle appears in this presentation as a zone of strongly divergent empirical β_T-values and hence as a zone of maximum mantle inhomogeneity, the lower mantle as a zone of fair homogeneity except perhaps for its innermost portion

(Fig. 6-11). A straightforward application of the W.-A. equation to region D yields an overall density increase of 1 gr./cm³ between 1,000 and 2,900 km for adiabatic conditions. This estimate would be slightly lowered by a super-adiabatic correction. Thus, when we start off with a subcrustal density of 3.3 gr./cm³, the density has to increase by more than 1 gr./cm³ between the M. discontinuity and the 1,000 km level to reach WIECHERT's density of an equivalent uniform mantle. In the best estimates now available a value of 4.6 ± 0.1 gr./cm³ is proposed for the 1,000 km level and a value of 5.5 ± 0.2 gr./cm³ for the bottom of the mantle (Fig. 6-12).

The bulk velocity drops sharply from 10.7 km/sec to 8.1 km/sec at the mantle-core interface (Fig. 6-6). This must be primarily taken as a density jump between a low-density mantle and a high-density core. If the incompressibility K_S is the same on both sides of the interface, the required increase of density would be 1.75-fold, i.e. from 5.5 gr./cm³ at the bottom of the mantle to 9.7 gr./cm³ at the top of the core [Eq. (3)].

We may presume that the core is well mixed and therefore homogeneous, at least in its outer portion. This justifies the use of the W.-A. equation. The result

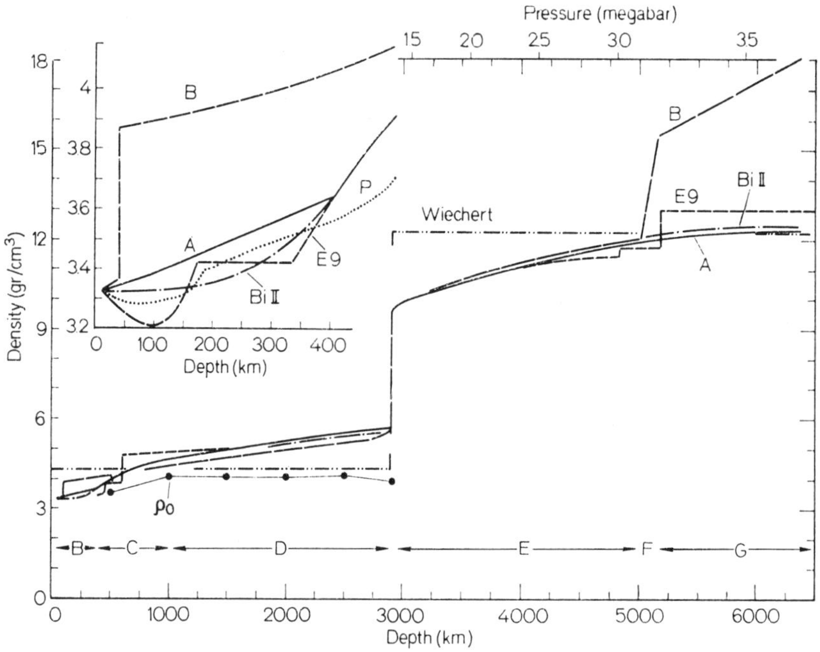

Fig. 6-12. Comparison of various density distributions for the Earth's interior. The straight lines give Wiechert's classical estimates for a high-density core and a low-density mantle, both assumed to be uniform. Bullen's models A and B evolved basically from the Jeffreys-Bullen distribution of P- and S-wave velocities, Anderson's model $E\,9$ from a combined interpretation of body wave data, surface wave dispersion and periods of free oscillations. Birch's model Bi II is based on an empirical velocity-density relation for upper mantle material, Clark and Ringwood's model P on a specific petrological model for the upper mantle, here their peridotite model. Large dots refer to calculated zero-pressure densities, model A, using Birch's formula for isothermal compression [Eq. (26)]. Regions of considerable complexity and uncertainty are the upper mantle (density minimum) and the transition from the outer to the inner core (central density). [From Bullen (1963), Birch (1964), Anderson (1967), and Clark and Ringwood (1964)]

is a 2 gr./cm³ density increase which brings ϱ up to about 11.5 gr./cm³ at 5,000 km depth. The steep increase of the P-wave velocity at the transition from outer to inner core is evidence of a major change in composition or phase, the latter being the more likely choice. However, the contribution of the inner core to the total mass and moment of inertia of the Earth is so small that the central density is quite uncertain.

Fig. 6-12 summarizes the results of four principal types of density calculations. Bullen's classical *model A* evolved from a piecewise integration of the W.-A.

equation through zones of presumed homogeneity, proceeding from a subcrustal density of 3.32 gr./cm³ on the basis of the Jeffreys-Bullen velocity distribution (Fig. 6-6). Excluded from the integration are the transition zones C and F of core and mantle. In this model the density at the Earth's center must exceed a value of 12.3 gr./cm³ but is otherwise freely adjustable. These upper and lower bounds fix in combination with mass and moment of inertia arguments the overall density increase in the C and F region.

BULLEN's alternative *model B* fits into the same distribution of seismic velocities. It evolved from the "pressure-compressibility" hypothesis, that K_S varies smoothly with increasing pressure throughout mantle and core; K_S is assumed in particular to be continuous at the mantle-core boundary. Using again a subcrustal value of 3.32 gr./cm³, the application of the W.-A. equation is restricted to the lower mantle, except for its innermost portion, and to the outer core.

The continuity of K_S yields first estimates for $\varrho = K_S/v_B^2$ and its fractional change

$$\frac{\Delta\varrho}{\varrho} = \frac{\Delta K_S}{K_S} - 2\frac{\Delta v_B}{v_B} \tag{22}$$

for inhomogeneous zones adjoining those of presumed homogeneity. Additional constraints arise from mass and moment of inertia arguments. BULLEN was able to derive in this way 18 gr./cm³ as a direct estimate for the density at the Earth's center. However, the periods of free oscillation favor model A rather than model B (Fig. 6-10).

Both models need amending in their uppermost portion where the now preferred Gutenberg velocity distribution differs most from the Jeffreys-Bullen distribution. Various refined density models for the upper mantle are emerging now from the study of surface wave dispersion and free oscillations (cf. ANDERSON's model E 9, Fig. 6-12).

BIRCH's calculations (1964) proceed from the assumption that inhomogeneities within the upper mantle arise primarily from phase transitions. Hence, he employed the W.-A. equation within the lower mantle, the empirical velocity-density relation (21) within the upper mantle, starting off with a subcrustal value of 3.32 gr./cm³ (BIRCH's 2nd solution). This fixes the parameter a in Eq. (21). The parameter b and the mass of the core are adjusted in such a way as to give acceptable values for mass and moment of inertia of the mantle.

The density calculations of CLARK and RINGWOOD (1964) are based on a specific upper mantle petrology, the *pyrolite* (cf. Sec. VI) or *eclogite* model. They begin at the M. discontinuity with the density of either peridotite (3.25) or eclogite (3.55) and make proper allowance for a gradual downward change of composition to pyroxene pyrolite and ultimately garnet pyrolite at 400 km depth.

The concurrent change of density is deduced directly from the differential equation of state (12) with experimentally determined values for incompressibility and thermal expansion under the presumed thermodynamical conditions. The resulting density distribution depends critically on the balance of self-compression and thermal expansion. The downward density gradient obviously becomes negative when $dT/dp > (K_T\,\alpha)^{-1}$. Inserting reasonable estimates, $K_T = 1.15$ megabar and $\alpha = 3.5 \cdot 10^{-5}\,°\mathrm{C}^{-1}$ (Tab. 6-4, 6-5) yields 25° C/kbar or 7.5° C/km as critical temperature gradient with $dp/dz = 0.33$ kbar/km (Sec. II, c). The actual gradient within the uppermost mantle most likely exceeds this critical value (Fig. 6-30) and a density minimum near 100 km depth as shown by the Clark-Ringwood and the Anderson E 9 model will be the consequence.

c) Gravity and Pressure within the Earth

The change of the gravitational acceleration g with depth follows, for a given density distribution, readily from Eq. (11) by numerical integration, starting with $g = 981$ gal ($=$ cm/sec²) at the surface. Because the downward increase of the

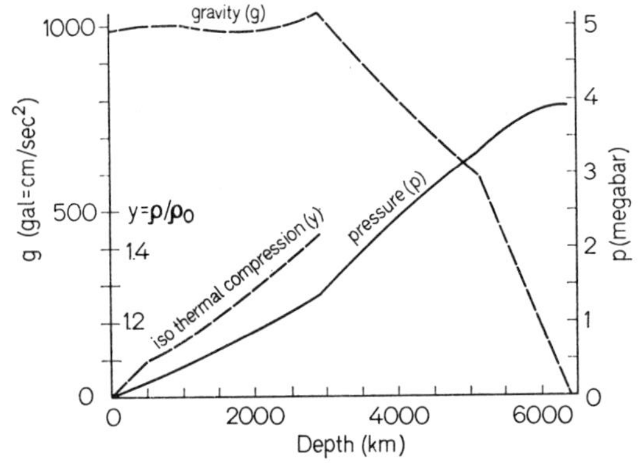

Fig. 6-13. Gravity and hydrostatic pressure as function of depth in BULLEN's model A. (From BULLEN, 1963.) The isothermal compression y has been obtained from BIRCH's equation of state, Eq. (26), approximating the isothermal bulk modulus K_T by the adiabatic modulus K_S from Fig. 6-14. Gravity remains nearly constant throughout the entire mantle. The pressure of 1.3 megabar ($= 1.3 \cdot 10^{12}$ dynes/cm²) at the bottom of the mantle compresses the material at this depth to about two thirds of its zero-pressure volume at the same temperature ($y = 1.5$)

first term $4\pi \varrho G$ is nearly the same as that of the second term $2g/(a-z)$, gravity remains remarkably constant throughout the entire mantle. It rises smoothly from 981 gal at the surface to a maximum of 1,060 gal at the mantle-core boundary before it drops to zero at the Earth's center.

The product $(g\varrho)$ determines the downward gradient of the hydrostatic pressure as is evident from (10). Hence, this gradient is roughly 0.6 kbar/km above the mantle-core boundary and amounts to 0.33 kbar/km in the uppermost mantle. The pressure reaches 1.3 megabar at the mantle-core interface and, after a steep ascent within the core, 3.7 megabar at the Earth's center (Fig. 6-13, Table 6-5).

d) Elastic Constants

After adopting one or the other density model the elastic constants for adiabatic compression and shear deformation, K_S and μ_S, follow readily from the P and S-wave velocities. The resulting values reflect an Earth's mantle of high rigidity and high incompressibility when compared with the elastic response of steel at zero pressure (Fig. 6-14). Poisson's ratio σ remains nearly constant, increasing from 0.27 beneath the M. discontinuity to 0.3 at the mantle-core boundary. These are acceptable values for solid matter.

The concurrent downward increase of seismic velocities and density implies that K_S and μ_S increase even faster with depth than ϱ throughout the mantle. There

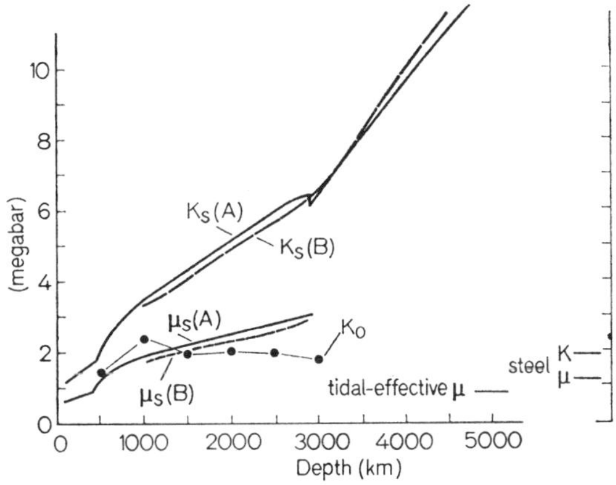

Fig. 6-14. Adiabatic incompressibility K_S and rigidity μ_S as function of depth, derived for Bullen's density models A and B (Fig. 6-12) and the Jeffreys-Bullen velocity distribution. The continuity of $K_S(B)$ at the mantle-core boundary reflects the pressure-compressibility hypothesis underlying the calculation of model B. Large dots refer to the zero-pressure values of $K_S(A)$, calculated from Eq. (24) with the isothermal compression shown in Fig. 6-13. The rapid increase of incompressibility in the lower mantle appears to be mainly a pressure effect which is removed by the reduction to zero pressure. The zero-pressure incompressibility of upper mantle material, on the other hand, increases nearly twofold between the M. discontinuity and 1,000 km depth. — The elastic constants of steel at zero-pressure are added for comparison, likewise the "tidal-effective" rigidity of the whole Earth as inferred from the overall yielding of the Earth's body to tidal forces (Sec. I, g)

may be a slight reduction of K_S at the mantle-core interface, but the incompressibility of the core increases again rapidly downward. A noticeable feature of the internal variation of rigidity is a *rigidity minimum* in the uppermost mantle, if we adopt Gutenberg's S-wave velocity distribution (Fig. 6-15). The reduction of μ relative to its subcrustal value would be small, however, and the rigidity remains well above 0.01 megabar, where true fluidity begins.

The core is impermeable for seismic shear waves and nothing definite can be concluded from seismic observations about its rigidity or lack of rigidity. Seismic energy, passing in the S-mode through the core, could be extinguished, for instance, by scattering or by high absorption in an elasto-viscous medium (Sec. II, f).

Evidence for a fluid core is (i) the yielding of the Earth to tidal forces (Takeuchi, 1950) and (ii) the amplitude of the 18.6-yearly forced nutation (Sec. III, a) of the Earth's axis of rotation (Jeffreys and Vicente, 1957). The observed deformation of the Earth's body as a whole is in both cases incompatible with a core having a rigidity greater than about 0.01 megabar, assuming that the seismically inferred rigidity of the mantle is also "tidal effective".

Hence, the core represents a nearly incompressible but highly deformable body ($K_S/\mu_S > 600$); only its innermost part could have higher strength. A likely explanation for the marked increase in the P-wave velocity by about 10% at 1,300 km from the Earth's center is indeed a sudden rise in rigidity (Fig. 6-6).

Neglecting the effect of accompanying changes of K_S and ϱ upon v_P the rigidity has to increase by $\frac{3}{2}K_S\varepsilon$ to produce a small fractional P-wave velocity increase $\varepsilon = \Delta v_P/v_P$ [Eq. (2)]. Setting $K_S = 12$ megabar and $\varepsilon = 0.1$ yields 1.8 megabar as the rigidity increase for the transition from the outer to the inner core.

e) Zero-Pressure Properties of the Mantle

Density and elastic constants of the deeper Earth's interior are strongly determined by the prevailing extreme pressure-temperature conditions. Both change with depth. Therefore it is of importance for conjectures about the possible composition of internal matter to reduce the observable in-situ properties to a constant reference pressure and temperature.

The reduction would be straightforward if a general *equation of state*, $\varrho = \varrho(p, T)$, valid for the interior of the Earth, could be formulated This is not yet possible and the reduction will be carried out along isotherms, yielding the zero-pressure density, ϱ_0, and the zero-pressure isothermal bulk modulus, $K_0 = \varrho_0(\partial p/\partial \varrho)_T$.

Density will be regarded as a continuous function of pressure. Hence, the prolongation of the isotherms from an elevated pressure p_1 to $p_2 = 0$ gives the zero-pressure properties of the metastable high-density phase if the material undergoes phase transitions between p_1 and p_2.

Highly compressed material becomes less and less capable of further compression under mounting pressure. Hence, the amount of isothermal compression, $y = \varrho/\varrho_0$, increases with pressure while the rate of compression, $y^{-1}\dfrac{dy}{dp} = K_T^{-1}$, decreases. Various equations of state, $f(p, y) = 0$, have been proposed to express this dependence of isothermal compression on pressure in best possible agreement with the experimental data.

BIRCH's (1952) *finite strain* equation of state has the approximate form

$$p = \tfrac{3}{2}K_0(y^{\frac{7}{3}} - y^{\frac{5}{3}}); \tag{23}$$

K_0 is readily identified as the zero-pressure value of K_T by setting $y = 1 + d\varrho/\varrho_0$ for $p \to dp$. Observing that $dy/dp = y/K_T$, a repeated differentiation of (23) toward p yields

$$K_T = \tfrac{1}{2}K_0(7y^{\frac{7}{3}} - 5y^{\frac{5}{3}}) \tag{24}$$

$$\left(\frac{\partial K_T}{\partial p}\right)_T = \beta_T = \frac{49y^{\frac{2}{3}} - 25}{21y^{\frac{2}{3}} - 15}. \tag{25}$$

After the elimination of K_0 from (23) and (24) it follows

$$y = \left(1 + \frac{2p}{3K_T - 7p}\right)^{\frac{3}{2}} \tag{26}$$

and thereby

$$\beta_T = 4\left(1 - \frac{35}{36}\frac{p}{K_T}\right). \tag{27}$$

MURNAGHAN's equation of state,

$$p = \frac{K_0}{m}(y^m - 1) \tag{28}$$

yields in a similar way $K_T = K_0 y^m$ and thereby

$$\begin{aligned} K_T &= K_0 + m\,p \\ \beta_T &= m \\ y &= \left(1 - \frac{m p}{K_T}\right)^{-\frac{1}{m}}. \end{aligned} \tag{30}$$

The in-situ properties at 2,000 km depth, for instance, are $K_T \approx K_S = 5.2$ megabar and $\varrho = 5.2$ gr./cm^3 with $p = 0.88$ megabar as ambient pressure. These values, when inserted into BIRCH's equation, yield $y = 1.3$, $K_T/K_0 = 2.5$, $\beta_T = 3.3$; hence $\varrho_0 = 4.0$ gr./cm^3 and $K_0 = 2.1$ megabar. The use of MURNAGHAN's equation gives about the same values, namely $y = 1.28$ and $K_T/K_0 = 2.3$, when $m = 3.3$. The nearly constant values which are found in this way for ϱ_0 and K_0 throughout the lower mantle suggest a remarkable homogeneity in phase and composition (Figs. 6-12 and 6-14).

f) Viscous Properties

A perfectly viscous medium has no strength and shows no resistance to deformation under shearing stress. The medium is said to "flow" or "creep" with internal friction limiting the spatial velocity gradient or the *rate of deformation*. The medium represents a *Newtonian fluid*, when the rate of deformation depends linearly on the applied stress, and a *non-Newtonian fluid*, when the stress-deformation rate relation is non-linear.

Consider two points P_1 and P_2 within deformable matter. The distance between them before deformation is s. Suppose a one-dimensional shearing stress σ acts upon the material, transverse to the line $P_1 P_2$, and displaces P_2 relative to P_1. In the case of perfect elasticity an instant and reversible small displacement

$$\Delta s = \frac{s}{\mu}\sigma \tag{31}$$

occurs in the direction of σ, $\Delta s/s$ representing the elastic shear and μ the *rigidity* of the material.

In the case of Newtonian viscosity P_2 starts to move with constant velocity v in the direction of σ and the viscous shear deformation increases irrevocably by $(v/s)\,\Delta t$ during the time increment Δt. The rate of shear v/s is a constant throughout the deformed medium and proportional to σ. Hence,

$$\frac{1}{s}\frac{\Delta s}{\Delta t} = \frac{v}{s} = \frac{\sigma}{\eta}, \tag{32}$$

defining η as *shear viscosity* of Newtonian fluids. An extension to three-dimensional stress and motion requires the introduction of a second viscous constant analogous to the two elastic constants of an isotropic elastic medium.

Viscous and elastic behaviour can be combined in various ways to describe the partially reversible and partially irreversible deformation of solids under one-dimensional shear stress. MAXWELL's formulation for an *elastoviscous* deformation is

$$\frac{2}{s}\frac{\Delta s}{\Delta t} = \frac{\sigma}{\eta_M} + \frac{\partial\sigma/\partial t}{\mu} = \left(\frac{1}{\tau_0} + \frac{\partial}{\partial t}\right)\frac{\sigma}{\mu}. \tag{33}$$

It establishes $\tau_0 = \eta_M/\mu$ as a characteristic time constant and η_M as the *Maxwell body viscosity*. The Maxwell body evidently behaves like a Newtonian fluid under quasi-stationary stress and like a rigid solid when the stress varies on a time scale which is much shorter than τ_0.

If the stress variations are sinusoidal in time with the period τ, the viscous term in (33) produces a phase lag between stress and deformation corresponding to the degree of anelasticity of the medium. The tangent of the phase lag,

$$\frac{\mu}{\eta_M} \cdot \frac{\tau}{2\pi} = Q_M^{-1}, \tag{34}$$

gives the specific dissipation of strain energy within the Maxwell body [cf. Eq. (7)].

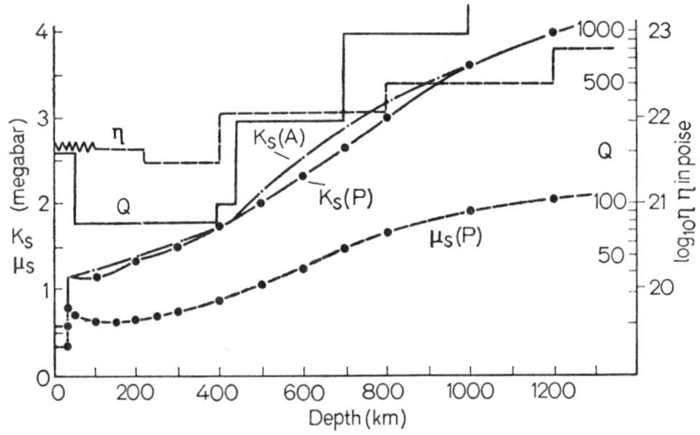

Fig. 6-15. Elastic and non-elastic properties of the upper mantle. The elastic constants $K_S(P)$ and $\mu_S(P)$ have been derived for GUTENBERG's upper mantle velocity distribution in combination with the peridotite density model of CLARK and RINGWOOD (Fig. 6-6 and 6-12). The incompressibility for BULLEN's model *A* is shown for comparison. GUTENBERG's *S*-wave velocity minimum in the upper mantle produces a rigidity minimum which more or less coincides with a maximum of the specific dissipation constant of seismic strain energy, Q^{-1} (model *G*, ANDERSON and ARCHAMBEAU, 1964), and a viscosity (η) minimum (MCCONNEL, 1965)

The hardly attenuated transmission of shear waves and the relatively slight damping of the torsional modes of free oscillations prove that the Earth's crust and mantle respond as a rigid body with little loss of energy when the period of oscillatory stress is less than one hour. A representative Q-value for the upper mantle is 100. In combination with a seismically effective rigidity of 10^{12} dynes/cm² this yields a viscosity of $5 \cdot 10^{16}$ *poise* (= dynes sec/cm²) for $\tau = 1$ hr, if the Earth is regarded as a Maxwell body.

The presence of regional deviations from hydrostatic equilibrium in crust and mantle and the slowness with which equilibrium conditions are restored clearly indicate that the secular deformation rate of mantle material is determined by much greater viscosities. The measurable post-glacial uplift of Fennoscandia, for instance, lags thousands of years behind the actual removal of the ice load, indicating an upper mantle viscosity of $3 \cdot 10^{21}$ poise (cf. JEFFREYS, 1959). This value certainly can not be interpreted as Maxwell-body viscosity, because the resulting time constant τ

for the uplift would be only of the order of years when the above cited value for Q^{-1} is inserted into Eq. (34).

By considering the various spatial harmonics of the Fennoscandian uplift separately, McConnel (1965) was able to infer tentatively the change of viscosity with depth. He concluded that the above cited value of $3 \cdot 10^{21}$ poise refers to a *viscosity minimum* between 100 and 400 km with higher viscosity above and below this depth range (Fig. 6-15). Hence, the mantle below 400 km seems to become with depth increasingly less capable of viscous flow. This conclusion is confirmed by its ability to support an equatorial bulge which is not in exact hydrostatic equilibrium (Sec. III, b). The present size of the bulge may be a relict from a faster rotation rate of the Earth millions of years ago and suggests a lower mantle viscosity near 10^{26} poise (McKenzie, 1966).

g) Strain Energy Dissipation

If a medium which is not perfectly elastic is set vibrating by an oscillatory force of the period τ, part of the strain energy is irrevocably lost because of internal friction, using this term in the broadest possible sense. The fractional loss of strain energy E_S per cycle defines the *specific dissipation constant*

$$Q^{-1} = \frac{\Delta E_S}{2\pi E_S} \tag{35}$$

of the medium as already introduced in Eq. (7). It embraces thermal losses due to conversion of mechanical energy into heat, scattering losses, and frictional losses due to stress-induced migration of lattice dislocations, including macroscopic sliding along internal surfaces. The dissipation constant is, as a measure of the degree of anelastic response, naturally related to the elastoviscous properties of the medium. This is exemplified by Eq. (34) in the case of a Maxwell body. The exact formulation of the functional dependence of Q from η thus depends on the theoretical concept which has been adopted to link stress and stress rate to deformation and deformation rate.

It has not yet been possible to devise a general elastoviscous theory for the Earth's interior which would hold over many orders in magnitude of the time constant of stress and of deformation. Hence, seismically inferred Q-values are not yet convertible into representative viscosity values for internal motion on geological time scales. We may just state qualitatively that internal zones with low seismic Q-values are more susceptible to viscous motion under stress than zones with high Q-values.

The most reliable Q-estimates for the Earth's interior are derived from the attenuation of the Love mode of surface waves [Eq. (7)] and from the damping of the torsional modes of free oscillations (Sec. I, g). Because the downward extent of their vibratory motion increases with increasing period τ, the effective Q-values when measured as a function of τ express implicitly the intrinsic Q-value as a function of depth z. The inversion of empirical $Q(\tau)$ data into a calculated $Q(z)$ profile is a solvable problem, when the intrinsic Q itself is not a function of τ (Anderson and Archambeau, 1964).

Love wave harmonics with periods between one and ten minutes yield Q-values of about 100, ($k=2.4\cdot10^{-5}\,\mathrm{km}^{-1}$ with $U=4.4$ km/sec and $\tau=300$ sec). There is clear evidence that Q increases toward longer periods, reaching 400 for $\tau=43$ minutes, the gravest torsional mode of free oscillations. Because Q seems also to increase toward short periodic Love waves we may speak of a $Q(\tau)$-minimum in the minute range. It has been interpreted by a corresponding $Q(z)$-minimum in the upper mantle between 50 and 400 km (ANDERSON and ARCHAMBEAU, 1964; Fig. 6-15). This establishes the upper mantle as a "soft" zone of high strain energy dissipation and low viscosity.

III. Gravity and Earth's Figure

a) Dimensions and Flattening of the Earth

The absolute dimensions of the Earth, i.e. its size, total mass and moments of inertia, have been inferred from geodetic measurements, gravity determinations, and astronomical data concerning the secular motion of the Earth's axis of rotation in space.

Geodetic triangulations at the Earth's surface yield in combination with star observations the absolute arc length of one degree on the meridian and thereby the geocentric distance r of the considered line element. The total mass M is then readily calculated from the surface gravity g according to the approximate formula $M=g\,r^2/G$, where G is the gravitational constant. For a correct determination proper allowance has to be made for the ellipticity of the actual Earth and the effect of centrifugal forces. In spite of these complications, the accuracy with which M is presently known is basically limited by the uncertainty in the experimental determination of G, amounting to about one part in thousand (Table 12-19).

The principal moments of inertia evolve from the precessional constant of the Earth's axis and the second zonal term in the expansion of the surface gravity field [Eq. (47)]. These moments of inertia are a direct expression of the internal distribution of mass (or density) [Eq. (37)]. They have revealed in particular, (i) that the Earth has a high-density core, and (ii) that the mass distribution in the mantle is not in perfect hydrostatic equilibrium (Sec. III, b).

It has been the rule to adopt certain fixed values for the various dimensions of the Earth according to the best available estimates and to define by means of them an *International Earth Ellipsoid* to which data concerning the actual shape and internal mass distribution may be referred. These values can be found in Table 19 and 20 of Chapter 12.

The Earth is of almost perfect symmetry with reference to both the polar axis of rotation and the equatorial plane. At a first approximation it has the spheroidal equilibrium figure of a rotating fluid of zero strength, implying that the (vector) sum of gravitational attraction due to internal masses and centrifugal acceleration is perpendicular to surfaces S of equal density. The centrifugal acceleration is directed outward (for a rotating observer), perpendicular to the axis of rotation, and given by $\omega^2 r\cos\varphi$; ω is the angular velocity of the Earth, r and φ denote the geocentric distance and latitude of an element ΔS of S.

The ratio of maximum centrifugal acceleration ($\varphi = 0°$) to gravity,

$$m = \frac{a^3\,\omega^2}{GM} \tag{36}$$

is about 1/300 [a: equivolumetric Earth radius, Eq. (40)]. This reflects the relative smallness of the centrifugal forces. The observation that the Earth's figure has been shaped by them imposes a definite upper limit upon the overall strength of the Earth's interior to resist deformation under the influence of longlasting stress (cf. Sec. II, f).

The most conspicuous feature of the centrifugal deformation is the *equatorial bulge* which makes the equatorial radius a_e longer than the polar radius a_p by 22 km or roughly by one part in 300. As a consequence, the moment of inertia about the polar axis,

$$C = \int_V \varrho\, \hat{r}^2 \cos^2 \hat{\varphi}\, d\hat{V} \tag{37}$$

is somewhat greater than the corresponding moment A about a principal axis in the equatorial plane; the integration is throughout the total volume V of the Earth, $\hat{r}$ and $\hat{\varphi}$ are the geocentric distance and latitude of the volume element $d\hat{V}$.

Comprehensive measures for the oblateness of the Earth's body are the *geometric flattening* (= geometric ellipticity)

$$f = \frac{a_e - a_p}{a_e} \tag{38}$$

and the *dynamical flattening* (= dynamical ellipticity or *precessional constant*)

$$H = \frac{C - A}{C}\,. \tag{39}$$

Both are well determined quantities. Their presently adopted values are 1/298.24 and 1/305.3, respectively.

If the ellipticity is to be disregarded in the course of some particular problem, a_e and a_p are properly combined to the radius

$$a = a_e \sqrt[3]{1-f} \tag{40}$$

of a sphere of equal volume. The principal moments of inertia are similarly joined to the moment of inertia

$$C_0 = \tfrac{1}{3}(C + 2A) \tag{41}$$

of an equivalent spherical Earth. Standard values for both quantities are $a = 6371$ km and $C_0 = 8.02 \cdot 10^{44}$ gr. cm^2.

The flattening of the Earth's body causes three principal effects (besides the obvious geodetic consequence that the arc length of one degree on the meridian increases from the equator to the poles): (i) Sun and Moon exert a torque upon the equatorial bulge, because the equatorial plane is inclined to the orbital planes of both celestial bodies; (ii) the surface gravity without centrifugal component decreases with latitude; (iii) the orbits of artificial satellites deviate from a simple ellipse. These effects have been utilized to determine the flattening of the Earth with great precision.

The torque due to the Sun tends to draw the equatorial bulge into the ecliptic plane. Because of the Earth's spin, a *precession* of the axis of rotation is the consequence, i.e. the Earth's axis moves in a cone about an axis normal to the ecliptic plane. A complete revolution takes place over about 26,000 years (EULER's *precession of the equinoxes*).

The torque due to the Moon acts similarly, but produces an additional oscillatory *nutation* of the Earth's axis during its precessional motion. This nutation has a principal period of 18.6 years, arising from the fact that the Moon's orbital plane is inclined to the ecliptic plane, the nodal line revolving completely once every 18.6 years.

The combined evaluation of precession and nutation yields (i) the mass of the Moon M_M in relation to the mass of the Earth M, and (ii) the dynamical flattening H of the Earth as introduced in Eq. (39). Standard values are $M/M_M = 81.375$ and $H = 0.003275$.

b) Low-Order Harmonics of the Geoid

The general definition of *hydrostatic equilibrium* states that the internal distribution of mass must be such that shear stresses are non-existent throughout the rotating or non-rotating body under consideration. Internal surfaces of equal density coincide with surfaces of equal gravity potential (see below). It is now clearly established that the internal mass distribution of the Earth, when viewed on a grand scale, is not in perfect hydrostatic equilibrium. It is also evident that these departures from equilibrium are not simply related to the distribution of oceans and continents, i.e. to the varying depth of isostatic compensation beneath the physical surface (cf. Sec. III, d).

Hence, the Earth's figure and gravity field contain certain large-scale features which are due to a *non-hydrostatic* distribution of mass within the solid mantle. — The core obviously has to be excluded because of its lack of supporting strength. — A comprehensive description of these non-hydrostatic anomalies in the global gravity field is provided by the low-order gravity or geoid harmonics which are obtained as follows:

The gravity vector $\boldsymbol{g}$ on or above the Earth's surface may be regarded as the gradient of a scalar potential U. Hence, $\boldsymbol{g} = grad\, U$ as the sum of gravitational attraction due to internal masses plus centrifugal acceleration stands normal on surfaces of constant potential. One distinguished equipotential surface is the *geoid* which envelops the Earth's body at sea level. This surface coincides with the mean free surface of the oceans. Its hypothetical continuation under land serves as zero mark for topographic elevations.

The International Earth Ellipsoid approximates the true geoid very closely. In fact, the low-order harmonics in the Earth's gravity field reveal that the distance between the geoid and the surface of the Earth Ellipsoid hardly exceeds 70 meters at the most (Fig. 6-16). Such geoid undulations when referred to the exact equilibrium figure of the rotating Earth are a direct expression of regional deviations from hydrostatic equilibrium within the Earth.

Let (r, φ, λ) be the geocentric spherical coordinates of the external point P on or above the surface of the geoid, λ denoting its longitude. The gravity potential at this point would be $\bar{U} = -G\,M/r$ if the Earth were a non-rotating sphere. The difference between the actual potential U and $\bar{U}$ at P shall be expressed by a series of spherical harmonics,

$$\bar{U} - U = \bar{U} \sum_{n=2}^{\infty} (a_e/r)^n \{J_n P_n (\sin \varphi) + \text{non-zonal terms}\} + \tfrac{1}{2} r^2 \omega^2 \cos^2 \varphi, \qquad (42)$$

with $P_n (\sin \varphi)$ denoting Legendre polynoms of the nth order. The last term accounts for the centrifugal acceleration. The non-zonal terms have the general form

$$J_n^m P_n^m (\sin \varphi) \cdot \sin (m \lambda - \delta_n^m);$$

$P_n^m (\sin \varphi)$ is a spherical function of the nth order and mth ($m = 1, 2, \ldots, n$) rank. These terms account for deviations from rotational symmetry of U with respect to the polar axis.

The even zonal harmonics $P_2, P_4, \ldots$ are symmetric with respect to the equatorial plane, while the uneven harmonics are non-symmetric. Hence, the uneven zonal and all non-zonal terms in the expansion of U are zero in the case of a rotating body in hydrostatic equilibrium. In general, the coefficients represent the internal mass or density distribution integrated throughout the volume of the geoid in the form

$$J_n \cdot M = \int_V (\hat{r}/a)^n \varrho P_n (\sin \hat{\varphi}) \, d\hat{V}. \qquad (43)$$

Suppose density variations as a function of latitude in the shell bounded by $r = r_1$ and $r_2 < r_1$ are (Cook, 1967)

$$\bar{\varrho} \sum_{1}^{\infty} \sigma_\nu P_\nu (\sin \varphi) \qquad (44)$$

and zero outside; $\bar{\varrho}$ denotes the mean density of the whole Earth. The orthogonality of Legendre's functions then allows a straight-forward evaluation of the volume integral (43), yielding

$$J_n = \frac{3\, \sigma_n}{(2n+1)(n+3)} \left[\left(\frac{r_1}{a}\right)^{n+3} - \left(\frac{r_2}{a}\right)^{n+3} \right] \qquad (45)$$

when the oblateness of the Earth is neglected. This reduces for a thin non-uniform shell to

$$J_n = \frac{3\, \sigma_n\, d}{(2n+1)\, r} \left(\frac{r}{a}\right)^{n+3} \qquad (45\text{a})$$

with $d = r_2 - r_1$ as its thickness and $r = (r_1 + r_2)/2$ as its mean radius.

In principle, the empirical values for the various zonal and non-zonal coefficients could be inferred from an analysis of surface gravity data, reduced to sea level (cf. Sec. III, c). However, their precise determination had to await the launching of artificial satellites, because the orbit of a not-too-distant satellite is critically determined by the global features of the Earth's gravity field.

The even zonal terms of U cause a slow rotation of the *lines of nodes* (intersection of orbital and equatorial plane) and *perigees* (position at closest approach). The uneven zonal terms cause a superimposed periodic motion of the lines of nodes, perigees and other orbital elements. Hence, the coefficients J_n can be found by observing the shifting orbits of a satellite at a single station.

The non-zonal coefficients J_n^m are responsible for perturbations within individual orbits and have to be investigated by a network of simultaneously observing stations.

In the practical determination the tracking data of numerous orbits are fitted to a limited set of coefficients, thus terminating the expansion of the gravity potential at some high order n. This leads, at least for the high-order terms, to a certain indetermination, since these terms do not form a converging series. In fact, the zonal and non-zonal coefficients above the 6th degree are all of the order 10^{-6} (Fig. 6-22).

Let $\bar{r}$ denote the mean radius of one particular surface $U=\text{const}$, here the geoid surface. Then its geocentric distance $r(\varphi, \lambda)$ may be expanded into a corresponding series of spherical harmonics:

$$r=\bar{r}\left[1-\sum_{2}^{\infty}\{N_n P_n(\sin\varphi)+\text{non-zonal terms}\}\right]. \tag{46}$$

Dominating in these expansions of the gravity potential (Eq. (42)] and geoid [Eq. (46)] are the second order zonal terms J_2 and N_2. They account for most of the centrifugal deformation of the Earth to a spheroid and have a straight-forward connection with the parameters f and m [Eqs. (36) and (38)].

A comparison of Eq. (43) with Eq. (37) suggests that the second order gravity term is related to the moments of inertia. This relation has the explicit form

$$J_2=H\frac{C}{M\cdot a_e^2} \tag{47}$$

and enables us to calculate the principal moments C and A for given values of H (from astronomical observations), $M\,a_e^2$, and J_2 (from gravity and satellite observations). The best available estimates are (cf. COOK, 1967)

$$J_2/H=0.3306$$

$$\begin{aligned} C &= 8.037\cdot 10^{44}\,\text{gr.cm}^2\\ A &= 8.011\cdot 10^{44}\,\text{gr.cm}^2 \end{aligned} \tag{47a}$$

with an accuracy of better than one part in thousand.

If the internal distribution of mass were in exact hydrostatic equilibrium, an intrinsic relationship would connect the dynamic and geometric flattening of the Earth (for a given rate of rotation and with an approximate knowledge of the radial distribution of mass). Accordingly, a *hydrostatic flattening* f_h can be calculated from empirical values of H and J_2 (cf. JEFFREYS, 1964) or, conversely, a *hydrostatic gravity coefficient* J_{2h} from empirical values of H and f (cf. COOK, 1967). The resulting differences $(f-f_h)$ or (J_2-J_{2h}) are of the order of 1/250 and presumably positive. Hence, the centrifugal deformation of the Earth appears to be slightly too large for the present-day rate of rotation (cf. Sec. II, f).

For $n>2$ gravity and geoid terms may be set equal, $N_n=J_n$, with corresponding identities for the non-zonal coefficients. The gravity potential of higher than second order is in this way readily converted into the undulating height of the geoid above the second-order figure of the Earth which — as noted above — is not an exact equilibrium figure. Furthermore, the fourth-order gravity term, J_4, has a sizeable "hydrostatic" component. Nevertheless, the resulting geoid contours may be taken to represent essentially the non-hydrostatic features of the Earth's gravity field. They provide a smoothed image of internal surfaces of equal density, the degree of smoothing depending on the postulated depth of these density imbalances.

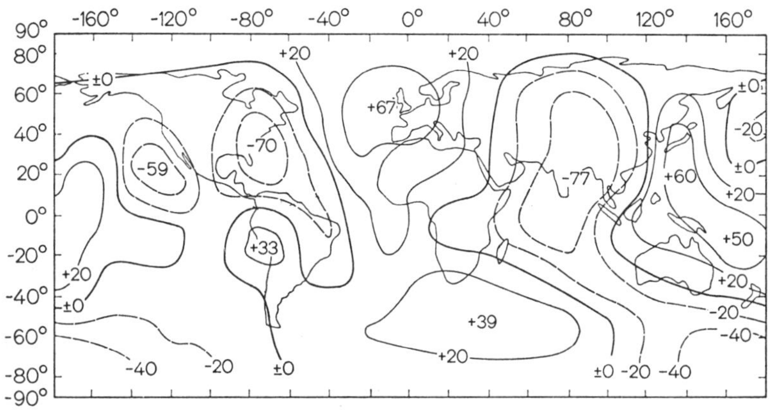

Fig. 6-16. Geoid contours in meters, representing the zonal and non-zonal gravity harmonics up to order 8, the zonal harmonics P_9, P_{10}, P_{12}, and the "resonance" harmonic P_{13}^{13}. These harmonics have been obtained from the "Doppler" tracking of artificial satellites. Zero reference is the Earth's spheroid with a geometric flattening $f = 1/298.28$. The probable "hydrostatic" flattening (see text) is close to 1/299.6. Hence, the contours reveal the centers of mainly nonhydrostatic mass imbalances within the Earth which may be connected with the upward and downward branches of mantle convection currents. Geoid highs: mass excess; geoid lows: mass deficit. There is no simple correlation between geoid contours and structural provinces of tectonic mobility or stability. (From Guier and Newton, 1965)

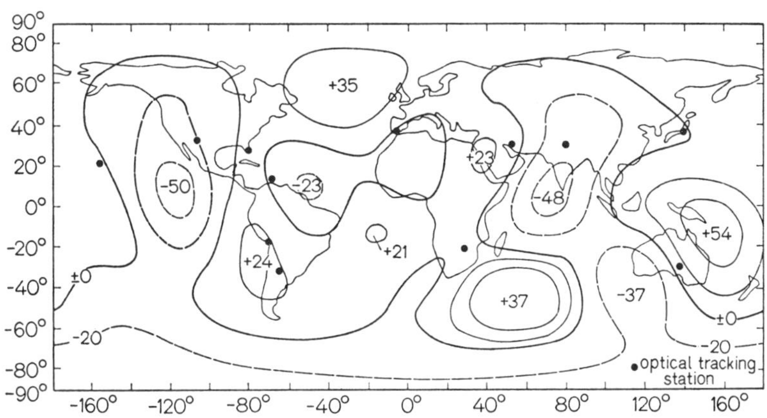

Fig. 6-17. Geoid contours in meters, comprising all gravity harmonics up to order 6 as obtained from optical tracking of artificial satellites. Zero reference is the spheroidal surface, given by the 2nd and 4th zonal harmonic. The general distribution of geoid highs (mass excess) and lows (mass deficit) agrees with that inferred from Doppler tracking (Fig. 6-16). (From Iszak, 1964)

A visual inspection of the geoid contours in Fig. 6-16/17 reveals a lack of correspondence between geoid highs and lows with the distribution of continents and oceans. Detailed studies have confirmed that there is indeed no significant correlation between the harmonics of the geoid surface and those of the Earth's topography, when the latter are likewise expanded into a series of spherical harmonics (MUNK and MACDONALD, 1960; GUIER and NEWTON, 1965; cf. also VENING-MEINESZ, 1958).

Suppose that the cause of the geoid undulations is an undulatory mantle-core interface, i.e. at the bottom of the mantle there is a thin non-uniform layer in which the density oscillates between 5.5 and 9.7 gr./cm³ (cf. Fig. 6-12). By inserting in Eq. (45a) the values $r = 3{,}470$ km, $\bar{\varrho}\,\sigma_n = 2.1$ gr./cm³ and $n = 6$ we obtain $d = 6.5$ km as the thickness of the non-uniform layer required to explain the sixth term $J_6 = 0.7$ 10^{-6} of the expansion of U. For $n \geqq 12$ the required thickness would have to be greater than 100 km to produce a zonal term of the order 10^{-7}. This is an unacceptable interpretation in view of the available seismological evidence for a smooth core surface. In summary, the low-order terms of the gravity field can originate from non-uniformities at the base of the mantle, those of higher order cannot.

Let us assume that the high-order terms originate from a non-uniformity between 140 and 200 km depth. For $n = 12$ and $J_n = 7 \cdot 10^{-7}$ we obtain $\bar{\varrho}\,\sigma_n = 0.005$ gr./cm³ which corresponds to lateral temperature variations of $\bar{\varrho}\,\sigma_n/(\varrho\,\alpha) \approx 50\,°\mathrm{C}$ ($\varrho = 3.3$ gr./cm³, $\alpha = 3 \cdot 10^{-5}$ °C⁻¹). They may be a direct consequence of convective motion. In this model geoid lows indicate ascending branches of convection which carry hot and light material to the surface. Similarly geoid highs would lie on top of descending branches, sucking in cold and heavy material.

Of particular interest is the distribution of shear stress which these hypothetical currents would exert upon the overlying rigid crust (RUNCORN, 1965). Geoid highs would become centers of maximum converging shear stress, leading to a compression of the crust, geoid lows to centers of diverging shear stress, producing a crustal dilatation. It may be significant that some active zones of mountain building, like the Andes, indeed coincide with geoid highs. The correlation between geoid lows and mid-oceanic ridges, which are assumed to be zones of crustal spreading (Sec. IV, d) is not so evident.

c) Gravity Anomalies

The ruggedness of the Earth's topography, as well as irregularly distributed masses below the surface down to, say, 60 km depth, produces a host of small-scale anomalies in the gravity field. Their magnitude is, however, small. Even prominent anomalies do not change the absolute surface value of gravity (≈ 981 gal) by more than a few parts in 10^5 (1 gal $= 1$ cm/sec²). The conventional unit in local gravity studies is therefore 1 milligal $= 10^{-3}$ gal.

The *half-width* L of anomalies (cf. Fig. 6-19) is a convenient measure for the depth at which the local mass irregularity (= deviation from a stratified distribution) is likely to occur. Except for this "half-width rule" the interpretation of gravity anomalies is an indeterminate problem. There always exists a family of equivalent mass distributions which cause the same surface anomaly of gravity. Any preference for one of them has to be guided by geological and petrological considerations, limiting the probable range of density variations as a function of depth.

The observed gravity g_P at a surface point P is subject to several reductions which remove the gravity effect of all known masses, including that of the Earth as a whole, approximated by the International Earth Ellipsoid. The *free air correction*, δg_f, reduces the local gravity field to some common reference surface, usually the geoid or sea-level surface.

If $g_{P'}$ denotes the gravity of the International Ellipsoid at the sea-level point P' perpendicularly below P, the difference

$$\Delta g_f = g_P - \delta g_f - g_{P'} \tag{48}$$

is the *free air anomaly* at P. Its correct determination obviously depends on the exact knowledge of the surface elevation h of P.

The *simple Bouguer correction*, δg_m, accounts for the gravity effect of stratified masses between P and P' at P, and the *topographic correction*, δg_t, for the gravity effect of the topographic highs and lows in the neighbourhood of P. The choice of an appropriate density for these masses above sea level is problematic and introduces some uncertainty into both corrections. Their sum, the *complete Bouguer correction* $\delta g_B = \delta g_m + \delta g_t$, is equal to the free air anomaly in the absence of internal mass irregularities.

Any remaining difference or *Bouguer anomaly*

$$\Delta g_B = \Delta g_f - \delta g_B \tag{49}$$

implies that local deviations from a stratified internal mass distribution exist. Bouguer anomalies obviously reflect in the first place geological subsurface conditions, e.g. the changing depth of a crystalline basement beneath a low-density cover of sediments. This uneven mass distribution might be known from other sources of information and its gravity effect can be incorporated into a *geological correction* δg_G (Fig. 6-18).

Bouguer anomalies which are not much altered by such corrections obviously arise from some yet unknown geological body of deviating density (Fig. 6-19) or they are a direct consequence of the isostatic adjustment of the underlying crust (Fig. 6-18).

d) Isostasy

Continental masses above sea level represent a load on the underlying material of the Earth, bounded by the (smoothed) geoid. Oceans may be regarded here as a "negative" load since their density is one third of the density of common rocks.

Suppose the underlying material has sufficient strength to carry these loads without deformation or re-distribution of mass: the density distribution below sea-level then remains undisturbed, i.e. a function of depth only, and the Bouguer anomaly (when corrected for local geological conditions) is zero. As a consequence, the Earth's interior is not in hydrostatic equilibrium (cf. Sec. III, b) down to some unspecified depth, the surface topography is said not to be in local *isostatic equilibrium* with its supporting substratum.

This applies to young oceanic islands and isolated mountains with a diameter of less than about 100 km which are — as a rule — without sizeable Bouguer anomalies. On the other hand, large topographic structures, in particular the major mountain systems (Alps, Andes, Himalaya), are clearly associated with Bouguer anomalies in such a way that the Δg_B-contours give a negative mirror image of the surface relief. Hence, beneath these mountains there exists a deficit in mass balancing the masses above sea level. This has been known since Bouguer made his classical gravity studies in the Andes around 1740.

If the internal mass deficit re-establishes hydrostatic equilibrium at some subsurface level, the isostatic compensation is said to be complete. Two basic concepts of complete *local isostatic* compensation have been advanced.

PRATT's concept (1855) assumes a deficit in mass extending from the physical surface down to a constant *depth of compensation*, H_0, below sea level, where hydrostatic equilibrium begins. Hence the density distribution below this depth is uniform or stratified, above this depth non-uniform and given by $(\varrho_0 - \Delta\varrho)$; ϱ_0 here is the mean density between the physical surface and the depth of compensation when the surface elevation h is zero, while $\Delta\varrho$ represents a variable density deficit when $h > 0$.

Treating gravity as a constant, it follows readily from the requirement of constant hydrostatic pressure at the level $z = H_0$ that

$$\Delta\varrho = \frac{\varrho_0 h}{H_0 + h} \tag{50}$$

[cf. Eq. (10)]. The slightly different Pratt-Hayford concept assumes a constant depth of compensation beneath the physical surface and sets $\Delta\varrho = \varrho_0 h/H_0$. If, for instance, $H_0 = 60$km and $\varrho_0 = 3$ gr./cm^3, the density deficit has to increase by 0.05 gr./cm^3 per kilometer increase of surface elevation to ensure complete isostatic compensation.

AIRY's concept (1855) explains the mass deficit as a projection or *root* of a floating crust into a substratum of higher density and zero strength. Hence, hydrostatic equilibrium begins beneath the undulating lower boundary of the *Airy crust* which is not necessarily identical with the seismically inferred crust.

Let $\bar{H}$ be the thickness of the Airy crust (below sea level) when the surface elevation $h = 0$. Its "freeboard" above the (hypothetical) surface of the substratum is $\bar{H}(\varrho_s - \varrho_c)/\varrho_s$ according to Archimedes' principle, when ϱ_c and ϱ_s denote the uniform densities of Airy crust and substratum. Floating equilibrium is preserved when any masses above sea level are compensated by an equivalent increment

$$\Delta H = \frac{h\,\varrho_m}{\varrho_s - \varrho_c} \tag{51}$$

of the thickness of the Airy crust, ϱ_m denoting the mean density of the surface load.

Setting $\varrho_c = \varrho_m = 2.67$ gr./cm^3 (granite) and $\varrho_s = 3.25$ gr./cm^3 (peridotite) yields $H = 4.6 \cdot h$, i.e. an increase of surface elevation by 1 km produces a root increment of nearly 5 km.

VENING-MEINESZ introduced the concept of *regional isostasy* in the sense that the compensating internal mass deficit is spread over a wider range than the surface load to be compensated (cf. HEISKANEN and VENING-MEINESZ, 1958). Regional isostasy obviously requires less redistribution of masses in the substructure than local isostasy, but it creates considerable shear forces in it. Hence, the substructure must have sufficient strength to withstand these forces without or with very slow viscous flow in the direction of local isostatic compensation.

The internal mass deficit, required by isostatic compensation, produces the gravity δg_i at P, which when subtracted as *isostatic correction* from the Bouguer anomaly yields the *isostatic* anomaly

$$\Delta g_i = \Delta g_B - \delta g_i. \tag{52}$$

The calculation of δg_i is similar to that of δg_t. The exact distribution of the assumed internal mass deficit for complete isostatic compensation and hence of δg_i itself depends on the particular concept of isostasy which is chosen and on the numerical values of the parameters $H_0, \varrho_0, \ldots$ [Eq. (50/51)] which are adopted.

The isostatic correction is negative in the case of a positive surface load and tends to cancel out the Bouguer correction. As a consequence, isostatic and free air anomaly are similar when the underlying crust is close to isostasy. Isostatic anomalies (but *not* free air anomalies) vanish completely when the isostatic compensation of surface loads is complete and according to the adopted concept of isostasy. A

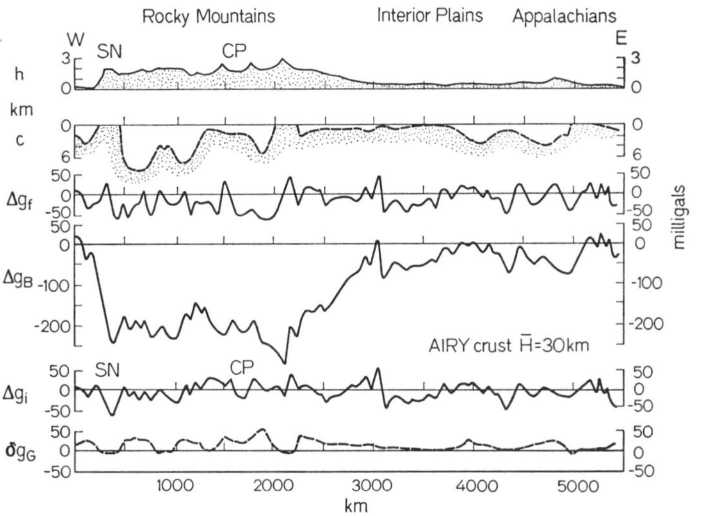

Fig. 6-18. Gravity anomalies along an east-west traverse in 38° latitude across the North American continent. The Bouguer anomaly Δg_B is nearly a negative mirror image of the topographic relief *h*, indicating overall isostatic equilibrium. The remaining isostatic anomaly Δg_i, calculated according to AIRY's concept, is therefore small, irregular, and similar to the free air anomaly Δg_f. The geological correction δg_G for the basement topography *c* offers no simple explanation for the main non-isostatic features which are therefore of deeper origin. For instance, the Sierra Nevada (*SN*) and the Colorado Plateau (*CP*) appear to be overcompensated ($\Delta g_i < 0$), i.e. the crustal mass deficit beneath them is larger than is required for isostatic compensation of the present-day surface elevation. (From WOOLLARD, 1966, profile *E*)

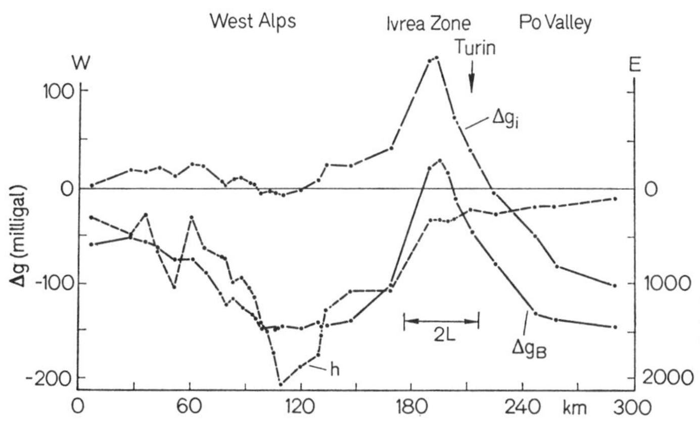

Fig. 6-19. Gravity profile across the West Alps in Europe. An uncompensated mass excess beneath the Ivrea zone causes a local Bouguer (Δg_B) and isostatic (Δg_i) gravity high. Its half-width *L* of about 20 km suggests that this Ivrea body is an uncompensated crustal lense of mafic to ultramafic material (Fig. 6-5). The topographic relief *h* of the West Alps appears to be in isostatic equilibrium ($\Delta g_i \approx 0$), while the adjacent Po valley does not. Its young filling with still unconsolidated sediments is isostatically too light for the underlying crust ($\Delta g_i < 0$). (From CORON, 1963)

positive isostatic anomaly implies that the negative correction δg_i (calculated under the assumption of complete isostatic compensation) is too large for the actually existing internal mass deficit. Hence, a surface load is *undercompensated* when $\Delta g_i > 0$ and *overcompensated* when $\Delta g_i < 0$ (Fig. 6-18/19).

Regional studies of isostatic anomalies should enable us to determine (i) the degree of isostatic compensation of surface features in one particular area, and (ii) how it is achieved. In practice, this is only partially possible because the regional features of isostatic anomalies are often obscured by uncompensated geological bodies of small dimensions (Fig. 6-18).

This does not invalidate the general statement that zones of recent mountain building have relatively small isostatic anomalies and are terefore more or less compensated. There are indications that we are confronted here by a combination of PRATT's and AIRY's concept, i.e. that mountain belts are supported by a root of light crustal material *and* by lower than average density in crust and uppermost mantle (Fig. 6—4). VENING-MEINESZ concept of regional isostasy seems to apply to those small-size structural surface features which lack a Bouguer anomaly and therefore local isostatic support.

IV. Geomagnetism

a) Basic Concepts

A magnet freely suspended above the ground is subject to a mechanical couple, tending to align it with the lines of force of a general Earth's magnetic field. It rotates, for instance, the compass needle in an approximate north-south direction. The magnetic field of the Earth's body thus evidenced is, in contrast to gravity, not a direct consequence of a fundamental physical law. Its most probable cause is fluid motions in the Earth's core which are thought to generate a self-sustaining system of electrical currents on the dynamo principle.

This dynamo has an external magnetic field which is weak by technical standards; its maximum flux density at the Earth's surface hardly exceeds 0.5 Gauss (Fig. 6-20). The Earth's magnetic field has been an unsteady phenomenon during the geological past. It is not unreasonable to assume that the Earth has been without a significant magnetic field for short intervals in its history.

The conventional units used in geomagnetism are the cgs-units *Oersted* (1 Oe = 79.6 Amp/m) for the magnetic *field strength* $\boldsymbol{F}$ and *Gauss* ($1\,\Gamma = 10^{-4}$ Volt-sec/m^2) for the *magnetic flux density* (= induction) $\boldsymbol{B}$ of the field; 1 gamma (γ) = 10^{-5} Gauss. The ratio

$$B/F = \mu_0\,\mu \tag{53}$$

defines the magnetic *permeability* μ of a medium, the vector difference

$$\boldsymbol{B} - \mu_0\,\boldsymbol{F} = 4\pi\,\mu_0\,\boldsymbol{I} \tag{54}$$

its *magnetization* $\boldsymbol{I}$ (magnetic moment per unit volume). The factor μ_0 is dimensionless and unity when B and I are measured in *Gauss* and F in *Oe*; the numerical values of B and F are then equal in a vacuum ($\mu = 1, I = 0$), which applies more or less to geomagnetic observations above the ground.

The *horizontal component* of the Earth's field is denoted usually by $\boldsymbol{H}$, the *downward component* by Z (Fig. 6.20). Their ratio Z/H defines the tangent of *inclination* of the field vector, i; the angle between the geographic meridian and the direction of $\boldsymbol{H}$ at some surface point, counted clockwise from north to east, defines the angle of *declination*, D. The quantities F (*total intensity*), H, Z, i, D, are referred to as *geomagnetic elements.*

Studies in geomagnetism have to consider (i) the time variations of the geomagnetic elements at a fixed site, and (ii) their spatial variability on a local, regional, or worldwide scale at a given instant of time. Experience has shown that the observed field $\boldsymbol{F}$ can be split into a quasipersistent part, the *main field* $\boldsymbol{F}_0$, and a fluctuating part, the *variation field* $\delta\boldsymbol{F}$. The elements of $\boldsymbol{F}_0$ vary smoothly on a secular time scale. Those of $\delta\boldsymbol{F}$ comprise various types of more or less "rapid" fluctuations which are superimposed on the main field without changing its elements by more than a few parts in thousand at the most.

The main field arises from sources inside the Earth as first proposed by GILBERT in 1600: "*magnus magnes ipse est globus terrestris*". The ultimate proof for GILBERT's postulate was given in 1838 by GAUSS' analytical presentation of the field in terms of spherical harmonics (Eq. (56)]. The variation field is primarily of extraterrestrial origin, but it has a secondary internal component due to sub-surface induction effects, as discussed in Sec. IV, g. Hence, the quasi-persistent main field as well as the variation field provide information about physical conditions inside the Earth.

b) The Dipole Field

The secular variability of the main field makes it necessary to reduce geomagnetic field observations to some fixed date, known as the *epoch* of the reduction. The variation field is removed by averaging the geomagnetic elements in time over some properly chosen interval (daily, monthly and yearly means). Determinations of this kind have been made on a continuous basis for nearly 150 years at many places around the globe.

During this time span the spatial configuration of the main field has been approximately that of a *magnetic dipole*, located at the Earth's geometric center and oriented nearly parallel to the polar axis of rotation. The axis of this fictitious dipole intersects the Earth's surface at the *geomagnetic poles*, which define a system of geomagnetic coordinates (Φ, Λ) analogous to the system of geographic coordinates (φ, λ), so that the geomagnetic meridian of zero longitude $(\Lambda = 0)$ passes through the geographic south pole. *Magnetic poles* are those points where the inclination of the main field is $\pm 90°$ $(H_0 = 0)$. Similarly the *magnetic equator* is defined as the line of zero inclination $(Z_0 = 0)$ while the *geomagnetic equator* is the great circle of zero geomagnetic latitude $(\Phi = 0)$.

Places on circles of geomagnetic latitude have the same dipole field, given in components by

$$H_{DP} = \frac{M \cos \Phi}{a^3}, \quad Z_{DP} = \frac{2\, M \sin \Phi}{a^3} \tag{55}$$

with M as the dipole moment and a as equivolumetric radius of the Earth; H_{DP} points in the direction of the geomagnetic meridian toward the geomagnetic pole of the northern hemisphere, which has by definition the polarity of a magnetic south pole (Fig. 6-20).

The moment and orientation of the dipole which most nearly approximates the observed main field $\boldsymbol{F}_0$ can be found as follows. The main field is irrotational above the ground, i.e. a magnetic field due to steady air-ground currents is not discernible. This makes it possible to derive $\boldsymbol{F}_0 = -\mathit{grad}\, V$ as gradient from a geomagnetic scalar potential V, which is Laplacian ($\Delta V = 0$). Its analytical approximation by a series of spherical harmonics is (in geographic coordinates)

$$V = a \sum_{n=1}^{\infty} (a/r)^{n+1} \sum_{m=0}^{n} P_n^m (\sin\varphi)\, [c_n^m \cos m\lambda + s_n^m \sin m\lambda] + \text{external terms} \qquad (56)$$

for $r \geqq a$, using the notations of Eq. (42). The radial dependence of the external terms is $(r/a)^n$, their latitude-longitude dependence is likewise $P_n^m \cos m\lambda$ and

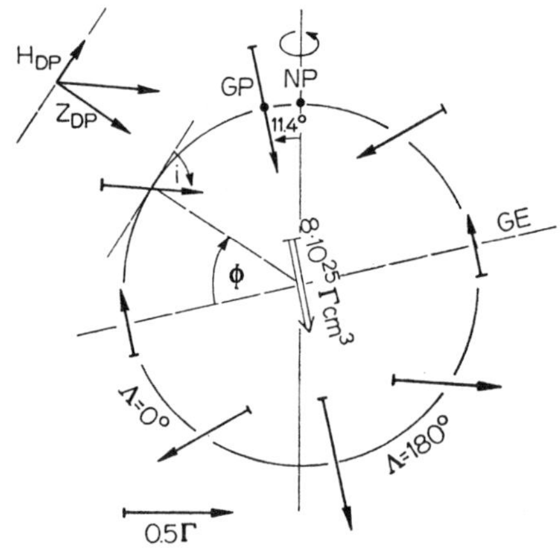

Fig. 6-20. Field of the centred geomagnetic dipole, shown for the 70° W meridional plane. *NP* geographic north pole; *GP* geomagnetic (south) pole; *GE* geomagnetic equator; Φ, Λ geomagnetic latitude and longitude; i inclination of the dipole field [$\operatorname{tg} i = 2 \operatorname{tg} \Phi$, Eq. (55)]

$P_n^m \sin m\lambda$. Internal terms of the first order $n = 1$ define the strength and orientation of the *best fitting centred dipole*, those of higher order the *non-dipole part* of the main field.

The numerical values of the internal and external coefficients can be computed by a least square method from empirical geomagnetic elements, averaged over a small surface element $a^2 \cos\varphi\, d\varphi\, d\lambda$. The vertical main field Z_0 as function of longitude and latitude yields (in essence) the difference between the external and internal coefficients, either one of the horizontal components their sum. This is readily verified by differentiating Eq. (56) with respect to r, φ and λ. Hence, combined analysis of the vertical and horizontal fields permits a separate determination of the internal and external terms. GAUSS' classical separation of the main field of 1835, and every separation since that epoch, gave small and insignificant external terms, thus proving the internal origin of the main field in an unbiased way.

The best fitting dipole of the 1965-epoch has a moment M of $8.01 \cdot 10^{25}\, \Gamma\, \text{cm}^3$ (ordinary magnets may have a moment of $10^3\, \Gamma\, \text{cm}^3$) with a geomagnetic pole at 88.6° N and 70.1° W in the arctic archipelago of Canada (LEATON *et al.*, 1965). The best fitting dipole moment of the 1835-epoch was significantly greater with $8.45 \cdot 10^{25}\, \Gamma\, \text{cm}^3$. This suggests a general decay of the dipole field during the last century. If it continues to decay at the present rate, the central dipole will have disappeared in 2,400 years.

In contrast, the dipole axis shows a remarkable secular stability. This state can be assumed also for the historic and geological past as evidenced by the fossil magnetization of baked pottery (*archaeomagnetism*) and rocks (*paleomagnetism*, Sec. IV, f). Hence, a centred dipole, roughly parallel to the polar axis, has been a predominant feature of the Earth at all times, only the polarity of the dipole must have changed from time to time after stable periods of variable length. During *normal* periods (like the present one) the dipole moment was antiparallel and during *reversed* periods parallel to the Earth's angular momentum of rotation (cf. Fig. 6-20). We conclude that the Earth's rotation has merely a stabilizing effect upon an intrinsic centred dipole without linear dependence of the dipole moment on the spin.

Turning to the physical cause of the dipole field, we observe that a dipole moment M is equivalent to a uniform magnetization M/V of the whole volume V of the Earth (for the 1965-dipole $I = 0.074\,\Gamma$). Since the deeper Earth's interior below the *Curie-isotherm* (Sec. IV, d) is practically non-magnetic, a significant contribution to the dipole field could arise only from a uniformly magnetized outermost shell of 30 km thickness at the most. The magnetization of this shell required to produce just one hundredth of the dipole field is $0.05\,\Gamma$. This value is improbable in view of the much lower magnetization of igneous and metamorphic rocks, not to mention the difficulty of explaining the required uniformity of the magnetization. Hence, rock magnetism has to be ruled out as a possible cause for the dipole field or any small fraction of it.

An alternative explanation is provided by zonal *electrical currents* which flow from east to west along circles of geomagnetic latitude. Crust and mantle appear to be relatively poor electrical conductors and the electromotive force to drive such currents in them would have to be extremely large (Sec. IV, g). A highly conductive and mobile Earth's core, consisting of molten iron, is therefore the most likely seat of the hypothetical current system which is responsible for the unsteady Earth's dipole.

Suppose these currents are concentrated in a thin shell of radius R. Then a sheet current density distribution

$$\frac{3}{4}\pi\frac{M}{R^3}\cos\Phi \tag{56a}$$

(in 10 amperes/cm) is equivalent to a centred dipole moment M (in $\Gamma\,\mathrm{cm}^3$). Assuming that this current flows at the very top of the core ($R = 3{,}470$ km), the equatorial density required to explain the 1965 dipole will be 0.46 million amperes per kilometer. The whole current, integrated between the poles, amounts to 3.2 billion amperes.

If this enormous system of *zonal currents* is to be generated and maintained like a self-exciting dynamo, it has to be combined with a second orthogonal system of *convective currents* in the core flowing in meridional planes. These will produce a tangential magnetic flux parallel to circles of latitude which is ten to one hundred times stronger than the intersecting flux produced by the zonal currents, but strictly confined to the interior of the core. Hence, this important convective current mode is undetectable by magnetic observations at the Earth's surface. Therefore the critical examination of the dynamo hypothesis relies on theoretical arguments.

Both current systems are thought to represent zonal and convective motions of a highly conductive core "fluid" across magnetic field lines. The motion could be initiated by a superadiabatic temperature gradient in the outer portion of the core. The electrodynamic couple between this motion and a weak initial magnetic field [Eq. (57)] builds up a secondary motion which in turn tends to amplify the initial one, and so on. The finite conductivity of the core fluid would limit the effectiveness of this dynamo process and also account for the intrinsic instability of the external dipole field.

c) Non-Dipole Field and Secular Variations

The presentation of the main field according to Eq. (56) leads to a fast converging series of spherical terms. This reflects the general observation that the main field is featureless in the intermediate half-width range between local anomalies

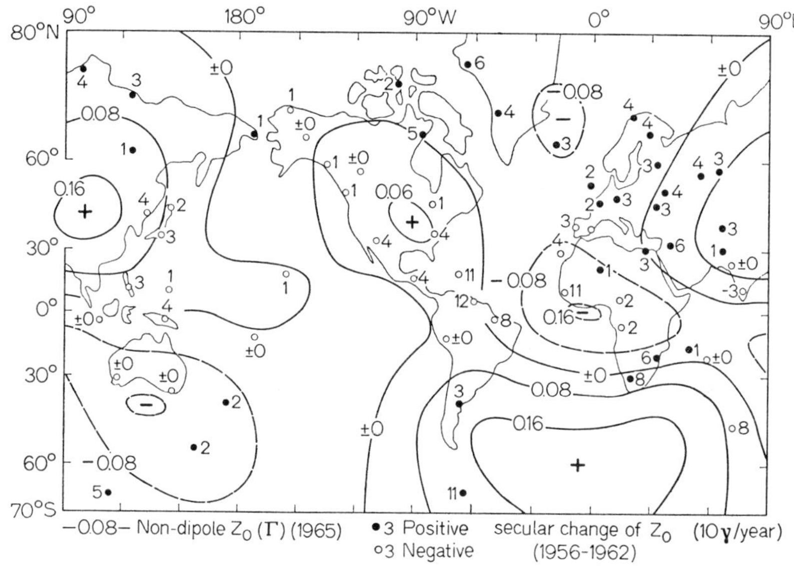

Fig. 6-21. Non-dipole part of the main field (epoch 1965) and secular variation (1956—62), shown for the vertical component Z_0. The contours of the nondipole Z_0 represent a smoothed image of a hypothetical electrical current systen in the surface of the highly conductive Earth's core. It is superimposed upon zonal east-west currents which produce the dipole part of the main field. Positive maxima are centres of clockwise current vortices, negative minima those of counter-clockwise vortices. These vortices do not correspond in number, position, and polarity to the highs and lows of the geoid, Fig. 6-16, thus diminishing the likelihood of a causal connection. The features of the nondipole field appear and dissolve on a secular time scale while drifting slowly in westerly direction (cf. text). Noteworthy are the rapid transitions from extremely high to nearly zero secular changes in South America and West Africa, furthermore the smallness of secular variations in the central Pacific area, including Australia. Non-dipole contours according to LEATON's analysis (1965), secular changes according to the tables of NAGATA and SAWADA (1963).

($L \lesssim 50$ km) and worldwide undulations ($L \gtrsim 1{,}000$ km), as is evident in Fig. 6-23. The Earth's magnetic field and gravity differ in this respect indicating that the non-dipole part of the magnetic field and the non-hydrostatic portion of the gravity field arise from basically different sources (Fig. 6-22).

Because the "half-width rule", cited in Sec. III, c for gravity anomalies, applies also to magnetic fields it follows that the upper mantle is free from sources contributing to the non-dipole field. On the other hand, its secular variability is incompatible with fixed sources in the solid lower mantle and suggests a connection with *non-convective* shifting motions in the core.

The secular variations of the dipole and non-dipole fields are altogether different. The latter seems to be growing at present at the expense of a slowly decaying dipole field, while the centres of maximum non-dipole Z_0 are drifting in a westerly direction

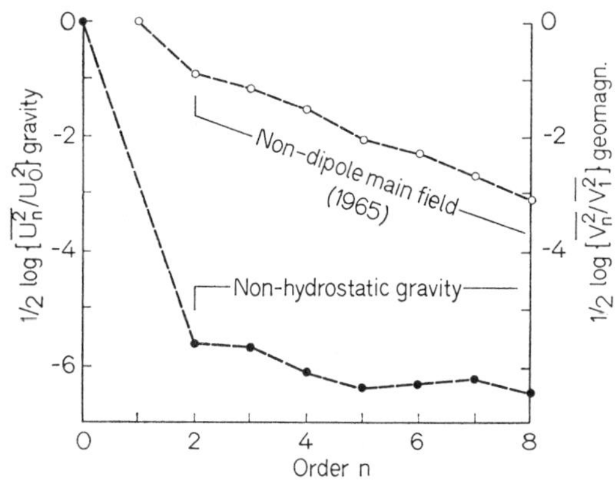

Fig. 6-22. The non-dipole part of the geomagnetic main field in comparison to the non-hydrostatic part of the gravity field. Shown are the relative contributions of the spherical harmonics to the total field as function of their order n. The non-dipole harmonics of the geomagnetic potential clearly diminish with increasing order, while the high-order gravity harmonics are not markedly reduced in size for $n > 3$. $\overline{V}_n$ and $\overline{U}_n$: Sum of the spherical terms of order n, squared and averaged over the surface of the Earth (excluding the hydrostatic portion of the even zonal gravity harmonics); U_0: gravity potential of the hydrostatic equilibrium figure of the Earth; $\overline{V}_1$: geomagnetic potential of the centred dipole [cf. Eq. (42) and (56)]. (Harmonic coefficients from LEATON *et al.*, 1965; GUIER and NEWTON, 1965; COOK, 1967)

(Fig. 6-21). BULLARD *et al.* (1950) found that the secular variations of the non-dipole geomagnetic elements, squared and averaged over the globe, can be minimized by displacing the meridians 0.18° westward each year. This *west-drift* of the non-dipole field accounts for about one-third of the observed secular variation.

A velocity estimate of the westward motion in the core fluid associated with the above mentioned drift requires a consideration of the interaction between motion and field. The magnetic field change in a highly conductive medium, moving with the velocity $\boldsymbol{v}$ across magnetic field lines is approximately given by

$$\frac{\partial \boldsymbol{F}}{\partial t} = \text{curl}\,(\boldsymbol{v} \times \boldsymbol{F}) \tag{57}$$

when field and motion have cosmic dimensions (cf. COWLING, 1957). The approximation implies that the electromagnetic diffusion of the field change through the medium is completely suppressed by eddy currents, i.e. the medium acts like a perfect conductor. As a consequence, the outward and inward magnetic flux through the surface of the conductor is invariant in time.

The velocity vector $\boldsymbol{v}$ obviously has no radial component at the core surface. Under the assumption that the core fluid is incompressible, Eq. (57) establishes a relation between the time derivative (secular variation) of the radial component of $\boldsymbol{F}$ and the tangential components of $\boldsymbol{v}$ at the inner surface of the core. Thus, the downward extrapolation of the observed secular variation field yields a hypothetical velocity field from which the zonal component may be extracted. KAHLE *et al.* (1967) obtained a north-south velocity profile for the motion of the core fluid along circles of (geographic) latitude, which shows a prominent westward motion between 20° N and 60° S with a maximum velocity of 6 km/year (=0.1°/year) at 20° S.

This westward velocity component implies that the top of the fluid core rotates slightly more slowly than the Earth's surface (one less full rotation in 3,600 years). BULLARD *et al.* (1950), on the other hand, argued that core and mantle are tightly coupled by electrical eddy currents induced in the lower mantle, counteracting any differential motion. Hence, a west drift in the outermost layers of the fluid core has to be balanced by an east drift further down.

Two special features of secular variations remain to be mentioned. The secular change of the geomagnetic main field is significantly lower in the central Pacific area than elsewhere. The same applies to the non-dipole part (Fig. 6-21). Paleomagnetic studies of Hawaiian lava flows show that this has been so for at least several hundred thousand years (DOELL and COX, 1965). This may reflect simply a regional peculiarity of the generating fluid motion in the core. But it can also imply that the inductive couple between outer core and lower mantle is here tighter than elsewhere or that the electromagnetic shielding of the lower mantle upon secular variations is unusually large (Sec. IV, g). These last explanations suggest that the mantle below the Pacific is electrically better conducting than elsewhere.

The second feature applies to the rate of secular variations. At many sites this rate undergoes from time to time a sharp change, usually within one or two years, although it is not known with certainty that it occurs simultaneously on a regional or continental scale. One possible explanation is that the inductive couple between core and mantle "slips" at irregular intervals, which may or may not cause detectable irregularities in the Earth's rotation rate for a surface observer. However, no definite correlation between the observed irregularities in the length of the day and geomagnetic secular variation has yet been found (cf. MUNK and MACDONALD, 1960; Chapter 11, 12).

d) Magnetic Anomalies

The usually weak and irregular magnetization of near-surface rocks causes minute local perturbations $\Delta \boldsymbol{F}$ in the quasi-persistent geomagnetic main field, known as magnetic *anomalies* (Fig. 6-23). There are only a few exceptional places where the main field is severely modified in strength and direction by local concentrations of highly magnetized iron ores.

Such anomalies arise from lateral variations of the internal magnetization $\boldsymbol{I}$. They do not disclose, at least not in a unique manner, the depth at which such deviations from a basically unknown layered distribution occur. Furthermore, the theoretical relations connecting $\Delta \boldsymbol{F}$ and $\boldsymbol{I}$ are complicated by the fact that the magnetization, a vector, may vary in strength *and* direction.

Suppose the surface anomaly is 2-dimensional and sinusoidal in the direction of the horizontal x-axis of rectangular coordinates, $\Delta Z \sim \sin(x/L)$; ΔZ denotes the vertical component of ΔF, $2\pi L$ the spatial wave length of the anomaly, L its half-width. The anomaly will arise from a variable downward magetization I_z in a layer of the thickness $2d$, which extends to infinity in both horizontal directions. Except for some arbitrary constant, I_z is then likewise sinusoidal and given by

$$\Delta I_z = \frac{\Delta Z \exp[(h+d)/L]}{4\pi \sinh(d/L)} \approx \frac{\Delta Z \cdot L}{4\pi\, d}; \tag{58}$$

h denotes the depth of the top face of the non-uniform layer beneath the level of observation. The approximation applies to thin layers when $(h+d)$ is small in comparison to L. Conversely, the non-uniformity increases exponentially, $\Delta I_z = (\Delta Z/2\pi)\exp(h/L)$, when $d \gg L$. According to the half-width rule we would set $h \approx L$ and thereby assume that $\Delta I_z \approx \frac{1}{2}\Delta Z$.

The ferromagnetic properties of rocks (Sec. IV, e) disappear above the Curie temperature which lies for the most abundant titano-magnetites between 250 and 500° C (Fig. 6-25) at atmospheric pressure. Experiments have shown that the

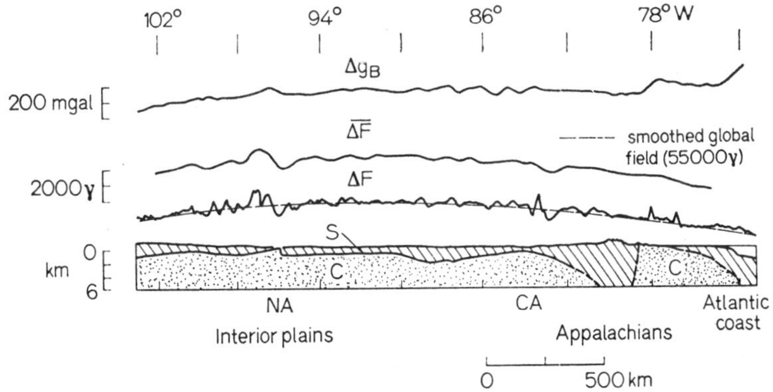

Fig. 6-23. Aeromagnetic profiles over the North American continent from the Atlantic coast (Washington, D.C.) to the foot of the Rocky Mountains (Denver, Colorado). Flight elevation 2 km to 3 km above sea level, thus diminishing the effect of very local features of shallow origin. The remaining spatial variations ΔF of the total field indicate (i) deep-seated intrusions of basic material, (ii) the relief of the crystalline basement (C) below nearly non-magnetic sediments (S). Distinct anomalies (also in gravity) above the Nemaha anticline (NA) and the Cincinnati arch (CA) which evidently bring highly magnetized material close to the surface. [Setting $\Delta Z = 800\,\gamma$ and $L = 10$ km in Eq. (58) gives $\Delta I_z = 250\,\gamma$ with $h = 1$ km and $d = 4$ km.] The running mean $\overline{\Delta F}$ (5-point average at 14 km intervals) brings out the long-wavelength content of the profile which is connected with deep crustal and possibly subcrustal structures. The wavelength for the smoothed Nemaha anomaly implies that its source can be as deep as 250 km/$2\pi \approx 40$ km, yielding for the Curie-isotherm (500° C) an even greater depth. The regional geothermal gradient is therefore low (Fig. 6-30). [Adapted from ZIETZ *et al.* (1966, Fig. 2 and 6, line 20)]

pressure dependence of their Curie temperature is small, i.e. in the order of 1°/kbar (SCHULT, 1968). Thus, the 500° C-isotherm within the Earth's interior may be regarded as the lower boundary for a significant magnetization of crustal and sub-crustal material (*Curie isotherm*). It may be reached 20 km below the surface in geothermal areas with high heat flow, but it can be more than 50 km deep where the geothermal gradient is small (Fig. 6-30). The magnetized surface layers of the

Earth have therefore a total thickness not exceeding a few tens of kilometers and this limits the half-width of magnetic anomalies to comparable values.

In summary, magnetic anomalies yield information (i) about geological bodies containing ferromagnetic minerals, (ii) about regional variations of the thermal

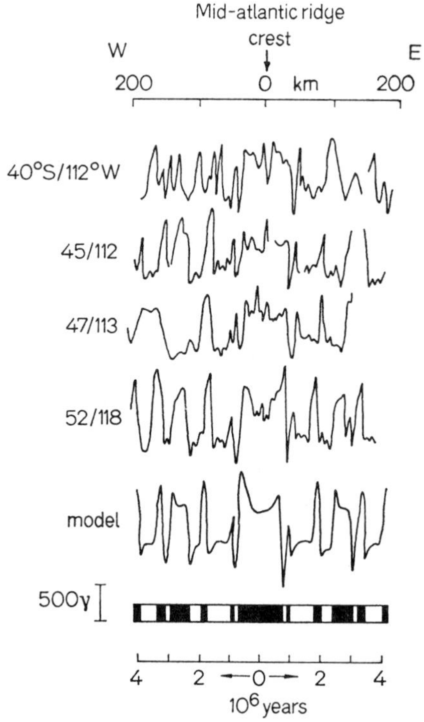

Fig. 6-24. Magnetic east-west profiles across the midatlantic ridge as observed with ship-towed magnetometers in the South Atlantic at various latitudes. The rugged total field anomalies which are characteristic for oceanic ridges show remarkably consistent features: Broad positive anomalies above the crest are flanked by sequences of nearly mirror-symmetric spikes when related to the center of the ridge. Hence, the sea floor of the ridge contains a system of north-south lamellae of highly magnetized material (presumably basalt) as indicated at the bottom. Black strips are magnetized in the direction of the present field, white strips are either non-magnetic or reversely magnetized with respect to the present field. (The model profile has been calculated for strips of 2 km thickness.) The Vine-Matthews hypothesis explains them by reversals of the geomagnetic dipole field in combination with sea floor spreading (cf. text). The time scale evolved from a comparison of these hypothetical reversals with those inferred from rockmagnetic studies, yielding a spreading-rate of 100 km in 2 million years (5 cm/year). (From HEIRTZLER *et al.*, 1968)

state influencing the depth of the Curie isotherm, and (iii) about the magnetic history of those geological structures which carry a remanent magnetization (Sec. IV, e). Fig. 6-23 and 6-24 give examples for these three types of conclusion.

The overall pattern of magnetic anomalies on land is complicated, any connection with continental geological structures not seldom being obscure or absent. In contrast, the world-wide system of oceanic ridges and fracture zones is accompanied by a regular pattern of long-stretched anomalies which seem to provide fundamental insight into the dynamics of the Earth's interior (Fig. 6-24). The mirror-symmetry of the ocean ridge anomalies led VINE and MATTHEWS (1963) to postulate a continuous upflow of hot basaltic melts at the centres of those ridges, where it is magnetized in the direction of the contemporary geomagnetic field while cooling down. The thus acquired magnetization is remanent (see below). Subsequently, the cooled and magnetized material is carried away from the centre of the ridge as a rigid block by subcrustal lateral flow, a hypothetical process commonly referred to as *sea floor spreading* (DIETZ, 1961; HESS, 1962).

It is reasonable to assume that the subcrustal lateral flow is symmetric and that it transports material which has been magnetized at the same time to equal distances on both sides of the ridge. Repeated reversals of the geomagnetic field (Sec. IV, f) in the course of this process lead to a mirror-image sequence of strips of normal and reversed magnetization which in turn explains the mirror-symmetry of the observed anomalies.

The now available absolute time scale for geomagnetic field reversals makes it possible to determine quantitatively the lateral flow or spreading velocity. It amounts to a few centimeters per year but seems to be subject to considerable regional variations (HEIRTZLER *et al.*, 1968).

e) Rock Magnetism

The magnetization of ordinary rocks is of *ferromagnetic* nature and controlled by the small fraction of ferromagnetic minerals which they contain. The term "ferromagnetism" is used here in the broad sense which includes the *ferrimagnetic* variety (two sets of opposite, but unequal atomic spin moments). Most prominent among those minerals are the ferrimagnetic members of the ternary system Fe_2O_3—FeO—$TiFeO_3$ (Fig. 6-25). Silicatic phases are either para- or diamagnetic and do not contribute significantly to the bulk magnetization.

The ferromagnetic magnetization of rocks is the vector sum of two basically different components,

$$\boldsymbol{I} = \boldsymbol{I}_i + \boldsymbol{I}_{rn}. \tag{59}$$

The first component is *induced* by and has the direction of the ambient field $\boldsymbol{F}$. Its intensity $\varkappa F$ is determined by the *susceptibility* $\varkappa$, a specific property of the material. The slope dB/dF [Eq. (53)] defines the (differential) magnetic permeability μ of the material for a given value of F. Hence, $\mu = 1 + 4\pi\varkappa$ (in cgs units).

Naturally occuring titano-magnetites have susceptibilities between 0.1 and 0.01 in weak fields ($F \leqq 1$ Oe), while the para- or diamagnetic susceptibility of silicatic minerals is of the order of 10^{-6}. The bulk susceptibility of rocks ranges accordingly from 10^{-6} to nearly 10^{-2}. Representative values are 10^{-5} (granite, gneiss, clastic sediments) and 10^{-3} (mafic igneous rocks). The alignment of elongated ferromagnetic minerals causes $\varkappa$ to become anisotropic (direction-dependent), a decrease in grain size usually effects a reduction of $\varkappa$. The susceptibility-temperature curve has a characteristic maximum just below the Curie point before its steep descent to zero.

The second component, $\boldsymbol{I}_{rn}$, is *remanent* and as such unaffected by the ambient field. It represents the magnetic memory of the rock for the geomagnetic field of the past. In particular, it preserves the direction of the paleomagnetic field at the time when the rock acquired its natural remanence which can occur in two principal ways.

Igneous rocks become magnetized when their ferromagnetic minerals pass through the *Curie temperature* during the cooling process. The direction of the contemporary geomagnetic field gives the preferred direction for the orientation of the newly formed elementary or Weiss' *domains* of *spontaneous magnetization*, a ferromagnetic speciality. The field strength determines the degree of alignment. This alignment is "frozen in" during the subsequent cooling while the spontaneous magnetization of the domains increases with decreasing temperature. The result is a stable *thermo-remanence TRM* of the rock. If several

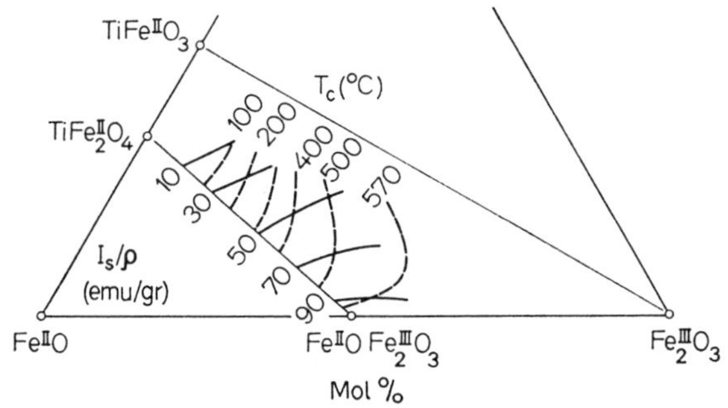

Fig. 6-25. Variation of Curie temperature (T_c) and saturation magnetization (I_s) of solid solutions between the binaries magnetite-ulvospinel ($TiFe_2O_2$) and hematite (αFe_2O_2)-ilmenite ($TiFeO_2$). The solid solutions, including magnetite and maghemite ($\gamma\, Fe_2O_2$) are ferrimagnetic; the magnetic spin moments of their Fe^{III} ions in octahedral and tetrahedral positions are opposite and self-cancelling which leaves the spins of the Fe^{II} ions responsible for the net moment of the solutions. Hematite and ilmenite are anti-ferromagnetic, i.e. the spin moments of their Fe and Ti-ions are self-cancelling. Substitution of Fe^{II} by Mg, Mn, Ni, ... tends to lower the Curie temperature. Data obtained from synthetic and natural titanomagnetites ($\varrho \approx 5.2$ gr./cm^3). (Adapted from Nagata, 1961)

ferromagnetic phases with different Curie temperatures are present, the mutual interaction of their partial TRM's can lead under special and seldom realized conditions to a total *TRM* at room temperature which is opposite to the direction of the ambient field (Uyeda, 1958).

The natural *TRM* of igneous rocks is several orders of magnitude greater than any isothermal remanence which a field of comparable strength can produce instantaneously without heating. However, the long-lasting isothermal action of the geomagnetic field over thousands of years may produce a sizeable *viscous remanence* which can obscure the original *TRM*. Fortunately, this viscous remanence is less stable and it can be destroyed by heating to moderate temperatures or by exposing the rock to alternating magnetic fields.

The ferromagnetic grains of igneous rocks retain their natural *TRM* during a later decomposition. If the sedimentation of detrital magnetite occurs under favorable conditions, the grains will align their intrinsic magnetic moments in the direction of the ambient field, thus giving the sediment a *detrital remanence DRM* in the direction of this field. A third and less important type of remanence results from reactions in the solid state, leading to the new formation of ferromagnetic phases (e.g. the reduction of hematite to magnetite). Directional stress and heating in the course of some subsequent tectonic or metamorphic process may produce a new piezo-magnetic or thermoremanence, thus altering the original *TRM* or *DRM* of the rock formation.

Even the most stable natural remanence will decay spontaneously on a geological time scale. Young rocks have, as a rule, a stronger remanence than old rocks of comparable composition and genesis. This makes it problematic to judge the intensity of the paleomagnetic field according to the present-day strength of the remanence which it produced.

In summary, the Precambrian basement and the deeper parts of the Earth's crust have a primarily induced magnetization ($I_{rn}/I_{in} \ll 1$, $I_{in} \approx 1-10\,\gamma$), while a predominant remanence is characteristic for young basic effusiva ($I_{rn}/I_{in} \gtrsim 1$, $I_{rn} \approx 30-300\,\gamma$). The ratio remanent to induced magnetization is known as the Königsberger *Q-ratio*.

f) Paleomagnetism

The stable or "non-viscous" part of the natural remanence of rocks faithfully reflects the local direction of the geomagnetic main field at some time instant during the geological past. Assuming that this field when averaged over secular changes has been that of a centred dipole, inclination and declination of the stable remanence yield immediately the paleo-geomagnetic latitude and longitude of the rock formation when it became magnetized (Fig. 6-20). Its present-day location fixes accordingly the polarity and position of the *paleo-geomagnetic* or *virtual pole* in geographic coordinates.

Careful comparative studies have shown that the restrictive assumption of a dipole field does not cast doubt upon the principal results of paleomagnetic research, *field reversals* and *polar wandering*. Rock formations which acquired their natural remanence during the same geological period in a certain area usually comprise two groups with opposing natural remanences. Hence, they yield identical pole positions but conflicting polarities of the paleo-pole.

Self-reversal of the remanence, as mentioned above, has to be ruled out as a general cause, leaving a polarity change of the centred dipole as the more likely explanation. Absolute datings of young effusiva of different continents established indeed that the polarity change of their remanence and thereby of the ancient field occurred simultaneously over the globe (Doell *et al.*, 1966). The oscillatory remanence, found in deep-sea cores, confirmed this result in a striking manner by yielding a matching sequence of normal and reversed geomagnetic epochs (Opdyke *et al.*, 1966). The last four million years have seen three well dated reversals at irregular intervals. Various attempts have been made to extend this paleomagnetic time scale deeper into the geological past (cf. Heirtzler, 1968).

The abrupt changes of the *DRM* polarity, found in deep-sea sediments, show that the polarity reversals of the dipole field must have been accomplished within a few thousand years, leaving the Earth without a significant field during this time. This time span is close to the free decay time of a dipole field, generated by electrical currents in a highly conductive core (Sec. IV, b).

A representative time constant for the stabilizing effect of self-induction is $L^2\,4\pi\sigma\mu$ with L as scale length of the decaying current system and σ as ambient conductivity in emu-cgs units (cf. Cowling, 1957). Assuming L to be a third of the core radius (1,100 km) gives an acceptable decay constant of 5,000 years for $\sigma = 10^{-6}$ emu (10^5 [ohm $\cdot$ m]$^{-1}$) and $\mu = 1$.

The position of the paleo-pole as seen from one particular continent has been changing in the course of the Earth's history in a systematic manner. The resulting *polar wandering curve* extends over half the globe. This movement of the virtual pole has to be apparent in the sense that it is the continent which moves relative to a fixed geomagnetic pole coinciding with the Earth's stable pole of rotation. This view also is favoured by theoretical arguments coupling the dipole axis with the Earth's axis of rotation and by paleoclimatic evidence.

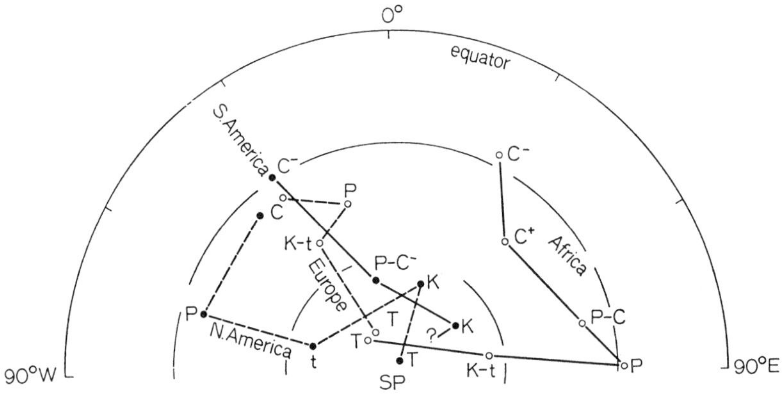

Fig. 6-26. Changing positions of the paleo-geomagnetic pole (cf. text) during the Earth's history, based on dated rocks with a non-viscous remanent magnetization. The statistical scatter of pole positions for one geological period is effectively reduced by combining the positions derived from rocks on the same continent or subcontinent. The resulting polar wandering curves are shown for the continents adjacent to the Atlantic ocean. They start from a Tertiary pole *T* near the present geographic south pole *SP* and diverge strongly for earlier periods (*K* Cretaceous; *t* Triassic; *P* Permian; C^+ and C^- upper and lower Carboniferous). This divergence excludes the hypothesis that the polar wandering represents a true shift of the geomagnetic pole of the ancient main field relative to the pole of rotation and implies continental drift. Reconstruction joints the paleozoic continents to a single supercontinent thus closing the Atlantic in accordance with WEGENER's original hypothesis. (Revised pole positions from CREER, 1968)

The consistency of the results from the same continent suggests that the continents have moved as rigid blocks, as already envisaged in WEGENER's classical theory of *continental drift*. A translatory displacement along a paleo-geomagnetic meridian causes a corresponding meridional shift of the paleo-pole. Similarly, a rotation of the continent around some fixed point *P* of observation leads to an apparent motion of the pole along a circle around *P*. Displacements along circles of paleo-geomagnetic latitude remain necessarily undetectable.

Polar wandering curves as obtained from the rock magnetization in different continents do not match, in particular not in their pre-jurassic sections (Fig. 6-26). This suggests that the continents have also moved relative to one another during the past. The discrepancy between the American and Euro-African curve, for instance, can be explained by a separation of these continental blocks commencing in the Jurassic period.

g) Electromagnetic Induction within the Earth

A time-varying magnetic field is reduced in amplitude and shifted in phase when penetrating electrically conducting matter. This shielding or *skin-effect* arises from the opposing action of eddy currents which are electromagnetically induced. Their strength depends on the ambient conductivity σ and on the magnitude of the time derivative of the magnetic flux (FARADAY's induction law). Hence, the degree of attenuation which the incident field suffers increases — for a given conductor — with the increasing lateral scale length and swiftness of the field fluctuation, i.e. with the frequency f and spatial uniformity of an oscillatory field.

A representative frequency-conductivity scale-length for this attenuation is the skin-depth

$$p = 1/\sqrt{2\pi\omega\sigma\mu} \tag{60}$$

($\omega = 2\pi f$), σ and μ to be measured in electromagnetic units. For geomagnetic induction problems, convenient units are (cph= cycles per hour) for f and $(\text{ohm} \cdot \text{m})^{-1}$ for σ. They yield ($\mu = 1$)

$$p = 30.2/\sqrt{f\sigma} \text{ [km]}.$$

Thus defined the skin-depth has a straightforward physical meaning when the conducting matter is uniform, occupying the lower half-space beneath a plane surface. Providing that p is small in comparison to the lateral scale length of an oscillatory magnetic surface field which originates from sources in the upper non-conductive half-space, p is that depth beneath the surface where the amplitude of the incident field is reduced to e^{-1} of its surface value ($e = 2.718\ldots$). The same applies to the amplitude of the associated electric field E and thus to the density $j = \sigma E$ (OHM's generalized law) of the induced current.

Two principal types of oscillatory magnetic fields pass in this way through the Earth's crust and mantle, testing their electrical conductivity. Secular variations of the quasi-persistent main field $\boldsymbol{F}_0$ diffuse *upward* from hypothetical sources in the core, while the transient geomagnetic variations $\delta\boldsymbol{F}$, comparatively "rapid" fluctuations of external origin, diffuse from the Earth's surface *downward* into the interior. In the first case we observe at the surface the attenuated variation field, not knowing its original strength and frequency spectrum. In the second case the situation is reversed, i.e. only the un-attenuated variation field is accessible to direct observations. It is possible in either case to come to certain conclusions about the internal distribution of conductivity which in turn may augment our knowledge about the internal thermal state.

The electric conduction in near-surface rock formations is mainly electrolytic through salty solutions filling pores and cracks. Their conductivity is thus quite variable. Representative values are 0.2 $(\text{ohm} \cdot \text{m})^{-1}$ for sediments and 0.005 for igneous and metamorphic rocks. Sea water in comparison has an average conductivity of 4, copper a conductivity of 10^8 $(\text{ohm} \cdot \text{m})^{-1}$ (Fig. 6-27).

Rocks become more and more insulating under pressure when their pores and cracks are closed. There is evidence that the Earth's crust and uppermost mantle are indeed poor conductors ($\sigma \leqq 10^{-3}$ $[\text{ohm} \cdot \text{m}]^{-1}$). At greater depth the conductivity rises again to values in excess of 1 $(\text{ohm} \cdot \text{m})^{-1}$ (see below), and it is not unreasonable to see in this rise a manifestation of the downward increase of temperature.

Olivine and other prospective constituents of the Earth's mantle are *semiconductors*, i.e. their conductivity increases with rising absolute temperature T according to the general formula

$$\sigma(T) = \sigma_0\, e^{-T_0/T}. \tag{61}$$

The parameters σ_0 and T_0 are pressure and composition dependent and apply to one particular mechanism of semi-conduction by impurities, electrons, or ions. Hence, σ_0 and T_0 may be regarded as constants within limited temperature ranges in which one particular conduction mechanism predominates. The temperature increment ΔT needed to double the conductivity within that range is evidently

$$\Delta T = \ln 2 \cdot T^2/T_0 \tag{62}$$

providing $\Delta T \ll T$. Inserting $T = 1{,}500°$ K and $T_0 = 31{,}200°$ K (HUGHES' value for olivine, Fig. 6-27) gives $\Delta T = 50°$ C. This illustrates the decisive influence of temperature upon the conductivity to be expected in the upper mantle. Eq. (61) implies that plots of $\log \sigma$ versus the inverse absolute temperature yield for semi-conducting materials segments of straight lines. This is indeed observed for many rock-forming minerals (Fig. 6-27).

The influence of pressure seems to be more complicated. HUGHES (1955) found at high temperatures (1,330—1,530° K) that the conductivity of olivine decreases at the constant rate of 4% per kbar (10^9 dynes/cm^2) for pressures up to 8 kbar. This he explained by a corresponding increase in T_0, i.e. in the activation energy necessary to mobilize charge carriers. HAMILTON (1965) found the opposite to be true at low temperatures (500—900° K).

The total differential change of conductivity under the combined influence of temperature and pressure can be written as

$$\frac{\Delta\sigma}{\sigma} = \frac{T_0}{T^2}\Delta T + \frac{\partial \ln \sigma}{\partial p}\Delta p, \tag{63}$$

thus avoiding the largely composition dependent parameter σ_0.

Inserting HUGHES' values and measuring Δp in (kbar) gives for $T = 1{,}500°$ C as ambient temperature

$$\frac{\Delta\sigma}{\sigma} = 0.014\,\Delta T - 0.04\,\Delta p. \tag{63a}$$

Hence, as already stated by HUGHES, a downward gradient of 1° C/km near the 1,500° C isotherm is sufficient to cancel the opposing effect of the pressure gradient (0.35 kbar/km). HAMILTON's values for $T = 900°$ K yield nearly the same temperature term but a reduced and positive pressure term.

The swiftness of secular variation reaching the Earth's surface puts an upper limit upon the inductive smoothing effect of conductive matter between the surface and the mantle-core boundary. The maximum permissible conductivity for the mantle as a whole turns out to be 50 (ohm · m)$^{-1}$ (ELSASSER, 1950), indicating that electrical semi-conduction in iron-magnesium silicates and oxides prevails throughout the mantle. Because the upper mantle is known to be less conductive (see below), the conductivity of the lower mantle may approach 1,000 (ohm · m)$^{-1}$ (Fig. 6-28). Estimates of the conductivity in the metallic core are of the order of 10^5 (ohm · m)$^{-1}$.

The downward diffusing geomagnetic variations provide more detailed information about the internal conductivity structure at shallow depth. The variation field $\delta \boldsymbol{F}$ as observed at the Earth's surface can be split into parts of *external* and *internal* origin in the same way as the quasi-persistent field (cf. comment to Eq. (56)). Beginning with SCHUSTER's analysis of the diurnal variations in 1889 it became clear that the internal part of geomagnetic variations thus deduced arises solely from electrical currents which are induced by the transient external part within the Earth. The surface ratio $\delta Z/\delta H$ of vertical to horizontal variations becomes in this way a sensitive indicator of the depth and distribution of these currents when set in relation to the lateral non-uniformity of the inducing field.

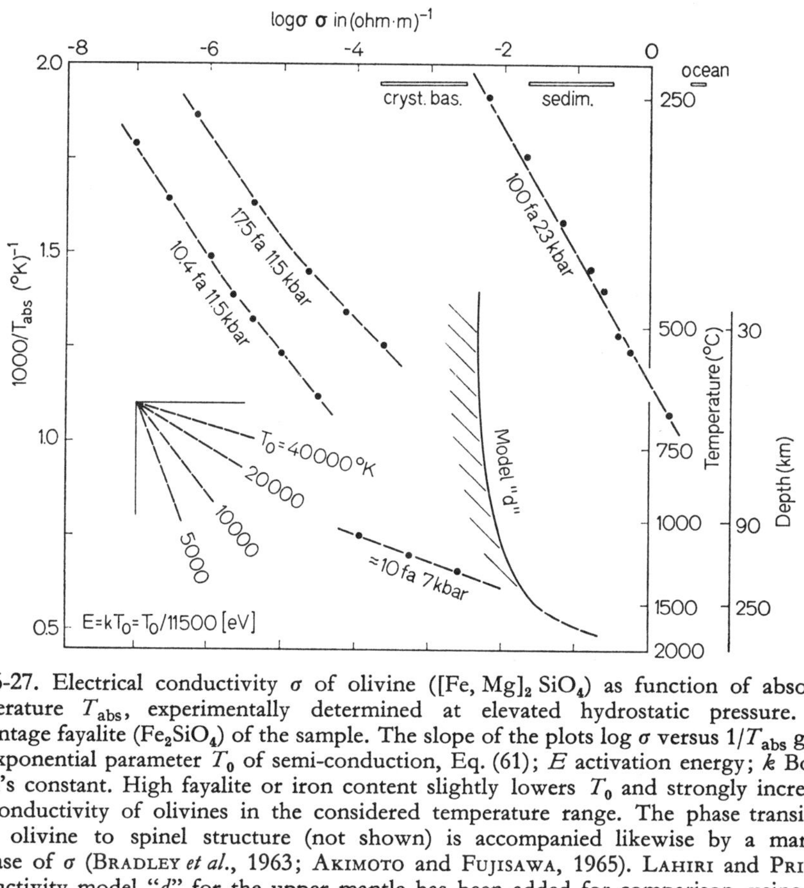

Fig. 6-27. Electrical conductivity σ of olivine ($[Fe, Mg]_2 SiO_4$) as function of absolute temperature T_{abs}, experimentally determined at elevated hydrostatic pressure. fa: percentage fayalite (Fe_2SiO_4) of the sample. The slope of the plots $\log \sigma$ versus $1/T_{abs}$ gives the exponential parameter T_0 of semi-conduction, Eq. (61); E activation energy; k Boltzmann's constant. High fayalite or iron content slightly lowers T_0 and strongly increases the conductivity of olivines in the considered temperature range. The phase transition from olivine to spinel structure (not shown) is accompanied likewise by a marked increase of σ (Bradley *et al.*, 1963; Akimoto and Fujisawa, 1965). Lahiri and Price's conductivity model "*d*" for the upper mantle has been added for comparison, using the temperature-depth relation for a normal continental heat flow (Figs. 6-28, 6-30). This model represents an upper limit for mantle conductivities except for regional uplifts of better conducting and possibly hot material (cf. text). The discrepancy of the slopes of model "*d*" and the experimental curves suggests that the downward rise of upper mantle conductivity is steeper than that indicated by model "*d*". (References: 100 fa Bradley *et al.*, 1963; 17.5 and 10.4 fa Hamilton, 1965; 10 fa Hughes, 1955)

The third observable quantity is the superficial intensity of the induced *earth currents* or the tangential component δE of the transient electric field driving them. (The electromagnetic unit of δE is equivalent to 1 millivolt/km). The surface ratio $\delta E/\delta H$ or the *impedance* of the variation field is — under certain conditions — a direct measure for the vertical distribution of the induced currents. A simplified model may explain how these ratios can be used to find the internal distribution of conductivity.

Consider a geomagnetic variation field of external origin which propagates in the form of a sinusoidal modulation with the speed ω/k in the positive x-direction of rectangular coordinates, z down. Its time-space dependence at the Earth's surface ($z=0$) is in complex notations $\exp(i[\omega t - kx])$; $2\pi/k$ represents the spatial wave length of the modulation, k^{-1} its lateral scale length, ω the angular frequency of harmonic time variation, observed at a fixed site. The conductivity in the upper half-space (air) up to te sources of the field will be zero, in the lower half-space (Earth's interior) a function of depth only, $\sigma(z)$.

The pertinent ratios for the surface field have the form ($i=\sqrt{-1}$)

$$\frac{\delta Z}{\delta H}=i k c \tag{64}$$

$$\frac{\delta E}{\delta H}=i \omega \mu c, \tag{65}$$

δE being normal to δH;

c is a complex-valued vertical scale length (cm) for the attenuated modulation in the conducting substratum and thus a function of ω, k, and $\sigma(z)$. The moduli of the complex expressions to the right give the amplitude ratios of the field components to the left, their arguments the respective differences in phase (time or space).

The vertical scale length c is independent of the spatial wave length if for the frequency considered $|kc| \ll 1$. In that case c indicates with its real and imaginary parts the centred depth of the in-phase and out-of-phase induced current distribution, when their phase is referred to δH. The connection of c with the vertical conductivity distribution is as follows: Set

$$c=h^* - i\tfrac{1}{2}p_a \tag{66}$$

and interpret

$$p_a=(2\pi\,\omega\,\sigma_a\,\mu)^{-\frac{1}{2}} \tag{66a}$$

as *apparent skin depth* for $z=h^*$ at the frequency considered (h^* and p_a are always positive and $dh^*/d\omega>0$). The resulting apparent conductivity versus depth profile, $\sigma_a(h^*)$, obtained from c as function of frequency, approximates the true vertical distribution $\sigma(z)$ very closely as a rule. — CAGNIARD's (1953) original definition of the apparent conductivity is different. It is based on the modulus of c, setting $p_a=\sqrt{2}\,|c|$, and leaves the argument of c unaccounted for. —

Eq. (64) illustrates the basic concept to derive the internal conductivity distribution from the $\delta Z/\delta H$ ratio, involving a wide spectrum of oscillatory or transient geomagnetic fields (CHAPMAN and PRICE, 1930; LAHIRI and PRICE, 1939). The required knowledge of the spatial configuration of the surface field (here contained in k) can evolve from simultaneous observations at different sites on a sufficiently large scale. (Most source fields for geomagnetic induction have global dimensions.) The ratio $\delta Z/\delta E=k/(\omega\mu)$ allows, at least in theory, an independent determination of this configuration.

Eq. (65) is the basic relation for combined magnetic and electric or *magnetotelluric* observations, to obtain $\sigma(z)$ (TIKHONOV, 1950; CAGNIARD, 1953). This method obviously does not depend on an exact knowledge of the surface field configuration, provided that its lateral scale length is sufficiently large in comparison to $|c|$ (see above).

The restrictive assumption of a stratified conductivity distribution which certainly does not apply to superficial rock formations and oceans limits a straightforward use of these relations. So far little progress has been made in treating the induction in a non-stratified Earth. However, superficial conductivity contrasts distort primarily the flow of superficial earth currents and hence the electric surface field. But the concurrent distortion of the magnetic surface field remains insignificant when the contribution of superficial earth currents to the internal part of $\delta \boldsymbol{F}$ is small for the region and frequency considered.

Particular attention has been paid to diurnal variations which originate from current vortices in the day-time high atmosphere and to the slow recovery phase of an equatorial ring current encircling the Earth after magnetic storms. The effective frequency spectrum of these world-wide disturbances extends from a few cycles

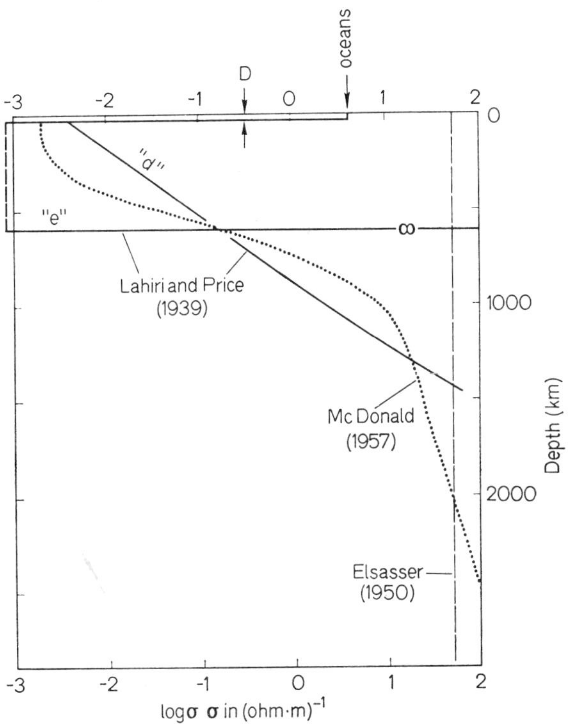

Fig. 6-28. Electrical conductivity as a function of depth in the Earth's mantle, deduced from the electromagnetic induction by time-varying geomagnetic fields of global extent. Models "*d*" and "*e*" of Lahiri and Price (1939) represent limiting distributions for the steep rise of conductivity in the upper mantle. They are compatible with the induced parts of diurnal and smoothed storm time variations when averaged over the globe. Both models incorporate a thin surface shell with an integrated conductivity equivalent to $D = 1{,}280$ m ("*e*") and $D = 500$ m ("*d*") sea water. It represents the oceans and continental surface layers or any other thin layer of comparable integrated conductivity at shallow depth. McDonald's (1957) distribution is a compromise between models "*d*" and "*e*" in the upper mantle and in accord with the frequency cut-off of secular variations in the lower mantle

to fractions of a cycle per day. Their lateral scale length is of the order of the Earth's radius and shielding by superficial earth currents should be small.

The world-wide magnitude of the Z/H-ratio reveals that these variations penetrate deeply into the Earth's mantle, inducing eddy currents at a scale depth of 600 km. Hence, the inductive shielding of the crust and uppermost mantle must be small for the indicated frequency range, i.e. the mean conductivity above, say, 250 km can hardly exceed 0.01 (ohm · m)$^{-1}$. At 600 km depth, where effective shielding sets in, the conductivity increases to at least 0.2 (ohm · *m*)$^{-1}$ with the tendency to rise (Fig. 6-28).

This remarkable change of the conductivity by two orders of magnitude occurs in the inhomogeneous *C* region of the mantle. It is therefore not readily explainable in terms of the downward increase of temperature and pressure without allowing for the additional influence of phase transitions and compositional changes upon σ.

h) Induction Anomalies

There exist considerable regional irregularities in the $\delta Z/\delta H$ ratio of diurnal variations which are not compatible with a purely downward change of the internal conductivity so far presumed. Faster variations with a reduced depth of penetration exhibit an even more complex behaviour in the same regions. It has been proven that it is indeed their internal or induced part which behaves anomalously under the influence of lateral internal conductivity gradients.

Geomagnetic *induction anomalies* of this kind have been encountered at many places, even though the depth of their origin is not always certain. The principal uncertainty arises in the case of fast variations from the largely unknown extent of induction in the outermost layers consisting of continental rock formations of variable conductivity and of oceans. Nevertheless, the upper mantle appears to be a highly non-uniform conductor and a causal connection with thermal imbalances is possible. This view is supported by the results of magneto-telluric and heat flow observations (Sec. V, a).

Current interest is focussed on induction anomalies of *bays* and similar geomagnetic variations in the frequency range of 1 cycle per hour. Their vertical scale depth is 250 km under ordinary continental conditions (LAHIRI and PRICE's model "*d*", Fig. 6-28), but it can be locally or regionally reduced to 100 km or less by unusually high conductivities at shallow depth, presumably in the uppermost mantle. This applies to parts of the Japanese island arc (RIKITAKE, 1959), to the Rio Grande rift valley in North America and to the adjoining Rocky Mountain system in general (SCHMUCKER, 1964; CANER *et al.*, 1967). Other prospective zones of high mantle conductivity are, among others, the Baikal Lake area (VANJAN and KHARIN, 1967), the Peruvian Andes (SCHMUCKER *et al.*, 1966), the Hungarian plane (ADAM *et al.*, 1964; WIESE, 1965), and parts of the Canadian Arctic archipelago (WITHAM, 1964).

A representative conductivity for these uplifts of highly conducting matter is 0.1 (ohm · m)$^{-1}$. It corresponds to an ambient temperature of 1,600° C for olivine ($\approx$10% fayalite) under a pressure of about 10 kbar (Fig. 6-27).

The situation is complicated near coastlines because bay-type disturbances do not seem to penetrate large and deep oceans to any extent. The consequence is pronounced coastal anomalies of the $\delta Z/\delta H$ ratio, observed, indeed, at most continental margins (PARKINSON, 1962). It is still undecided whether the deep conductivity structure beneath the ocean is basically different from that under continents. Magnetic and magnetotelluric observations in offshore California, finding high conductivities at shallow depth beneath the ocean floor, suggest that this might be the case (FILLOUX, 1967).

V. The Thermal State

a) Terrestrial Heat Flow and Volcanism

The inner heat of the Earth is observable in the volcanic discharge of melts and steams, in hot springs, in contact and regional metamorphosis, and — most important of all — in the rise of temperature with increasing depth common to deep mines and boreholes. The last-mentioned general *geothermal gradient* dT/dz in combination with the *thermal conductivity* K of the rock formations causes a continual upflow of internal heat to the Earth's surface ($z=0$) from which it escapes into space as *terrestrial heat flow* at the rate of

$$q_0 = K\left(\frac{\partial T}{\partial z}\right)_{z=0} \tag{67}$$

per unit surface area [Eq. (74)].

Rocks at low temperatures have an effective thermal conductivity of about 6 mcal/(° C cm sec), 1 mcal $=10^{-3}$ cal. Outside volcanic areas the geothermal gradient measures 15 to 35° C/km, which implies a *conducted* heat flow of 1 to 2 μcal/(cm² sec), 1 μcal $=10^{-6}$ cal. A representative global average *is 1.5 μcal/(cm² sec)* or 1 cal/m² per minute. Integrated over the whole surface of the Earth this gives, a total outflow of internal heat of *$760 \cdot 10^{10}$ cal/sec* (Table 6-6).

A small fraction of the current heat flow can be explained by the loss of initial heat retained by the Earth's body since its formation more than four billion years ago (cf. Sec. V, f). Its major part must come from secondary heat sources which are considered in the following sections.

The daily influx of energy into the sunlit high atmosphere measures, for vertical incidence, 20,000 cal/m² per minute and thus is four orders of magnitude greater than the thermal energy outflow from the Earth's interior. Hence, the large-scale thermal balance of the atmosphere and of well-mixed oceans is hardly influenced by the terrestrial heat flow.

Heat loss through volcanism is insignificant. Volcanoes are estimated to bring to the Earth's surface one cubic kilometer of lava per year. If lava cools from the temperature T to $(T-\Delta T)$, the heat loss per unit volume is

$$E/V = \varrho\, c_p\, \Delta T \tag{68}$$

for an unrestricted change of volume, c_p denoting the specific heat at constant pressure. A representative value for rocks is $c_p=0.2$ cal/(gr. °C). After adding the latent heat of solidification $L^*=100$ cal/gr., the total energy release is $(6.7+2.8)\cdot 10^{17}$ cal when one cubic kilometer of lava cools from 1,200° C to room temperature ($\varrho=$ 2.8 gr./cm³). The volcanic heat balance may be regarded as stationary on a global scale, assuming that lava cools at the same rate at which hot melts are brought up at other places. This implies that internal heat is lost by volcanism at the relatively slow rate of *$3\cdot 10^{10}$ cal/sec.*

A heat flow of *1.2 μcal/(cm² sec)*, prevailing at a distance from centers of recent tectonic or magmatic activity, is regarded as *normal* for continental and oceanic regions (cf. von Herzen, 1967). Because of the bias of high heat-flow values, the mean heat flow in either region exceeds the normal one by 0.3. The median of all determinations on land as well as at sea is about 1.3 (cf. Lee and Uyeda, 1965).

The equality of the continental and oceanic terrestrial heat flow became evident after BULLARD, MAXWELL, and REVELLE (1956) succeeded in making the first reliable heat flow determinations on the sea floor. It implies that the integrated secondary heat production beneath surface areas of equal size is basically the same for continents and oceans in spite of the great difference in crustal thickness. As seen in Sec. V, f, about one half of the normal continental heat flow can be referred to the radiogenic heat production in the crust, while the normal oceanic heat flow must be almost entirely of subcrustal origin. Because the radiogenic heat production in ultramafic subcrustal material is small, the required heat sources beneath the oceanic crust are spread over a wide depth range. As a consequence, the oceanic substructure will be "hotter" than the continental substructure at comparable depth (Figs. 6-29 and 6-30).

Detailed studies have revealed some coherence between regional deviations from normal heat flow and geological structural provinces. Stable continental shields tend to have a slightly subnormal heat flow around 1.0 μcal/(cm^2 sec). Many zones of recent mountain building have higher-than-normal heat flow, reflecting the presence of deep-seated hot magma lenses. Very high heat flow (>2.0) is characteristic of mid-oceanic ridges, but it is also found on continents, as for instance in the western United States and in Hungary. Heat flow lows (<0.8) accompany the flanks of mid-oceanic ridges, even though a distinct geographical association of heat flow highs and lows has not yet been proven.

There is a remarkable lack of correlation between the rate of the surface heat flow and the thickness of the crust. Hence, regional heat flow variations reflect thermal imbalances in the upper mantle rather than varying degrees of crustal radiogenic heat production. This view is supported by geomagnetic observations. They indicate that various well-established continental heat flow highs are also zones of high electrical mantle conductivity and therefore prospective zones of higher-than-normal mantle temperature (cf. Sec. V, f).

b) Non-Radiogenic Heat

The thermal balance of the Earth's interior involves two types of continuous heat production which are not connected with the decay of radioactive elements. *Frictional heat* results from a conversion of strain energy when internal matter does not respond with perfect elasticity to stress. This applies in particular to the energies associated with earthquakes and Earth's tides (Sec. II, g). In the Earth's early history, the kinetic energy of incoming bits of planetary matter contributed in a similar way to the initial heating of the Earth's nucleus formed by gravitational aggregation.

The dissipation of tidal strain energy leads to a reduced and, more important, to a delayed response of the Earth and its oceans to tidal forces [cf. Eq. (34)]. As a result, the tidal bulge is not in line with the direction of the gravitational pull causing it and is hence subject to a mechanical couple. It tends to reduce the Earth's angular velocity Ω, while the lost angular momentum is transferred to the orbital motion of the attracting celestial body (cf. Sec. VI).

MUNK and MACDONALD (1960) give a contemporary total tidal deceleration of the Earth of (ratio solar to lunar torques due to ocean tides 1:3.4)

$$\dot{\Omega} = 5.8 \cdot 10^{-22} \text{ radians/sec}^2$$

besides a recently observed small acceleration of unknown origin. This value corresponds to a lengthening of the day by 2.2 milliseconds per century. Hence, rotational energy is lost presently at a rate of [Eq. (115)]

$$\dot{E}_{\text{rot}} = \Omega \cdot \dot{\Omega}\, C = 82 \cdot 10^{10} \text{ cal/sec} \tag{69}$$

with $\Omega = 2\pi/(\text{sidereal day})$ and C from Eq. (47a).

Only one tenth of this energy can be associated with the irreversible tidal deformation of the Earth's interior, the rest must be dissipated by bottom friction and internal wave motion in the oceans and marginal seas.

The estimate for the solid Earth is based on a tidal-effective dissipation constant of $Q^{-1} = 0.01$. Assuming that these $8.2 \cdot 10^{10}$ cal/sec are dissipated in the "soft" zone of the upper mantle between 50 and 350 km depth (Fig. 6-15), the rate of heat production by tidal friction becomes here 0.6 cal/km³ per second or one order of magnitude less than the expected rate of ambient radiogenic heat production (Table 6-1).

The release of stored strain energy by earthquakes is locally concentrated in areas of outstanding seismicity. South America north of the 37th parallel and the Japan-Kamchatka region alone contribute one third of the estimated global seismic energy output of 10^{12} cal/sec (GUTENBERG and RICHTER, 1949).

Latent heat is released when molten matter solidifies or when it transfers, in response to changing conditions of state, to a less closely packed solid phase. The thermal energy associated with endothermic chemical reactions falls into the same category of internal heat production. A representative value for the latent heat of solidification of molten magmas or lavas is 100 cal/gr.

Solidification, phase transitions and chemical reactions are accomplished, wherever they may occur, within a limited span of time, yielding no more than a transitory contribution to the thermal balance of the Earth's interior. This does not withstand the fact that the latent heat of phase transitions imposes a formidable barrier against thermal convection, as discussed in Sec. V, i.

c) Radiogenic Heat

The decay energy of radioactive elements represents by far the most effective source of internal heat. Short-lived isotopes may have been largely responsible for the initial heating of the primitive Earth. On a geological time-scale, however, radiogenic heat is derived almost exclusively from the slow decay of four isotopes with half-lives of the order of 10^9 years. They are the uranium isotopes U^{238} and U^{235}, decaying into Pb^{206} and Pb^{207}, thorium Th^{232}, decaying into Pb^{208}, and the potassium isotope K^{40}, decaying into Ca^{40} (β-decay) and Ar^{40} (K-electron capture).

Let c_i denote the concentration and ε_i the decay energy per unit mass and time of the ith radioactive isotope, contained in internal matter of the bulk density ϱ. Then radiogenic heat is produced at the rate

$$A = \varrho \sum_i \varepsilon_i c_i \tag{70}$$

per unit volume, the summation being over all isotopes. Measuring ϱ in (gr./cm³) and the concentrations in (ppm = grams per metric ton of internal matter) yields

$$A = \varrho\, [(22.4 + 1.0)\, c_{\text{U}} + 6.3\, c_{\text{Th}} + 8 \cdot 10^{-4}\, c_{\text{K}}]\, \frac{\text{cal}}{\text{km}^3 \text{ sec}} \tag{70a}$$

with the decay energies of Table 12—5.

The heat produced by uranium is given separately for U^{238} (22.4) and U^{235} (1.0), using their present isotope abundances of 99.27 gr. U^{238} and 0.72 gr. U^{235} per 100 gr. U. Heat production by potassium is based on an abundance of 0.012 gr. K^{40} per 100 gr. K.

While radiogenic heat is conducted from the Earth's interior to the surface, its sources decay exponentially with time. The pertinent time constant [Eq. (88)] for the conduction of heat from the upper mantle to the surface is of the order of 10^9 years and thus comparable to the half-lives of the four isotopes. Hence, the contribution of radiogenic heat sources in the upper mantle to the present-day surface heat flow will reflect a mean production rate $\tilde{A}(t)$ for the time t since this conduction started.

Table 6-1. *Radiogenic heat production in rocks and stony meteorites* [Compiled according to the tables of CLARK *et al.* (1966) and WAKITA *et al.* (1967)]

Rock type	Concentrations (ppm)			Density (gr./cm³)	Heat production rate (cal/km³/sec)	
	c_U	c_{Th}	c_K		A[a]	$\tilde{A}$ (4.5 AE)[b]
Granitic igneous rocks	4.7[c] (535)[d] 2.3—5.5[e]	20 (156) 11—22	36,000[f] 18,000—42,000	2.65	710 370—800	1,080
Granodiorites	2.6 (156) 1.5—3.0	9.3 (92) 6.5—11.0	20,000[f] 12,000—23,000	2.70	360 230—430	570
Mafic igneous rocks	0.9 (236) 0.3—1.0	2.2 (169) 0.8—2.5	15,000[g] 5,000—17,000	2.80	130 50—150	240
Dunites, peridotites (mostly nodules)	0.01—0.05 (25)	0.03—0.12 (21)	30—180 (19)	3.25	1.5—6.8	2.2—10.5
Eclogites (excl. pipe eclogites)	0.04—0.20 (13)	0.2—0.5 (13)	200—2,400 (4)	3.4	8—33	12—57
Carbonaceous chondrites	0.01—0.02 (3)	0.07 (1)	410—730 (10)	3.3	3.3—5.0	6.0—9.8
Ordinary chondrites	0.01—0.02 (20)	0.04—0.06 (8)	840 (38)	3.3	3.8—5.0	8.6—10.4

[a] Eq. (70a).
[b] Eq. (73).
[c] Mean value.
[d] Total number of individual analyses and average values.
[e] Range of centred 50%.
[f] Calculated for $c_U/c_K = 1.3$.
[g] Calculated for $c_U/c_K = 0.6$.

Let c_i^0 be the initial concentration of the i-th isotope at the time $t=0$ and λ_i its decay constant. Then

$$\tilde{A}(t) = \varrho \sum_i \varepsilon_i c_i^0 \tilde{g}_i(\lambda_i t) \tag{71}$$

defines a *weighted mean production rate* for the time interval t with

$$\tilde{g}(\lambda t) = \frac{e^{-\lambda t}}{\sqrt{\lambda t}} \int_0^{\sqrt{\lambda t}} e^{u^2} du \tag{72}$$

as weighting factor. Its implication will be shown in Sec. V, e. The integral appearing in Eq. (72) is tabulated as DAWSON's integral. For $\lambda_i t \ll 1$ the weighting factor becomes unity, while it approaches zero as $(2\lambda_i t)^{-1}$ for $\lambda_i t \gg 1$.

Using the decay constants of Tables 12—5 and $t = 4.5$ billion years as the age of the Earth, the $\tilde{g}$-factors are 0.645 for U^{238}, 0.135 for U^{235}, 0.86 for Th, and 0.272 for K^{40}. Hence,

$$\tilde{A}\,(4.5\ \text{AE}) = \varrho\,[(29+11)\,c_U + 6.8\,c_{Th} + 23\cdot 10^{-4}\,c_K]\,\frac{\text{cal}}{\text{km}^3\,\text{sec}} \tag{73}$$

(10^9 *years* $= 1 AE$) when the present-day concentrations $c_i = c_i^0 \exp(-\lambda_i t)$ and the isotope abundances, quoted above, are inserted. The concentrations are measured again in (ppm).

A comparison with Eq. (70a) shows clearly that the fast-decaying isotopes U^{235} and K^{40} have contributed significantly to the heat which has been conducted from greater depth to the surface in the course of the Earth's history.

The concentrations of uranium, thorium, and potassium in a particular type of rock are to some degree correlated, as we know from geochemical studies (cf. Chapters 90 and 92, vol. II). In common igneous rocks thorium is three or four times more abundant than uranium, potassium about 10^4 times more abundant than uranium. These ratios imply that the four isotopes here considered have generated comparable amounts of radiogenic heat when averaged over the age of the Earth according to Eq. (73).

Data for various rock types and stony meteorites are listed in Table 6-1. High grade metamorphic rocks like granulites have less uranium and thorium than their non-metamorphic igneous equivalents and hence a smaller heat production.

It is evident that heat-producing elements are inversely related to dark minerals. Therefore, radiogenic heat production will be high in the granitic upper crust, considerably lower in the mafic lower crust, and much lower in the presumably ultramafic upper mantle. A representative present-day production rate for crustal material is $A = 200$ cal/(km³ sec). A peridotitic upper mantle can be expected to have generated heat at a weighted mean rate of 5 cal/(km³ sec) for $t = 4.5$ AE.

d) Heat Transfer

The transfer of thermal energy within the Earth's interior is accomplished by *conduction*, *radiation*, and *advection*. Advection is a transport of heated matter by viscous flow and is considered separately (Eq. [80]).

The amount of conducted and radiated heat passing through a unit area of cross-section per unit time defines the *heat flow vector*

$$\boldsymbol{q} = -K \operatorname{grad} T. \tag{74}$$

Pointing in the direction of falling temperature T, its value depends for a given temperature gradient on the *effective thermal conductivity* K of the medium, a scalar when the thermal properties are isotropic.

The effective conductivity of igneous and metamorphic rocks at room temperature ranges from about 4 to 8 mcal/(° C cm sec) for granitic and basaltic material. The conductivity of ultramafic material is about twice as high (Clark, 1966a).

In non-metallic crystallized solids the conduction of heat implies the propagation of thermal energy by elastic lattice vibrations. In metals heat is spread mainly by the kinetic energy of free electrons. *Lattice conduction* predominates in the Earth's crust and mantle, *metallic conduction* in the Earth's core. The transmittancy of a lattice for thermal energy diminishes when its vibrations become very intense. In fact, the thermal *lattice conductivity* K_l may be regarded as being proportional to the inverse absolute temperature,

$$K_l = C_l / T \tag{75}$$

or, to a higher degree of approximation, $K_l \sim T^{-\frac{5}{4}}$ (Lubimova, 1958).

The parameter C_l is, for a given composition, dependent on the prevailing pressure, increasing when homogeneous solid matter is compressed. Representative zero-pressure values are 2 cal/(cm sec) for crustal and 3.5 cal/(cm sec) for ultramafic subcrustal material. They have been deduced from CLARK's tables by subtracting from the measured effective conductivities the radiative component K_r with a proper choice of the transparency [cf. Eq. (77)].

The second non-convective mode of internal heat transfer results from the emission and absorption of electromagnetic radiation. Its importance for the internal thermal balance was first recognized by CLARK (1957) and LUBIMOVA (1958).

Integrated over all wavelengths, radiation energy is emitted at a rate of (sT^4) per unit surface area of heated matter according to the STEFAN-BOLTZMANN law; $s = 1.3 \cdot 10^{-12}$ *cal/(cm² sec °K⁴)*. The degree of transmission of radiated energy through the surrounding matter depends on its *transparency* ε^{-1} as function of the wavelength in the spectral range considered. This transparency, a length, refers to the mean free path of a quantum of radiation, measuring a few millimeters or less in olivine crystals at low temperature (see below). Its reciprocal is the *opacity* ε.

By combining the emission and transmission laws a formal *radiative thermal conductivity* can be introduced, given for one particular wave length by

$$K_r = C_r/T \tag{76}$$

with

$$C_r = A\varepsilon^{-1} s T^4$$

(CLARK, 1957). When added to the lattice conductivity from Eq. (75), it yields the *effective thermal conductivity*

$$K = (C_l + C_r)/T \tag{77}$$

of internal matter at the elevated absolute temperature T. The dimensionless parameter A in Eq. (76) is about 15 for material with an optical refraction index of 1.7.

CLARK, studying olivine samples, found a spectrum of transmission limited to infrared wavelengths between 0.3 and 12 microns. Within this range radiation energy is dissipated (i) by dispersion, (ii) by absorption bands associated with the presence of transition elements such as iron, and (iii) by a general background absorption, due to free electrons in the conduction band of the semi-conductors here considered. These free electrons reduce the transparency and hence the radiative heat transfer while providing for the electrical conductivity σ of the material.

If we ignore band absorption, the opacity can be considered to rise linearly with increasing conductivity,

$$\varepsilon(T) = \varepsilon_0 + a\,\sigma(T), \tag{78}$$

where $\sigma(T) \sim \exp(-T_0/T)$ in accordance with Eq. (61); the zero-temperature term ε_0 accounts for the absorption by dispersion. Experimentally determined transparencies of iron-magnesium silicates show considerable scatter when averaged over the transmitted spectrum at room temperature. CLARK (1957) gives 10 to 30 cm⁻¹ as a reasonable estimate of ε_0 and he sets $a = 220\,\Omega$ on the basis of theoretical arguments.

With these values heat radiation becomes the predominant mode of (non-convective) heat transfer below the 800° C isotherm, causing a steep rise of the effective thermal conductivity with depth in the upper mantle (Table 6-2). This rise levels off near the base of the upper mantle, where the electrical conductivity exceeds

Table 6-2. *Thermal (K) and electrical (σ) conductivity in crust and upper mantle*

K: Calculated from Eq. (76)-(78) with $\varepsilon_0^{-1} = 0.1$ cm, $A = 15$ and $a = 220\ \Omega$. σ: Lahiri and Price model "*d*" (Fig. 6—28).

z (km)	T (° C)	σ $(\Omega$ m$)^{-1}$	ε (cm^{-1})	C_l (cal/cm sec)	C_r (cal/cm sec)	K (mcal/° C cm sec)
0	0	>0.01	10	2	0.01	7
30	500				0.72	3.5
100	1,000	>0.01	10		5	7
400	1,800	0,04	10,1	3.5[a]	36	19
1,000	2,500	3	17		68	26

[a] Value for ultramafic material at zero pressure.

1 $(\Omega \cdot \text{m})^{-1}$ and where absorption by free electrons becomes decisive. The lower crust appears to be a zone of minimum effective conductivity acting as a zone of thermal insulation for the deeper interior.

e) Basic Equation of Heat Diffusion

The state of *thermal equilibrium* within a closed surface S implies that the outflow of conducted heat through S is balanced by an equal amount of heat which is generated within or advected to the volume enclosed by S. Accordingly, the basic differential equation for stationary thermal equilibrium condition is

$$\text{div}\ \boldsymbol{q} = A \tag{79}$$

with q as heat flow vector [Eq. (74)]; A is either the production rate per unit volume of fixed heat sources within S or, in the case of advection,

$$A = \varrho\, c_p (\boldsymbol{v} \cdot \text{grad}\ T) \tag{80}$$

with $\boldsymbol{v}$ as velocity vector of the material in motion, transporting heat into S.

If $A = A(z)$ within the Earth is a function of depth z alone for $b \leqq z \leqq a$ and zero for $z > a$, the integration of Eq. (79) gives

$$q(b) = A_a^b (a - b) \tag{81}$$

for a plane Earth's surface and

$$q(r_b) = A_a^b \frac{r_b}{3}\left[1 - \left(\frac{r_a}{r_b}\right)^3\right] \tag{82}$$

for a spherical Earth's surface of radius r_0 ($r = r_0 - z$, $r_b = r_0 - b$, $r_a = r_0 - a$);

$$A_a^b = \frac{1}{a-b}\int_b^a A(z)\, dz \tag{83}$$

and

$$A_a^b = \frac{3}{r_b^3 - r_a^3}\int_{r_a}^{r_b} A(r)\, r^2\, dr \tag{84}$$

are the *mean heat production rates* for the depth range *a—b* of stationary heat production.

Assuming that K and A are constants for $z_1 \leqq z \leqq z_2$, the change of temperature with depth in this range is

$$T(z) = T(z_1) + \frac{z - z_1}{K}\left[q(z_1)\cdot f - A(z - z_1)\frac{2f+1}{6}\right]. \tag{85}$$

The factor

$$f = \frac{r_0 - z_1}{r_0 - z}$$

accounts for the curvature of the Earth's surface. The heat flow through the lower boundary $z = z_2$ is, for $z_2 \ll r_0$,

$$q(z_2) = q(z_1) - A(z_2 - z_1) \tag{86}$$

in virtue of Eq. (81).

Because T and q are continuous functions of depth, the temperature distribution within a multi-layered model can be found by applying Eq. (85) to each layer successively, beginning with given surface values of T and q.

Measuring q in (μcal/cm² sec), A in (cal/km³ sec), K in (mcal/°C cm sec), and z in (km) gives ($f = 1$)

$$T(z) = T(z_1) + \frac{z}{K}\left[100\, q(z_1) - A\,\frac{z - z_1}{200}\right] {}^\circ\mathrm{C}. \tag{85a}$$

If the outflow of conducted heat exceeds the rate at which heat is generated or advected, the ambient temperature within S drops according to the loss of heat, given by Eq. (68) for unrestricted thermal contraction. Heat is transferred in this way from regions of high temperature into those of low temperature until complete equalization or a stationary state is reached. The governing equation for the *non-stationary thermal state*, combining Eq. (79) with (68), is

$$\operatorname{div} \boldsymbol{q} = A - \varrho c_p \frac{\partial T}{\partial t} \tag{87}$$

where $\boldsymbol{q}$, A, and T are now functions of time t.

The *time constant* of this diffusion process is

$$\tau(L) = L^2 \frac{\varrho c_p}{\pi K}, \tag{88}$$

a function of the linear dimensions L of the volume from which heat diffuses away. Conversely, when t denotes the time after diffusion has started from a heated region, the characteristic *scale length* for its lateral extent is

$$p(t) = \sqrt{\frac{\pi K t}{\varrho c_p}}. \tag{89}$$

Measuring t in (10^4 years) and L in (km) gives

$$\tau(L) = 2L^2 \varrho / K \quad (10^4 \text{ years}) \tag{88a}$$

$$p(t) = 0.7\sqrt{Kt/\varrho} \quad (\text{km}) \tag{89a}$$

with ϱ in (gr./cm³), K in (mcal/°C cm sec) and $c_p = 0.2$ cal/(gr. °C).

A non-stationary state will asymptotically approach thermal equilibrium if the heat sources are inexhaustible and the heat sinks have an infinite capacity on a time scale which is large in comparison to $\tau(L)$.

It may be added that $\tau(L)$ is also the time constant for the heat transfer by thermal convection, L referring then to the linear dimensions of the convection cells (Sec. V, i). Conversely, $p(t)$ gives the maximum size of convection cells which can be set in motion by a superadiabatic state during the time t.

Straigthforward analytical solutions of Eq. (87) are limited to thermal models of simple geometry, involving either plane or spherical isothermal surfaces. The following examples have been formulated on the basis of a treatise by CARSLAW and JAEGER (1959).

1. Consider a subsurface intrusion of a magma at the time $t=0$. Its initial temperature shall exceed that of the surrounding rocks by T_0 degrees and the total volume shall have uniform properties K, ϱ, c_p. If the intrusion is in form of an infinite plane sheet of thickness d, the temperature on its mid-plane will be reduced to

$$T(t) \approx \frac{1}{2} T_0 \frac{d}{p(t)} \tag{90}$$

for $t \gg \frac{1}{4}\tau(d)$. If the intrusion has the form of a sphere of radius a, its central temperature after a time $t \gg \frac{1}{2}\tau(a)$ will be:

$$T(t) \approx \frac{\pi}{6} T_0 \left[\frac{a}{p(t)}\right]^3 . \tag{91}$$

Evidently, the temperature excess of the intrusion is reduced in either case to about one half of its initial value after a time $t=\tau$. The additional diffusion of latent heat L^* from solidification of material can be included by replacing T_0 in Eqs. (90) and (91) by (T_0+L^*/c_p).

2. Consider the outflow of heat from a heated half-space with uniform properties K, ϱ, c_p, bounded by the $z=0$ plane of rectangular coordinates; z is down, pointing into the half-space. The boundary is kept at some fixed temperature, T_c, i.e. heat which is conducted to it is immediately dispersed. This model may be used to approximate the non-stationary outflow of heat from the Earth's interior, provided that $p(t)$ is at the considered time t small in comparison to the Earth's radius.

The initial temperature in the heated half-space shall be (T_0+T_c) at $t=0$. Afterwards heat shall be generated at the exponentially decreasing uniform rate $A_0 \exp.(-\lambda t)$, $t>0$. Then the surface heatflow through the $z=0$ plane as a function of time is

$$q(t,0) = \frac{K T_0}{p(t)} + \frac{2}{\pi} \tilde{A}(t)\, p(t) \tag{92}$$

with

$$\tilde{A}(t) = A_0 \frac{e^{-\lambda t}}{\sqrt{\lambda t}} \int_0^{\sqrt{\lambda t}} e^{u^2}\, du \tag{93}$$

as *weighted mean production rate*, Eq. (71). The concurrent downward rise in temperature is approximately

$$T(t,z) \approx T_c + \frac{z}{K}\left[q(t,0) - \frac{1}{2} A_0 e^{-\lambda t} z\right] \tag{94}$$

for $z \ll p(t)$, revealing a formal correspondence to Eq. (85). During the final state of internal heating $(\lambda t \gg 1)$ the difference $(T_0 - T_c)$ approaches zero as $1/\sqrt{\lambda t}$.

If the secondary heat is generated by numerous sources of different decay constants λ_i, the total surface heat flow $q(t,0)$ can be obtained by superimposing the production rates as indicated by Eq. (71).

f) Thermal Balance and Temperature Distribution

As a starting condition it is assumed that the sources of heat below the Earth's surface are in *thermal equilibrium* with the observed surface heat flow q_0. Thus ignoring the Earth's thermal history q_0 is used to derive with the aid of Eq. (82) mean production rates A_a^b for selected depth ranges $a-b$. The calculations refer to *normal* oceanic and continental regions (Sec. V a) with $q_0 = 1.2$ µcal/(cm² sec).

The mean production rate from the surface down to the Earth's center follows then as

$$A_{6,371}^{0} = 5.7 \text{ cal/(km}^3 \text{ sec)} \tag{95}$$

or, when the core is excluded, as

$$A^0_{2,900} = 6.7 \text{ cal/(km}^3 \text{ sec)}. \tag{96}$$

Both values are of little significance because the time constant for the conduction of heat from the core and lower mantle to the Earth's surface exceeds the total age of the Earth. Assuming a transparency of $\varepsilon^{-1} = 0.1$ cm, the mean thermal properties of the (upper) mantle are

$$K = 17 \text{ mcal/(° C cm sec)} \qquad c_p = 0.26 \text{ cal/(° C gr.)} \tag{97}$$

according to Table 6-2 and 6-4. They yield, when inserted into Eq. (88), the time constants $\tau = 60$ AE for the mantle as a whole ($\varrho = 4.5$ gr./cm³, $L = 2{,}900$ km) and $\tau = 4.5$ AE for the upper mantle alone ($\varrho = 3.9$ gr./cm³, $L = 860$ km). These estimates imply that the conducted outflow of generated and initial heat since the formation of the Earth has been limited to the crust and upper mantle. Furthermore, the scale length of thermal convection cells which could turn over at least once during the lifetime of the Earth is likewise limited to about 1,000 km (Sec. V, e).

The exclusion of the lower mantle gives

$$A^0_{1,000} = 14 \text{ cal/(km}^3 \text{ sec)}$$

as the required mean production rate of the upper mantle and crust. An attempt is now made to subtract from this combined estimate that portion which originates from the crust. To first order [Eq. (86)] the heat flow through the M. discontinuity at the depth d is

$$q_{MD} = q_0 - A^0_d \cdot d,$$

yielding in combination with Eq. (82) the mean production rate $A^d_{1,000}$ for the upper mantle alone.

The radiogenic heat production in granodiorites and mafic igneous rocks, as listed in Table 6-1, may be regarded as the range of acceptable values of crustal heat production A^0_d. The assumption of a granodioritic crust yields $q_{MD} \approx 0$ for the continental crust ($d = 32$ km), implying that no heat can be produced in a mantle of uniform temperature. A gabbroic crust with $q_{MD} = 0.78$ µcal/(cm² sec) leaves a mantle production rate of 9.2 cal/(km³ sec). Excluding the extreme conditions mentioned which are unlikely for geological reasons, a crustal heat production of $A^0_d = 200$ cal/(km³ sec) is assumed. It gives

$$\left.\begin{aligned} q_{MD} &= 0.56 \text{ µcal/(cm}^2 \text{ sec)} \\ A^d_{1,000} &= 6.7 \text{ cal/(km}^3 \text{ sec).} \end{aligned}\right\} \tag{98a}$$

for continents.

The thin and presumably basaltic crust beneath oceans obviously contributes little to the mean production rate of crust and mantle. Inserting $A^0_d = 130$ cal/(km³ sec) and $d = 6$ km yields

$$\left.\begin{aligned} q_{MD} &= 1.12 \text{ µcal/(cm}^2 \text{ sec)} \\ A^d_{1,000} &= 13.2 \text{ cal/(km}^3 \text{ sec)} \end{aligned}\right\} \tag{98b}$$

which is twice as high as the corresponding production rate in the mantle beneath continents.

The time constant $\tau(L)$ of heat diffusion through the continental crust is 10 million years, indicating that it is in thermal equilibrium outside areas of recent volcanism and magmatism. This value has been obtained with $L=30$ km, $K=5$ mcal/(° C cm sec), $\varrho=2.8$ gr./cm³, and $c_p=0.2$ cal/gr. The same applies to the upper mantle down to, say, 200 km depth on a time scale which is still short in comparison to the half-lives of radiogenic heat sources. Setting $L=200$ km, $K=15$ mcal/(° C cm sec), $\varrho=3.5$ gr./cm³, and $c_p=0.25$ cal/gr. gives $\tau(L)=230$ million years. Hence, any initial heat from this depth range has escaped a long time ago.

This justifies a straightforward application of the stationary heat flow equation (85) to get the internal temperature distribution down to the indicated depth. Examples for such calculations are shown in Figs. 6-29 and 6-30. It is evident from ELSASSER's curves that only an effective radiative heat transfer can keep the upper mantle temperatures below the melting curve of ultramafic material.

CLARK and RINGWOOD calculated their temperature curves on the assumption that 0.5 μcal/cm² sec of the observed surface heat flow originates from quasi-stationary sources below 400 km depth. The heat production in the upper mantle beneath continents is taken as zero. Beneath oceans a distribution is chosen such that oceanic and continental temperature curves merge at 400 km depth. This requires that the heat production in the oceanic mantle decreases with depth from 27 cal/(km³ sec) at the base of the crust to 6.8 at 400 km depth. As a result, the steep ascent of the oceanic temperature curve levels off near 150 km. At this depth the oceanic temperature exceeds the continental temperature by nearly 200° C. The resulting imbalance causes a deflection of the heat flow near continental margins, transferring heat from the oceanic mantle to the continental crust. Model calculations for the resulting upbending of the isotherms at coastlines have been carried out by MACDONALD (1963).

VON HERZEN (1967) gives particular attention to the thermal state beneath areas of unusually high or low heat flow (Fig. 6-30). Because thermal equilibrium is presumed, his results do not apply to local heat flow anomalies which are connected with subsurface intrusions of hot material and refer to extended regions ($L \gtrsim 100$ km) of excessive or deficient heat flow.

It is unlikely that regional patterns of heat flow variations reflect merely changing rates of crustal heat production. This conclusion is obvious for the thin and "unproductive" oceanic crust, but it can be applied also to the continental crust (cf. Sec. V, a). They imply the existence of pronounced regional differences in the subcrustal heat flow, which ranges in Fig. 6-30 from more than 2 μcal/(cm² sec) in oceanic high heat flow areas to less than 0.4 in continental low heat flow areas. These discrepancies are hardly explainable in terms of a regionally intensified or reduced heat transfer because of differences in effective conductivity in the uppermost mantle. They almost certainly imply that heat is advected to and from the base of the crust by thermal convection currents in an elasto-viscous mantle.

Consider an upgoing branch of thermal convection in the upper mantle ($\varrho=3.6$ gr./cm³, $c_p=0.24$ cal/gr.), flowing with a velocity of 0.1 cm/year. The amount of heat which is advected by this flow in the presence of a geothermal gradient of 2.2° C/km is equivalent to a non-advected heat generation of 60 cal/(km³ sec) as is readily inferred from Eq. (80). When integrated over a vertical column of 200 km length, this flow augments the subcrustal heat flow by 1.2 μcal/(cm² sec) as required for geothermal areas. It is presumed that the convection is stationary on a sufficiently long time scale to warrant a quasi-stationary conductive transfer of the advected heat to the Earth's surface. Downgoing branches of

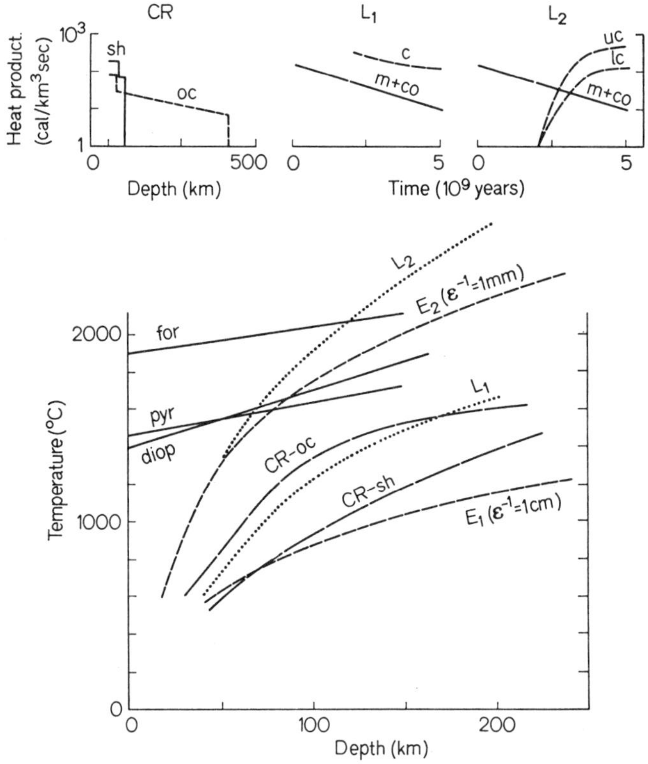

Fig. 6-29. Temperature models for the Earth's upper mantle, yielding a stationary (E, CR) or non-stationary (L) present-day surface heat flow of 1.2 cal/cm²/sec (1.0 cal/cm²/sec, model $CR-sh$). Heat transfer is non-convective, i.e. radiative and by lattice conduction [Eq. (77)]. — Models E_1 and E_2 have zero heat production within the considered depth range and demonstrate the effect of *radiative transfer* in dependence of the infrared transparency e^{-1} of upper mantle material (Elsasser, 1966). — Model $CR-oc$ (oceans) and $CR-sh$ (continental shields) include *stationary heat sources*, concentrated within the crust below continental shields but extending down to 400 km depth below oceans (upper graph). The resulting thermal imbalance in the upper mantle is evident from the diverging temperature curves $CR-oc$ and $CR-sh$, derived for an effective infrared transparency of about 2 mm (Clark and Ringwood, 1964). — Models L_1 and L_2 additionally regard the *thermal history* of an Earth, formed by cold accretion 5 billion years ago. An un-differentiated primitive Earth for 2 billion years (upper graphs); then instantaneous (L_1) or gradual (L_2) differentiation into crust (*c*), mantle (*m*), and core (*co*), leading to a gradually decreasing (L_1) or gradually increasing (L_2) heat production in the upper (*uc*) and lower (*lc*) continental crust (Lubimova, 1967). Model temperatures E_2 and L_2 rise above the indicated solidus curves of forsterite (Davis and England, 1964), diopside (Boyd and England, 1963), and pyrolite (Clark and Ringwood, 1964). The resulting melting is not in agreement with seismological evidence for a basically rigid upper mantle

convection in the upper mantle cause heat sinks with negative valued production rates, thus lowering the subcrustal heat flow.

The mean heat flow for the whole Earth is 1.5 μcal/(cm² sec) (Sec. V, a). When the normal heat flow of 1.2 is regarded as representative for the outflow of radiated and conducted internal heat, the excess of 0.3 gives an estimate of the effectiveness of heat transport by mantle convection. Integrated over the surface of the mantle it yields $150 \cdot 10^{10}$ cal/sec versus $430 \cdot 10^{10}$ cal/sec for conducted heat, obtained from the subcrustal heat flow values in Eq. (98a) and (98b). The total outflow of heat ($760 \cdot 10^{10}$ cal/sec) minus the sum of conducted and advected heat is balanced by radiogenic heat production in the crust, providing $180 \cdot 10^{10}$ cal/sec (Table 6-6).

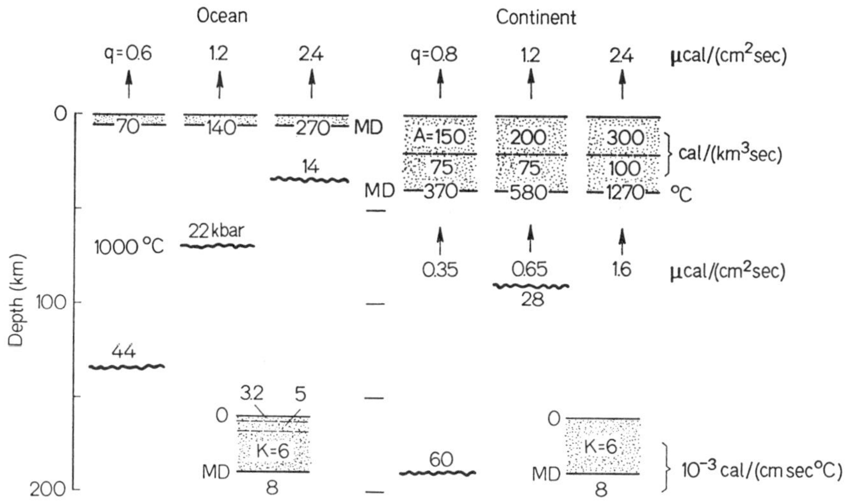

Fig. 6-30. Temperature at the base of the crust (*MD* Mohorovičić discontinuity) and depth of the 1,000° C isotherm (wavy line) beneath oceanic and continental regions of low, normal, and high heat flow q, here regarded as stationary [Eq. (85)]. Zero radiogenic heat production A in the mantle above the 1,000° C isotherm and in the oceanic crust. About one half of the normal continental heat flow is assumed to arrive from crustal heat sources. Heat flow anomalies are explained (i) by a deviating subcrustal heat flow and, on continents, (ii) by a deviating heat production in the crust. Hydrostatic pressure at the 1,000° C isotherm. (Adapted from von Herzen, 1967)

These estimates are decidedly different when the normal subcrustal heat flow likewise contains a significant advective component. It is reasonable to assume that normal areas would lie above horizontal branches of thermal convection cells in the mantle. Hence, any advective heat production is determined here by the horizontal temperature gradient dT/dx, i.e. by the temperature difference between up and downgoing branches. For a first estimate, let $dT/dx = 0.1°$ C/km and let the horizontal flow of heated subcrustal matter ($\varrho\, c_p = 0.85$ cal/° C cm³) be 100 km thick. Then a quasi-stationary flow velocity of 1 cm/year would account for one half of the subcrustal heat flow beneath continents and a velocity of 2 cm/year to account for that under oceans. These are acceptable lateral velocities for thermal convection currents, indicating that even in normal areas the subcrustal heat may be partially advected rather than conducted.

The attempt is made in conclusion to consider the *time-dependent* outflow of *radiogenic* and *initial heat* from the deeper regions of the upper mantle. The use of Eq. (92) presumes that the conducted heat flow to be evaluated is observed on a surface of constant temperature. Hence, by applying it to the subcrustal heat flow q_{MD}, a certain error is made because the temperature at the M. discontinuity, T_{MD}, has not been constant throughout the Earth's history.

Using the upper mantle properties of Eq. (97), the scale length of the diffusion of Eq. (89) is $p(t)=860$ km after $t=4.5$ billion years of cooling. Two extreme assumptions are made: (i) The subcrustal heat flow as given by Eq. (98) results exclusively from the outflow of initial heat $\varrho c_p(T_0+T_{MD})$; $T_{MD}=580°$ C for continents and 150° C for oceans. Setting $\tilde{A}(t)=0$ in Eq. (92) gives

(oceans) $T_0=5{,}700°$ C

(continents) $T_0=2{,}800°$ C.

These temperatures imply that, even after 4.5 billion years of cooling, the mantle temperature is still above the melting curve which is an unacceptable supposition (Table 6-4).

(ii) The initial heat is T_{MD} and the subcrustal heat flow is entirely due to secondary heating by radiogenic heat sources. Their weighted mean production rates follow readily from Eq. (92) as with T_0

$$\begin{aligned} &\text{(oceans)} && \tilde{A}(t)=20.5 \text{ cal/(km}^3\text{ sec)} \\ &\text{(continents)} && \tilde{A}(t)=10.2 \text{ cal/(km}^3\text{ sec)}, \end{aligned} \tag{99}$$

representing upper limits for the required heat production in the upper mantle. A correction for the neglected sphericity of the Earth would increase these estimates by about 15%.

As a compromise between these extreme assumptions it is assumed that the initial temperature within the primeval mantle was uniformly 1,500° C. Hence, $T_0=920°$ C for the continental and $T_0=1{,}350°$ C for the oceanic mantle. These values imply that, after 4.5 billion years of cooling, the initial heat contributes 0.18 μcal/(cm² sec) to the continental and 0.27 μcal/(cm² sec) to the oceanic heat flow [Eq. (92)]. The balance comes from radiogenic heat sources with weighted mean production rates of

$$\begin{aligned} &\text{(oceans)} && \tilde{A}(t)=15.5 \text{ cal/(km}^3\text{ sec)} \\ &\text{(continents)} && \tilde{A}(t)=7.0 \text{ cal/(km}^3\text{ sec)}. \end{aligned} \tag{100}$$

The value for the mantle beneath continents now matches the heat production in typical ultramafic rocks and carbonaceous chondrites (Table 6-1, last column). The choice of a higher initial temperature, implying a further reduction of $\tilde{A}(t)$, is not sensible because it leads to temperatures above the melting point.

The evaluation of Eq. (94) with the production rates of Eq. (100), the heat flow values of Eq. (98) and $\lambda t=0.85$ yields $T=1{,}660°$ C at $z=400$ km depth under continents and 2,270° C at the same depth under oceans. The second value still reaches the probable melting temperature at the indicated depth (Table 6-4). An adequate reduction can be obtained by assuming that radiogenic heat sources are concentrated in the uppermost oceanic mantle with a corresponding depletion of its lower part.

In order to demonstrate the geochemical consequences, the production rates of the various models are converted into uranium concentrations with the aid of Eqs. (70a) and (73). Table 6-3 gives the resulting concentrations for a reasonable choice of present-day abundance ratios. Comparable values have been obtained by various authors, who considered the outflow of heat from more elaborate thermal models, using numerical methods to solve Eq. (87) (cf. LUBIMOVA, 1967; MACDONALD, 1964).

Model I in Table 6-3 is stationary and thus not based on any particular mode of subcrustal heat transfer. In models II and III the advection heat of is excluded. The uranium concentrations obtained reflect in all models the adopted scale length of heat diffusion. The choice of a less effective thermal conduction would reduce this scale length and thereby increase the calculated concentrations.

In models I and III the total mass of uranium beneath one square centimeter of surface area is the same for continents and oceans. In model III it amounts to 18 gr./cm^2 when the lower mantle with its uncertain rate of radiogenic heating is excluded.

Table 6-3. *Uranium concentrations (ppm) beneath areas of normal heat flow (1.2 μcal/cm² sec)*

Model	Continents		Oceans	
	crust	mantle	crust	mantle
I	1.3	0.03	0.9	0.06
II	1.3	0.03[a]	0.9	0.06[a]
III	1.3	0.02[a]	0.9	0.045[a]
Abundance ratio U:Th:K	1:3.5:13,000	1:4:10,000	1:2.5:16,000	1:4:10,000

[a] Add 15% for neglected sphericity.

Model I: Stationary heat production in crust and upper mantle; Eq. (98).
Model II: Exponentially decaying heat production, zero initial heat; Eq. (99).
Model III: Exponentially decaying heat production and outflow of initial heat; Eq. (100).
Thermal properties for model II and III: Eq. (97).

Assuming that it has the uranium concentration of the upper continental mantle, we obtain 29 gr./cm^2 which corresponds to a mean concentration of 0.036 ppm for mantle and crust together.

The uranium content of the continental mantle, model III, is in good agreement with the concept of an upper mantle of peridotitic or chondritic composition (cf. Table 6-1). The more than twice as high uranium concentration of the oceanic mantle implies that it is in a less differentiated state and therefore less depleted of the lithophile radioactive isotopes than the continental mantle. According to subcrustal seismology, the mantle material beneath oceans and continents has about the same elastic constants. There is no indication for a marked and systematic difference in density either (cf. Sec. III, b).

CLARK and RINGWOOD (1964) suggest that matter equivalent to a mixture of peridotite and basalt in the ratio 3:1 can reconcile the results of seismology and heat flow. This hypothetical primeval mantle material, called *pyrolite*, has the density and elasticity of peridotite, but an abundance of radioactive isotopes which is about four times greater. Basalts are derived from pyrolite by fractional melting.

The thermal models which have been considered so far are deficient in two principal aspects: (i) The thermal conductivity is treated as a temperature-independent property of the material, thus ignoring the T^3-dependence of its radiative component [Eq.

(76)]. (ii) Radiogenic heat sources are treated as if fixed in place in the course of the Earth's history. In reality, however, their present-day concentration in the continental crust is the result of a differentiation process, depleting the deeper interior of long-lived radioactive isotopes. Hence, in addition to conduction, radiation, and advection, an upward migration of radiogenic heat sources has controlled the early thermal history of the Earth. Both effects have been taken into account by LUBIMOVA (1967) in deriving the temperature curves L_1 and L_2 of Fig. 6-29.

The lower boundary of the zone from which heat is conducted to the surface moves downward with a velocity $p/t = 100$ m/(10^6 years). This is readily seen by differentiating Eq. (89) with respect to time and by setting $t = 4.5$ AE. Hence, the thermal state of the lower mantle will remain unchanged on a geological time scale, i.e. its present-day heat balance is largely determined by the unknown amount of initial heat which it has retained since the formation of the Earth.

The primeval Earth during and just after its formation has been heated up (i) by the dissipation of gravitational energy of formation [Eq. (113)] (ii) by the decay energy of short-lived isotopes, and (iii) by the dissipation of rotational energy from a faster spinning Earth which was slowed down by tidal friction (Table 6-6). A quantitative treatment of these sources of initial heat is hardly possible. As a consequence, the thermal state of the deep interior beneath the zone from which heat has been conducted to the surface is uncertain. Temperature estimates for these regions can be obtained only indirectly by a consideration of melting curves (Sec. V, g) and from the distribution of the electrical conductivity (Sec. IV, g).

g) Melting Curves

The melting of rocks and minerals is connected with a reduction of density. Their melting temperature T_m increases accordingly with pressure p. An exact knowledge of the *melting curve* $T_m(p)$ of matter within the Earth can give us maximum temperatures in the Earth's mantle, which is solid, and minimum temperatures in the Earth's core which is, at least in its outer portion, liquid.

Basaltic rocks melt at atmospheric pressure between 1,000 and 1,200° C. At the base of the crust these melting temperatures will be 50 to 100° C higher. At 75 km depth melting can be assumed to occur around 1,400° C (YODER and TILLEY, 1962).

The melting curves of various silicates have been determined experimentally for pressures up to 50 kbar (cf. CLARK's Table 15-4, 1966b). Within this range T_m does not rise linearly with pressure but behaves more or less according to the Simon equation

$$T_m = T_0 (p/p_0 + 1)^{\frac{1}{c}} \tag{101}$$

(T_0: melting temperature at zero pressure in °K); p_0 and c are composition-dependent adjustable constants, $2 \leqq c \leqq 4$ for many substances.

KRAUT and KENNEDY (1966) report a linear relationship between T_m and the isothermal compression of the material:

$$T_m = T_0 \left(1 + C \frac{\varrho_R - \varrho_{0R}}{\varrho_R}\right) \tag{102}$$

(ϱ_R: density at room temperature and pressure p, ϱ_{0R}: density at room temperature and zero pressure, C: an adjustable constant).

Gilvarry (1956a, 1966) has shown that Eq. (101) is, in combination with Grüneisen's theory of solids, a direct consequence of the generalized Lindemann melting law and he regards Eq. (102) as an approximation for low compression.

A differentiation of the melting laws with respect to pressure gives

$$\frac{dT_m}{dp} = T_0 \frac{(1+p/p_0)^{\frac{1}{c}}}{c(p+p_0)} \tag{103}$$

$$\frac{dT_m}{dp} = T_0 \frac{C}{y_R K_{TR}} \tag{104}$$

where K_{TR} denotes the isothermal bulk modulus at room temperature and $y_R = \varrho_R/\varrho_{0R}$. Hence, the initial slope of the melting curve ($y_R = 1$, $p = 0$) fixes $c \cdot p_0$ and C/K_{TR}. In the case of silicates it ranges from 5° C/kbar for forsterite to more than 10° C/kbar for diopside and albite according to Clark's table. Their zero-pressure melting temperatures range from 1,120° C for albite to 1,900° C for forsterite, zero-pressure bulk moduli at room temperature from 0,5 megabar for albite to 1.2 megabar for forsterite.

The values for *forsterite*, when inserted above, yield with $c = 2$

$$p_0 = 220 \text{ kbar} \qquad C = 2.76$$

as constants of the melting laws; $y_R(p, K_{TR})$ will be derived from Eq. (26), using the approximation $K_{TR}(p) = K_{TR}(0) + 4p$. The resulting melting temperatures of forsterite at 100 and 400 km depth (30 and 130 kbar) can be found in Table 6-4 together with the melting point gradients $dT_m/dz = \varrho g\, dT_m/dp$. Olivine with a certain iron content will have a considerably lower melting temperature, but it is not possible to say how much it will change.

The melting curve of *iron* has an initial slope of only 3° C/kbar, starting at $T_0 =$ 1,532° C. Accordingly, we may indeed expect that the temperature at the mantle-core interface is below the melting point of iron but above the melting point of Mg-Fe silicates. Inserting $K_{TR} = 1.7$ megabar for $p = 0$ and setting $c = 3$ gives the constants

$$p_0 = 200 \text{ kbar} \qquad C = 2.84.$$

Extrapolating with these parameters the melting curve of iron to the mantle-core interface ($p = 1.37$ megabar, $y_R \approx 1.5$) yields melting temperatures above 3,000° C (Table 6-4). Even though the effect of phase transitions upon the melting curve is ignored, 3,000° C may be regarded as a lower limit for the temperature at the base of the mantle.

If the transition from the outer to the inner core is marked by the solidification of iron, the melting point gradient must exceed the geothermal gradient dT/dz somewhere between 2,900 and 5,000 km depth.

h) Thermal Expansion and Specific Heat

The response of matter to a differential increase of temperature at constant pressure is expressed by the coefficient of *thermal volume expansion*

$$\alpha = -\frac{1}{\varrho}\left(\frac{\partial \varrho}{\partial T}\right)_p \tag{105}$$

and by its *isobaric specific heat* c_p, accounting for the differential increase of intrinsic energy. If, in the case of solids, the volume V rather than the pressure is kept constant, the differential change of intrinsic energy E_i in relation to the change in pressure defines the GRÜNEISEN *ratio*

$$\gamma = V\left(\frac{\partial p}{\partial E_i}\right)_V = \frac{\alpha K_T}{\varrho c_v}. \tag{106}$$

The second identity follows from $(\partial p/\partial T)_V = K_T\alpha$ and $dE_i = \varrho V c_v dT$ with c_v as specific heat at constant volume. The GRÜNEISEN ratio, a dimensionless number usually between 1 and 2, is basic for the theory of solids and thereby for the equation of state of the Earth's interior (cf. GILVARRY, 1956b). It connects the specific heats of solids according to

$$c_p = c_v(1+\alpha\gamma T) \tag{107}$$

and the isothermal with the adiabatic bulk modulus [Eq. (13)]. The geophysical significance of these thermal parameters arises from their determining influence upon convection (Sec. V, i).

The thermal expansion coefficient of common silicatic minerals ranges from about 1.5 to $3\cdot 10^{-5}\,°C^{-1}$ at room temperature and atmospheric pressure (cf. SKINNER, 1966). It is almost twice as high at 800° C and is reduced by increasing pressure. A representative value for the Earth's crust is $2.5\cdot 10^{-5}\,°C^{-1}$.

The pressure derivative of α is obviously related to the temperature derivative of isothermal compression. The relation is readily found by differentiation:

$$\left(\frac{\partial\alpha}{\partial p}\right)_T = -\alpha\frac{\delta_T}{K_T}; \tag{108}$$

$$\delta_T = -(\alpha K_T)^{-1}\cdot\left(\frac{\partial K_T}{\partial T}\right)_p$$

is a dimensionless parameter of similar theoretical implications as β_T, introduced in Eq. (17) (BIRCH, 1952, 1968; ANDERSON, 1967). In accordance with high-pressure research we may set $\beta_T\approx\delta_T$ which implies that the product (αK_T) is independent of pressure.

The temperature derivative of α shall be given in terms of the temperature derivatives of c_v and γ, both to be inferred from the theory of solids. After differentiating Eq. (106) with respect to T at constant pressure we obtain

$$-\left(\frac{\delta(\alpha^{-1})}{\delta T}\right)_p = \delta_T - 1 + \frac{\varepsilon_T}{\alpha T} \tag{109}$$

where

$$\varepsilon_T = \frac{\partial\ln c_v}{\partial\ln T} + \frac{\partial\ln\gamma}{\partial\ln T}.$$

The first term of the dimensionless parameter ε_T can be found by differentiating Eq. (110) with respect to T. It shows that $\partial(\ln c_v)/\partial(\ln T)$ decreases from 0.39 for $T=\Theta/2$ to 0.1 for $T=\Theta$, disappearing as $\Theta/(2T)$ for $\Theta/T\to 0$. The temperature derivative of the GRÜNEISEN ratio is related to the temperature derivative of β_T and presumably is near zero. Hence, at sufficiently high temperatures $\varepsilon_T=0$, yielding ANDERSON's approximation which has been used to derive Eq. (20).

The specific heat of one mole of solid matter increases with temperature to a limiting value of $3R\approx 6$ cal/(mole °C) for all substances. The temperature dependence of c_v at low temperatures is likewise independent of composition when the temperature is referred to the DEBYE *temperature* Θ. It is obtained by fitting experimental

$c_v(T)$-curves to the DEBYE *function* $D(\Theta/T)$ so that

$$m c_v(T) = 3 R D(\Theta/T) \tag{110}$$

for a given mean atomic weight m. This function increases from zero at low temperatures ($T \ll \Theta$) to unity at high temperatures ($T \gg \Theta$).

Silicates with $m \approx 20$ have a specific heat of 0.2 cal/(° C gr.) at room temperature and atmospheric pressure, corresponding to $\Theta = 850°$ K. The specific heat is reduced by isothermal compression. Hence, the in-situ DEBYE temperature of compressed iron-magnesium silicates within the Earth will be somewhat higher, even though it should stay well below the ambient temperature under mantle (p, T) conditions (cf. BIRCH, 1952).

Table 6-4. *Specific heat, thermal expansion, and melting temperature*
Crust: mafic igneous rocks. Core: iron.

Depth	Spec. heat	Expansion coeff.	Grüneisen ratio	Adiabat. gradient	Temp.	Melting temperature			
						Simon Eq.		Kraut-Kennedy Eq.	
(km) z	(cal/° C gr.) c_v	(10^{-6} ° C^{-1}) α	γ	(° C km) τ_0	(° C) T	(° C) T_m	(° C/km) dT_m/dz	(° C) T_m	(° C/km) dT_m/dz
crust	0.20	25	0.9			1,200			
100	0.26[a]	35[a, b]	1.2	0.41	1,000	2,040[c]	1.5	2,040[c]	1.4
400	0.26	28	1.2	0.53	1,800	2,470	1.4	2,400	1.1
1,000	0.27	22	1.5	0.54	2,500				
2,900	0.27	(6)	(0.6)	(0.21)	3,500				
2,900	0.11	(7)	(1)	(0.6)	3,500	(3,300)	0.8	(3,200)	0.5

[a] Olivine ($m = 22$, $\Theta = 850°$ K).
[b] Calculated for isothermal compression at 800° C ($\alpha_0 = 40 \cdot 10^{-6}$ ° C^{-1}).
[c] Forsterite.

Starting with the thermal expansion coefficient α_0 at zero pressure the integration of Eq. (108) gives the reduced coefficient α at any elevated pressure. BIRCH (1968) has shown that this integration can be avoided by assuming that the equation of state has the general form

$$p = K_0(T) \cdot F(y)$$
$$y(T, p) = \varrho(T, p)/\varrho(T, 0)$$

(cf. Sec. II, e). Differentiating it with respect to temperature at constant pressure and observing that $dF/dy = K_T/(y K_0)$, it follows without further assumptions that

$$\alpha = \alpha_0 \left(1 - \frac{\delta_0 p}{K_T}\right) \tag{111}$$

with δ_0 denoting the zero pressure value of δ_T.

The thermal expansion coefficients in Table 6-4 have been derived from this relation with $\delta_0 = 4$, inserting K_S from Table 6-5 instead of K_T. No attempt has been made to integrate Eq. (109) to account for the dependence of α on temperature.

Hence, the reported values are presumably too low for the ambient temperature in the mantle. The same applies to the GRÜNEISEN ratio and the adiabatic gradient, calculated according to Eq. (106) and (15) for the indicated temperatures. VERHOOGEN (1951) obtained comparable values by using an intrinsic relationship between γ and the change of the elastic constants with density and hence with depth. The temperatures to which the adiabatic gradients are referred represent reasonable estimates.

i) Thermal Convection

Thermal convection is intrinsically related to the concept of an *adiabatic temperature* distribution $T_0(z)$. Its derivative with respect to depth z is the adiabatic gradient $\tau_0 = \alpha g\, T/c_p$ (Eq. 15, 13, 107). Consider a depth range $z_1 \leqq z \leqq z_2$, in which $\alpha g/c_p$ is constant and in which the temperature distribution $T(z)$ is adiabatic, $T = T_0$. Then, by integration

$$T_0(z) = T_0(z_1) \exp\left[\frac{\alpha g}{c_p}(z - z_1)\right]. \tag{112}$$

In the case of a general distribution, set $T(z_1) = T_0(z_1)$ and subtract $T_0(z)$ from $T(z)$. The temperature in this range is *superadiabatic* where $T > T_0$ and *subadiabatic* where $T < T_0$.

The superadiabatic state in a gravitational field implies an *energetically unstable density* distribution. Buoyancy forces tend to convert it into a stable one by convecting hot and light material upwards into regions of lower temperature and vice versa. In the case of a viscous fluid the imbalance becomes critical, i.e. *convection* sets in, when a characteristic dimensionless parameter, the RAYLEIGH *number*, exceeds a certain threshold value. Suppose the fluid is bounded at the planes $z = z_1$ and $z = z_1 + d$ of constant but different temperatures T_1 and T_2. If

$$\Delta T = T_2 - T_1[1 + \exp(\alpha g d/c_p)]$$

denotes the superadiabatic temperature difference between the upper and lower boundary, the RAYLEIGH number is proportional to $\Delta T d^3$ and inversely proportional to the viscosity of the medium.

The Earth's upper mantle is in a highly superadiabatic state. The expected rise in temperature with depth is far greater than that inferred from the adiabatic gradients in Table 6-4. Inserting into Eq. (112) $T_0(z_1) = 600°$ C as temperature at the base of the crust and using $\alpha g/c_v = 2.5 \cdot 10^{-4}$ (km^{-1}) gives an adiabatic temperature of 840° C at 1,000 km depth. This temperature is definitely too low for any reasonable thermal model, and the superadiabatic temperature difference ΔT between top and bottom of the upper mantle is presumably of the order of 1,000° C.

This difference is considered to be adequate to initiate thermal convection provided that the upper mantle viscosity is below 10^{21} poise and the size of the cells of the order of 1,000 km. The viscosities quoted in Sec. II, f, are slightly above this limit but the difference is not regarded as significant enough to prevent convection in the upper mantle (cf. review articles by TOZER, 1966, and KNOPOFF, 1966).

The conduction and radiation of heat is effective enough to establish thermal equilibrium conditions down to, say, 200 km depth on a geological time scale (Sec. V, f). Hence, the importance of thermal convection for the thermal state arises from the advection of heat from the deeper interior. This implies that the convection transports material across the heterogeneous zone between 400 and 700 km which is marked by seismic discontinuities (Sec. I, e).

A truly convecting medium cannot retain chemical inhomogeneities. Hence, the concept of thermal convection throughout the upper mantle implies that the zone between 400 and 700 km depth is chemically uniform and marked by a phase transition of Mg-Fe silicates from the olivine into the spinel structure. The latent heat of this phase transition represents a partial barrier for convection (VERHOOGEN, 1965; GEBRANDE, 1967): Material in a downgoing branch of convection is heated up by the release of latent heat when it changes at the phase boundary to the more closely packed phase. Conversely, material in an upgoing branch is cooled down by the consumption of latent heat while transferring into the less closely packed phase. As a consequence, the boundary between the phases dissolves into an extended region in which a gradual transition from one phase to the other occurs. Because both phases co-exist in this region the ambient temperature gradient is that of the phase transition curve. It must exceed the adiabatic gradient to sustain convection.

The prevailing temperature and pressure gradients in the upper mantle suggest that the temperature gradient in the transition zone from olivine into spinel is superadiabatic. Hence, large-scale convection which includes the deeper parts of the upper mantle is possible and also probable. The resulting consequences for the thermal state have been discussed in Sec. V, f.

Assuming a temperature $T(z_1)=2{,}500°$ C for the base of the upper mantle, the evaluation of Eq. (112) with $\alpha g/c_v = 1.3 \cdot 10^{-4}$ (km^{-1}) gives 3,300° C as adiabatic temperature for the base of the lower mantle. The actual temperature may be somewhat higher but it is reasonable to assume that superadiabatic temperatures in the lower mantle are low. This implies in conjunction with a high ambient viscosity that the lower mantle is at present not in convective motion but that it may have been so during the early history of the Earth.

VI. Conclusions

This summary reviews the principal results of the preceding sections and considers their bearing on the *chemical composition* and *energy balance* of the Earth's interior.

Crustal seismology has confirmed the geochemical conclusion (Chapt. 7, IV) that the *upper* third of the *continental crust* is of *granitic* to *granodioritic* composition. This is evident from a comparison of experimentally determined with observed P and S-wave velocities (Fig. 6-31).

The central part of the continental crust to about 20 km depth is characterized by a negligible velocity gradient. In this depth range the opposing effects of increasing pressure and temperature upon the seismic properties of fairly homogeneous matter are self-cancelling. Toward the base of the continental crust a gradual change to *gabbroic* composition is probable. In regions with a discernible intra-crustal Conrad-discontinuity this transition will be abrupt. The thin crust beneath oceans may be regarded as a solidified layer of basaltic melts. We may assume that the entire Earth's crust originated ultimately from the underlying upper mantle which underwent fractional melting from time to time in the course of the Earth's history.

The *upper mantle* presumably consists of *peridotite* with a minor gabbro component beneath oceans in order to account for the required rate of suboceanic radiogenic heat production (Sec. V, f). Eclogite, containing the high pressure modifications of gabbro minerals, is not a likely main constituent of the upper mantle. The phase transition *gabbro-eclogite* would be spread over a considerable depth range under the ambient (p, T)-conditions at the base of the crust (Green and Ringwood, 1966).

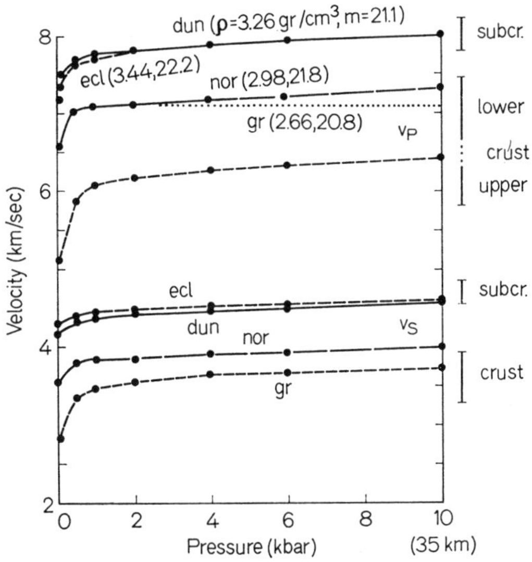

Fig. 6-31. Velocity of ultrasonic compressional (v_P) and shear (v_S) pulses in rocks, measured under static pressure up to 10 kbar in the laboratory. The shown specimens have representative velocities for granitic (*gr*), basaltic-gabbroic (*nor*), peridotitic (*dun*), and eclogitic (*ecl*) composition. Density and mean atomic weight m in parenthesis. The range of seismic crustal and subcrustal velocities is indicated at the margin. Smooth velocity increase with pressure above 1 kbar. A concurrent rise in temperature would tend to lower the P-wave velocity at a rate of about 0.5% per 100° C. This reduction is indicated by the dotted line for the sample *nor*; it refers to a linear temperature increase to 600° C at 10 kbar, corresponding to conditions at the base of the continental crust (Fig. 6-30). *gr* granite, Barre/Vermont; *nor* norite, Bushveld/Transvaal; *dun* dunite Mt. Dun/New Zealand; *ecl* eclogite, Healdsburg/California. [Data from Birch (1960, 1961) and Simmons (1964)]

This disagrees with the first-order seismic discontinuity which in general separates crustal and subcrustal matter. Further objections against the interpretation of the M. discontinuity as universal phase boundary arise from the observation that the regional variations in subcrustal temperature appear to be unrelated to those of subcrustal pressure (Fig. 6-30).

The mantle regions B and D are zones of fairly uniform density and compressibility, extrapolated to zero pressure (Table 6-5). The properties of region B *(uppermost mantle)* are typical for peridotite in its low-pressure phase. Region D *(lower mantle)* has with the possible exception of its lowest part the properties of the high-pressure phase of shocked olivine rocks (dunite) with 20% *fayalite* (Fig. 6-32). In fact, the density-pressure curve between 1,000 and 2,500 km depth follows closely the $p-\varrho$ curve (Hugoniot) of a shocked low-iron dunite sample except for a shift toward slightly higher densities. The variation of density with temperature is not considered, but we may assume that the temperature increases along the Hugoniots in the same way as it increases with depth in the lower mantle.

Static pressure experiments in the kilobar range have shown that the *phase transition* of *olivine* into the more closely packed *spinel* structure occurs with a density increment of 10% (Akimoto *et al.*, 1965). This increment is insufficient to explain the zero-

Table 6-5. *State and properties of the Earth's interior*[a]

Accuracy: $\lesssim 5\%$ (except η and Q^{-1}). ϱ_0, K_0: Zero-pressure properties, calculated for isothermal compression, Eqs. (24) and (26), $K_T \approx K_S$.

Region (Bullen)	Depth (km)	Pressure (megabar)	Density (gr./cm³)		Elastic constants (megabar)			Dissip. const.	Viscosity (10^{21} poise)	Gravity (gal)
	z	p	ϱ	ϱ_0	K_S	K_0	μ_S	Q^{-1}	η	g
A	crust	0	2.65 2.8		0.55 1.0		0.34	≈0	rigid	981
		0.01								985
B	30 100 200 400	 0.03 0.06 0.13	3.3 3.2—3.3 3.3—3.4 3.6	3.3 3.2 3.2 3.3	1.2 1.1—1.2 1.3—1.4 1.7	1.2 1.0 1.1 1.2	0.72 0.62—0.7 0.65—0.7 0.9	1/80 1/80 1/80 1/100	 3 2	 988 993 1,000
C	600	0.21	4.1	3.7	2.6	1.8	1.3	1/300	10	1,003
D	1,000 2,000 2,900	0.39 0.88	4.6 5.2 5.5	4.0 4.0 3.8	3.5 5.2 6.4	2.1 2.1 1.8	1.9 2.5 3.0	1/1,000 1/1,500	 10^5	1,000 1,004
		1.37								1,060
E	2,900 4,000	 2,4	9.7 11,3		6.3 10.2				fluid	780
F	5,000	3,2	12,1		12.6					≈500
G	inner core	3.3 3.6—3.7	12—14	(8)	(14)	(2.6)	(2)		rigid?	390—460 0

[a] Based on Jeffreys-Bullen distribution of P and S-wave velocities, including Gutenbergs low-velocity layer in the upper mantle (Fig. 6-6). Densities in regions B and G according to surface wave dispersion and free oscillations, Fig. 6—12.

pressure density of shocked olivine rocks in the metastable state, suggesting that they change in the megabar range, at least partially, into an assemblage of even more closely packed *oxides* of Mg, Fe, and Si. The same may apply to the lower mantle as first proposed by Birch (1952).

The ambient (p, T) conditions in the transitional mantle *region C* are presumably of the right order to explain its non-uniformity by a gradual phase transition of olivine into spinel (Ringwood, 1956 and 1966). If there is an additional change of composition, it will increase the mean atomic weight of the mantle material between region B and D by 2. This value has been inferred with Birch's empirical density-velocity relation, Eq. (21), and corresponds to a 20% increase of the fayalite content of olivines. Recently discovered discontinuities in region C support this concept of a slightly inhomogeneous mantle composition.

The seismic discontinuity at 2,900 km depth almost certainly represents a compositional boundary. RAMSEY's hypothesis which regards it as a phase boundary finds no support in shock wave experiments. The density-pressure curve of the core is in good agreement with the Hugoniots of iron and iron alloys but incompatible with the Hugoniots of the prospective main constituents of the mantle.

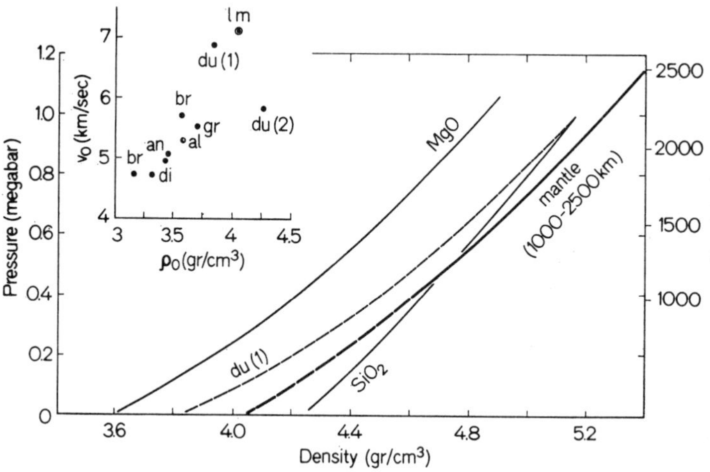

Fig. 6-32. Shock-wave compression of materials which possibly compose the deeper layers of the Earth and the pressure-density curve for the lower mantle as inferred from seismology and model calculations (Fig. 6-12, 6-13). The selected depth range has fairly uniform zero-pressure properties ϱ_0, K_0 (Table 6-5); this justifies an isothermal extension (dashed) of the $p-\varrho$ curve to zero pressure, setting $dp/d\varrho \approx K_0/\varrho_0$ for $p=0$ [Eq. (14)]. The experimental shock wave data have been transposed to metastable $p-\varrho$ *Hugoniots* which are shown. They extend the properties of the shocked high-pressure phase to zero pressure, yielding its zero-pressure density ϱ_0 and bulk velocity v_0. The insert displays properties thus obtained for shocked rock samples of varying composition in comparison to the zero-pressure properties of the lower mantle (*lm*), approximating its zero-pressure adiabatic bulk velocity by $(K_0/\varrho_0)^{\frac{1}{2}}$. *gr* granite; *al* albitite; *an* anorthosite; *di* diabase; *br* bronzitite; *du* (1) dunite (92% olivine with 12% fayalite); *du* (2) dunite (90% olivine with 55% fayalite). — The $p-\varrho$ Hugoniot and zero-pressure properties of the sample *du* (1) resemble those of the lower mantle most closely. This establishes olivine as possible main constituent. Olivine with 20% fayalite (mean atomic weight: 22) should have about the right zero-pressure density 4.05 gr./cm³. (Shockwave data from MCQUEEN *et al.*, 1967)

MCQUEEN and MARSH (1966) confirmed AL'TSHULER's result that shocked pure iron between 1 and 2 megabar is about 1 gr./cm³ denser than would be expected in the Earth's core. Shocked iron-nickel alloys have even higher densities. Hence, if *iron* is the main constituent of the core, it must be mixed with some light element such as *silicon*, at least in its outer portion. BALCHAN and COWAN (1966) found that an Si-content of *14 to 20 wt.* % would suffice. There is no geophysical evidence to support the assumption that the core contains appreciable amounts of nickel. This assumption was derived from the relatively high cosmic abundance of nickel and thermodynamical considerations about partition of Ni between metal and oxides.

The energy balance, presented in Table 6-6, is intrinsically connected with the dynamics of the Earth's interior, involving differentiation, convection, expansion and compression. The dominant term is the loss of *potential energy of position*, E_{gr}, which the internal matter suffered when it aggregated during the formation of the Earth.

Consider a growing Earth nucleus of spherical symmetry, attracting with its gravity field planetary matter from infinite dispersion. If r and g denote its radius and surface gravity,

Table 6-6. *Energy balance of the Earth's interior*

E^*_{gr}: Gravitational energy (cold, undifferentiated Earth). E_{gr}: Gravitational energy (present-day Earth). E_S: Stored strain energy of isothermal compression. $\dot{E}_S$: Rate of strain release by earthquakes. E^*_R: Radiogenic heat production (U, Th, K) in the undifferentiated Earth from $t=0$ until $t=1.5$ AE after formation (abundances according to model III, Table 6—3). E_R: Radiogenic heat production in the differentiated Earth from $t=1.5$ AE until the present time (see above). $\dot{E}_R$: Radiogenic heat production rate. qF: Contribution of the depth region to the present-day integrated surface heat flow. E_{rot}: Energy of current rotation. $\dot{E}_{rot}$: Tidal deceleration. E_{DP}: Magnetic potential energy of the centred dipole. $\dot{E}_{DP}$: Secular decay of the centred dipole.

	Energies (10^{28} cal)						
	E^*_{gr}	E_{gr}	E_S	E^*_R	E_R	E_{rot}	E_{DP}
Whole Earth	5,680[a]	5,970	320	120	115	5.1	$1.8 \cdot 10^{-10}$
Crust		40	0		24	0.05	
Upper mantle		2,320	9		47	2.4	
Lower mantle		2,680	127		(44)	2.1	
Core		930	184		0	0.6	

	Current 1st time derivatives (10^{10} cal/sec)					Current 2nd time derivatives (10^{-7} cal/sec^2)	
	$\dot{E}_R$	qF	$\dot{E}_S$	$\dot{E}_{rot}$	$\dot{E}_{DP}$	$\ddot{E}_R$	$\ddot{E}_{rot}$
Whole Earth	850	−760	−100	−82	$-5 \cdot 10^{-3}$	−430	+10
Crust	180	−180				−110	
Upper mantle	350	−580[b]				−170	
Lower mantle	(320)	(−150)				(−150)	

[a] BIRCH (1965).

[b] 320 cal/sec conducted radiogenic heat, 110 cal/sec conducted initial heat, 150 cal/sec advected heat (possibly from the lower mantle).

any newly attracted mass element ϱdV in its outermost shell will have lost the potential energy $\varrho g r dV$. The integration over the volume of the Earth or some part of it gives the relevant *gravitational energy* which has been converted into other forms of energy. The values in Table 6-6 have been obtained by considering ϱ in the core, g in the mantle, and $g \cdot \varrho$ in the crust as constants. Then, in virtue of Eq. (11),

$$E_{gr} = g_0 \begin{cases} \frac{3}{5} R M_{co} \\ \frac{2}{3} a M [1-(R/a)^3] \\ a M_c \end{cases} \tag{113}$$

(M, M_{co}, M_c: mass of the Earth, the core and the crust; a, R: radius of the Earth and the core; g_0: mean gravity in the mantle).

Differentiation within the primeval Earth implies a further irreversibles loss of potential energy. BIRCH (1965) estimates that the *formation of the core* released $290 \cdot 10^{38}$ cal of gravitational energy more than 80% of which has been converted into heat. This energy, raising the ambient temperature of the whole Earth by roughly 1,500° C, exceeds the total amount of radiogenic heat from the decay of uranium, thorium and K^{40}. The current balance of gravitational energy may be regarded as stationary ($\dot{E}_{gr}=0$) provided that the differentiation process is completed and that at the Earth's surface the mountains are eroded at the same rate as new mountains are built up.

A small fraction of the lost potential energy is stored in form of compressional *strain energy* E_S within the Earth's body, mainly in its highly compressed core. Earthquakes receive their energy from the subsequent build-up of shearing strain within the uppermost 700 km. Below this depth the mantle appears to have sufficient strength to withstand the shearing forces which arise from the slightly non-hydrostatic equatorial bulge of the Earth (Sec. III, b).

The strain energies in Table 6-6 have been calculated under the assumption that the compression occurred at constant temperature according to Eq. (23). If V_0 denotes the zero-pressure volume of the considered depth region and $V=V_0/y$ its present-day volume,

$$E_s=\int_V^{V_0} p\, d\hat{V}=\tfrac{9}{8}\, y\, V K_0 (1+y^{\frac{4}{3}}-2y^{\frac{2}{3}}). \tag{114}$$

This relation has been evaluated with the values of Table 6-5, using $K_0=2.6$ megabar and $y=1.5$ for the core.

The *radiogenic heat* production E_R refers to the indicated thermal model with the additional assumption that the lower mantle of the differentiated Earth has the same abundances of U, Th, and K^{40} as the upper continental mantle. The *kinetic energy of rotation*,

$$E_{\text{rot}}=\tfrac{1}{2}\,\Omega^2\, C \tag{115}$$

[cf. Eq. (69)] is small as against the other energies so far considered, but it is still large in comparison to the energy of the geomagnetic dipole field. Assuming that the dipole moment M originates from zonal electrical currents at the surface of the core according to Eq. (56a), the magnetic field energy is

$$E_{DP}=\tfrac{1}{2}\, M^2/R^3. \tag{116}$$

The current in- and outflow of energy ($\dot{E}$) is mainly determined by the balance between radiogenic heating and terrestrial heat flow. The production rates adopted imply stationary conditions in the crust, a cooling of the upper mantle by 1.7° C in 10 million years, and a heating of the lower mantle by 1.5° C in 10 million years. These values have been obtained from Eq. (87), setting $\varrho c_p=1$ cal/(° C cm³) in the upper and 1.4 in the lower mantle. Heat transfer across the boundaries of the lower mantle is excluded (div $q=0$).

The loss of rotational energy by tidal friction ($\dot{E}_{\text{rot}}$) deserves particular attention because it represents the only *extra-terrestrial* term in the energy balance. The concurrent secular acceleration of the Moon increases its distance from the Earth by 10^{-7} cm/sec in accordance with KEPLER's third law (cf. MUNK and MACDONALD, 1960). This in turn reduces the amount of tidal deceleration and energy dissipation at the indicated rate $\ddot{E}_{\text{rot}}$).

The geomagnetic dipole field shows the greatest relative energy change $\dot{E}/E$, reflecting its basic instability. The strain energy E_S, on the other hand, represents in this balance the most stable form of potential internal energy. If the cosmological hypothesis is correct and the gravitational constant G is slowly decreasing in time, strain energy is continuously released within an expanding Earth. Hence, this hypothesis has far-reaching consequences for the energetic state of the Earth's interior and its dynamics (cf. JORDAN, 1966).

Acknowledgments

I wish to thank Professor G. ANGENHEISTER, H. W. BARTELS, Dr. K. M. CREER, and Dr. O. HARTMANN for a searching reading of the manuscript or parts of it.

Symbols

a	Earth's radius (6,371 km); Eq. (40).
A	Heat production rate.
A_a^b	Mean of A in the depth range a–b; Eq. (83/84).
$\tilde{A}(t)$	A weighted mean of A; Eq. (71).
AE	10^9 years (aeon).
A, C	Principal moments of inertia; Eq. (47a).
C_0	Mean of the principal moments; Eq. (41).
c_p, c_v	Specific heat at constant pressure and volume.
c_i	Current concentration (by weight) of a radioactive element "i".
E	Energy (Table 6-6); also electric field.
f	Geometric flattening of the Earth; Eq. (38).
F	Geomagnetic field vector; Sec. III, a.
g	Gravity.
G	Gravitational constant.
gal	cm/sec
h	Surface elevation (Sec. III).
H	Dynamical flattening, Eq. (39); also thickness of the AIRY crust and horizontal component of F.
I	Magnetization.
K_S, K_T	Adiabatic and isothermal bulk modulus (incompressibility).
K_0	Zero-pressure value of K_T.
K	Effective thermal conductivity.
L	A scale length, half-width
m	Mean atomic weight.
M	Mass of the Earth ($5.976 \cdot 10^{27}$ gr.); also its magnetic dipole moment.
p	Hydrostatic pressure; also a scale length of thermal and electromagnetic diffusion (skin-depth).
q, q_0, q_{MD}	Heat flow, surface heat flow, heat flow through the M. discontinuity.
Q^{-1}	Dissipation constant of strain energy; Eq. (35).
r	Distance from the Earth's center.
t	Time.
T	Travel time of a seismic pulse; also the (absolute) temperature.
T_m	Melting temperature.
v_P, v_S	Velocities of compressional and shear waves; Eq. (2).
v_A	Apparent surface velocity of a seismic pulse; Eq. (8).
v_B	Bulk velocity; Eq. (3).

y	Isothermal compression; Sec. II, e.
z	Depth beneath the Earth's surface.
Z	Vertical component of F.
α	Coefficient of thermal expansion.
α_0	Zero-pressure value of α.
β_T	Isothermal pressure derivative of K_T; Eq. (17).
γ	10^{-5} Gauss; also Grüneisen ratio.
Δ	Distance from the epicenter of an earthquake (measured on a great circle).
ε	Opacity (ε^{-1} transparency).
λ	Decay constant of a radioactive element; also east longitude.
η	Viscosity; Eq. (32).
μ	Magnetic permeability.
μ_S	Adiabatic rigidity (shear modulus).
ϱ	Density.
ϱ_0	Zero-pressure value of ϱ.
σ	Electrical conductivity.
τ	Period or a time constant; Eq. (88).
τ_0	Adiabatic gradient; Eq. (15).
φ	Geographic latitude.
ω	Angular frequency.
Ω	Mean angular velocity of the Earth's rotation.

References

Adam, A., A. Wallner u. H. Wiese: Elektrische Leitfähigkeitsanisotropien des Untergrundes im Spiegel magnetotellurischer und geomagnetischer Messungen. Gerlands Beitr. Geophys. **73**, 310 (1964).

Akimoto, S., and H. Fujisawa: Demonstration of the electrical conductivity jump produced by the olivine-spinel transition. J. Geophys. Res. **70**, 443 (1965).

— —, and H. Katsura: The olivine-spinel transition in Fe_2SiO_4 and Ni_2SiO_4. J. Geophys. Res. **70**, 1969 (1965).

Anderson, D. L.: Latest information from seismic observations. In: *The earth's mantle* (T. F. Gaskell, ed.). London and New York: Academic Press 1967.

—, and C. Archambeau: The anelasticity of the earth. J. Geophys. Res. **69**, 2071 (1964).

Anderson, O. L.: Equation for thermal expansivity in planetary interiors. J. Geophys. Res. **72**, 3661 (1967).

Balchan, A. S., and G. R. Cowan: Shock compression of two iron-silicon alloys to 2.7 megabars. J. Geophys. Res. **71**, 3577 (1966).

Birch, F.: Elasticity and constitution of the earth's interior. J. Geophys. Res. **57**, 227 (1952).

— The velocity of compressional waves in rocks to 10 kilobars. J. Geophys. Res. **65**, 1083 (1960); **66**, 2199 (1961).

— Density and composition of mantle and core. J. Geophys. Res. **69**, 4377 (1964).

— Energetics of core formation. J. Geophys. Res. **70**, 6217 (1965).

— Thermal expansion at high pressures. J. Geophys. Res. **73**, 817 (1968).

Bolt, B. A.: The velocity of seismic waves near the earth's center. Bull. Seismol. Soc. Am. **54**, 191 (1964).

Boyd, F. R., and J. L. England: Effect of pressure on the melting of diopside and albite in the range up to 50 kilobars. J. Geophys. Res. **68**, 311 (1963).

Bradley, R. S., A. K. Jamil, and D. C. Munro: Electrical conductivity of fayalite and spinel. Nature **193**, 965 (1962).

Bullard, E., C. Freedman, H. Gellman, and J. Nixon: The westward drift of the earth's magnetic field. Phil. Trans. Roy. Soc. London, Ser. A **243**, 67 (1950).

—, A. E. Maxwell, and R. Revelle: Heat flow through the deep sea floor. Advan. Geophys. **3**, 153 (1956).

BULLEN, K. E.: Introduction to the theory of seismology, 3rd edit. London: Cambridge University Press 1963.

CAGNIARD, L.: Basic theory of the magneto-telluric method of geophysical prospecting. Geophysics **18**, 605 (1953).

CANER, B., W. H. CANNON, and C. E. LIVINGSTONE: Geomagnetic depth sounding and upper mantle structure in the Cordillera region of western North America. J. Geophys. Res. **72**, 6335 (1967).

CARSLAW, H. S., and J. C. JAEGER: Conduction of heat in solids, 2nd edit. London: Oxford University Press 1959.

CHAPMAN, S., and A. T. PRICE: The electric and magnetic state of the interior of the earth, as inferred from terrestrial magnetic variations. Phil. Trans. Roy. Soc. London, Ser. A **229**, 427 (1930).

CLARK, S. P.: Radiative transfer in the earth's mantle. Trans. Am. Geophys. Union **38**, 931 (1957).

— Thermal conductivity. In: *Handbook of physical constants* (S. P. CLARK, ed.). New York: The Geological Society of America 1966a.

— High-pressure phase equilibria. In: *Handbook of physical constants* (S. P. CLARK, ed.). New York: The Geological Society of America 1966b.

—, Z. E. PETERMAN, and K. S. HEIER: Abundances of uranium, thorium, and potassium. In: *Handbook of physical constants* (S. P. CLARK, ed.). New York: The Geological Society of America 1966.

—, and A. E. RINGWOOD: Density distribution and constitution of the mantle. Rev. Geophys. **2**, 35 (1964).

COOK, A. H.: Gravitational considerations. In: *The earth's mantle* (T. F. GASKELL, ed.). London and New York: Academic Press 1967.

CORON, S.: Aperçu gravimétrique sur les Alpes occidentales. AGI Série XII, Fasc. 2, Centre Nat. Rech. Sci. (1963).

COWLING, T. G.: Magnetohydrodynamics. New York: Interscience Publishers 1957.

CREER, K. M.: Arrangement of the continents during the paleozoic era. Nature **219**, No. 5149, 41 (1968).

DAVIS, T. C., and J. L. ENGLAND: The melting of forsterite up to 50 kilobars. J. Geophys. Res. **69**, 1113 (1964).

DIETZ, R. S.: Continent and ocean basin evolution by spreading of the sea floor. Nature **190**, 854 (1961).

DOELL, R. R., and A. COX: Paleomagnetism of Hawaiian lava flows. J. Geophys. Res. **70**, 3377 (1965).

— G. B. DALRYMPLE, and A. COX: Geomagnetic polarity epochs: Sierra Nevada data 3. J. Geophys. Res. **71**, 531 (1966).

DOHR, G., u. K. FUCHS: Statistical evaluation of deep crustal reflections in Germany. Geophysics **32**, 951 (1967).

ELSASSER, W.: The earth's interior and geomagnetism. Rev. Mod. Phys. **22**, 1 (1950).

— Mechanics of the upper mantle. In: *Advances in earth science* (P. M. HURLEY, ed.). Cambridge, Mass.: The M. I. T. Press 1966.

FILLOUX, J. H.: Oceanic electric currents, geomagnetic variations and the deep electrical conductivity structure of the ocean-continent transition of central California. Ph. D. thesis, University of California, San Diego (1967).

FUCHS, K., ST. MÜLLER, E. PETERSCHMITT, J.-P. ROTHÉ, A. STEIN u. K. STROBACH: Krustenstruktur der Westalpen nach refraktionsseismischen Messungen. Gerlands Beitr. Geophys. **72**, 149 (1963).

GEBRANDE, H.: Der Einfluß von Phasenumwandlungen auf Konvektionsströme im Erdmantel. Z. Geophysik **33**, 297 (1967).

GIESE, P., C. PRODEHL u. C. BEHNKE: Ergebnisse refraktionsseismischer Messungen 1965 zwischen dem Französischen Zentralmassiv und den Westalpen. Z. Geophysik **33**, 215 (1967).

GILVARRY, J. J.: The Lindemann and Grüneisen laws. Phys. Rev. **102**, 308 (1956a).

— Grüneisen Parameter for a solid under finite strain. Phys. Rev. **102**, 331 (1956b).

Gilvarry, J. J.: Lindemann and Grüneisen laws and a melting law at high pressure. Phys. Rev. Letters **16**, 1089 (1966).

Green, D. H., and A. E. Ringwood: An experimental investigation of the gabbro to eclogite transformation and its petrological applications. Geochim. Cosmochim. Acta **31**, 767 (1967).

Guier, W. H., and R. R. Newton: The earth's gravity field as deduced from the Doppler tracking of five satellites. J. Geophys. Res. **70**, 4613 (1965).

Gutenberg, B.: Wave velocities at depth between 50 and 600 kilometers. Bull. Seismol. Soc. Am. **43**, 224 (1953).

— Wave velocities in the earth's core. Bull. Seismol. Soc. Am. **48**, 301 (1958).

—, and C. F. Richter: Seismicity of the earth. Princeton, N. J.: Princeton University Press 1949.

Hamilton, R. M.: Temperature variation at constant pressure of the electrical conductivity of periclase and olivine. J. Geophys. Res. **70**, 5679 (1965).

Heirtzler, J. R., G. O. Dickson, E. M. Herron, W. C. Pitman, and X. Le Pichon: Marine magnetic anomalies, geomagnetic field reversals and motions of the ocean floor and continents. J. Geophys. Res. **73**, 2119 (1968).

Heiskanen, W. A., and F. A. Vening-Meinesz: The earth and its gravity field. New York: MacGraw-Hill 1958.

Herrin, E., and J. Taggart: Regional variations in Pn velocity and their effect on the location of epicenters. Bull. Seismol. Soc. Am. **52**, 1037 (1962).

Herzen, R. P. von: Surface heat flow and some implications for the mantle. In: *The earths' mantle* (T. F. Gaskell, ed.). London and New York: Academic Press 1967.

Hess, H. H.: History of ocean basins. In: *Petrologic studies: A volume to honor A. F. Buddington.* New York: Geological Society of America 1962.

Hughes, H.: The pressure effect on the electrical conductivity of peridotit. J. Geophys. Res. **60**, 187 (1955).

Ibrahim, A. K., and O. W. Nuttli: Travel time curves and upper mantle structure from long period S waves. Bull. Seismol. Soc. Am. **57**, 1063 (1967).

Izsak, I. G.: Tesseral harmonics of the geopotential and corrections to station coordinates. J. Geophys. Res. **69**, 2621 (1964).

Jeffreys, H.: The earth, 4th edit. London: Cambridge University Press 1959.

— Some normal earthquakes. Geophys. J. **6**, 493 (1962).

— On the hydrostatic theory of the figure of the earth. Geophys. J. **8**, 196 (1964).

—, and R. Vicente: The theory of nutation and the variation of latitude. Monthly Notices Roy. Astron. Soc. **117**, 142, 162 (1957).

Jordan, J., R. Black, and C. C. Bates: Patterns of maximum amplitudes of Pn and P waves over regional and continental areas. Bull. Seismol. Soc. Am. **55**, 693 (1965).

Jordan, P.: Die Expansion der Erde. Braunschweig: Vieweg 1966.

Kahle, A. B., E. H. Vestine, and R. H. Ball: Estimated surface motions of the earth's core. J. Geophys. Res. **72**, 1095 (1967).

Knopoff, L.: Thermal convection in the earth's mantle. In: *The earth's mantle* (T. F. Gaskell, ed.). London and New York: Academic Press 1967.

Kraut, E. A., and G. C. Kennedy: New melting law at high pressures. Phys. Rev. **151**, 668 (1966).

Kuo, J. T., and M. Ewing: Spatial variations of tidal gravity. In: *The earth beneath the continents* (J. S. Steinhart and T. J. Smith, ed.). Washington D. C.: American Geophysical Union 1966.

Lahiri, B. N., and A. T. Price: Electromagnetic induction in non-uniform conductors, and the determination of the conductivity of the earth from terrestrial magnetic variations. Phil. Trans. Roy. Soc. London, Ser. A **237**, 509 (1939).

Leaton, B. R., S. R. C. Malin, and M. J. Evans: An analytical representation of the estimated geomagnetic field and its secular change for the epoch 1965. J. Geomagn. Geoelect. **17**, 187 (1965).

Lee, W. H. K., and S. Uyeda: Review of heat flow data. In: *Terrestrial heat flow* (W. H. K. Lee, ed.). Washington D. C.: American Geophysical Union 1965.

Lehmann, I.: *S* and the structure of the upper mantle. Geophys. J. **4**, 124 (1961).

LIEBSCHER, H.-J.: Reflexionshorizonte der tieferen Erdkruste im Bayrischen Alpenvorland, abgeleitet aus Ergebnissen der Reflexionsseismik. Z. Geophysik **28**, 162 (1962).
LUBIMOVA, E. A.: Theory of thermal state of the earth's mantle. In: *The earth's mantle* (T. F. GASKELL, ed.). London and New York: Academic Press 1967.
LUBIMOVA, H. A.: Thermal history of the earth with consideration of the variable thermal conductivity of its mantle. Geophys. J. **1**, 115 (1958).
MACDONALD, G. J. F.: The deep structure of continents. Rev. Geophys. **1**, 587 (1963).
— Dependence of the surface heat flow on the radioactivity of the earth. J. Geophys. Res. **69**, 2933 (1964).
MCCONNEL, R. K.: Isostatic adjustment in a layered earth. J. Geophys. Res. **70**, 5171 (1965).
MCDONALD, K. L.: Penetration of the geomagnetic secular field through a mantle with variable conductivity. J. Geophys. Res. **62**, 117 (1957).
MCKENZIE, D. P.: The viscosity of the lower mantle. J. Geophys. Res. **71**, 3995 (1966).
MCQUEEN, R. G., and S. P. MARSH: Shock-wave compression of iron-nickel alloys and the earth's core. J. Geophys. Res. **71**, 1751 (1966).
— —, and J. N. FRITZ: Hugoniot equation of state of twelve rocks. J. Geophys. Res. **72**, 4999 (1967).
MELCHIOR, P.: The earth tides. Oxford: Pergamon 1966.
MÜLLER, ST., and M. LANDISMAN: Seismic studies of the earth's crust in continents, 1; evidence for a low-velocity zone in the upper part of the lithosphere. Geophys. J. **10**, 525 (1966).
MUNK, W. H., and G. J. F. MACDONALD: The rotation of the earth. London: Cambridge University Press 1960.
NAGATA, T.: Rock magnetism, 2nd ed. Tokyo: Maruzen 1961.
—, and M. SAWADA: Annual mean values of geomagnetic elements since 1955. J. Geomagn. Geoelect., Suppl. **15**, 1 (1963).
NUTTLI, O.: Seismological evidence pertaining to the structure of the earth's upper mantle. Rev. Geophys. **1**, 351 (1963).
OPDYKE, N. D., B. GLASS, J. D. HAYS and J. FOSTER: Paleomagnetic study of Antarctic deep-sea cores. Science **154**, 349 (1966).
PAKISER, L. C., and J. S. STEINHART: Explosion seismology in the western hemisphere. In: *Research in geophysics*, **2**, *Solid earth and interface phenomena*. Cambridge, Mass.: The M. I. T. Press 1964.
—, and I. ZIETZ: Transcontinental crustal and upper-mantle structure. Rev. Geophys. **3**, 505 (1965).
PARKINSON, W. D.: The influence of continents and oceans on geomagnetic variations. Geophys. J. **6**, 441 (1962).
PEKERIS, C. L., and H. JAROSCH: The free oscillations of the earth. In: *Contributions in geophysics in honor of Beno Gutenberg*. Intern. Ser. Monogr. Earth Sci. **1**. London: Pergamon 1958.
PRESS, F.: Seismological information and advances. In: *Advances in earth science* (P. M. HURLEY, ed.). Cambridge, Mass.: The M. I. T. Press 1966.
— Earth models obtained by Monte Carlo inversion. J. Geophys. Res. **73**, 5223 (1968).
RIKITAKE, T.: Anomaly of geomagnetic variations in Japan. Geophys. J. **2**, 276 (1959).
RINGWOOD, A. E.: The olivine-spinel transition in the earth's mantle. Nature **178**, 1303 (1956).
— The chemical composition and origin of the earth. Mineralogy of the mantle. In: *Advances in earth science* (P. M. HURLEY, ed.). Cambridge, Mass.: The M. I. T. Press 1966.
RUNCORN, S. K.: Changes in the convection pattern in the earth's mantle and continental drift: Evidence for a cold formation of the earth. Phil. Trans. Roy. Soc. London, Ser. A **258**, 228 (1965).
SACKS, I. S.: Residuals in travel time from teleseisms on the Andean stations. Carnegie Inst. Yearbook **65**, 47 (1966).
SCHMUCKER, U.: Anomalies of geomagnetic variations in the southwestern United States. J. Geomagn. Geoelect. **15**, 193 (1964).
—, S. E. FORBUSH, O. HARTMANN, A. A. GIESECKE, M. CASAVERDE, J. CASTILLO, R. SAL-

GUEIRO, and S. DEL POZO: Electrical conductivity anomaly under the Andes. Carnegie Inst. Yearbook **65**, 11 (1966).

SCHULT, A.: The effect of pressure on the Curie temperature of magnetite and some other ferrites. Z. Geophysik **34**, 505 (1968).

SIMMONS, G.: Velocity of shear waves in rocks to 10 kilobars, 1. J. Geophys. Res. **69**, 1123 (1964).

SKINNER, B. J.: Thermal expansion. In: *Handbook of physical constants* (S. P. CLARK, ed.). New York: The Geological Society of America 1966.

SMITH, T. J., J. S. STEINHART, and L. T. ALDRICH: Lake superior crustal structure. J. Geophys. Res. **71**, 1141 (1966).

TAKEUCHI, H.: On the earth tide of the compressible earth of variable density and elasticity. Trans. Am. Geophys. Union **31**, 651 (1950).

TIKHONOV, A. N.: (Determination of electrical parameters of deep layers within the earth.) Dokl. Nauk USSR **73** (2), 295 (1950).

TOZER, D. C.: Towards a theory of thermal convection in the mantle. In: *The earth's mantle* (T. F. GASKELL, ed.). London and New York: Academic Press 1967.

TUVE, M. A., H. E. TATEL, and P. J. HART: Crustal structure from seismic exploration. J. Geophys. Res. **59**, 415 (1954).

UYEDA, S.: Thermoremanent magnetism as a medium of paleomagnetism, with special reference to reverse thermoremanent magnetism. Japan. J. Geophys. **2**, 1 (1958).

VANJAN, L. L., and E. P. KHARIN: (Magnetic depth sounding in the Lake Baikal area.) In: *Regional geophysical investigations in Siberia.* Novosibirsk: Academy of Sciences of the USSR, Siberian department 1967.

VERHOOGEN, J.: The adiabatic gradient in the mantle. Trans. Am. Geophys. Union **32**, 41 (1951).

— Phase changes and convection in the earth's mantle. Phil. Trans. Roy. Soc. London, Ser. A **258**, 276 (1965).

VINE, F., and D. H. MATTHEWS: Magnetic anomalies over ocean ridges. Nature **199**, 947 (1963).

WAKITA, H., H. NAGASAWA, S. UYEDA, and H. KUNO: Uranium, thorium, and potassium contents of possible mantle materials. Geochemical J. **1**, 183 (1967).

WIESE, H.: Geomagnetische Tiefentellurik. Deut. Akad. Wiss. Berlin Geomagn. Institut Potsdam, No. 36 (1965).

WITHAM, K.: Anomalies in geomagnetic variations in the Arctic Archipelago of Canada. J. Geomagn. Geoelect. **15**, 227 (1964).

WOOLLARD, G. P.: Regional isostatic relations in the United States. In: *The earth beneath the continents* (J. S. STEINHART and T. J. SMITH, ed.). Washington D. C.: American Geophysical Union 1966.

YODER, H. S., and C. E. TILLEY: Origin of basalt magmas, an experimental study of natural and synthetic rock systems. J. Petrol. **3**, 342 (1962).

ZIETZ, I., E. R. KING, W. GEDDES, and E. G. LIDIAK: Crustal study of a continental strip from the Atlantic ocean to the Rocky Mountains. Bull. Geol. Soc. Am. **77**, 1427 (1966).

CHAPTER 7

K. H. WEDEPOHL

COMPOSITION AND ABUNDANCE OF COMMON IGNEOUS ROCKS

Only rock types of major abundance contribute to a detectable degree to chemical cycles at and near the surface of the Earth; rare species tell us about the limiting conditions in the natural environment. A handbook of this type has to be restricted to a condensation of more general information. Therefore the abundant rock types must be considered exclusively. This section will not contribute new names or introduce ncw aspccts to the petrographic nomenclature or review the extensive and partially controversial literature. It will act only as a reference on rock names for authors and readers of Part II of this Handbook.

Nomenclature should be free of rock names related to a special genesis. The system must try to meet the most common use of a term on an international scale. It should satisfy the requirements of a petrographer to describe a natural environment consisting of units larger than a cubic meter. Magmatic processes commonly produce only special combinations of the major rockforming minerals in a certain range of variation. The principle of natural selection is that of approaching mixtures with lowest possible melting temperatures.

Some secondary process of probable magmatic origin may change a primary rock composition. Using strict rules of nomenclature the resulting rock is a metamorphic one. But a great number of people still quote spilitic rock types (see Table 7-1 c) for instance as a magmatic group. Another process allowing deviation from primary mineral-associations is that of spatial separation into monomineralic or polymineralic segregations through gravitative settling, convection etc. These are mechanical actions producing local inequilibrium. Rapid consolidation of a rock melt often prevents the early products of crystallisation from equilibrating with the residual melt. Zoning of mixed crystals, partially resorbed crystal relicts and interstitial glass are definite indications of inequilibrium. The whole range of equilibrium formations and inequilibrium relicts must be covered by rock nomenclature. Therefore, limits of variability are determined more for convenience than by results from high temperature experiments.

In the historical course of defining rock types chemical and mineralogical systems have each prevailed from time to time. A third aspect of rock terminology is a geometric one called "texture" in English-speaking countries (French and German petrographers use the term "structure" instead). This factor includes relative grain size, shape and arrangement of mineral constituents and gives information about the sequence of mineral formation. Because properties of this type are mainly important from petrological aspects, but less so from the geochemical point of view, they will not be considered in this article in detail.

The quantitative mineral composition, called the mode, is the main basis of a modern nomenclature of magmatic rocks. In several cases — natural glass and rocks with some glassy portion or very fine-grained matrix — chemical composition must be considered in nomenclature. An often recommended method is the use of norm minerals which are computed following well known rules (for instance CIPW-rules: CROSS, IDDINGS, PIRSSON and WASHINGTON, 1902, and modern textbooks). Computers can be easily programmed for these rules (VITALIANO et al., 1965; JUNG, SCHULZ, 1965; HEY et al., 1966; etc.).

If we plot the names of common magmatic rocks into a graphical system considering all the combinations of 3 or 4 abundant rock forming minerals there must be gaps due to the above described principle of selection.

I. Nomenclature of Igneous (Magmatic) Rocks

To make the system simple we only use two subgroups for all magmatic types.

a: *Intrusive rocks* as products of subsurface consolidation of melts (plutonic rocks etc.).

b: *Volcanic rocks* consolidated after extrusion of melts on the surface or intrusion near the surface (hypabyssal rocks) of the Earth (a melt in this sense consists at least partially of molten material).

Pyroclastic rocks are listed and described in Chapter 8 under sedimentary rocks.

In some descriptions of geologic settings and certain geochemical literature only collective terms are used such as:

granitic-rhyolitic rocks
intermediate rocks
basaltic-gabbroic rocks
peridotitic rocks
alkalic rocks.

The old collective terms as acid, basic and ultrabasic rocks should be avoided because they are chemically misleading (rocks with more than 66% SiO_2 have been called: acid; those between 45 and 50% SiO_2: basic, and rocks with less than 45% SiO_2: ultrabasic). Another "chemically sounding" term probably cannot be replaced by a short, generally accepted name; that is "*alkalic rock*" for a mineral association containing feldspathoids or other alkali minerals such as acmite. These rocks have a molar proportion of alkalies to silicium larger than 1:3 (SHAND, 1930). No group of magmatic rocks is more restricted to such special geologic environments or consists of so many varieties under specific names as that of the alkalic rocks. The number of names is not proportional to the abundance of this group.

BECKE's and HARKER's rock classification based on geographic predominance (calc-alkaline = Pacific-rocks, alkaline = Atlantic rocks) can be misleading and does not solve the problem of regional associations of magmatic rock types related to special tectonic environments.

We avoid duplication of a modern literature search by STRECKEISEN (1966, 1967) for convenient rock definitions on the base of abundant mineral proportions if we adopt in general his suggestions. A reasonable number of modal analyses from common magmatic rock types has been published and give information about the predominance of certain mineral compositions and their ranges of variation. We know from the statistics of analytical errors and from other distributions of data that it is inconvenient to consider the rare and extreme values and therefore favor only those of higher probability. The problem is to draw appropriate limits of variability in mineral composition of rock types.

STRECKEISEN has used data on modal rock analyses as a base of his nomenclature. Data came from literature compilations (JOHANNSEN, 1931—1939; TRÖGER, 1935; SHAND, 1943) and from publications about several magmatic provinces. According to suggestions of P. NIGGLI and E. TRÖGER the proportion of the light rockforming minerals offers a convenient basis for a nomenclature system. It can be easily plotted into twin triangles having a joint base line. The four corners of this figure are occupied by the four major groups of light minerals:

Q SiO_2-minerals (mainly quartz)
A alkali feldspars (potassium feldspars, anorthoclase, and albite (up to ~10% anorthite)
P plagioclase with more than ~10% anorthite
F feldspathoid minerals (including melilite, contrary to the suggestions of STRECKEISEN) and zeolites.

Except for including the non-albitic plagioclases, these groups are mainly identical with the minerals of BOWEN's so called "residual system": $NaAlSiO_4$—$KAlSiO_4$—SiO_2. Through correlation with plagioclase which usually coexists with abundant dark or mafic (*M*)[1] silicates, the pyroxenes, amphiboles etc. will be indirectly represented in the graphical plot (Fig. 7-1). A formal representation of the mafic minerals graphically would require a third dimension. The resulting twin tetrahedron makes the use of the system unwieldly. Terminology of a relative small number of magmatic rocks, predominantly mafic types, is based on proportions of dark silicates and cannot be specifically described by the three minerals of triangle *AQP* or *AFP* (Fig. 7-1). Selectivity of natural magmatic processes makes our system simple.

In the suggested system the break between plagioclase with less than and more than 10 mole-% $CaAl_2Si_2O_8$ (into albite as alkali feldspar and into "plagioclase") cannot strictly consider the different ranges of naturally occuring mixtures between the K—Na—Ca—feldspar molecules. Compositions of mixed crystals have been plotted for instance by TUTTLE and BOWEN (1958) in a triangular diagram of the ternary system. The ranges of miscibility are temperature dependent. And the temperature of a rock melt is highly influenced by its content of gases like H_2O etc. In the mentioned diagram of TUTTLE and BOWEN an alkali feldspar with a maximum of about 13 mole-% anorthite coexists in a trachytic dry melt with plagioclase containing more than 13 mole-% $CaAl_2Si_2O_8$. Dry rhyolite, phonolite or wet trachyte melts have lower melting temperatures producing alkali feldspar with a maximum of 7% anorthite and a related coexisting plagioclase. The alkali feldspars from wet rhyolite or phonolite melts must be even lower in anorthite (less than 4%).

Some petrographers divide all the magmatic rocks between melanocratic and leucocratic types depending on the prevalency of either dark (60—100 vol.-%) or light minerals. Leucocratic rocks have less than 30% mafic minerals. The intermediate group with 30—60% dark minerals is often called mesocratic. The portion of dark

[1] *M*: Pyroxenes, amphiboles, olivines, magnetites, ilmenites, titanites, garnets, biotites, apatites and other accessories.

minerals (M) which is equivalent to the color index can usually be estimated from visual inspection.

Fig. 7-1 represents the limiting conditions for light-colored minerals from common intrusive and volcanic rocks. In plotting the vol.-% of the light minerals, belonging to the groups A, Q, P or A, F, P, their sum must be equal to 100%.

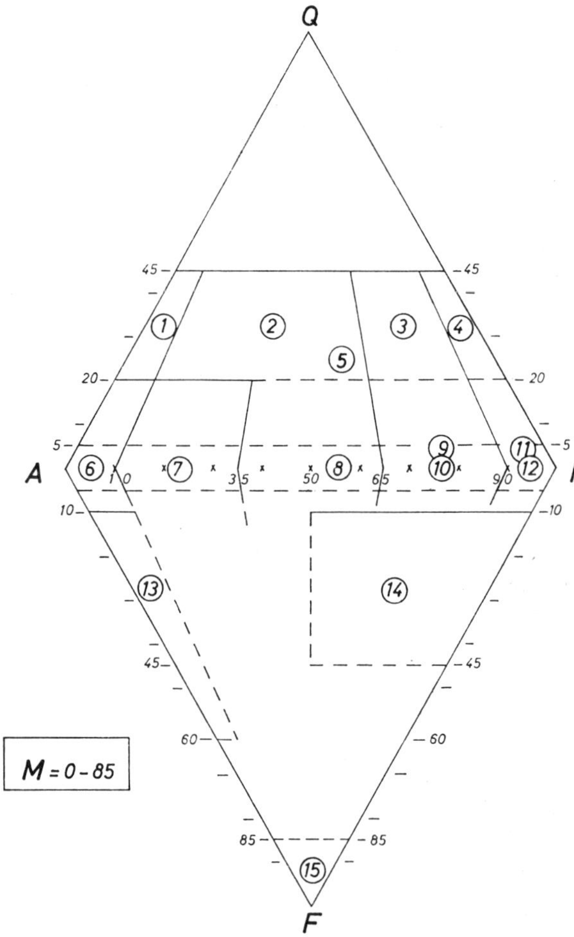

Fig. 7-1
A: alkali feldspar Q: SiO_2 P: plagioclase F: feldspathoids For rock numbers see Tables 7-1a to 7-1c

This figure is a copy of STRECKEISEN's original drawing with respect to most lines limiting ranges of composition. Only those rock types are mentioned by name which are geochemically important because of their abundance. This selection by the present author also has disadvantages but, since our petrographic handbooks list more than one thousand names of magmatic rocks, limitations are required.

The textbook by TURNER and VERHOOGEN (1960) represents an example of describing most of the important magmatic processes with a relative small number of rock terms.

The position of the following limits has been somewhat modified from STRECKEISEN's suggestions (the latter in brackets). Compositions are given in terms of volume percent.

1. In the diagrams all rock compositions between 0 and 85% (90%) dark or mafic minerals (pyroxenes, amphiboles, olivines, biotites etc.) can be plottet.

2. A quartz content of 45% (60%) is the maximum for most granites, granodiorites, quartzdiorites and the related volcanic rocks.

3. A feldspathoid content of 10% is the minimum and that of 45% (60%) is the maximum for most essexites, theralites and the related volcanic rocks.

4. The main fields of alkali syenites, syenites, monzonites, monzogabbros, gabbros and related volcanic rocks cover the area from 5% quartz to 5% (0%) feldspathoids.

5. The minimum content of feldspathoid minerals in the sum of feldspars and feldspathoids of ijolites, nephelinites, leucitites, melilitites is 85% (90%). Nephelinites, leucitites etc. with less than 85% (90%) *F* and much more than 45% *F* are called tephritic-or phonolitic nephelinites etc. respectively. Melilites are counted as feldspathoids (mafic minerals).

6. The area of essexites (including theralites, nepheline tephrites, leucite tephrites, nepheline basanites and similar rocks) extends to the *P*—*F*-borderline. Essexites are the alkali-feldspar-rich varieties and theralites the plagoclase-rich rocks of this field (14). Tephrites etc. with more than 50% alkali feldspar can be called phonolitic tephrites etc.

Broken lines in Fig. 7-1 indicate that the range of variability is here somewhat larger than the numbered areas. Transitional rock types outside of the areas of common composition should not get special names. Combinations of rock terms with "quartz-bearing" or "quartz-poor" etc. describe their special mineral content. A latite with 10% quartz is a "quartz-bearing latite". A syenite with 8% nepheline is a "nepheline bearing syenite" respectively. A quartz monzonite = adamellite (5a) is a granite relatively poor in quartz and alkali feldspar.

Rock types containing feldspathoids are much less common than quartz-bearing species or those lacking in both of these mineral groups. Therefore the lower triangle is less important than the upper. The lesser importance is visible in the restriction of composition in which abundant rock types occur.

Most of the rock species of the right side of the *A*—*P*—*F*-triangle have pyroxene- (and even plagioclase-) contents comparable to basalts and gabbros, which they often accompany; therefore they are included in the more general terms "basaltic" and "gabbroic" rocks.

The recommended rock terms are listed in Tables 7-1a to c including information about the proportion of mafic (*M*) minerals such as pyroxenes, amphiboles, olivines, biotites, garnets etc. This part of the tables makes unnecessary a third dimension in Fig. 7-1. The last column of Tables 7-1a to c contains additional information about the character of dark minerals etc. on which some terms like basanite, norite depend.

The European use of the term "porphyry" for an altered rock type conflicts with its use as a textural description (of larger crystals in a finer groundmass) in the United States. Therefore it is recommended that "porphyry" be used only in connection with "quartz-" as a rock name or in connection with a rock name of Table 7-1b for describing its texture. "Porphyritic" is synonymous with "porphyry".

European petrographers call a spilitic and/or saussuritic basalt a "diabase"; in America the latter name is restricted to basalts (mainly of dikes and sills) having an ophitic texture (euhedral plagioclase in pyroxene mesostasis). This difference can only be overcome by avoiding both definitions, calling a basalt with a special texture only a basalt and calling a spilitic basalt a spilite.

Table 7-1a. *Intrusive rocks with less than 85 vol.-% mafic constituents* (*M*)

Number of rock in Fig. 7-1*	Rock name	Common synonym	*M* in vol.-%	Special remarks
1a	alkali granite		0—20	
	[granite pegmatite		<5]	
2a	granite		5—20	
3a	granodiorite		5—25	biotite ≧ hornblende
4a	quartz diorite	tonalite (dark) trondhjemite (light)	15—40	
5a	quartz monzonite	adamellite	5—25	
6a	alkali syenite		0—25	
7a	syenite		10—35	
8a	monzonite		15—45	
9a	monzodiorite		20—50	
10a	monzogabbro		25—60	
11a	diorite		25—50	andesine, biotite, amphibole etc.
12a	gabbro		35—65	labrador-bytownite, pyroxene etc.
		norite	35—65	orthopyroxene > clinopyroxene
	anorthosite		0—10	
13a	nepheline syenite		0—85	
		foyaite	0—30	
		malignite	30—60	
		shonkinite	60—85	
14a	essexite (nepheline as *F*)		30—60	
		theralite		$\frac{A}{P}$(theral.) > $\frac{A}{P}$(essex.)
	teschenite (analcite as *F*)		30—60	
15a	ijolite (nepheline as *F*)		30—60	
	melteigite (nepheline)		60—85	
	urtite (nepheline as *F*)		0—30	
	Intrusive rocks with 85—100 vol.-% mafic constituents (*M*)			
	peridotite		85—100	olivine (30—90%)
		Iherzolite		ol., ortho- and clinopyrox.
	dunite		100	olivine ≫ pyroxene
		kimberlite	100	phlogopite-bearing
	griquaite		100	garnet > pyroxene, olivine
	pyroxenite		85—100	
	hornblendite		85—100	

* The subclassification into a, b, c as *intrusive*, volcanic and altered rock types is not used in Fig. 7-1.

Table 7-1b. *Volcanic rocks*

Number of rock in Fig. 7-1*	Rock name	Common synonym	M in vol.-%	Special remarks
1b	alkali rhyolite		0—12	
		pantellerite		alkali pyroxene
2b	rhyolite rhyolitic obsidian	liparite	0—15	
3b	rhyodacite		0—20	
4b	dacite		5—25	
5b	quartz latite	dellenite	0—20	
6b	alkali trachyte		0—20	
7b	trachyte		5—25	
8b	latite		5—35	
	minette		30—50	lamprophyre: pyroxene and biotite
		vogesite		hornblende
9b	latite andesite	doreite mugearite	15—40	potassium feldspar, olivine
10b	latite basalt	trachybasalt	40—60	
	spessartite		40—75	lamprophyre: hornblende > pyroxene
11b	andesite		20—40	
	kersantite		20—40	lamprophyre: biotite>pyroxene
12b	tholeiitic basalt	diabase**	40—75	hypersthene in norm
	alkali olivine basalt		40—75	nepheline in norm
	picrite basalt	oceanite	40—75	>20% olivine
	camptonite		40—75	lamprophyre: pyroxene and hornblende
13b	phonolite		0—25	
14b	nepheline basanite		30—70	>5% olivine
	nepheline tephrite		30—70	<5% olivine
	leucite basanite		30—70	>5% olivine
	leucite tephrite		30—70	<5% olivine
15b	nephelinite		40—75	
		ankaratrite	60—85	>50% pyroxene
	leucitite		40—75	
	melilitite		40—75	
	alnöite		40—75	lamprophyre: biotite > pyroxene; melilite

* The subclassification into a, b, c as intrusive, *volcanic* and altered rock types is not used in Fig. 7-1.

** Ophitic basaltic rock in U.S.A.

Table 7-1 c. *Altered rock types of the spilite-keratophyre-association (characterized by replacement of olivine by serpentine, of pyroxene and calcic plagioclase by chlorite, epidote and albite etc.)*

Number of rock in Fig. 7-1*	Rock name	Common synonym	*M* in vol.-%	Special remarks
3c, 4c	quartz porphyrite		0—30	
6c, 7c	quartz keratophyre		0—20	> 15% quartz
	keratophyre		10—25	
11c	porphyrite		20—40	
12c	spilite	diabase (European petrographers)	40—75	

* The subclassification into a, b, *c* as intrusive, volcanic and *altered* rock types is not used in Fig. 7-1.

Furthermore terms such as granophyre, dolerite etc. should be avoided as rock names because they only indicate a texture (granophyric: irregular intergrowth of quartz patches in a base of feldspar), coarseness (doleritic) etc.

Some volcanic rocks contain major or minor portions of glass. A "norm" calculation of potential minerals from a chemical analysis of the glass can often help in adequately locating the rock in the *A—P—Q—F*-system of Fig. 7-1. Several rock names are in use for specific or more general compositions of the glass. For instance the term "limburgite" means a nepheline basanite (14b) with glass, which on crystallizing may contain nepheline and plagioclase. Obsidian and pitchstone are more general names of natural glasses with less than or more than 3% water respectively. They always should be combined with a prefix from Table 7-1b to indicate the potential mineral composition after crystallization. Rhyolitic obsidian is the most common of these types.

Names of the so-called lamprophyre association are included in Table 7-1b although they appear as often heavily altered dike rocks intermediate between volcanic and intrusive types. This alteration seems to be connected with their special genesis which is characterised by a relative high concentration of potassium and volatile constituents (H_2O, CO_2 etc.). These rock compositions point to the alkali olivine basalt-type of original magma.

Rocks of the spilite-keratophyre association mostly from geosynclinal environments are compiled in Table 7-1c. Type and degree of deviation from the "normal" rocks as placed into Table 7-1b leads to the assumption of a metamorphic alteration on the final rock composition. Low grade replacement of calcic feldspar of basaltic rocks etc. by albite is called "spilitization". "Saussuritization" is epidote formation from calcic plagioclase and Ca-bearing solutions[1]. Both processes including chloritization etc. are mainly responsible for the formation of rocks of Table 7-1c. Serpentinization of olivine rocks is a comparable reaction. Since rocks formed in these processes are not strictly of magmatic origin they are mentioned both here and under metamorphic rocks for completeness.

[1] The rock name "spilite" is often restricted to basaltic rocks with visible albitization. In this publication it replaces the name "diabase" of English and German petrographers, even for rocks without abvious sodium affinities, to avoid confusion due to the different meaning of "diabase" in America.

II. Chemical Composition of Common Igneous Rocks

The average chemical composition of abundant igneous rocks is of interest to the geochemist. Up to 1954 the best known averages for various rock types were those computed by R. A. DALY. The 1933 edition of DALY's book includes all of the improvements he added to his early concept. Since DALY computed these averages, more reliable data have become available on representative rocks; also analytical methods and rock terminology have been improved.

In 1954 NOCKOLDS published his evaluation of a literature survey on analyses from fresh and representative magmatic rocks. Except for having an upper limit of 10% quartz (instead of 5%) in syenites, intermediate and gabbroic rocks, NOCKOLDS uses definitions similar to those of STRECKEISEN (1966). A few rocks in his averages do not resemble the standard rock composition of Tables 7-1 a and b. Only the more common types from NOCKOLDS' collection were selected for Tables 7-2 to 7-6. His computations of mineral norms are included in these tables.

A plot of the *AQP* or *AFP* minerals on Fig. 7-1 from norm data of Tables 7-2 to 7-6 can give information on the average rock composition in cases of a reasonable agreement between norm and modal minerals.

III. Rock Nomenclature and Current Trends in Igneous Petrology

In the introduction to this article we have cautioned, as a primary rule of nomenclature, to avoid rock names which are related to special petrogenetic processes. Progress in understanding natural processes can change views on the origin of a special rock unit but its names should be preserved nevertheless.

Some disadvantages arising from this rule must be tolerated. Rocks originating from the same or a similar process can have different names. Other rocks with the same name might be derived from different processes. We will quote a few examples.

Many granites and granodiorites when plotted in fields 2—3 of Fig. 7-1 are close to the borderline between these two rock types. The reason for a major abundance of granitic rocks of these compositions is their coincidence with the mixture of lowest melting point. TUTTLE and BOWEN (1958) and WINKLER and VON PLATEN (1961) have plotted the distribution of normative albite, *K*-feldspar, quartz compositions from more than 1,000 granitic rocks (in a broad sense) for comparison on an isobaric triangular diagram showing the melting relations in the same system. The average composition has only a slightly higher *K*-feldspar proportion than the minimum melting temperature at 3,000 atm water pressure. Under increasing pressure mixtures of minimum melting temperature become richer in albite. LUTH, JAHNS and TUTTLE (1964) have investigated this system up to 10,000 kilobars pressure. VON PLATEN (1965), WEIL and KUDO (1968) have investigated the influence of calcium on the above mentioned system. The norm of average granites from Table 7-2 gives an albiteanorthite ratio of 4.7. The low melting composition related to such a melt in VON PLATEN's experiments still lies in the field containing 53% of this total of 1,200 plotted granitic rocks. The same author has furnished information about the increase of alkali feldspar in granitic mixtures of lowest melting point under growing HCl partial pressure.

Table 7-2. *Averages of granitic rocks.* (After Nockolds, 1954)
Effusive rocks include obsidians; rock number of Table 7-1a, b in [brackets]; number of analyses used for average in (brackets).

	Alkali granites [1a] (48)	Alkali rhyolites [1b] (21)	Granites [2a] (72)	Rhyolites [2b] (22)	Quartz monzonites [5a] (121)	Quartz latites [5b] (58)	Grano-diorites [3a] (137)	Rhyodacites [3b] (115)	Quartz diorites [4a] (58)	Dacites [4b] (50)
SiO_2	73.86	74.57	72.08	73.66	69.15	70.15	66.88	66.27	66.15	63.58
TiO_2	0.20	0.17	0.37	0.22	0.56	0.42	0.57	0.66	0.62	0.64
Al_2O_3	13.75	12.58	13.86	13.45	14.63	14.41	15.66	15.39	15.56	16.67
Fe_2O_3	0.78	1.30	0.86	1.25	1.22	1.68	1.33	2.14	1.36	2.24
FeO	1.13	1.02	1.67	0.75	2.27	1.55	2.59	2.23	3.42	3.00
MnO	0.05	0.05	0.06	0.03	0.06	0.06	0.07	0.07	0.08	0.11
MgO	0.26	0.11	0.52	0.32	0.99	0.63	1.57	1.57	1.94	2.12
CaO	0.72	0.61	1.33	1.13	2.45	2.15	3.56	3.68	4.65	5.53
Na_2O	3.51 [a]	4.13	3.08	2.99	3.35	3.65	3.84	4.13	3.90	3.98
K_2O	5.13 [a]	4.73	5.46	5.35	4.58	4.50	3.07	3.01	1.42	1.40
H_2O^+	0.47	0.66	0.53	0.78	0.54	0.68	0.65	0.68	0.69	0.56
P_2O_5	0.14	0.07	0.18	0.07	0.20	0.12	0.21	0.17	0.21	0.17
qz [b]	32.2	31.1	29.2	33.2	24.8	26.1	21.9	20.8	24.1	19.6
or	30.0	27.8	32.2	31.7	27.2	26.7	18.3	17.8	8.3	8.3
ab	29.3	35.1	26.2	25.1	28.3	30.9	32.5	35.1	33.0	34.1
an	2.8	2.0	5.6	5.0	11.1	9.5	16.4	14.5	20.8	23.3
c	1.4	—	0.8	0.9	—	—	—	—	—	—
$CaSiO_3$	—	0.1	—	—	—	0.2	—	1.3	0.3	1.3
$MgSiO_3$	0.6	0.3	1.3	0.8	2.5	1.6	3.9	3.9	4.9	5.3
$FeSiO_3$	1.1	0.6	1.7	—	2.2	0.8	2.9	1.3	4.1	2.8
ac	—	—	—	—	—	—	—	—	—	—
mt	1.2	1.9	1.4	1.9	1.9	2.5	1.9	3.0	2.1	3.3
il	0.5	0.3	0.8	0.5	1.1	0.8	1.1	1.4	1.2	1.2
ap	0.3	0.2	0.4	0.2	0.5	0.3	0.5	0.3	0.5	0.3

[a] Pegmatites mainly differ in averages by their higher potassium and slightly lower sodium contents (~6.3% K_2O) etc.

[b] The following abbreviations are used for normative minerals: qz = quartz; or = K-feldspar; ab = albite; an = anorthite; c = corundum; lc = leucite; ne = nepheline; ac = acmite; mt = magnetite; il = ilmenite; ap = apatite; cc = calcite.

Table 7-3. *Averages of intermediate rocks.* (After Nockolds, 1954)
Rock number of Tables 7-1a, b in [brackets]; number of analyses in (brackets).

	Alkali syenites [6a] (25)	Alkali trachytes [6b] (15)	Syenites [7a] (18)	Trachytes [7b] (24)	Monzonites [8a] (46)	Latites [8b] (42)	Monzo-diorites [9a] (56)	Latite andesites [9b] (38)	Diorites [11a] (50)	Andesites [11b] (49)
SiO_2	61.86	61.95	59.41	58.31	55.36	54.02	54.66	56.00	51.86	54.20
TiO_2	0.58	0.73	0.83	0.66	1.12	1.18	1.09	1.29	1.50	1.31
Al_2O_3	16.91	18.03	17.12	18.05	16.58	17.22	16.98	16.81	16.40	17.17
Fe_2O_3	2.32	2.33	2.19	2.54	2.57	3.83	3.26	3.74	2.73	3.48
FeO	2.63	1.51	2.83	2.02	4.58	3.98	5.38	4.36	6.97	5.49
MnO	0.11	0.13	0.08	0.14	0.13	0.12	0.14	0.13	0.18	0.15
MgO	0.96	0.63	2.02	2.07	3.67	3.87	3.95	3.39	6.12	4.36
CaO	2.54	1.89	4.06	4.25	6.76	6.76	6.99	6.87	8.40	7.92
Na_2O	5.46	6.55	3.92	3.85	3.51	3.32	3.76	3.56	3.36	3.67
K_2O	5.91	5.53	6.53	7.38	4.68	4.43	2.76	2.60	1.33	1.11
H_2O^+	0.53	0.54	0.63	0.53	0.60	0.78	0.60	0.92	0.80	0.86
P_2O_5	0.19	0.18	0.38	0.20	0.44	0.49	0.43	0.33	0.35	0.28
qz	1.7	—	2.0	—	—	0.5	2.0	7.2	0.3	5.7
or	35.0	32.8	38.4	43.9	27.8	26.1	16.7	15.6	7.8	6.7
ab	46.1	54.0	33.0	28.8	29.3	27.8	31.9	29.9	28.3	30.9
an	4.2	3.3	10.0	9.7	15.8	19.2	21.1	22.2	25.8	27.2
ne	—	0.6	—	2.0	—	—	—	—	—	—
$CaSiO_3$	3.0	2.1	3.0	4.2	6.3	4.5	4.5	4.1	5.6	4.2
$MgSiO_3$	2.4	1.6	5.0	3.2	8.0	9.7	9.9	8.5	15.3	10.9
$FeSiO_3$	2.1	—	2.1	0.5	4.1	2.4	5.4	3.0	8.5	5.3
Mg_2SiO_4	—	—	—	1.4	0.8	—	—	—	—	—
Fe_2SiO_4	—	—	—	0.2	0.4	—	—	—	—	—
mt	3.3	3.3	3.3	3.7	3.7	5.6	4.9	5.3	3.9	5.1
il	1.2	1.4	1.5	1.2	2.1	2.3	2.1	2.4	2.9	2.4
ap	0.5	0.4	1.0	0.5	1.0	1.2	1.0	0.8	0.8	0.7

Table 7-4. *Averages of gabbroic-basaltic rocks.* (After NOCKOLDS, 1954)
Rock number of Tables 7-1a, 7-1b in [brackets]; number of analyses used for averages in (brackets).

	Gabbros [12a] (160)	Tholeiitic basalts [12b] (137)	Alkali olivine basalts [12b] (96)
SiO_2	48.36	50.83	45.78[a]
TiO_2	1.32	2.03	2.63
Al_2O_3	16.81	14.07	14.64
Fe_2O_3	2.55	2.88	3.16
FeO	7.92	9.00	8.73
MnO	0.18	0.18	0.20
MgO	8.06	6.34	9.39[a]
CaO	11.07	10.42	10.74
Na_2O	2.26	2.23	2.63[a]
K_2O	0.56	0.82	0.95[a]
H_2O^+	0.64	0.91	0.76
P_2O_5	0.24	0.23	0.39
qz	—	3.5	—
or	3.3	5.0	6.1
ab	18.9	18.9	18.3
an	34.2	25.9	24.7
ne	—	—	2.3
$CaSiO_3$	8.0	10.3	10.8
$MgSiO_3$	14.0	15.8	7.1
$FeSiO_3$	7.4	11.2	2.9
Mg_2SiO_4	4.3	—	11.5
Fe_2SiO_4	2.5	—	5.0
mt	3.7	4.2	4.6
il	2.4	3.8	5.0
ap	0.6	0.5	1.0

[a] The mean of literature data from TURNER, VERHOOGEN (1960) is higher in SiO_2 (48.4), Na_2O (3.2), K_2O (1.3), mainly compensated by lower MgO.

Table 7-5. *Averages of peridotitic and anorthositic rocks.* (After NOCKOLDS, 1964)
Rock number of Tables 7-1a, 7-1b in [brackets]; number of analyses used for averages in (brackets).

	Peridotite (23)	Anorthosite [12a] (9)		Peridotite (23)	Anorthosite [12a] (9)
SiO_2	43.54	54.54	qz	—	1.4
TiO_2	0.81	0.52	or	1.7	6.7
Al_2O_3	3.99	25.72	ab	4.7	39.3
Fe_2O_3	2.51	0.83	an	7.5	45.9
FeO	9.84	1.46	ne	—	—
MnO	0.21	0.02	$CaSiO_3$	3.9	0.3
MgO	34.02	0.83	$MgSiO_3$	14.8	2.1
CaO	3.46	9.62	$FeSiO_3$	2.6	1.2
Na_2O	0.56	4.66	Mg_2SiO_4	49.1	—
K_2O	0.25	1.06	Fe_2SiO_4	9.6	—
H_2O^+	0.76	0.63	mt	3.7	1.2
P_2O_5	0.05	0.11	il	1.5	0.9
			ap	0.1	0.3

Table 7-6. *Averages of alkalic rocks.* (After NOCKOLDS, 1954)
Rock number of Tables 7-1a, b in [brackets]; number of analyses used for averages in (brackets).

	Nepheline syenites [13a] (80)	Phonolites [13b] (47)	Essexites [14a] (15)	Nepheline tephrites [14b] (8)	Leucite tephrites [14b] (31)	Ijolites [15a] (11)	Olivine nephelinites [15b] (21)	Olivine leucitites [15b] (11)	Olivine melilitites [15b] (10)
SiO_2	55.38	56.90	46.88	44.82	47.05	42.58	40.29	43.64	37.08
TiO_2	0.66	0.59	2.81	2.65	1.54	1.41	2.90	2.54	3.31
Al_2O_3	21.30	20.17	17.07	15.42	16.05	18.46	11.32	10.82	8.08
Fe_2O_3	2.42	2.26	3.62	4.28	3.49	4.01	4.87	5.11	5.12
FeO	2.00	1.85	5.94	6.61	5.78	4.19	7.69	5,89	7.23
MnO	0.19	0.19	0.16	0.16	0.17	0.20	0.22	0.15	0.18
MgO	0.57	0.58	4.85	7.27	6.20	3.22	13.28	13.86	16.19
CaO	1.98	1.88	9.49	10.32	10.80	11.38	12.99	10.66	16.30
Na_2O	8.84	8.72	5.09	5.30	2.35	9.55	3.14	2.16	2.30
K_2O	5.34	5.42	2.64	1.26	5.38	2.55	1.44	4.09	1.36
H_2O^+	0.96	0.96	0.97	1.56	0.60	0.55	1.08	0.72	1.89
P_2O_5	0.19	0.17	0.48	0.35	0.59	1.52	0.78	0.63	0.96
CO_2	0.17	—	—	—	—	0.38	—	—	—
Cl	—	0.23	—	—	—	—	—	—	—
SO_3	—	0.13	—	—	—	—	—	—	—
or	31.1	31.7	15.6	7.8	22.2	10.0	—	6.9	—
ab	32.0	36.2	14.7	12.6	—	—	—	—	—
an	2.8	1.7	16.1	14.5	17.5	—	12.8	6.1	7.5
lc	—	—	—	—	7.4	3.9	6.5	13.8	6.5
ne	23.3	18.7	15.3	17.3	10.8	43.7	14.2	9.9	10.5
Ca_2SiO_4	—	—	—	—	—	—	1.6	—	12.8
$CaSiO_3$	2.1	2.9	11.6	14.3	13.6	18.5	17.2	17.8	10.7
$MgSiO_3$	1.2	1.4	8.2	10.4	9.3	8.0	13.1	14.5	8.6
$FeSiO_3$	0.8	0.9	2.4	2.5	3.2	2.4	2.2	1.1	0.8
Mg_2SiO_4	0.1	—	2.8	5.5	4.3	—	14.1	14.1	22.3
Fe_2SiO_4	0.1	—	0.8	1.5	1.7	—	2.7	1.2	2.5
mt	3.5	3.3	5.3	6.3	5.1	5.8	7.2	7.4	7.4
il	1.4	1.2	5.3	5.0	2.9	2.7	5.5	4.9	6.2
ap	0.4	0.3	1.2	0.8	1.3	3.6	1.8	1.5	2.3
cc	0.4	—	—	—	—	0.9	—	—	—

Discriminating between gabbros, norites, anorthosites and the different basalts is impossible by means of the classification of Fig. 7-1 where they all plot in field 12. For these types nomenclature must be mainly based on the pyroxenes (see Tables 7-1a, b).

Major emphasis has been given to the problem of the origin of basaltic melts. From studies of the regional distribution of basalts the identification of at least two principal magma types has been proposed. This is not the place to review modern research on field relations and melting experiments of basalts and rocks assumed to be their parental material. From the great number of pertinent publications only a few comprehensive papers, which cover the field in some detail, are cited; these include lists of more than 350 references: YODER and TILLEY (1962), GREEN and RINGWOOD (1966). At first inspection of Table 7-4 chemical differences between the two principal types, the so called tholeiitic basalts and the alkali olivine basalts, do not seem to be very large. If one plots the sum of alkalies against silica the samples cover areas which can be divided by a straight line[1]. The higher alkali to silica ratio of the alkali olivine basalt is usually represented by normative nepheline whereas higher silica to alkali gives normative quartz (see Table 7-4). CHAYES (1966) has evaluated more than 600 analyses of Cenozoic basaltic volcanics to show that less than 25 percent of this number yield norms containing neither ne nor qz (for abbreviations of normative minerals see footnote on Table 7-2). He critically reviews the historical line of basaltic nomenclature and proposes the names alkaline and subalkaline basalts for the ne-normative and qz-normative rocks respectively. For the basalts of minor importance with norms lacking ne and qz he classifies these into two groups by means of a discriminant based on normative $CaMgSi_2O_6$, $(Mg, Fe)SiO_3$ and $(Mg, Fe)_2SiO_4$. In this arrangement most analyzed basalts which would be called tholeiitic by YODER and TILLEY (1962) are classified as subalkaline. All others are classified as alkaline, including some which would certainly not be called alkali olivine basalt by YODER and TILLEY because their norms contain $(Mg, Fe) SiO_3$. As already mentioned the critical feature in the definition is the nature of the pyroxenes. Diopsides and augites with a tendency to be rich in sodium and/or in aluminium are the only representatives of this mineral group among the alkali olivine basalts. From the composition range of the modal pyroxene, it follows that hypersthene is a significant normative mineral in all subgroups of tholeiitic basalt. Modal pigeonite and/or hypersthene commonly occur in this rock group. Modal olivine is not a specific mineral for distinguishing the principal magma types. Mentioning this mineral in only one name of the principal types appears to be misleading but is justified as long as norms are considered and the subgroup of tholeiitic olivine basalt is excluded.

Besides the basaltic types already mentioned KUNO (1960) considers the so called non-porphyritic high-alumina basalt as product of a special primary magma. This view is not accepted by YODER and TILLEY (1962) or CHAYES (1966).

From the point of view of the origin of the different principal magmas, it is important to observe that a continuum exists between all basaltic types. This fact excludes the possibility of there being fundamentally different starting materials for the two so-called primary magmas. Seismic evidence in connection with volcanism, assumed temperature gradients in the earth, experiments on basaltic melts and isotope studies make it highly probable that the upper mantle is the source of the principal

[1] This line is characterized by the equation: $SiO_2 = 2.78\ (\%\ Na_2O + \%\ K_2O) + 38.8$.

magmas. Peridotites of the mantle are assumed to be the potential starting material for partial melting. Separation into two principal magmas can be caused by fractional melting, fractional precipitation and remelting under pressures of the upper mantle as YODER and TILLEY (1962) and GREEN and RINGWOOD (1966) etc. deduce from their high pressure-high temperature experiments.

A group of elements like K, Ba, La-Nd, U, Th, P etc. which do not readily substitute in solid solution in the major minerals of the peridotite in the mantle are increasinglye nriched from the different subgroups of tholeiitic basalts to those of alkali olivine basalts. Tholeiitic basalts from deeply submerged volcanic features of the oceans have the lowest concentrations of the mentioned elements and are called by ENGEL et al. (1965) primitive tholeiitic basalts. From the so called "primitive" tholeiitic basalt to alkali olivine basalt the ratio of Fe^{3+}/Fe^{2+} increases in average by a factor of two. CHAYES, VELDE (1965) find in their statistical evaluation that basaltic lavas from the oceanic islands are much richer in titania than those from the landward flanks of the ocean deeps.

The use of the term "tholeiitic" after Tholey in the Saar area (Western Germany) can be misleading. If contrary to CHAYES (1966), the adjective "tholeiitic" is to be preserved for a fundamental group of basaltic rocks (from a historical point of view), these never should be called "tholeiites". JUNG (1958) has presented enough petrological information about the rock from the type locality to discredit it as a typical representative of "tholeiitic basalts".

For more detailed information on the rapidly developing concepts of igneous petrology the reader should consult the special literature. Granitic and basaltic rocks have been considered here as examples to demonstrate that the needs in rock terminology for more or less special definitions must be balanced by the needs of petrology to analyse certain processes.

IV. Abundance of Common Intrusive Rocks

Estimates of the composition of the Earth's crust suffer heavily from the fact that only a third of its volume (in the continents) can be inspected directly. The material of the lower crust of the continental areas, covering about a third of the Earth's surface, and that of the total crust under the oceans can only be investigated from indirect measurements (see Chapter 6). The average thickness of the continental crust has been found by seismic reflections and refractions to be 35 km and that of the oceanic crust to be 5 to 6 km. The continents extend oceanwards to approximately the 500 m depth-contour (OLIVER, EWING and PRESS, 1955). We estimate that through deep erosion in shield regions and steep folding of mountains no more than the upper third of the continental crust is exposed randomly and locally. This includes only 30 volume percent of the Earth's crust.

From heat balance computations based on surface heat flow and average heat production of ^{238}U, ^{232}Th and ^{40}K in common igneous rocks, the maximum thickness of granitic rocks over large areas in the continents is estimated to be about 12 km They probably represent more or less the upper continental Earth's crust. In deep exposures, no regular gradient has been observed from granites or chemical equivalents into dioritic or gabbroic rocks of the lower crust. Not long ago seismic evidence from the continental crust was interpreted as the product of velocities through two or more distinct rock layers. Recent seismological work interpreted by Birch (1955), Tatel and Tuve (1955) and others destroyed this simple concept. Why should the invisible crust be built up more simply than the visible one, which is a complicated "mosaic structure" of folded and faulted blocks? One reason for the uniform pattern of the seismic velocities is their relative small range of variability with rock composition in the crust. A systematic correlation of velocity with composition is mainly found at higher pressures, after the disturbing effect of porosity has been eliminated. Tatel and Tuve (1955) describe two different types of seismic behavior in the continental lower crust which they base upon high quality seismic data. In the gradational type the velocity of compressional waves increases slowly with depth from the upper crustal value of 6 to 7 km/sec. The Mohorovičić discontinuity is marked by an abrupt change of velocity from 7 to 8 km/sec for the mantle. In the second type of behavior the crust appears to consist mainly of a 6 km/sec layer to over 20 km depth. In a 6 km thick zone of the lower crust the compressional velocity increases from 6 km/sec to that of the mantle without an abrupt transition. The increase of compressional velocities above 6 is probably related to a change in composition from granitic to dioritic and gabbroic (7 km/sec) rocks and not to an influence of pressure. Peridotite, the probable upper mantle material, has velocities of 7.8—8.2 km/sec. Although much of the evidence for an intermediate layer between upper crust and mantle is indirect it seems likely that such a layer is present over larger areas. The nature of the transition zone between upper and lower crust is still unknown[1]. Seismic refraction measurements have proved that the crust under ocean basins is on the average no more than one sixth as thick as that of the continents. A large part of it corresponds in compressional wave velocities with gabbroic rocks (similar to those of the continental lower crust). The absence of granitic rocks in the oceanic crust is confirmed through the rarity of granitic xenoliths in oceanic lavas.

Until more information is available one may assume that $\geqq 50$ percent of the Earth's crust (lower continental crust, oceanic crust) consists of rocks chemically equivalent to basalts or gabbros. The object of this section is to report estimations of rock abundances in the upper continental crust mainly based on direct observations. This is the portion of the Earth's crust which has more often taken an active part in geological processes than the layer underneath. Folding and faulting has exposed deeper units of the upper crust to erosion and introduced the products into cycles of exogenic processes. After mechanical and chemical segregation the eroded matter accumulated as sediments. After this it became potential source material for metamorphic rocks. Partial melting has recycled some of the erosion products into magmatic sources for rock formation.

If the subsediment units of the upper crust acquire random positions in relation to the Earth's surface through folding, faulting, plutonism and erosion their exposed surfaces are assumed to be proportional to their volume. Portions of abundant magmatic rock types in the total volume of an upper crust layer can be estimated from geologic maps of continental shields and mountain belts. Lava flows and sediments

[1] Lambert and Heier [Geochim. et Cosmochim Acta **31**, 377 (1967)] have demonstrated that pyroxene granulites are lower in U, Th and K than their chemical equivalents existing under lower pressure.

which still rest conformably on the Earth's surface must be excluded from this type of compilation. They will be considered in a different section (V).

The classical evaluations of geologic maps for rock abundances were published by SEDERHOLM (1925) and by DALY (1933). A compilation like this relies on map coverage, rock classification, quality of mapping and quality of rock determination. Comparison between different areas of the world is needed to compute

Table 7-7. *Percentages of areas occupied by specific intrusive rock types*

<table>
<tr><th>Author</th><th>SEDERHOLM (1925)</th><th>DALY (1933)</th><th>MOORE (1959)</th><th>DALY (1933)</th><th>GROUT (1938)</th><th>RUDMAN et al. (1965)</th></tr>
<tr><td>Mapped area</td><td>Finland 191,400 km²</td><td>Cordillera U.S.A. 14,100 km²</td><td>Cordillera U.S.A. 44,800 km²</td><td>Appa-lachians 3,950 km²</td><td>Canadian Shield</td><td>Basement Mid-western U.S.A. (sampled through wells)</td></tr>
<tr><td>Granites (incl. quartz monzonites)</td><td rowspan="3">86.8</td><td>46.7
37.3</td><td>34.6
19.1</td><td rowspan="3">88</td><td rowspan="3">(<50)
87
(>25)</td><td rowspan="3">87</td></tr>
<tr><td>Granodiorites</td><td></td><td></td></tr>
<tr><td>Quartz diorites</td><td>1.0</td><td>33.7 [a]</td></tr>
<tr><td>Diorites</td><td></td><td>4.1</td><td>1.9</td><td>0.8</td><td></td><td></td></tr>
<tr><td>Gabbros</td><td>13.2</td><td>6.9</td><td>10.7</td><td>10.8</td><td>13</td><td>13</td></tr>
<tr><td>Peridotites</td><td></td><td>1.4</td><td></td><td></td><td></td><td></td></tr>
<tr><td>Syenites, alkalic rocks</td><td></td><td>2.0</td><td>0.1</td><td>0.3</td><td></td><td></td></tr>
<tr><td>Anorthosites</td><td></td><td>0.9</td><td></td><td></td><td></td><td></td></tr>
</table>

[a] The author interprets this value as an overestimation at the expense of granite, granodiorite due to sampling in an area with regional prevalences (quartz diorite occurs predominantly west and granite-granodiorite east of a boundary line through the Western U.S.).

a representative section of the upper earth's crust. A large portion of the data for these computations dates back more than 30 years. It is highly regrettable that the vast amount of data from drilling operations and modern mapping which is stored and only partially published by geological survey institutions, oil companies etc. is not evaluated for this purpose.

Data from SEDERHOLM (1925), DALY (1933), MOORE (1959), GROUT (1938) and RUDMAN et al. (1965) are used for Table 7-7[1]. Estimations of the proportions of igneous rocks in the U.S.S.R. by S. P. SOLOVJOV (1952) published in "Gosgeolizdat" are only known to the present author from a reference (48.7% intrusive granites; 13.5% acid effusive rocks; 37.4 basic and ultrabasic intrusive and effusive types). Because effusive rocks seem to be overestimated in SOLOVJOV's compilation the portion of basic rocks is probably too large.

Comparing the values reported by DALY (1933) with those by MOORE (1959) from the same gross area, some increase is to be observed of quartz diorites at the expense of granodiorite (and granite). Diorites, syenites, alkalic rocks and anorthosites appear to be overestimated in the smaller sampling area of DALY.

[1] Additional information on North America is presented by A. E. J. ENGEL in: Nature **140**, 145 (1963).

A good chemical average of the upper crust exposed throughout Earth's history can probably be derived from its sediment layer. Decomposition of magmatic rocks, transport of matter and formation of sediments must be a closed system except for H_2O, CO_2, O_2, S, Cl and some minor elements. The author has compared averages of a great number of elements in igneous rocks (standard composition as indicated by Table 7-7) with those in sedimentary rocks (for compositions see Chapter 8 of this book). Because several elements are characteristic of granitic, gabbroic, peridotitic or alkalic rocks respectively, balance computations of this type can give additional information about their influence on sediment composition. They confirm in general the proportions of Table 7-7. Information from Table 7-7 and from the forementioned balance computations are used for Table 7-8 to give an a tentative average of magmatic rocks constituting the upper Earth's crust.

Table 7-8. *A tentative standard section of the upper continental Earth's crust (intrusive rocks)*

Granites, quartz monzonites	44 vol.-%
Granodiorites	34 vol.-%
Quartz diorites	8 vol.-%
Diorites	1 vol.-%
Gabbros	13 vol.-%
Peridotites	<0.5 vol.-%
Syenites, alkalic rocks	<0.5 vol.-%
Anorthosites	<0.5 vol.-%

It is surprising that all columns of Table 7-7 agree within a small range in the abundance of granitic rocks: 85—88% (granites, granodiorites, quartz monzonites and quartz diorites) and of gabbroic rocks: 7—13%. The ratio of granitic to gabbroic rocks is 6.4. WICKMAN (1954) reported several ratios computed from balances with sedimentary rocks which should be comparable to this. He gives a maximum of 2.83 and a minimum of 1.43. For several reasons the present author cannot confirm these results. All igneous rock types besides granitic and gabbroic species are unimportant in abundance.

V. Abundance of Common Volcanic Rocks

It has been mentioned in the foregoing section that volcanic rocks conformable to the Earth's surface must be excluded from estimations of the abundance of igneous rocks. They usually form a relatively thin layer of lava flows with a large exposed surface covering the upper Earth's crust locally.

The largest known masses of volcanic flows are those in the Parana basin of Brazil, Argentina, Uruguay and those of India on the Deccan Peninsula, the so-called Siberian traps, the Karroo-volcanics and the effusive rocks of the northwestern United States of America. Basalt masses of unknown size are expected to cover vast areas on the ocean floor and to be interlayered with deep sea sediments. The islands of Hawaii and Iceland are large composite volcanoes but are only small portions of the basalt masses of the oceans.

Continental basalt plateaus of Mesozoic to Recent age cover areas in the range of 10^6 km^2 as that of the Parana basin. They cover a total of about 2—3 percent on the continental surface. These plateaus are usually a composite of numerous single flows with not more than about 15 m average thickness. On the Deccan peninsula local total thicknesses of nearly 2,000 m are observed. The thickness of the Parana basalt plateau varies around 400 m (SCHNEIDER, 1966). In places of plateau basalt coverage these are assumed to have an average portion of about 3 percent in the upper continental Earth's crust. From this estimation the total volume of Mesozoic to Recent volcanics is not more than (3 percent of 3 percent) about 0.1 percent of the volume of the upper continental crust. The time span of volcanic production we have considered is 5—10 percent of that of the total rock-forming period of the earth. If volcanism has not changed in intensity during earth's history one can assume a total production of about 2 percent of the rocks of the upper continental crust. The volcanics older than Mesozoic occur mainly covered and folded in the mapped areas used for the data in Table 7-7. They are counted as intrusives. As the main result of this section we have to admit that young volcanic rocks conformable on the Earth's surface represent only a negligible mass if we compare it with the mass of the upper continental crust. In the oceans this portion increases to a larger but unknown amount.

The geologic history of the sampling area has a major influence on the relative abundance of common volcanic rock types. The circum-Pacific volcanism with its higher portion of andesites and rhyolites differs extremely from the oceanic flows with their definite predominance of basalts. Table 7-9 contains information about the relative abundance of volcanic rocks in general and of basalt types in different geologic settings.

Regional and temporal variations of magma types make a general estimation of abundances on the base of available geologic data rather uncertain especially if oceanic areas have to be included. Enormous masses of tholeiitic basalts occur in the previously mentioned continental floods as well as in the old plateau type accumulations (Canadian Shield etc.). Cenozoic volcanism has produced remarkable amounts of alkali olivine basalts on the continents (more than 75% of the Tertiary volcanics in Czechoslovakia, Hungary, Germany, France, Spain etc. for instance). Elevated volcanoes along major rift systems in the oceans include representatives of both types of the so-called principal magma types (Hawaii, Iceland etc.).

Volcanism in the circum-Pacific belts is furthermore characterized by its enormous production of pyroclastics. There the ratio of pyroclastics to lava is >10 (SAPPER, 1917). On the continents this ratio usually varies between 0.1 and 1. In the latter environment only the rare alkalic lavas produce a high fraction of pyroclastics by means of their abnormal gas content.

ENGEL et al. (1965) present a diagram in which they plot the estimated decreasing amount of basaltic production on the continents during Earth's history. This production is assumed to consist predominantly of tholeiitic basalt with a total portion of 2% alkali olivine basalt almost exclusively restricted to Mesozoic and Cenozoic volcanism.

Summing up the available information, tholeiitic basalts are by far the most abundant volcanic rocks on earth, followed by andesites. Rhyolites, dacites, trachytes, latites and alkalic rocks range around one percent of the total mass of volcanics.

Table 7-9. *Relative abundance of common volcanic rocks in percentage of total area covered by volcanis (column 1, 2, 3), in percentage of total sample number (column 4, 5), in percentage of total area covered by basalts (column 6), in percentage of total volume of basalts (column 7)*

Author	DALY (1933)	DALY (1933)	SUGIMURA et al. (1963)	CHAYES (1966)	CHAYES (1966)	SUGIMURA et al. (1963)	ENGEL et al. (1965)
Sampling area	volcanic rocks on Earth	volcanic rocks, Cordillera (U.S.A.)	Tertiary and Pleistocene volcanic rocks, Japan 175,000 km²	346 basalt samples, circum oceanic areas	219 basalt samples, from oceanic islands outside Hawaii	Pleistocene basalts, Japan 3,800 km²	continental basalts of all ages
Tholeiitic basalts	} 82	} 32	} 45	82[a]	15[a]	94	98
Alkali olivine basalts				2[a]	49[a]	6	2
Andesites	16	44		n.d.	n.d.	n.d.	n.d.
Trachytes, Alkalic rocks	n.d.	<0.5		n.d.	n.d.	n.d.	n.d.
Dacites	n.d.	0.9	} 55	n.d.	n.d.	n.d.	n.d.
Rhyolites	2	23		n.d.	n.d.	n.d.	n.d.

[a] CHAYES (1966) has counted basalts with quartz in the norm as "subalkaline" (tholeiitic) and those with nepheline in the norm as "alkaline" (alkali olivine basalts) leaving a rest with $ne+qz=0$. n.d. — not determined.

VI. Average Composition of the Upper Continental Earth's Crust

This section mainly deals with the proportion of igneous rocks in the arbitrarily defined upper third of the continental crust. If we include the cover of sedimentary and metamorphic rocks the computed averages will not differ very much from those based only on considering magmatic rocks. Weathering and sediment formation as well as regional metamorphism in connection with the respective source rocks are assumed as being closed systems for a large number of chemical elements.

Chemical averages for large earth units can be compared with the cosmic abundances to indicate certain processes of earth formation. Relative accumulation of characteristic elements in the continental crust informs us of its specific history as a derivative product of the mantle.

Some older estimates on the composition of the upper crust are based mainly on averages of large numbers of rock analyses without considering the abundance of the rocks they represent. In Table 7-10 the figures reported by CLARKE, WASHINGTON (1924) are the arithmetic mean of 5159 analyses and therefore give too much statistical weight to rare rock types. Abnormal objects are often sampled as a matter of interest.

Table 7-10. *Composition of the "average igneous rock"*

	CLARKE, WASHINGTON (1924): arithmetic mean of 5159 analyses of magmatic rocks (%)	SEDERHOLM (1925): average of rocks from Finland by area (%)	GROUT (1938): average of rocks from the Canadian Shield by area (%)	SHAW *et al.* (1967): average of statistically weighted rocks from the Canadian Precambrian Shield (%)	This paper: average by area computed from Table 7-10 and rock analyses reported by NOCKOLDS (1954) (%)
SiO_2	59.12	67.58	63.08	64.93	66.4
TiO_2	1.05	0.41	0.81	0.52	0.7
Al_2O_3	15.34	14.66	16.75	14.63	14.9
Fe_2O_3	3.08	1.27	2.38	1.36	1.5
FeO	3.80	3.29	2.91	2.75	3.0
MnO	0.12	0.04	0.02	0.068	0.08
MgO	3.49	1.69	1.78	2.24	2.2
CaO	5.08	3.40	4.07	4.12	3.8
Na_2O	3.84	3.07	3.64	3.46	3.6
K_2O	3.13	3.56	3.07	3.10	3.3
H_2O^+	1.15	0.79	0.79	0.79	0.6
P_2O_5	0.30	0.11	0.22	0.15	0.18
CO_2	0.10	0.12	0.39	1.28	n.d.

KNOPF (1916) has considered areal distribution of rock types for his computation of the average igneous rock, but he has overestimated the portion of andesites, basalts and rhyolites by regarding equal surface exposures of intrusives and young volcanics as representing equal rock masses.

SEDERHOLM (1925), GROUT (1938), BARTH (1961) and SHAW (1967) have published averages of igneous rocks weighted for their abundance. In the last column of Table 7-10 the values of the present author are added which consider the common igneous rocks in proportions derived from Table 7-8 and averages of rock types reported by NOCKOLDS (1954).

The last column agrees reasonably well with the two sets of estimates to Finland and Canada. This average is even more close to one of the Canadian Shield computed by SHAW et al. (1967). Areal averages of rocks from Southern Norway (BARTH, 1961), including large masses of metamorphic and sedimentary rocks, mainly differ in higher Al, Mg and Na figures from column 4 of Table 7-10. Most of the newly computed figures of Table 7-10 range between the related values of the two inner columns.

The fictive average igneous rock (comparable with the upper earth's crust) must mainly consist of the major minerals of the granitic rocks. This "rock" is similar to a granodiorite as GOLDSCHMIDT (1954) has often mentioned.

Mineral composition of the upper Earth's crust beneath the sediment cover can be computed from the abundance of common rock types (Table 7-8) and an approximate average mineral content of these rocks (LARSEN, 1942). Some of LARSEN's figures have been corrected to a minor degree by the present author partially on the base of granitic norm values of NOCKOLDS (1954) (Tables 7-2, 7-3) partially from literature data on large gabbro masses (HESS, 1960; TURNER, VERHOOGEN, 1960; etc.) and on granitic rocks.

Table 7-11. *Approximate average mineral composition of common intrusive rocks* (LARSEN, 1942) *with minor corrections as mentioned in the text*

	In vol.-% of:					
	granite	grano-diorite	quartz diorite	diorite	gabbro	upper crust as in Table 7-8
Plagioclase	30	46	53	63	56	41
Quartz	27	21	22	2	—	21
Potassium feldspar (incl. microperthite)	35	15	6	3	—	21
Amphibole	1	13	12	12	1	6
Biotite	5	3	5	5	—	4
Orthopyroxene	—	—	—	3	16	2
Clinopyroxene	—	—	—	8	16	2
Olivine	—	—	—	—	5	0.6
Magnetite, ilmenite	2	2	2	3	4	2
Apatite	0.5	0.5	0.5	0.8	0.6	0.5

The figures of the last column of Table 7-11 computed by the present author, must differ from those which CLARKE and WASHINGTON (1924) reported from norm calculations based on an arithmetic mean of numerous analyses. We have already mentioned that these analyses represent oversampled rare rock types. This is obvious in the higher portion estimated for mafic minerals.

Plagioclases are by far the most abundant minerals not only of the upper continental crust but probably of the total Earth's crust.

Specific information about other than the reported elements in common igneous rocks can be found in the respective sections of volume II of this Handbook.

References

BARTH, T. F. W.: Abundance of the elements, areal averages and geochemical cycles. Geochim. et Cosmochim. Acta **23**, 1—8 (1961).

BIRCH, F.: Physics of the crust, p. 101—118. In: POLDERVAART 1955.

CHAYES, F.: Alkaline and subalkaline basalts. Am. J. Sci. **264**, 128—145 (1966).

—, and D. VELDE: On distinguishing basaltic lavas of circumoceanic and oceanic — island type by means of discriminant functions. Am. J. Sci. **263**, 206—222 (1965).

CLARKE, F. W., and H. S. WASHINGTON: The composition of the Earth's crust. U.S. Geol. Survey. Profess. Papers **127** (1924).

CROSS, W., J. P. IDDINGS, L. V. PIRSSON, and H. S. WASHINGTON: Quantitative classification of igneous rocks. Chicago 1903.

DALY, R. A.: Igneous rocks and the depths of the Earth. New York: McGraw-Hill Book Co. 1933.

ENGEL, A. E. J., C. G. ENGEL, and R. G. HAVENS: Chemical characteristics of oceanic basalts and the upper mantle. Bull. Geol. Soc. Am. **76**, 719—734 (1965).

GOLDSCHMIDT, V. M.: Geochemistry. Oxford: Clarendon Press 1954.

GREEN, D. H., and A. E. RINGWOOD: The genesis of basaltic magmas. In: Petrology of the upper mantle. Dept. Geophys. Geochim. Australian Natl. Univ. Publ. **444**, 117—205 (1966).

HESS, H. H.: Stillwater igneous complex, Montana. Geol. Soc. Am., Mem. **80** (1960).

HEY, M. H., R. W. LE MAITRE, and B. C. M. BUTLER: A versatile computer programme for the recalculation of rock and mineral analyses. Mineral. Mag. **35**, 788 (1966).

JOHANNSEN, A.: A descriptive petrography of the igneous rocks. Chicago 1931—1939.

JUNG, D.: Untersuchungen am Tholeyit von Tholey (Saar). Beitr. Mineral. u. Petrog. **6**, 147—181 (1958).

—, u. H. SCHULZ: Beschreibung von Algol-Programmen zur Berechnung der Niggli-Werte und der CIPW-Norm. Neues Jahrb. Mineral. Abhandl. **103**, 256—272 (1965).

KNOPF, A.: The composition of the average igneous rock. J. Geol. **24**, 620—622 (1916).

KUNO, H.: High-alumina basalt. J. Petrology **1**, 121—145 (1960).

LARSEN, E. S.: In: F. BIRCH, J. F. SCHAIRER, and H. C. SPICER, Handbook of Physical Constants. Geol. Soc. Am., Spec. Papers **36**, 3 (1942).

LUTH, W. C., R. H. JAHNS, and O. F. TUTTLE: The granite system at pressures of 4 to 10 kilobars. J. Geophys. Research **69**, 759—773 (1964).

MOORE, J. G.: The quartz diorite boundary line in the Western United States. J. Geol. **67**, 198—210 (1959).

NOCKOLDS, S. R.: Average chemical compositions of some igneous rocks. Bull. Geol. Soc. Am. **65**, 1007—1032 (1954).

OLIVER, J., M. EWING, and F. PRESS: Crustal structure of the Arctic regions from the Lg phase. Bull. Geol. Soc. Am. **66**, 1063—1074 (1955).

PLATEN, H. v.: Kristallisation granitischer Schmelzen. Beitr. Mineral u. Petrog. **11**, 334—381 (1965).

POLDERVAART, A. (ed.): Crust of the Earth. Geol. Soc. Am., Spec. Papers **62** (1955).

RUDMAN, A. J., C. H. SUMMERSON, and W. J. HINZE: Geology of basement in Midwestern United States. Bull. Am. Ass. Petrol. Geologists **49**, 894—904 (1965).

SAPPER, K.: Katalog der geschichtlichen Vulkanausbrüche. Straßburg 1917.

SCHNEIDER, A. W.: Personal communication about estimations by V. LEINZ 1966.

SEDERHOLM, J. J.: The average composition of the Earth's crust in Finland. Bull. comm. géol. Finlande **70**, 3—20 (1925).

SHAND, S. J.: Eruptive Rocks. London 1943.

SHAW, D. M., G. A. REILLY, J. M. MUYSSON, G. PATTENDEN, and F. E. CAMPBELL: An estimate of the chemical composition of the Canadian Precambrian shield. Can. J. Earth Sci. **4**, 829—853 (1967).

STRECKEISEN, A.: Die Klassifikation der Eruptivgesteine. Geol. Rundschau **55**, 478—491 (1966).

— Classification and nomenclature of igneous rocks. Neues Jahrb. Mineral. Abhandl. **107**, 144—240 (1967).

SUGIMURA, A., T. MATSUDA, K. CHINZEI, and K. NAKAMURAK: Quantitative distribution of late Cenozoic volcanic materials in Japan. Bull. volcanol. **26**, 125—140 (1963).

TATEL, H. E., and M. A. TUVE: Seismic exploration of the crust. p. 35—50. In: POLDERVAART 1955.

TRÖGER, W. E.: Spezielle Petrographie der Eruptivgesteine. Berlin 1935.

TURNER, F. J., and J. VERHOOGEN: Igneous and Metamorphic Petrology, second ed. New York-Toronto-London: Mc. Graw-Hill Book Co. 1960.

TUTTLE, O. F., and N. L. BOWEN: Origin of granite in the light of experimental studies in the system $NaAlSi_3O_8$—$KAlSi_3O_8$—SiO_2—H_2O. Geol. Soc. Am., Mem. **74** (1958).

VITALIANO, C. J., R. D. HARVEY, and J. H. CLEVELAND: Computer program for norm calculation. Am. Mineralogist **50**, 495—498 (1965).

WEDEPOHL, K. H.: Geochemie. Berlin: W. de Gruyter & Co. 1967 (Sammlung Göschen 1224—1224b).

WEIL, D. F., and A. H. KUDO: Initial melting in alkali feldspar-plagioclase-quartz systems. Geol. Mag. **105**, 325—337 (1968).

WICKMAN, F. E.: The "total" amount of sediments and the composition of the "average igneous rock". Geochim. et Cosmochim. Acta **5**, 97—110 (1954).

WINKLER, H. G. F., u. H. v. PLATEN: Experimentelle Gesteinsmetamorphose. V. Experimentelle anatektische Schmelzen und ihre petrogenetische Bedeutung. Geochim. et Cosmochim. Acta **24**, 250—259 (1961).

YODER, H. S., and C. E. TILLEY: Origin of basalt magmas: An experimental study of natural and synthetic rock systems. J. Petrology **3**, 342—532 (1962).

CHAPTER 8

K. H. WEDEPOHL

COMPOSITION AND ABUNDANCE OF COMMON SEDIMENTARY ROCKS

Sediments and their diagenetic consolidation products cover the magmatic and metamorphic crustal rocks over much more than 50 percent of the Earth's surface. This blanket is formed from the weathering products and residues of magmatic, metamorphic and sedimentary rocks after transport and accumulation in water, air or ice. During Earth's history, an increasing amount of recycled sedimentary materials has been incorporated into new sediments. Compared with the composition of the primary magmatic rocks, average sediments are distinctly higher in water, carbon (mainly as carbon dioxide), chlorine, and in the ratio of ferric to ferrous iron. Chemical weathering of high temperature silicates mainly proceeds in reactions with water, carbon dioxide and oxygen. Large amounts of former sediments presently occur as metamorphic rocks. Except for a loss of water, carbon dioxide and minor volatiles and some reduction of the ratio of ferric to ferrous iron, regional metamorphism usually does not alter the bulk chemical composition of the former sediments. Only high grade metamorphism can be associated with chemical differentiation of the parental sediments through partial melting, metasomatism etc. At this stage, remaining indicators of a sedimentary origin may be lost. The quantity of original sediments camouflaged in metamorphic and magmatic rocks should be considered but cannot be easily estimated from balance computations.

As mentioned in previous chapters of this handbook, rock terms needed for geologic interpretation should be restricted to types of major abundance. If differences between species are too small to be readily identified the system becomes too complicated to be easily applied by non-specialists. If greater sophistication is needed for a certain problem subclassification of a simple basic system may be desirable. The properties selected as a basis for discrimination must be easily definable and of importance for genetic interpretations. Mineral composition and texture (as a function of grain size, pore size, sorting, roundness etc.) are the most important properties for categorizing sedimentary rocks. A restriction of a rock name to a certain genetic process is not advisable. But terminology must allow the description of these processes clearly and adequately according to modern concepts. There are a few exceptions to the common use of nongenetic terms worth mentioning. The medium of deposition, water, ice or air has to be considered sometimes in the discrimination of special sediment types. The medium of transport can often be recognized from the degree of sorting of grain sizes, degree of rounding, scratches and facets on pebbles etc.

Most magmatic and metamorphic rocks can be identified after inspection with the polarizing microscope, although for some a chemical analysis may also be needed. As grain size is often the critical property of sedimentary rocks this must usually be analyzed by special equipment. Size and shape of voids (pore space) can also be an important property. With decreasing grain size of the mineral constituents to be identified the petrographic microscope must be replaced by x-ray diffraction instruments, electron microscopy, thermal analysis etc.

As under conditions of magmatic and metamorphic processes, certain principles of natural selection will exclude a great number of mineral combinations during the formation of sedimentary rocks. Temperatures of magmatic and metamorphic processes more easily allow equilibration than do conditions in the environment of sediment formation. Therefore one finds detrital minerals from rather different conditions of formation in the sedimentary environment; several minerals with a large temperature range of formation or with special resistance against weathering can be easily incorporated into sediments after they are set free by decomposition of magmatic or metamorphic rocks. As common minerals of this type quartz, dioctahedral mica, chlorite and alkali feldspars may be mentioned. The list of resistant accessories includes: andalusite, anatase, apatite, barite, brookite, cassiterite, chromite, columbite, corundum, kyanite, garnet, monazite, rutile, sillimanite, sphene, spinel, staurolite, topaz, tourmaline, wolframite, xenotime, zircon. The occurrence of amphiboles, pyroxenes, magnetite, olivine, epidote and ilmenite as detrital minerals indicates limited decomposition.

I. Nomenclature of Sedimentary Rocks

The system of classification of sedimentary rocks must be more heterogeneous than that of magmatic or metamorphic rocks. Mineral composition is the accepted basic property for classification. Although the system favored in this article will be mainly used for geochemical purposes, the general trend in petrographic nomenclature should not be ignored. A rock group, once established, cannot be expanded or reduced without causing confusion. Redefinition of terms must be undertaken with care. It is sometimes desirable to sharpen a term rather than to alter drastically its meaning.

For certain sedimentary rocks physical properties of texture based on grain size and distribution are the basis for classification and for others, composition of detrital minerals or composition of authigenic minerals. Detrital minerals can indicate a special source or provenance area of the accumulated matter. Authigenic minerals characterize the conditions of deposition or diagenesis.

For simplicity all sedimentary rocks are grouped into two main classes (a, b).

a: *Clastic* (fragmental) *sediments and sedimentary rocks* (including pyroclastic deposits).

b: *Chemical and biochemical sediments and sedimentary rocks.*

The most serious problem in this classification is the possible mingling between the two classes a and b of sedimentary rock types. This will be considered under section c *clastic sediments and sedimentary rocks with specific chemical constituents or organic residues*. Intermediate types are named after the prevailing component. It is convenient, to mention a subordinate portion (10—50%) of one sediment type in another by a prefix to the main rock name such as: argillaceous limestone and calcareous shale; arenaceous limestone and calcareous sandstone. In the case of carbonaceous or bituminous sediments fractions even smaller than 10% coal or bitumen are mentioned in the name. Detailed information on the nomenclature of various mixtures within class a or b and of mixtures between the two is given in sections a, b and c.

a) Clastic Sediments and Sedimentary Rocks

They are the products of physical accumulation of matter in air, water or ice and are conveniently classified into groups on the basis of particle dimensions. Subgroups are distinguished by special mineral constituents, composition of the cement, filling of pore space, and other properties like sorting or roundness.

The bulk chemical composition of a clastic sedimentary rock depends on the mineral grains and the void-filling cement. Because pore space in gravel and sand may amount to more than 30%, there can be a large proportion of cement minerals (quartz, calcite, clay minerals, iron oxides, dolomite etc.) which are usually formed after deposition of the sediment in contrast to the detrital minerals of the framework.

The pore space of argillaceous rocks behaves differently from that of coarser clastic rocks and of limestones. The proportion of voids to framework in unconsolidated marine clays not compressed by a heavy load is much larger than that of average sandstone but permeability is smaller in the former. Permeability is a function of the amount of connected pore space and of capillaries. Marine sediments with a large proportion of pore space and minor permeability usually contain an appreciable amount of seawater trapped during deposition. During diagenesis, the composition of this seawater changes as a consequence of reactions with the sedimentary minerals, formation of dolomite and chlorite — decrease of Mg in pore solution — and bacterial sulfur reduction. The absolute salinity of the solution may increase by compaction of the mineral framework, accompanied by filter-pressing. For detailed information see the special publications by VON ENGELHARDT (1960), RITTENBERG et al. (1955) and others. HAMILTON (1959) reports a decrease of porosity from more than 60 vol.-% to less than 30% by the load of a sediment cover from zero to 1,000 m thickness in young marine clays.

Convenient rock classifications based on grain size must be a product of agreement among a great number of experienced investigators. In clastic rocks the predominant grain size and the mineral composition can be closely correlated if the ranges of size fractions are carefully selected. The grain size classes of the proposed system often coincide with fractions separated by natural processes. Incomplete sorting of grains can indicate a special environment. Two critical grain sizes are used for the characterization of sorting in a certain bed: $\mathcal{Q}_3$ and $\mathcal{Q}_1$. By definition (TRASK, 1932) 75 weight percent of a sample has a smaller grain size than $\mathcal{Q}_3$ and 25 percent a smaller grain size than $\mathcal{Q}_1$. The quantity $So = \sqrt{\mathcal{Q}_3/\mathcal{Q}_1}$ is called the sorting coefficient and gives a measure of the spread of the particle size frequency curve. Sediments with an *So*-ratio smaller than 1.4 are described as very well sorted, those in the range 1.4—1.5 well sorted and those > 1.87 poorly sorted etc. Quartz sandstones often have:

Table 8-1. *Classification of clastic sediments and -sedimentary rocks*

Predominant grain size (mm in diameter)	Name of sediment	Name of sedimentary rock	Common synonyms	Rock composition
Larger than 2	*gravel* (containing rounded boulders, larger than 200 mm in diameter or pebbles, smaller than 200 mm i.d.) *rubble* (angular fragments)	*conglomerate* (rounded fragments), *breccia* (angular fragments)	rudaceous or psephitic deposits	polygenetic rock fragments, quartz pebbles common, $\geqq 30\%$ sand-, silt-, clay- or lime cement
0.06 to 2	*sand*	*sandstone* (quartz sandstone, quartz arenite: more than 95% quartz) *arkose* *greywacke* (or graywacke)	arenaceous or psammitic deposits	detrital quartz (75 vol.-% or more) detrital quartz (75 vol.-% or less), feldspar (25 vol.-% or more) detrital quartz (60 vol.-% or less) feldspar (25 vol.-% or more) chlorite and mica (15 vol.-% or more) (rock fragments)
0.002 to 0.06	*silt* (loess: unstratified aeolian calcareous silt)	*siltstone*		detrital quartz, dioctahedral mica, chlorite
Less than 0.02	*clay,* mud	claystone, mudstone, *shale, slate* (well developed fissility)	argillaceous, lutaceous or pelitic deposits	dioctahedral mica (illite), quartz
Unassorted assemblage of coarse and fine material	*till,* unstratified deposit containing specific constituents mentioned in the fifth column *tilloid* (without faceted or striated pebbles or cobbles)	*tillite*	(glacial) boulder argillite, conglomeratic mudstone, diamictite	angular boulders and pebbles (faceted or striated) in a fine grained structure-less matrix representing 80—90% of the total deposit

So = 1.2. Graywackes are usually less well sorted with *So* about 1.7 (FÜCHTBAUER, 1959).

Grain size classes specified for Table 8-1 are based on the suggestions published by CORRENS (1939), TWENHOFEL (1950), PETTIJOHN (1957), FÜCHTBAUER (1959), HUCKENHOLZ (1963b), HATCH, RASTALL (1965).

In Table 8-1 silt and siltstone have been specified as a special class following common rules. Silt is found usually as an abundant constituent of clays and shales; therefore the silt and clay classes are separated only by a broken line in Table 8-1. PETTIJOHN (1957) estimates that the silt fraction of argillaceous deposits averages 60%.

The term "loess" is usually applied to an unstratified (windblown) calcareous sediment mainly composed of silt grains. The meaning of the term "greywacke" has often been debated in English-speaking countries. A term is doubtless needed for a special sandstone "with low maturity" having usually less than 50% quartz but appreciable portions of feldspars, phyllosilicates (chlorites, mica) and rock fragments. If one refers to the type locality, in the Harz mountains in Germany, there is no ambiguity in the term "greywacke". For petrographic descriptions of greywackes from the type locality see HELMBOLD (1952), HUCKENHOLZ (1959), MATTIAT (1960).

1. Pyroclastic Deposits — Volcanic Tuffs

Explosive volcanic eruptions project fragments of country rocks and splash lava into the atmosphere or, if submarine, into the water. After settling the finer matter usually forms a layered sediment because of the change of grain size (and composition) with time. The proportion of pyroclastic ejecta to the mass of fluid magma erupted into flows increases with the gas pressure in the lava. The amount of country rocks occluded as xenoliths in the pyroclastics usually decreases with increasing volume erupted from a volcanic vent. The ejected volume which can be discharged by one volcanic explosion ranges up to a cubic kilometer. The chemical composition of pyroclastics is often more variable locally and temporally than that of either pure magmatic or sedimentary rocks.

As a consequence of this variability the terminology of pyroclastic deposits is complex. It is mainly based on the predominating grain size of the material representing more than 50 percent of the total volume. The composition of the parent lava of the ejecta should be included as an adjective in a combined name, as in "rhyolitic ash", "trachytic lapilli" etc. For conventions used in igneous rock terms see Chapter 7. Tuffs are compacted pyroclastic material of small grain size, the indurated equivalents of volcanic ash and dust deposits. Large portions of pyroclastic material consist of glass due to the quenching of a melt during the eruption ("vitric tuffs"). If the glass shards are hot enough during deposition to weld together the so called "welded tuffs" are formed. As a metastable product, glass easily undergoes devitrification and decomposition during diagenesis and contributes to the cementation of the deposit. "Palagonite" is a special tuff, which is formed from basaltic glass mainly by hydration etc. Tuffs that consist chieflly of crystalline rock fragments are called "lithic tuffs".

The grain size classification of pyroclastic deposits by WENTWORTH and WILLIAMS (1932) does not follow the system for normal clastic sediments as listed in Table 8-1. These authors suggest that the following fractions should be separated: fine ash, and dust (lithified: tuff) $<$ 0,25 mm $<$ coarse ash (lithified: tuff) $<$ 4 mm $<$ lapilli (small fraction: cinder) $<$ 260 mm $<$ bombs, blocks (lithified: agglomerate).

The fine grained pyroclastics may be transported by wind and water over large distances and then become major or minor parts of non-volcanic sediments (formation of so called "tuffites" for instance).

b) Chemical and Biogenic Sediments and Sedimentary Rocks

They have been accumulated predominantly by chemical precipitation or sedimentation of organic residues in water. This section includes products of widely different origins. Most chemical precipitates are transformed during diagenesis because of their moderate to high solubility. A large proportion of the mineral grains constituting a rock of this type is not an original product of precipitation. Calcium carbonate sediments which accumulated in seawater at elevated temperatures usually consist in part of the metastable minerals aragonite and high-magnesium calcite. Organisms with rock forming skeletons or tests use calcite, high-magnesium calcite and aragonite in their structures. Some skeletal materials are exclusively aragonite or calcite and some are in part aragonite and in part calcite. The factors responsible for the differences have been studied by LOWENSTAM (1954) and several other investigators. Evidence from diagenesis of recent sediments together with the rarity of these minerals in older deposits prove that these compounds are easily transformed into calcite through dissolution and precipitation in pore solutions. Primary calcite is less soluble under ordinary conditions and has a higher resistance against diagenetic dissolution and recrystallization (foraminifera shells etc.).

Preservation of shells in limestones and of primary textures of oolites in oolitic rocks does not indicate that they still contain the original minerals of deposition. In many cases it is difficult to recognize the original composition of a certain chemical sediment from the consolidated rock. The proportion of calcium carbonate from inorganic precipitation and from shell debris often cannot be estimated in a limestone. Therefore classifications based on the original constituents are inadequate. A classification system by FOLK (1959) has been used to some extent recently. In this system the following carbonate constituents of a limestone are distinguished: I) fine-grained idiomorphous carbonate grains (<0.004 mm in diameter) as "micrite", II) coarse grained crystalline carbonates (>0.01 mm in diameter) as "sparry calcite" or "sparite" and III) pellets, ooids, detrital grains and skeletal debris as "allochems". Recent investigations by means of strontium analyses (FLÜGEL, WEDEPOHL, 1967, etc.) have proved that all of the three mentioned types of constituents are often composed of diagenetic calcite. Distinguishing between I and II is of little help in genetic considerations. Constituents of type III may be partially disintegrated into I (and II) to be biomicrite, biospar or related particles. The origin of I or II from pellets, certain skeletons etc. can rarely be determined except in recent sediments where the production of the sediment particles can be observed. Preserved structures of organic residues (coral reefs, algae, stromatolites, shell accumulations) and inorganic products of precipitation (oolitic textures) are most valuable indicators for special environments. The oolitic texture is almost always a primary feature characteristic of calcium carbonate precipitation in shallow, often agitated waters. The uniformity of size of the oolites is evidence for the sorting action of streaming water. The particles float until a certain accretionary size is acquired by precipitation. Chemical indicators of an environment of deposition are rarely preserved during diagenesis of limestones and dolomites. Several features of diagenetic solution and reprecipitation, such as granoblastic textures, stylolitic seams etc. can often be observed but are not present in every chemical sediment that has undergone gross secondary alteration.

Table 8-2. *Classification of chemical and biochemical sediments and sedimentary rocks*

Predominant mineral	Name of sediment	Name of sedimentary rock	Common synonym	Rock composition and specific properties
Calcite (and/ or aragonite)	lime mud, aragonite mud, *calcareous ooze*, *globigerina ooze*, *pteropod ooze* (ooze: more than 50% $CaCO_3$ of organic origin) chalk	*limestone* (accretionary and detrital)		limestone: more than 85% carbonate minerals (magnesian limestone: 5—10% dolomite; dolomitic limestone: 10—50% dolomite); argillaceous limestone: 15—50% clay; bituminous limestone: usually less than 10% bituminous matter, predominantly hydrocarbons
Dolomite	(primary dolomitic sediments are rare)	*dolomite*	dolostone	more than 50% carbonate, less than 10% calcite (calcitic dolomite: 10—50% calcite) bituminous dolomite: usually less than 10% bituminous matter, predominantly hydrocarbons
Gypsum anhydrite		bedded *gypsum* bedded anhydrite (mainly dehydration product of gypsum)		
Halite (and magnesium potash salts as polyhalite, kieserite etc.)		bedded *rocksalt*		
Silica (quartz, opal, α-cristobalite)	*radiolarian-*, *diatomaceous ooze* radiolarian-diatomaceous earths (siliceous ooze has more than 30% silica of organic origin), silica sinters	*chert*, radiolarite, diatomite etc.	flint	more than 30% silica of organic origin or from chemical precipitation
Carbonaceous matter (degraded plant remains)	*peat*, lignite, *brown coal*	bituminous[a] coal, *coal*, anthracite (and various transitional ranks)		coal is generally well jointed, brilliant in luster (gravity: 1—1.8)

[a] Term commonly used for coals which burn freely with flame although they contain no bitumen.

Some limestones are predominantly composed of fragmental or detrital matter which has been mechanically transported and deposited. For these clastic-looking rocks the use of terms such as lime sandstones, lime siltstones, calcirudites, calcarenites, calcilytites etc. has been suggested. Because diagenetic reprecipitation has often obscured the primary grain size, the application of rock terms related to a predominating size fraction cannot help to elucidate the history of fine grained limestones. The mechanically deposited limestones can show the same textures and structures as do non-carbonate clastic sediments (stratification, cross-bedding, graded bedding etc.).

Cherts are rather common silica-rich sedimentary rocks. The origin of silica from either radiolaria, diatomaceous and sponge residues or from chemical precipitation cannot be easily determined in cherts. The solubility of silica is high enough so that many of the thin walled skeletons and needles of siliceous organisms are lost from dissolution and reprecipitation in pore fluids of cherts.

Normal seawater of the ocean deeps (having 2.5—5 ppm Si) is undersaturated with respect to quartz and colloidal silica. Surface water contains even less silica in solution. Except through organic production, silica can only reach saturation if hot springs with higher concentrations locally mix with seawater. Because of the sediment's ambiguous history, the term "chert" is suggested for consolidated rock as a neutral name with no implications as to the origin of the silica.

Phosphate rocks — usually of complex composition — can have a biogenetic (remains of vertebrate bones, guano, coprolites) or an inorganic origin (often interbedded with dark shales).

Dolomite, bedded gypsum, bedded anhydrite, bedded halite and the rarer evaporites with higher solubility are always inorganic precipitates. Dolomite and anhydrite are usually secondary products from reactions with calcium carbonate or gypsum respectively. During metamorphism under pressure (load) gypsum is often dehydrated to form anhydrite.

The only sediment type of Table 8-2 which has an exclusively biogenic origin is represented by degradation products of plant and animal tissues. The resistance to oxidation and decomposition of organic materials varies. Least resistant are proteins, sugars and starches; cellulose and fats decompose less readily. Chitin, resins and waxes can remain unchanged over geologic periods. Peat usually consists of 70—90 percent of organic residues largely of cellulose (with little protein). Coalification follows burial by partial decomposition of the predominating plant remains without access of air (prevented by the water table) in normal cases under increased pressure and temperature. The degree of coalification is used (beside names indicating a special origin of the organic matter) to classify the coal into ranks with increasing carbon content and decreasing portions of volatile hydrocarbons. According to CLARKE (1924), the carbon from lignite to anthracite increases from 73 to 94% C, compensated by decrease of oxygen from 20.5 to 2.7% O and of hydrogen from 5.2 to 2.8% H (nitrogen remaining almost constant at 1.5% N). Modern coal petrography considers characteristic components, which are often segregated texturally into bands such as: vitrain, fusain, clarain and durain. These main constituents are made up of macerals as organic units like minerals. — Bedded coal is not abundant but carbonaceous components of normal sedimentary rocks are widely distributed. Even in commercially valuable areas coal usually forms not more than a percent of the oridinary sedimentary rocks with which it is interbedded.

The proportion of higher plants in the total organic production is comparable with the marine phyto- and zooplankton contribution to the worldwide carbon cycle. Most of this matter is already oxidized in the aerated ocean water beneath the levels of organic production and does not reach the sea bottom. Predominantly in shallow seas some degraded matter from the large organic primary production remains to be incorporated into the sediment. Because of the wide distribution of primitive organic production these degraded substances are much more widely distributed in different sediments than is the carbonaceous matter. The remains of phyto- and zooplankton are richer in fatty and protein substances than is peat. During degradation, various types of hydrocarbons form which are much more mobile than carbonaceous substances. Therefore, the so called bitumen (bitumen: all hydrocarbons soluble in carbon bisulfide) as a mixture of degradation products, has often moved from its primary site of deposition and has accumulated in reservoir rocks with reasonable permeability. Because of the disperion of primary organic residues and the mobility of secondary products, bitumen occurs only in rare cases as a separate deposit. It is usually a constituent of shales, sandstones or carbonate rocks and will be listed in Table 8-3.

Rare types of sedimentary deposits, such as iron rich sedimentary rocks (chamositic-, sideritic-, cherty-, residual ironstones), some bauxites, borates, nitrates (caliche) etc. are not considered in the compilation of Table 8-2.

c) Clastic Sediments and Sedimentary Rocks with Specific Chemical Constituents or Organic Residues

(Table 8-3)

The proportion of chemical or biogenic components and clastic constituents depends on the rate of chemical precipitation in relation to detrital accumulation. Limestones often contain 10 percent argillaceous material demonstrating that a ratio of 10 in speed of chemical to detrital accumulation is rather common. As we know from recent samples of deep ocean sediments, the carbonate to clay ratio is influenced not only by the biogenic calcium carbonate precipitation but also by dissolution of the shells of dead plankton at the bottom. This again is a function of calcium carbonate undersaturation of the deep ocean water, of the shell size, of protection of the carbonate grains by organic matrix etc.

By convention we call the limestones containing 15 to 50 percent clay or sand portions: argillaceous or arenaceous limestones respectively. A shale or sandstone with 10 to 50 percent calcium carbonate is a calcareous shale or sandstone. In some countries the unconsolidated mixture of 25 to 75 percent carbonate and 75 to 25 percent clay is called "marl" with intermediate compositions as clayey marl, marl clay, marly clay etc. In other countries the name "marl" either does not connote any particular composition or is restricted to calcareous freshwater deposits. Therefore this term should be abandoned. It should be mentioned that a Recent foraminiferal sediment with 30% $CaCO_3$ and more has conventionally been called globigerina ooze. This convention leads to a discrepancy in properties between the unconsolidated sediment and the sedimentary rock. A requirement of at least 50% $CaCO_3$ for a calcareous mud to be called globigerina-, pteropod- etc. ooze would be more appropriate (see Table 8-2).

It has already been mentioned that most organic residues do not reach the ocean bottom, but are oxidized in the aerated water column through which the dead matter must settle. Therefore the carbon content of a sediment is only distantly related to the primary organic production in the surface waters (which depends on temperature, availability of nutrients). RONOV (1958) reports an average of 0.67% C for clays and

Table 8-3. *Classification of clastic sediments and sedimentary rocks with specific chemical constituents or organic residues*

Predominant detrital sediment	Specific chemical or biogenic constituents	Name of sediment	Name of sedimentary rock	Specific accessories
Clay	10—50% $CaCO_3$	calcareous clay	calcareous shale	
	10—50% $CaMg(CO_3)_2$	dolomitic clay	dolomitic shale	
	dark colored from organic residues (mainly of hydrocarbons)	bituminous clay	bituminous shale, black shale, oil shale (kerogenite, yielding oil on slow destillation)	V, Mo, U, Cu, As[a] enriched
	dark colored from plant residues (coal)	carbonaceous clay	carbonaceous shale	Ge, Mo, As[a] enriched
	MnO_2-coating on sediment particles, manganese nodules rare or common	pelagic clay (red clay)		>1,000 ppm Mn (Mo, Cu, Co, Ni[a] enriched)
	10—50% silica from organic residues or precipitation	siliceous clay	siliceous-, cherty shale	
	red colored argillaceous and arenaceous sediments (ferric oxide coating on particles)		red beds	decolorized zones, haloes etc. from local reduction (containing C, V, U etc.)
Sand	10—50% $CaCO_3$	calcareous sand	calcareous sandstone	
	10—50% $CaMg(CO_3)_2$	dolomitic sand	dolomitic sandstone	
	dark colored from organic residues (mainly hydrocarbons)	bituminous sand, oil sand (oil productive)	bituminous sandstone	
	dark colored from plant residues (coal)	carbonaceous sand	carbonaceous sandstone	

[a] For average composition of nearshore deposited, non-bituminous clays and shales see Turekian, Wedepohl (1961).

shales (more than 9,000 samples). The average carbon in rocks of oil basins is three times as high as in rocks of areas barren of oil. A literature compilation demonstrates that samples with more than 15% C are rare. Some authors distinguish between bituminous deposits deposited under an aerated water column, as gyttja, and those formed under stagnant water (euxenic conditions) with bacterial sulfate reduction, as sapropels. The specific elements which may be precipitated from seawater by hydrogen sulfide are those whose sulfides have very low solubilities, (Hg, Ag, Cu, Fe etc.). Such elements cannot form detectable concentrations in a sediment unless detrital accumulation is very slow.

Table 8-4. *Average chemical compo-*

	Sandstones (253)[d] CLARKE (1924) (%)	*Sandstones*[b] (from platforms) (3,700)[d] VINOGRADOV, RONOV (1956) (%)	*Greywackes* (61[c])[d] PETTIJOHN (1963) (%)	*Shales* (mainly from geosynclines) (277)[d] CLARKE (1924), GOLDSCHMIDT (1933), MINAMI (1935), SHAW (1956) (%)	*Shales* (from platforms) (6,800)[d] VINOGRADOV, RONOV (1956) (%)
SiO_2	78.7	70.0	66.7	58.9	50.7
TiO_2	0.25	0.58	0.6	0.78	0.78
Al_2O_3	4.8	8.2	13.5	16.7	15.1
Fe_2O_3	1.1	2.5[a]	1.6	2.8	4.4[a]
FeO	0.3	1.5[a]	3.5	3.7	2.1[a]
MnO	0.03[a]	0.06[a]	0.1	0.09	0.08
MgO	1.2	1.9	2.1	2.6	3.3
CaO	5.5	4.3	2.5	2.2	7.2
Na_2O	0.45	0.58	2.9	1.6	0.8
K_2O	1.3	2.1	2.0	3.6	3.5
H_2O^+	1.3	3.0	2.4	5.0	5.0
P_2O_5	0.08	0.10[a]	0.2	0.16	0.10[a]
C/CO_2	n.d./5.0	0.26[a]/3.9	0.1/1.2	0.6[a]/1.3	0.67[a]/6.1
S/SO_3	n.d./0.07	n.d./0.7	0.1/0.3	0.24[a]/n.d.	n.d./0.6

[a] Data are from literature compilations different from those listed at head of the column.
[b] Psammitic rocks in general.
[c] Similar to an average of 70 greywackes from the Harz mountains (Germany), the type locality.
[d] Number of samples used for average in brackets.

Another sediment type which acquires its specific chemical properties by a very small rate of detrital deposition (mm per thousand years or less) is the clay of the abyssal ocean areas far off the continents. This sediment was called "red clay" by its early investigators, but often the material is not red in color; therefore, other names came into use. The term "pelagic clay" has been applied instead but does not represent an ideal alternative because "pelagic" pertains primarily to special communities of marine organisms.

The term "porcelanite", which is used for cherty shales, has additional meanings (contact metamorphosed calcareous shale, fused shale from burning coal) and therefore should be abandoned.

Names for rarer deposits, anhydrite sandstones for instance, can be formed corresponding to those listed in Table 8-3.

II. Average Composition of Common Sedimentary Rocks

For several abundant types, enough analytical data have become available so that averages may be computed (Table 8-4). The sequence of the columns is the same as that of rock names in Tables 8-1 to 8-3. The number of samples used for the average is noted in brackets. Some of the listed averages are more representative of their rock class than others. There as been a tendency to favor rarer types for investigation

sition of selected sedimentary rocks

Tillites (68)[d] GOLDSCHMIDT (1933) (%)	*Limestones* (93)[d] WEDEPOHL (unpubl.) (%)	*Carbonate rocks* (from platforms) (1,500—8,300)[d], VINOGRADOV, RONOV (1956) (%)	*Cherts* (10)[d] CRESSMANN (1962) (%)	*Pelagic clays* (430)[e] LANDERGREN (1964) (%)
58.9	6.9	8.2	89.9	54.9
0.79	0.05	n.d.	0.2	0.78
15.9	1.7	2.2	3.7	16.6
3.3	0.98	1.0[a]	2.3	} 7.7
3.7	1.3	0.68[a]	n.d.	
0.10	0.08	0.07[a]	0.1	(2.0)
3.3	0.97	7.7	0.5	3.4
3.2	47.6	40.5	0.3	0.72
2.1	0.08	n.d.	0.7	1.3
3.9	0.57	n.d.	0.7	2.7
3.0	0.84	n.d.	1.2	(9.2)
0.21	0.16	0.07[a]	0.9	0.72
n.d./0.6	38.3	0.23[a]/35.5	(0.3)/n.d.	computed carbonate free
0.08/0.09	0.11/0.02	n.d./3.1	n.d.	n.d.

[e] Weighted average for the three oceans, sea salt free samples (35 samples of sea salt containing Pacific clays, reported by GOLDBERG and ARRHENIUS, 1958, contain about 6 to 7% NaCl. This is a high value compared with 3—4% NaCl in Atlantic clays, analysed by BEHNE, 1953.)

and analysis. The sampling of unaltered sedimentary rocks is often restricted to quarries, where they are well exposed, and, therefore, to samples of commercial value. Some types, such as sandstones in the general sense, have a rather wide range of chemical variation. Fine grained rock types usually contain mineral constituents averaged from different source areas, because of the large distances of travel before deposition. Limestone as one of the listed rock types, has definitely an age dependent chemical composition. Daly had already published in 1909, and after him several investigators have confirmed, that the average magnesium content increases with the geological age of the samples (VINOGRADOV, RONOV, 1956, etc.). This increase depends on the probability of a reaction between limestones and migrating magnesium-bearing solutions.

For the average mineral composition of some of the mentioned common rock types, we can list tentative data. Note that several of the data of Table 8-5 are from sources and on samples different from those for Table 8-4 (chemical composition).

The difficulty of a quantitative mineral determination increases with decreasing grain size. Every specialist in the x-ray phase analysis of clay particles knows this problem, which has several causes: deviations from ideal lattice structure (mixed-layering etc.), variability in chemical composition of clay minerals, incomplete randomness of grain orientation during x-ray exposures, etc. Therefore, only tentative data from a great number of shales are presented. Pelagic clays have an even higher proportion than shales of the finest grain size fraction. A number of semiquantitative data on fractions of samples from both oceans have been recently published by GRIF-

Table 8-5. *Average mineral composition of selected sedimentary rock types in percent* (The important mineral of a group is indicated in parentheses abbreviated.)

	Sandstones HUCKENHOLZ (1963a)	Greywackes HUCKENHOLZ (1963a)	Shales[a] WEDEPOHL (1967)	Limestones[b] (computed from Table 8-4)
Quartz	82	37	20	
Plagioclase Potassium feldspar	5	28	10—15	
Muscovite (M)/illite (I)	8 (M) (with Chlorite)	29 (C) (with Chlorite, Kaolinite)	45—55 (I)	14
Chlorite (C) Kaolinite			14	
Calcite Dolomite	3	3	3	84 2
Accessories	2	3	1	

[a] Data on the clay minerals are mainly averages of WEAVER's (1967) 70,000 x-ray analyses of American shales, which are comparable with the author's averages of European Paleozoic and Mesozoic shales. The quartz content is only from European shales, which contain on the average more plagioclase than K-feldspar. The average sodium of shales from Table 8-4: 1.2% Na_2O indicates about 10% albite beside 0.25 soluble NaCl. The figure for illite includes expanded clay minerals.

[b] By computing the mineral constituents from the analysis of 345 limestones which CLARKE (1924) has published (7.5% silicates, 36% dolomite), one realises that a large number of dolomites has been included in the composite of American limestones. The average carbonate rocks from the Russian platform (Table 8-4, VINOGRADOV, RONOV, 1956) contain about 12% silicates and 35% dolomite. It seems to be reasonable to assume that between a quarter and a third of all carbonate rocks consists of dolomites.

FIN and GOLDBERG (1963), GOLDBERG and GRIFFIN (1964) and BISCAYE (1965). There is poor agreement between the latter two publications on chlorite in the Atlantic Ocean, demonstrating the experimental difficulties. In comparison with the data on shales (Table 8-5) the pelagic clay samples have, in general, the following features: less quartz, more kaolinite and montmorillonite (tentatively: 20 percent of every mineral), comparable illite and chlorite. Zeolites (phillipsite etc.) are specific constituents of numerous pelagic sediments especially those from the Pacific. There are regional differences with respect to almost every mineral constituent within and between both large oceans.

Chemical constituents characteristic for bituminous, carbonaceous and pelagic clays mainly occur in the minor element concentration range. Compilations of a great number of literature data on bituminous and pelagic argillites were published by the present author in 1964 and 1967 respectively. Some elements of major importance are listed in Table 8-3.

III. Rock Nomenclature and Current Trends in Sedimentary Petrology

There is a general tendency in rock classifications to avoid names which are specifically restricted to a single process of formation, as already mentioned. This tendency is a consequence of the inconvenience of changing names after dubious genetic relations become clarified. Genesis and diagenesis of sediments, environment of origin, transport and deposition of sediment constituents are major objects of petrologic investigation.

A certain framework of genetic names cannot be excluded from the terminology of rocks. Names such as clastic, chemical and biochemical sediments, pyroclastic deposits etc. have genetic implications. The characteristic features of transported minerals can be easily observed in coarse grained sedimentary rocks and permit their ready classification as products of mechanical accumulation. With decreasing grain size of clastic sediments, specific genetic features and mineral composition determinable with the petrographic microscope become more and more obliterated.

Not until the second third of this century methods have been developed to prove that fine grained clastic rocks consist predominantly of detrital minerals. X-ray diffraction methods for powdered samples became the prominent tool of clay mineral analysis after HADDING (1923) had published x-ray evidence for the occurrence of quartz, kaolinite and muscovite in clays. The identification of the minerals was the prerequisite for theories about their origin. After numerous investigations of clay minerals species, structures and occurrences had been published in the thirties CORRENS (1939) divided the clay constituents into four genetic groups as follows: Two groups of detrital minerals, classed as: weathering residua and weathering products. (Weathering comprises all processes of low temperature rock alteration at the earth's surface). A third group contains authigenic (mainly diagenetic) minerals and a fourth one, biogenetic products.

The proportion of detrital to authigenic minerals of argillaceous rocks in general has been extensively debated. For some time authigenic clay mineral formation has been assumed to produce the largest fraction of modern sediments. But recent experimental results have clearly demonstrated that the predominant fraction of clays and shales must be of detrital origin. Numerous potassium-argon ages of marine argillaceous deposits are evidence for the provenance of clay micas. Illites from Recent Atlantic clays have an average Paleozoic age and considerably predate deposition (HURLEY et al., 1963). The regional pattern of clay mineral distribution in the Atlantic and Pacific Ocean sediments is mainly controlled by the predominance of certain clay minerals on the neighboring continents (GRIFFIN, GOLDBERG, 1963; GOLDBERG, GRIFFIN, 1964; BISCAYE, 1965).

These studies on provenance and environment rely on advanced methods of clay mineralogy. A detailed treatment of these methods has been presented by BROWN (1961) in a second edition of a book edited by BRINDLEY (1951). The book lists as many as eight groups of clay minerals. A class of interstratified clay minerals is included. These so called "mixed layer" minerals are regularly or randomly composed of layers structurally equivalent to the "primitive" clay minerals.

RONOV and coworkers (1966) have studied chemical differentiation in different geotectonic zones, such as platforms and geosynclinal basins. Both zones contain comparable proportions of major rock types (see Table 8-6b). But chemical weathering on platforms contributes higher rates of mobile material into continental run off.

The portion of authigenic minerals in clastic sediments often increases with decreasing rate of detrital accumulation. Data about absolute rates of Recent sediment formation are mainly the result of measuring the decay of nuclides, including ^{14}C,

^{230}Th, ^{231}Pa, as a function of depth in a sediment core. The methods, based on the decay of ^{230}Th and ^{231}Pa, involve many difficulties and several assumptions have to be made which are outside the scope of this article. Only a few ranges of clay accumulation will be quoted: 0.2—6.6 mm/thousand years (^{230}Th method: GOLDBERG, GRIFFIN, 1964) or 1—70 mm/thousand years (^{14}C method: TUREKIAN, STUIVER, 1964) for the Central South Atlantic. Clays on the abyssal plains of the Central South Pacific accumulate more slowly (GOLDBERG, KOIDE, 1962) and normal shelf sediments form more rapidly than the central South Atlantic deposits. For shelf sediments, a range of centimeters per thousand or hundred years has been reported.

Some authigenic minerals characteristic of slowly accumulating sediments have been mentioned in section II and Table 8-3: the MnO_2 and the phillipsite from pelagic clays. Precipitation of heavy metal sulfides in stagnant waters by bacterial sulfate reduction indicates a sapropelic (euxenic) environment. These sulfides become prominent sediment constituents only under conditions of slow detrital deposition (WEDEPOHL, 1964).

Some remarkable chemical changes within sediments are the result of reactions with pore solutions. Seawater must be trapped during sediment deposition and relative depletion of Mg^{2+}, SO_4^{2-} compared with sea salt composition has been generally observed in pore solutions from diagenetic reactions (VON ENGELHARDT, 1960; WHITE, 1965). Bacterial activity leads to a reduction of the sulfate concentration. The resulting sulfide is usually fixed as iron-sulfide. Interstitial waters in recent sediments usually contain potassium in seawater proportions. Feldspar alteration can cause a slight increase of the potassium concentration. Pore solutions from sediments which have been under deep burial often have lost some potassium. The complex problem of the low temperature geochemistry of potassium has been recently stressed by WEAVER (1967). This author presents evidence for a decrease of illites in shales with geologic age.

The study of chemical sediments by comparison with phase relations in experimental systems dates back to the fundamental investigations of VAN'T HOFF (1905). An up-to-date critical evaluation of data and theories about the composition and origin of marine salt deposits has been presented by BRAITSCH (1962). Phase studies prove that the present composition of certain salt deposits can be a product of low temperature metamorphism. But in accordance with common practice, rocks like anhydrite are listed as sedimentary in Table 8-2.

Modern thinking about the origin of limestone has been highly influenced by the observation of recent carbonate formation, mainly in the favored environment of tropical and subtropical marine waters. The occurrence of aragonite and magnesian calcite (up to 30% $MgCO_3$) as predominant constituents of sediment-forming animal shells, but metastable with respect to conditions in seawater, is an important object of investigation. During genesis of carbonate sediments the calcium carbonate phases dissolve in the sequence of their decreasing solubility and reprecipitate as calcite. Final consolidation must be connected with recrystallization of almost all primary constituents of carbonate sediments including calcite. Otherwise the average Sr concentration in limestones cannot be as low as about 400—500 ppm Sr. Primary calcium carbonates of biogenic or inorganic marine origin contain more than 1,000 ppm Sr (for references see: GRAF, 1960; FLÜGEL and WEDEPOHL, 1967, etc.).

OXBURGH, SEGNIT and HOLLAND (1959) have proved that partition of strontium between calcite and solution favors the solution ($K < 1$).

If a major portion of rock constituents has formed after sedimentation one expects a change of grain size distribution and texture in the rock during diagenesis. Therefore the significance of these properties in terminology and genetic considerations is doubtful; we do not emphazise a detailed classification of carbonate rocks on the basis of grain size or texture, except for recognizable organic residues, oolites and coarse grained detrital constituents.

Observation of recent carbonate formation and experiments in sea water-simulating systems leads to the classification of dolomite mainly as a diagenetic mineral. Diagenetic dolomite forms by reactions between either seawater or pore solutions with calcium carbonate under conditions of elevated salinity and temperature. In connection with his recent studies on phase relations in the Na—Ca—Mg—CO_3—Cl—SO_4—H_2O system, USDOWSKI (1967) has quoted papers on dolomite formation in lagoonal environments and the problem of the restricted primary deposition of this mineral. This author estimates that between 40 and 60 percent of dolomite has formed by reactions between consolidated limestones and pore solutions.

Numerous calculations about low temperature equilibria in chemical systems of major importance to the seawater-sediment environment have been published by GARRELS, CHRIST (1965) etc. They especially consider p_H, $E_h(p_{CO_2}, p_S)$ — relations in systems containing Fe, Cu, Mn, Pb, Ni, Co, W, U, V, Au.

IV. Abundance of Common Sediments and Sedimentary Rocks

The total amount of sedimentary matter in the Earth's crust must be ultimately derived from igneous rocks. The total mass of sediments has been computed from balances between magmatic rocks and their weathering products (sediments and seawater cations). GOLDSCHMIDT (1933) and CORRENS (1948) have estimated the sediment-mass as 170 kg per square centimeter of the earth's surface. For this kind of estimation, one must assume that the amount of sodium in the oceans balances the difference in sodium between igneous rocks and the sedimentary weathering products. If sedimentary rocks have an average density of about 2.4 g/cm^3, the normal thickness of the sediment layer, equally spread over the total earth's surface, is 700 m (using the quoted figure of 170 kg/cm^2 estimated by GOLDSCHMIDT, 1933). By considering more recent data on sodium abundances in magmatic (3.3% Na_2O) and sedimentary rocks (1.45% Na_2O) one ends up with about 900 m for the average thickness of the sediment layer.

This type of balance computation does not consider the sodium in salt deposits and that in pore solutions trapped in sediments and sedimentary rocks. BRAITSCH (1963) estimates that the total sodium from salt deposits of the world is not more than about 0.1% Na_2O in the sediment-mass. RONOV's (1968) figure for the abundance of evaporites is 1% of all sedimentary rocks. There are only tentative data available from which the pore volume of average sedimentary rocks and the sodium concentration of typical pore solutions can be estimated. We assume that 0.5% Na_2O is a reasonable figure for the average soluble sodium from pore solutions and salt deposits in all sedimentary rocks. By using this estimate for a mass balance, we get an average sediment thickness of 1,200 m or a total mass of sedimentary rocks on Earth of 1.5×10^{18} metric tons. RONOV (1968) mentions 1850 m as the average sediment thickness from his balance computation.

A sediment mass of 1.5×10^{18} tons still might be too low because the amount of sedimentary rock types which are chemically very similar to average magmatic rocks has not been considered in balance computations. Greywackes, as common rock types, and tillites belong to this group. In Table 8-6 references to publications are quoted from which it must be assumed that greywackes amount to (a minimum of?) 15 percent of the total mass of sedimentary rocks.

Seismic evidence for sediment thickness in typical sections of the Earth's crust is not yet substantial enough to produce a very reliable figure for the total mass of sedimentary matter. The areas of sediment deposits are to be divided into: pelagic (53%), hemipelagic and shelf (18%), young folded belts (8%) and shields (21%). Most uncertain is the average thickness of sediments on the bottom of the oceans. In 1964 M. and J. EWING stated that in large areas of the Atlantic, Pacific and Indian Oceans the seismic measurements found only 0.5 to 1 km of material which was clearly sedimentary (1.5—2.5 km/sec compressional wave velocity). On the basis of additional information from the recent literature, not more than the lower limit of 0.5 km should be taken as the average for pelagic sediments. Tentative values for the average thickness of sediments in other typical sections of the crust are mainly taken from KAY (1951), KUENEN (1950) and POLDERVAART (1955) as: hemipelagic and shelf: 4,000 meters; young folded belts: 5,000 meters; shield areas: 500 meters. Using these values one gets an average sediment thickness on Earth of 1,500 meters (comprising a total of 1.7×10^{18} tons). This estimate is only slightly higher than that from the author's balance computations. The recent rivers of the world deliver 3.9×10^9 tons dissolved (LIVINGSTONE, 1863) and about 10^{10} tons of suspended material into the oceans each year. Within 3 billion years about 3.10^{19} tons of sediment material should have been deposited if the present rate of river transport is typical. Most of the deposited material has been eroded and recycled for river transport again and again. Ten cycles must be assumed for the average sedimentary material if no appreciable portion has been extracted from this reservoir to form metamorphic rocks. There are only a few tentative estimates available for the mass of metamorphic rocks on earth. RONOV (1968) assumes that about 60% of the upper continental crust is composed of metamorphic rocks. If the major portion of metamorphic rocks has formed from sedimentary deposits by maintaining their bulk chemical composition this rock mass cannot be estimated from chemical balances. BARTH's (1962) assumption that almost all granitic rocks have formed from primeval sedimentary rocks must be considered with caution until the origin of a great number of granitic rocks has been established by $^{87}Sr/^{86}Sr$ ratios, etc.

Our present knowledge of the abundance of chemical elements in sedimentary matter suffers less from a lack of precise analytical data on common rock species than from our limited information about the abundance of sedimentary rock types.

Table 8-6 presents several sets of data on the abundance of common sedimentary rock types and recent sediments. These are estimated from planimetry of geologic

Table 8-6. *Abundance of sedimentary rock types*

a) *Environment of sediment deposition* (estimated as fraction of present continental sediment volume of $5 \cdot 10^8$ km^3).
Geosynclinal areas: 75 vol.-% (10 km average thickness).
Platform areas: 25 vol.-% (1,8 km average thickness) RONOV (1968).

b) *Abundance of rock types as fraction of sediment mass on the continents.*

Author	% limestones, dolomites	% sandstones, graywackes, arkoses	% shales, clays
LEITH, MEAD[a] (1915) G	22	32	46
CLARKE (1924) C	5	(15)	<80
SCHUCHERT (1931) G	20		
GOLDSCHMIDT, (1933) C	8.6		
v. ENGELHARDT (1934) C	8.6	(12)	80
KUENEN[b] (1941) G	29	14	57
KRYNINE (1948) G		40	
WICKMAN (1954) C	9.5	(8)	83
RONOV (1968) G	22[c] (evaporites: 1%)	25[c]	52[c]
This investigation C	7.5—8	(≧15)	≦77

[a] North America, Europe, Asia.
[b] East Indies.
[c] With small differences between platforms and geosynclines.

G: Estimated from geological maps and sampling (handling sandstones as a grain size class).
C: Estimated from chemical balances (considering quartz sandstones only).

c) *Abundance of sandstone types in percent of total sandstone class.*

Author	% (quartz) sandstones	% arkoses	% greywackes
KRYNINE (1948)	22.5	32.5	45
PETTIJOHN (1960, 1963)	34	15	46
MIDDLETON (1960)	34	16	50

d) *Abundance of deep sea sediment types as fraction of total covered ocean area*

Author	% pelagic clay	% foraminifera (and pteropod) ooze[d]	% diatomaceous and radiolaria ooze[d]
VAUGHAN (1924)	50	38	10
SVERDRUP et al. (1942)	37.8	48.7	13.5
RONOV (1968)	39	44	17

[d] It must be considered that in former compilations those beds containing organic carbonates or silica as low as 30% are counted as "oozes". Requiring a minimum of 50% calcareous residues in foraminifera ooze the fraction of this sediment type is probably less than 30%.

Table 8-6 (Continued)

e)

	Pelagic clay from the deep oceans as fraction of total sediment mass
Kuenen (1950) G	>50%
Poldervaart (1955) G	23%
Goldberg, Arrhenius (1958) C	13%
Wedepohl (1960) C, G•	10—20%
Ronov (1968) G•	27% [e]

[e] Including second layer of seismic velocities.

G: Estimated from geological maps and sampling (handling sandstones as a grain size class).

C: Estimated from chemical balances (considering quartz sandstones only).

G•: Estimated from geophysical and geological data.

maps and geologic sampling (*G*). For deep sea deposits, seismic evidence for layer thickness has been used (*G*•). The remaining values are computed from material belances between the average chemical composition of common magmatic and sedimentary rocks in the upper continental crust (*C*). All computations rely on an assumption of the composition of the upper continental crust. The assumption of the present author is explained in Chapter 7 of this handbook as: 44% granites, quartz monzonites, 34% granodiorites; 8% quartz diorites; 13% gabbros.

Values on the abundance of the sandstone class, as derived from chemical balance computations (*C*), have been listed in brackets in cases where greywackes were excluded. As already explained, the overall abundance of greywackes cannot be computed because the chemical composition of greywackes is similar to that of magmatic rocks representing the upper crust. From geological sampling Krynine (1948); Middleton (1960) and Pettijohn (1963) estimate that greywackes comprise 45—50 percent of the sandstone class.

The abundance of greywackes has additional important consequences arising from their close resemblance to granitic rocks in chemical composition. Partial melting of meta-greywackes under high grade metamorphic conditions will produce the largest possible fraction of melt in comparison with metamorphic rocks of other composition. From the first part of Table 8-6 we recognize an appreciable discrepancy between the two different methods of estimating the abundance of limestones. The present mass of carbonate rocks is at least twice the amount which can be derived from weathering of present day crustal rocks. The surplus could be a recycled product of earlier crustal composition.

In future research extra attention should be paid to the change of chemical composition of sedimentary matter in relation to environment and time. Ronov and coworkers, Strakhov (1967), Weaver (1967) and others have already contributed numerous observations on different chemical trends in sedimentary lithogenesis due to environment. — The oldest sedimentary rocks probably will allow us to recognize early stages of evolution of the Earth's crust.

References

BARTH, T. F. W.: Die Menge der Kontinentalsedimente und ihre Beziehung zu den Eruptivgesteinen. Neues Jahrb. Mineral., Monatsh. **3/4**, 59 (1962).

BEHNE, W.: Untersuchungen zur Geochemie des Chlor und Brom. Geochim. et Cosmochim. Acta **3**, 186 (1953).

BISCAYE, P. E.: Mineralogy and sedimentation of recent deep-sea clay in the Atlantic Ocean and adjacent seas and oceans. Bull. Geol. Soc. Am. **76**, 803 (1965).

BRAITSCH, O.: Entstehung und Stoffbestand der Salzlagerstätten. Berlin-Göttingen-Heidelberg: Springer 1962.

— Evaporite aus normalem und unverändertem Meerwasser. Fortschr. Geol. Rheinland u. Westfalen **10**, 151 (1963).

BROWN, G.: The x-ray identification and crystal structures of clay minerals. London: Mineralogical Soc. 1961.

CLARKE, F. W.: Data of geochemistry, fifth ed. U.S. Geol. Survey, Bull. **770** (1924).

CORRENS, C. W.: In: T. F. W. BARTH, C. W. CORRENS u. P. ESKOLA, Die Entstehung der Gesteine. Berlin: Springer 1939.

— Die geochemische Bilanz. Naturwissenschaften **35**, 7 (1948).

CRESSMAN, E. R.: Nondetrital siliceous sediments, chapt. T in: M. FLEISCHER (ed.), Data of geochemistry. US. Geol. Survey, Profess. Papers **440**-T (1962).

ENGELHARDT, W. v.: Die Geochemie des Bariums. Chem. Erde **10**, 187 (1934).

— Der Porenraum der Sedimente. Berlin-Göttingen-Heidelberg: Springer 1960.

EWING, M., and J. EWING: Distribution of oceanic sediments. Studies on Oceanography 525 (1964).

FLÜGEL, H. W., u. K. H. WEDEPOHL: Die Verteilung des Strontiums in oberjurassischen Karbonatgesteinen der Nördlichen Kalkalpen. Ein Beitrag zur Diagenese von Karbonatgesteinen. Contr. Mineral. and Petrol. **14**, 229 (1967).

FOLK, R. L.: Practical petrographic classification of limestones. Bull. Am. Assoc. Petrol. Geologists **43**, 1 (1959).

FÜCHTBAUER, H.: Zur Nomenklatur der Sedimentgesteine. Erdöl u. Kohle **12**, 605 (1959).

GARRELS, R. M., and C. L. CHRIST: Solutions, minerals and equilibria. New York: Harper & Brothers 1965).

GOLDBERG, E. D., and G. O. S. ARRHENIUS: Chemistry of Pacific pelagic sediments. Geochim. et Cosmochim. Acta **13**, 153 (1958).

—, and J. J. GRIFFIN: Sedimentation rates and mineralogy in the South Atlantic. J. Geophys. Research **69**, 4293 (1964).

—, and M. KOIDE: Geochronological studies of deep sea sediments by the ionium-thorium method. Geochim. et Cosmochim. Acta **26**, 417 (1962).

GOLDSCHMIDT, V. M.: Grundlagen der quantitativen Geochemie. Fortschr. Mineral. Krist. Petrogr. **17**, 112 (1933).

GRAF, D. L.: Geochemistry of carbonate sediments and sedimentary carbonate rocks, part I—IVb. Illinois State Geol. Survey, Circ. **297**, **298**, **301**, **308**, **309** (1960) (of special importance: part III Minor element distribution).

GRIFFIN, J. J., and E. D. GOLDBERG: Clay mineral distribution in the Pacific Ocean. In: The sea, vol. 3, p. 728. New York: Interscience Publ. 1963.

HADDING, A.: Eine röntgenographische Methode kristalline und kryptokristalline Substanzen zu identifizieren. Z. Krist. **58**, 108 (1923).

HAMILTON, E. L.: Thickness and consolidation of deep-sea sediments. Bull. Geol. Soc. Am. **70**, 1399 (1959).

HATCH, F. H., and R. H. RASTALL: Petrology of the sedimentary rocks. (4th ed. revised by J. T. GREENSMITH). London: Murby & Co. 1965.

HELMBOLD, R.: Beitrag zur Petrographie der Tanner Grauwacke. Heidelberger Beitr. Mineral. u. Petrog. **3**, 243 (1952).

HUCKENHOLZ, H. G.: Sedimentpetrographische Untersuchungen an Gesteinen der Tanner Grauwacke. Beitr. Mineral. u. Petrog. **6**, 261 (1959).

— A contribution to the classification of sandstones. Geol. Fören. i Stockholm Förh. **85**, 156 (1963a).

HUCKENHOLZ, H. G.: Der gegenwärtige Stand in der Sandsteinklassifikation. Fortschr. Mineral. **40**, 151 (1963b).
HURLEY, P. M., B. C. HEEZEN, W. H. PINSON, and H. W. FAIRBAIRN: K-Ar values in pelagic sediments of the North Atlantic. Geochim. et Cosmochim. Acta **27**, 393 (1963).
KAY, M.: North American geosynclines. Mem. Geol. Soc. Am. **48**, 143 p. (1951).
KRYNINE, P. D.: The megascopic and field classification of the sedimentary rocks. J. Geol. **56**, 130 (1948).
KUENEN, P. H.: Geochemical calculations concerning the total mass of sediments in the Earth. Am. J. Sci. **239**, 161 (1941).
— Marine geology. New York: John Wiley & Sons 1950.
LANDERGREN, S.: On the geochemistry of deep-sea sediments. Repts. Swed. Deep-Sea Exped. **10**, Spec. Inv. **5**, 59 (1964).
LEITH, C. K., and W. J. MEAD: Metamorphic geology. New York: Holt 1915.
LIVINGSTONE, D. A.: Chemical composition of rivers and lakes, chapt. G in: M. FLEISCHER (ed.), Data of geochemistry. U.S. Geol. Survey, Profess. Papers **440** G (1963).
LOWENSTAM, H. A.: Factors affecting the aragonite: calcite ratios in carbonate-secreting marine organisms. J. Geol. **62**, 284 (1954).
MATTIAT, B.: Beitrag zur Petrographie der Oberharzer Kulmgrauwacke. Beitr. Mineral. u. Petrog. **7**, 242 (1960).
MIDDLETON, G. V.: Chemical composition of sandstones. Bull. Geol. Soc. Am. **71**, 1011 (1960).
MINAMI, E.: Selen-Gehalte von europäischen und japanischen Tonschiefern. Nachr. Ges. Wiss. Göttingen, Math.-physik. Kl. **4**, 1, 143, (1935).
OXBURGH, U. M., R. E. SEGNIT, and H. D. HOLLAND: Coprecipitation of strontium with calcium carbonate from aqueous solutions. Abstracts GSA-meeting Pittsburgh 95 A (1959).
PETTIJOHN, F. J.: Sedimentary rocks, 2nd ed. New York: Harper & Brothers 1957.
— Chemical composition of sandstones — excluding carbonate and volcanic sands, chapt. S in: M. FLEISCHER (ed.), Data of geochemistry. US. Geol. Survey, Profess. Papers **440**-S (1963).
POLDERVAART, A.: Chemistry of the Earth's crust. Geol. Soc. Am. Spec. Pap. **62**, 119 (1955).
RITTENBERG, S. C., K. O. EMERY, and W. L. ORR: Regeneration of nutrients in sediments of marine basins. Deep-Sea Research **3**, 23 (1955).
RONOV, A. B.: Organic carbon in sedimentary rocks. Geochemistry (U.S.S.R.) (English Transl.) **5**, 510 (1958).
— Common tendencies in the chemical evolution of the Earth's crust, ocean and atmosphere. Geochemistry Internat. **4**, 713 (1964).
— Probable changes in the composition of seawater during the course of geological time. Sedimentology **10**, 25 (1968).
—, YU. P. GIRIN, G. A. KAZAKOV, and M. T. ILYUKHIN: Sedimentary differentiation in platform and geosynclinal basins. Geochemistry Internat. **3**, 595 (1966).
—, and A. A. YAROSHEVSKY: Chemical composition of the Earth's crust. Geochemistry (U.S.S.R.) **11**, 1285 (1967).
SCHUCHERT, C.: Geochronology or the age of the Earth on the basis of sediments and life. Natl. Research Council Bull. **80**, 10 (1931).
SHAW, D. M.: Geochemistry of pelitic rocks. III. Major elements and general geochemistry. Bull. Geol. Soc. Am. **67**, 919 (1956).
STRAKHOV, N. M.: Principles of lithogenesis, ed. by S. I. TOMKEIEFF, and J. E. HEMINGWAY. New York: Plenum Press 1967.
SVERDRUP, H. U., M. W. JOHNSON, and R. H. FLEMING: The oceans. New York: Prentice Hall 1942.
TRASK, P. D.: Origin and environment of source sediments of petroleum. Houston (Texas) 1932.
TUREKIAN, K. K., and M. STUIVER: Clay- and carbonate accumulation rates in three South Atlantic deep-sea cores. Science **146**, 55 (1964).
—, and K. H. WEDEPOHL: Distribution of the elements in some major units of the Earth's crust. Bull. Geol. Soc. Am. **72**, 175 (1961).

TWENHOFEL, W. H.: Principles of sedimentation. New York: McGraw-Hill Book Co. 1950.

USDOWSKI, H. E.: Die Genese von Dolomit in Sedimenten. Mineralogie und Petrographie in Einzeldarstellungen. Berlin-Heidelberg-New York: Springer 1967.

VAN'T HOFF, J. H.: Zur Bildung der ozeanischen Salzablagerungen. Braunschweig: Vieweg & Sohn 1905.

VAUGHAN, T. W.: Oceanography in its relations to other earth sciences. J. Wash. Acad. Sci. **14**, 313 (1924).

VINOGRADOV, A. P., and A. B. RONOV: Composition of the sedimentary rocks of the Russian platform in relation to the history of its tectonic movements. Geochemistry (U.S.S.R.) (Engl. Transl). **6**, 533 (1956).

WEAVER, C. E.: Potassium, illite and the ocean. Geochim. et Cosmochim. Acta **31**, 2181 (1967).

WEDEPOHL, K. H.: Spurenanalytische Untersuchungen an Tiefseetonen aus dem Atlantik. Geochim. et Cosmochim. Acta **18**, 200 (1960).

— Untersuchungen am Kupferschiefer in Nordwestdeutschland; ein Beitrag zur Deutung der Genese bituminöser Sedimente. Geochim. et Cosmochim. Acta **28**, 305 (1964).

— Environmental influences on the chemical composition of shales and clays. Phys. and Chem. Earth **8** (1968) (in press).

WENTWORTH, C. K., and H. WILLIAMS: The classification and terminology of the pyroclastic rocks. Natl. Acad. Sci.-Natl. Research Council Rept. Comm. Sed. Bull. **89**, 19 (1932).

WHITE, D. E.: Saline waters of sedimentary rocks. In: Fluids in subsurface environments — a symposium. Mem. Am. Assoc. Petrol. Geologists **4**, 342 (1965).

WICKMAN, F.: The "total" amount of sediments and the composition of the "average igneous rock". Geochim. et Cosmochim. Acta **5**, 97 (1954).

Chapter 9

K. R. Mehnert

COMPOSITION AND ABUNDANCE OF COMMON METAMORPHIC ROCK TYPES

The geochemical investigation of metamorphic rocks is generally not as advanced as that of magmatic and sedimentary rocks. The reason for this may be sought in their relatively complex origin. Evidently in metamorphic rocks the number of independent genetical parameters is much greater than in normal magmatic or sedimentary rocks where general trends of formation predominate.

Hence the natural complexity of metamorphic rock formation has already to be considered in their *classification* and *nomenclature*. Two essentially different systems of nomenclature can thus be distinguished:

a) According to the "educt", i.e. the parent rock from which the metamorphic rock originated. This educt can be designated by the prefix "meta-", as for instance in the term "meta-graywacke", or in the often used comprehensive term "metabasite", and many others.

b) According to the "product", i.e. the rock type as it is portrayed by the modal composition and structure of a metamorphic rock without any consideration of its origin. For instance, a rock may be described as a sericite-chlorite schist, or a biotite-plagioclase gneiss, etc., without the respective parent rock being quoted or even known.

In the beginning of any petrographical or geochemical investigation metamorphic rocks are generally classified purely according to their "phenotype", i.e. their present composition and structure according to b. Only after some thorough investigation, mainly with the microscope, rocks can be classified according to their "genotype", i.e. the educt from which they originated, according to a.

Besides this general scheme of nomenclature there are many special petrographical terms. The following definitions can give but a brief summary of the most important conceptions. For detailed descriptions including their exact limitation or possible transitions reference should be made to textbooks of petrology.

General Terms

Gneiss: Metamorphic rock exhibiting parallel structure and consisting of feldspar ($>20\%$ of the whole rock) and generally of quartz beside some mafic minerals like biotite, muscovite, hornblende, pyroxene, cordierite, sillimanite, garnet, etc.

Schist: Metamorphic rock exhibiting parallel structure generally free of or poor in feldspar ($<20\%$ of the whole rock) consisting mainly of quartz and/or mafic minerals (as above).

Fels: Metamorphic rock without parallel structure.

Special Terms

Hornfels: Generally fine-grained metamorphic rock without parallel structure typically occurring in the innermost zones of contact aureoles (see p. 286).

Phyllite: Generally fine-grained sericite-chlorite schist mainly consisting of sericite, chlorite and quartz, often carbonate-bearing.

Amphibolite: Metamorphic rock composed of common hornblende and plagioclase.

Granulite[1]. Metamorphic rock composed essentially of a fine-grained mosaic of feldspar with or without quartz. Mg-Fe-minerals, if present, are predominantly anhydrous. Granulites typically contain lenticular (or elongate) grains or aggregates of grains.

Eclogite: Equigranular rock mainly composed of omphacite (Mg-Ca pyroxene rich in Na) and garnet (Mg-Ca-Al garnet rich in Mg) often containing kyanite, enstatite or hypersthene.

I. General Trends in Metamorphism

In principle every magmatic or sedimentary rock can be transformed into a metamorphic rock and hence occur in every stage of increasing or decreasing metamorphism. In addition, metamorphic rocks themselves can in turn be subjected to another or even several processes of metamorphism whereby their structure and composition may again be altered and adopt every stage of metamorphism ("polymetamorphic" rocks).

From a petrological as well as a geochemical point of view it is thus expedient first to investigate rather simple cases of metamorphism, e.g. the same parent rocks undergoing varying metamorphic conditions or different parent rocks under the same conditions.

In the simplest case transformation takes place without any alteration of the chemical and mineralogical composition of the parent rock, simply through the formation of a new metamorphic structure. This process can lead either to the formation of a parallel structure from a previously massive rock or, vice versa, to the formation of a massive structure from a previously schistose rock.

Recrystallization is generally accompanied by a coarsening of the grain size, and particular minerals even grow preferentially as compared with the other minerals of the matrix and hence form the so-called *porphyroblasts.* This process of preferential crystal growth during metamorphism is called *blastesis.*

In this context the following problem is interesting geochemically: According to petrographic experience recrystallization or even blastesis hardly occur in rock-forming silicates solely due to a rise in temperature. There are for instance rocks, e.g. hornfelses and granulites, that must have been exposed to relatively high temperatures in the course of geological time. In spite of this they have preserved their originally fine-grained structure. On the other hand, there are metamorphic rocks

[1] Other definitions of "granulite" at present in use, for instance to describe fine-grained muscovite granites, are less widespread. They should be abandoned in favor of the definition given above.

that obviously endured but a relatively small increase in temperature and yet show strong recrystallization. This apparently contradictory behavior may be explained by the action of hydrothermal solutions which were probably infiltrated during the process of metamorphism and thereby caused extensive recrystallization even in rocks affected only by low-temperature metamorphism. On the other hand, no coarsening of the grain size could happen where hydrothermal solutions were lacking even if the reaction temperature was relatively high.

These solutions obviously quit the rock after metamorphism leaving no other traces than the well-known microscopical inclusions of fluids within the minerals. It can thus happen that a rock shows approximately or even exactly the same composition before and after metamorphism, whereas it possessed an entirely different bulk composition during the transformation process itself. Hence the presence and action of hydrothermal fluids during metamorphism can as a rule be postulated in spite of the impossibility to detect them directly by geochemical methods.

Another type of metamorphism is represented by a transformation of the mineral content without changing the chemical composition. Consequently this kind of transformation is called *isochemical metamorphism*. It is simply characterized by a redistribution of the elements present in the mineral phases of the parent rock ("paleosome") and the formation of new mineral phases ("neosome") according to the bulk composition of the whole rock. In principle those phases are formed that are stable under the prevailing PTX-conditions.

According to Eskola (1921) all minerals that are stable under the same PTX-conditions comprise a *mineral facies*. Petrographic experience has shown that metamorphic rocks can be adequately classified and even mapped geologically according to their respective mineral facies. Because of the generally complex composition of metamorphic rocks this comprehensive mineral facies principle is more adequate than the use of the stability relations of only one mineral as a petrogenetic reference.

In order to study the limits of stability of co-existent mineral phases it is essential to realize that the reactions amongst rock-forming silicates are generally rather slow. Consequently minerals unstable in natural rocks can indeed be retained as *relics* even under conditions where they should properly have been transformed according to the thermodynamic stability limits.

Hence in order to establish the limits of a mineral facies in nature the first appearance of a newly formed, so-called "critical" mineral has to be fixed, even if it were present in very small amounts. The disappearance, however, of a mineral cannot be taken on principle as a criterion of the limits of a facies, unless equilibrium has evidently been achieved.

In addition, it should be noted here that many rock-forming minerals crystallize as members of a rather complex mixed-crystal series. Generally only minerals showing quite a specific composition are stable within limited PT-domains and hence are characteristic of a particular mineral facies or sub-facies. Since most metamorphic rocks in spite of their complex chemical composition are composed of but a few yet rather complex groups of minerals (micas, garnets, pyroxenes, amphiboles, etc.) it is of great importance to investigate even small geochemical differences between such co-existing minerals. Only then can these complex minerals be utilized for the exact investigation of a specific mineral facies or sub-facies.

Equilibrium between co-existing mineral phases is rather quickly established if the rock was affected not only by hydrothermal pore fluids but also by internal movements caused for instance by tectonic processes. Unilateral pressure (shearing stress) was occasionally assumed to constitute an additional parameter in the formation of minerals. Nevertheless it can now be taken for granted that shearing stress is only effective in the more rapid establishment of equilibrium by some sort of "stirring effect". Locally, however, pressure can evidently increase considerably due to tectonic causes, especially in orogenic zones, and thereby minerals can be formed even at rather shallow depths that normally occur only under high confining pressure, i.e. at great depth.

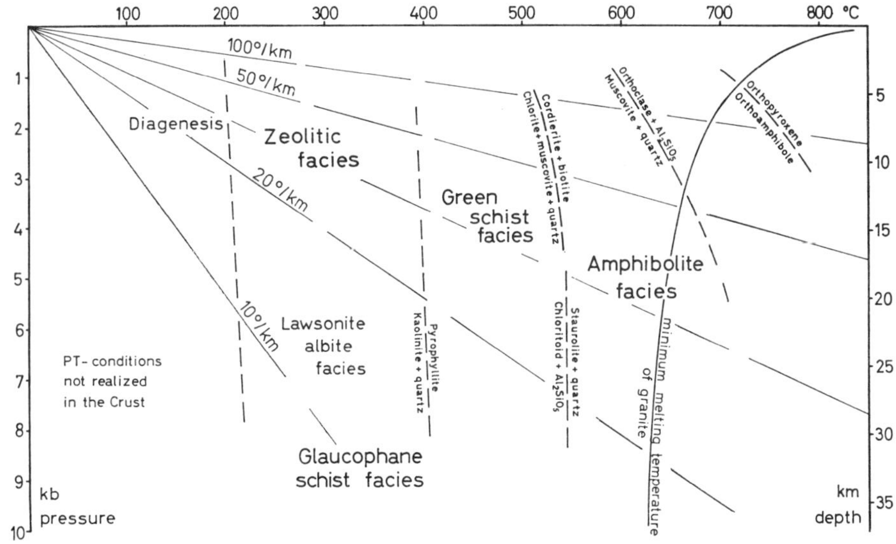

Fig. 9-1. Distribution of mineral facies in a temperature-pressure diagram of the upper crust according to Winkler (1967)

The other main parameter of metamorphism is temperature. The normal geothermal gradient of the upper crust of the Earth is around 30°/km, yet according to Birch (1955), Tuttle (1955), Schuiling (1957), Clark (1960), and o hers, it can vary between 5° and 50°/km. Locally the gradient can evidently even rise to 150—200°/km for instance around magmatic "heat domes". It could be demonstrated in several recent publications that a relatively steep geothermal gradient of about 50—150°/km has to be assumed in many metamorphic zones of orogenic belts (Miyashiro, 1961; Zwart, 1962; Den Tex, 1963; Green, 1963, a.o.). Here the "isograds", i.e. the lines of equal grades of metamorphism from the lowest to the highest mineral facies, are rather closely aligned and hence prove the geothermal gradient to have been extraordinarily steep during the main phase of metamorphism (see Fig. 9-1).

Metamorphism by definition begins below the zone of weathering, sedimentation and diagenesis. Hence for geological reasons all geochemical alteration processes occurring at or near the Earth's surface, like oxidation and hydration, are generally not included within the concept of metamorphism *sensu stricto*. As Winkler (1967) has pointed out metamorphism begins with the formation of minerals that are not stable under sedimentary conditions, i.e. minerals of the so-called zeolitic facies, like laumontite and prehnite. These minerals are formed at around 220° C at a H_2O-pressure of 2 kb, and hence the initial stages of metamorphism can be taken to occur at low to medium pressures at about 200—220° C as shown in Fig. 9-1. This temperature of the very beginning of the lowest mineral facies, the zeolitic facies, is appreciably higher than hitherto assumed. The subsequently following greenschist facies, which is widely distributed in regional metamorphic terrains of the epizone of mountain belts, begins at about 390—420° C at 2 kb H_2O-pressure. At 530—550° C the amphibolite facies is reached.

In the high-temperature range of the amphibolite facies rocks begin to show incipient melting phenomena. This process of regional rock melting is called *anatexis*. The "ternary" eutectic of the granite-system quartz-albite-orthoclase for instance lies at about 640—650° C at H_2O-pressures of 4—5 kb (TUTTLE and BOWEN, 1958; LUTH, JAHNS and TUTTLE, 1964). It is important to note that sedimentary rocks of psammitic or pelitic composition as well as their corresponding metamorphic rocks begin to melt around 680—750° C at 2 kb pressure.

On the other hand, rocks can remain unmolten under these or still higher temperatures, if fluxes like water or other volatiles are deficient or lacking altogether. In this way, for instance, metamorphic rocks of the granulite facies are supposed to have been formed.

If the rocks are rich in Mg-Al-Fe components, their melting interval is, even in the presence of water, so high that they will remain unmolten under the PT-conditions of the upper crust. To this group belong rocks of sedimentary origin as well as metabasites or ultrabasites. Under very high pressures, generally not existent in the upper crust but in the lower crust and the upper mantle, they are transformed to rocks belonging to the eclogite facies.

In principle metamorphism *sensu stricto* ends as anatexis begins. The incipient stages of anatexis are often still included in the concept of "*ultrametamorphism*". Here molten and unmolten portions of the rocks are thoroughly mixed and thus the so-called *migmatites* are formed. In advanced stages the molten masses collect and finally unite to larger bodies which can intrude other, unmolten masses or the adjacent rocks outside the anatectic rock series. At this stage, when the neosomatic melt definitely leaves the rock system whence it originated, the process of anatexis ends by definition, and the *magmatic* sequence of rock formation begins. It is thus that a new *petrogenetic cycle* can be initiated leading through the magmatic, sedimentary, metamorphic and anatectic stages eventually back again to the beginning.

II. Isochemical Series of Metamorphism

The geochemical investigation of transformation processes of increasing metamorphism presupposes that sufficient analyses of metamorphic rocks of exactly the same parent rock are available, and that at different stages of transformation. This apparently simple demand is in fact relatively difficult to fulfil.

At first it has to be ascertained, as far as it is possible, that the educt and the product (see p. 272) of the transformation process correspond to one and the same rock before metamorphism. This is, for instance, demonstrable in the case of sedimentary rocks, where a certain layer or horizon persists in its composition to such an extent that its pre-metamorphic state may with certainty be assumed to be invariable. Magmatic parent rocks, too, may often be sufficiently homogeneous to fulfil this presupposition.

In practice, however, most sedimentary or magmatic rocks reveal small heterogeneities when investigated by sensitive geochemical methods. If, hence, differences between the parent rock and its final metamorphic product were to be detected, the primary heterogeneity of the former, as well as that of the latter, have at first to be accurately known.

Another difficulty arises from the fact that the transformation of rocks generally proceeds selectively. Metamorphic processes hence progress irregularly, i.e. according to the — slightly variable — primary composition of the rock. This in effect means that some parts of the rock adjust more readily to the changing conditions than others which thus form the so-called *resisters*. For instance, meta-pelitic layers of sediments usually attain their metamorphic character more easily than do meta-arenitic layers. Sandstones, and especially quartzites, retain their primary character even under metamorphic conditions when intercalations of former clay already are completely transformed.

The geochemical differences actually found in a series of metamorphic rocks can hence either be derived from primary heterogeneities or can be due to selective metamorphism. Furthermore they can be interpreted by selective metamorphism *following* primary heterogeneities. A clue to this intricate problem may be found in the following consideration: If the differences existing between the parent rock and its metamorphic counterpart are greater than the statistically determined heterogeneity of the parent rock, then it is rather probable that the differences are due to metamorphic reactions. This conclusion evidently rests on the statistical comparison of co-genetic sample series resulting from increasing metamorphism (see Table 9-1 and 9-2).

The last and probably main difficulty in the geochemical investigation of metamorphic rock series arises from the fact that as a rule the metamorphic reactions take place by the simultaneous change of several, mutually independent parameters, such as temperature, total pressure, partial H_2O-pressure, partial pressure of other volatiles, and the rather variable concentration of chemical elements by primary origin or by migration, i.e. addition or removal of mobile components. Strictly speaking, in geochemical investigations only one of these parameters should undergo a controllable change and the others should remain constant.

In reality this is, of course, hardly ever met with. In order to solve this rather complex problem it is thus necessary to lump together several of these parameters which according to all geological experience normally go together. We can thus establish a "normal series" of increasing regional metamorphism whereby all special developments have at first to be disregarded.

Thereby, however, the question naturally arises which rock types in actual fact represent the "normal series" of increasing regional metamorphism. The classical conception, proposed for instance by Rosenbusch or Grubenmann-Niggli, utilized as the normal series of increasing metamorphism of pelitic rocks the sequence:

clay → shale → slate → sericite-chlorite schist → mica schist → biotite gneiss ± cordierite, garnet, sillimanite ("kinzigite").

This sequence in fact is almost isochemical, as the examples of chemical analyses of Table 9-1 demonstrate, except for a decrease of the H_2O content, especially in the initial stages. Further differences between the analyses can however be of primary as well as of secondary origin. In order to distinguish between these two possibilities the statistical variance of each type of rocks should be determined, so as to eliminate incidental differences between the types.

Table 9-1. *Mean values of typical examples of metamorphic rocks of pelitic origin*

	1 Shales	2 Slates	3 Phyllites	4 Mica schists	5 Kinzigites
SiO_2	60.15	60.18	61.78	60.42	59.64
TiO_2	0.76	0.67	0.73	0.91	0.75
Al_2O_3	16.45	17.70	16.54	16.41	17.42
Fe_2O_3	4.04	2.45	2.43	2.85	2.42
FeO	2.90	4.67	5.45	6.84	5.84
MnO	tr	0.16	0.07	0.09	0.13
MgO	2.32	3.01	2.35	3.76	3.82
CaO	1.41	1.52	1.94	0.96	1.82
Na_2O	1.01	1.62	1.89	1.60	1.98
K_2O	3.60	3.79	4.14	3.25	3.32
H_2O^+	3.82	3.19	2.32	2.39	2.23
P_2O_5	0.15	0.19	0.27	0.21	0.04
CO_2	1.46	0.30			
C	0.88	0.36			
SO_3	0.58	0.10			
S		0.08	0.01	0.14	

1: Average of 51 Paleozoic shales (CLARKE, 1924, p. 552).
2: Average of 13 slates (NANZ, 1953, p. 53/54, No. 3, 4, 8, 10—14, 16—18, 23, 25).
3: Average of 4 phyllites (SIMONEN, 1953, p. 32, No. 3, 4, 7, 9).
4: Average of 3 mica schists (SIMONEN, 1953, p. 50, No. 47—49).
5: Average of 6 kinzigites (SIMONEN, 1953, p. 51, No. 55—59. 62).

Table 9-2. *Composition of phyllites, mica schists and sillimanite*

	Phyllites ($n_1 = 6$ anal.)			Mica schists ($n_2 = 17$ anal.)			Sillimanite gneisses ($n_3 = 10$ anal.)		
	$\bar{x}_1$	s	$s_{\bar{x}}$	$\bar{x}_2$	s	$s_{\bar{x}}$	$\bar{x}_3$	s	$s_{\bar{x}}$
SiO_2	60.21	3.25	2.60	60.06	5.78	2.75	59.26	3.94	2.44
TiO_2	1.32	0.16	0.13	1.12	0.22	0.10	1.11	0.26	0.16
Al_2O_3	19.03	1.89	1.51	19.89	3.50	1.66	19.74	2.85	1.77
Fe_2O_3	4.09	1.19	0.95	2.62	0.99	0.47	2.20	1.15	0.71
FeO	4.59	1.03	0.82	4.46	1.05	0.50	4.52	1.04	0.64
MgO	1.31	0.53	0.42	1.99	1.01	0.48	2.70	0.88	0.55
CaO	1.37	0.25	0.20	1.52	0.87	0.41	2.22	0.97	0.60
Na_2O	1.73	0.36	0.29	1.54	0.83	0.39	1.87	0.54	0.33
K_2O	3.29	0.35	0.28	4.23	0.90	0.43	4.12	0.93	0.58
H_2O	2.89	0.84	0.67	2.41	0.81	0.39	2.02	0.48	0.30
P_2O_5	0.28	0.03	0.02	0.26	0.08	0.04	0.16	0.06	0.04

Only a few serious attempts in this direction have until now been made and, as a rule, they concern only 2 or 3 stages of the normal series mentioned above.

Table 9-2 lists the mean values of 33 chemical analyses of phyllites, mica schists and sillimanite gneisses from the Central Pyrenees after ZWART (1959). From these mean values ($\bar{x}$), their standard deviations (s) and 95% confidence limits ($s_{\bar{x}}$), the standard deviations of the differences (s_d) and the test value t_x were calculated. The latter in conjunction with STUDENT's t-distribution determines with a 95% probability whether the differences thus found are statistically significant (+) or not (−).

The data of Table 9-2 demonstrate that only a few significant changes can be observed during this type of progressive metamorphism from phyllites to mica schists and to sillimanite gneisses. For instance a decrease of the Fe_2O_3 content is evident. The slight increase in K_2O from phyllites to mica schists is probably due to the accidental influx of alkali (see p. 284) since this increase is found to be retrograde in the following stage of metamorphism.

Even more pronounced details of progressive metamorphism were given by SHAW (1956) when presenting statistical calculations on the basis of 155 carefully chosen chemical analyses from the literature (Fig. 2). Group A comprises low-grade metamorphic schists of pelitic origin, whilst group B contains high-grade schists of the same origin. The differences recognized as statistically significant are shown by full lines, whilst those of only insignificant differences by interrupted lines.

It can thus be seen that SiO_2 and Al_2O_3 as well as MgO, Na_2O and K_2O exhibit no significant changes. The significant decrease of Fe_2O_3 is coupled with a likewise significant increase in FeO. If Fe^{II} and Fe^{III} are added, the changes cancel each other. That means: only a reduction of Fe^{III} to Fe^{II} took place with a small decrease in total oxygen. CaO and CO_2 both decrease. Hence dissolution and removal of $CaCO_3$ obviously are geochemical significant processes during increasing regional metamorphism.

As a result it can thus be stated that no general change in composition during progressive regional metamorphism can be observed. The only obvious changes are a decrease of H_2O and $CaCO_3$, as well as a reduction of Fe^{III} to Fe^{II}. It can hence be

gneisses from the Central Pyrenees. (After H. J. ZWART, 1959.)

Phyllites/mica schists ($t_{95}=2.08$)				Mica schists/sillimanite gneisses ($t_{95}=2.06$)			
$\bar{x}_2-\bar{x}_1$	s_d	t_x	sign./ not sign.	$\bar{x}_3-\bar{x}_2$	s_d	t_x	sign./ not sign.
−0.15	6.53	0.05	−	−0.80	5.19	0.39	−
−0.20	0.21	2.01	−	−0.01	0.23	0.11	−
+0.86	3.19	0.66	−	−0.15	3.28	0.12	−
−1.47	1.05	2.95	+	−0.42	1.06	1.00	−
−0.13	1.05	0.26	−	+0.06	1.05	0.01	−
+0.68	0.92	1.56	−	+0.71	0.96	1.86	−
+0.15	0.77	0.41	−	+0.70	0.91	1.94	−
−0.19	0.74	0.54	−	+0.33	0.74	0.78	−
+0.94	0.80	2.46	+	−0.11	0.91	0.30	−
−0.48	0.82	1.23	−	−0.39	0.71	1.38	−
−0.02	0.07	0.58	−	−0.10	0.23	1.08	−

taken for granted that under normal conditions regional metamorphism takes its course in a geochemically "conservative" manner. Strictly "isochemical" reactions are but rare, especially within the domain of a single mineral. The bulk composition of the whole rock, however, remains generally unchanged.

This is also true with regard of the trace elements, as SHAW demonstrated in 1954. Statistical tests on 63 samples of the Littleton formation of New Hampshire indicated

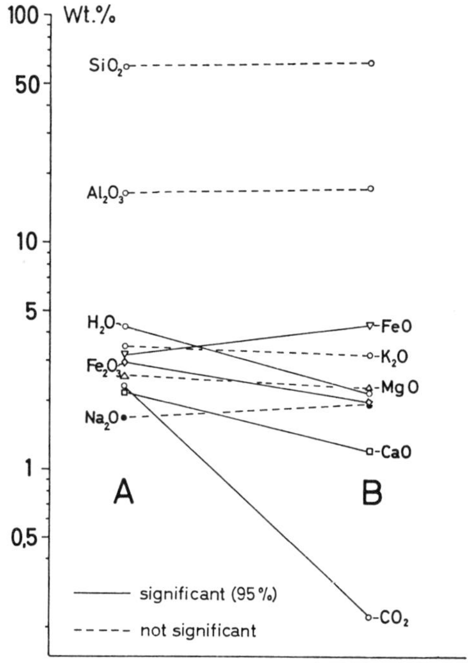

Fig. 9-2. Diagram showing the mean values (*A*) of 85 pelitic rocks of low-grade metamorphism compared with the mean values (*B*) of 70 pelitic rocks of high-grade metamorphism. (After SHAW, 1956, p. 928.) The differences of *A* and *B* are statistically "significant" (solid lines) or "not significant" (dotted lines) according to STUDENT's *t*-test (see text p. 279)

a rather wide range in the original composition of the rocks. However, the mean values of low-, medium- and high-grade metamorphic rocks showed that the distribution of the trace elements remained essentially the same during progressive metamorphism. Table 9-3 demonstrates this "conservative" behavior of the trace elements. Only Ni and Cu slightly decrease, and Li and Pb increase in the order of about 100%, probably because of potash metasomatism as mentioned above. Metasomatism of course can always interfere with otherwise geochemically "conservative" regional metamorphism.

In this way the metasomatic influence can become stronger and stronger and finally supplant the isochemical character of regional metamorphism completely. Whilst for instance GOLDSCHMIDT (1911), BILLINGS (1937, 1938), SHAW (1954, 1956), TAYLOR (1955) and ZWART (1959) emphasized that regional metamorphism normally progresses in the "conservative" manner demonstrated above, FERSMAN (1929) and

Wenk (1954) pointed out that definite changes in composition can occur, but that the bulk composition of the whole rock series investigated can nevertheless remain unchanged. They hence explain the metasomatic alterations by internal migrations within a more or less extensive rock mass due to differences in PTX-conditions.

Moreover, many authors described metamorphic reactions as taking place in a generally open system and hence emphasized systematic geochemical changes with increasing grade of metamorphism. The most evident changes include the decrease of the H_2O content (Yoder, 1955) and of the oxygen content (Eugster, 1959) with increasing depth.

Table 9-3. *Average composition of low, medium- and high-grade meta-pelites of the Devonian Littleton formation, New Hampshire.* (After Shaw, 1954.)

	Low-grade rocks (13 samples)		Medium-grade rocks (20 samples)		High-grade rocks (30 samples)		Average of all rocks	
	mean	*s*	mean	*s*	mean	*s*	mean	*s*
Ga	20.8	7.5	15.9	5.14	19.8	5.84	18.8	6.34
Cr	116	38.7	113	33.5	109	31.2	112	33.1
V	109	44.0	125	27.0	120	38.0	119	3.85
Li	54.7	22.0	108	157	127	67.0	106	104
Ni	80.5	33.4	63.7	23.7	57.4	19.8	64.2	25.5
Co	16.8	4.0	19.4	6.74	18.0	7.01	18.2	6.59
Cu	23.1	15.5	23.8	23.3	12.5	12.5	18.3	18.0
Sc	11.3	4.3	11.9	3.10	15.6	9.56	13.5	7.35
Zr	191	68.0	213	78.7	203	69.3	204	72.7
Y	38.8	11.5	37.9	12.6	51.7	24.2	44.7	20.0
Sr	524	416	731	244	760	260	705	310
Pb	16.1	9.30	23.3	13.7	27.3	10.7	23.7	12.3

Of special interest are the systematic changes in the direction of a generally granitic composition by increasing metamorphism as shown for instance by Lapadu-Hargues (1945), Reynolds (1946), Chao (1951), Higazy (1954), Engel and Engel (1958, 1960). This kind of transformation is called *granitization* and is discussed in detail on p. 289.

In conclusion it can hence be said that on the one hand "normal series" of increasing metamorphism can be established in which the transformations on the whole take place isochemically, and where migrations are restricted to relatively small domains.

On the other hand there are allochemical "normal series" characterized by an extensive migration of mobile components within the earth's crust. Obviously both types can occur side by side and even merge with each other to a certain degree.

a) Retrogressive Metamorphism

In principle, every process of increasing metamorphism must finally decline and terminate in a *retrograde* or *retrogressive* phase, if the rock involved is actually found at the Earth's surface. This latter kind of metamorphism hence takes place at generally declining PT-conditions whereby the mineral paragenesis is adapted more or less to the new conditions.

A chemical change in the bulk composition does not necessarily occur so that retrogressive metamorphism can proceed geochemically just as "conservatively" as does progressive metamorphism (see p. 279).

If this evolution from high to low temperature takes place rapidly, generally little or hardly any mineralogical change occurs. Metamorphic rocks thus can preserve their highest grade ever attained completely, even over geological ages, like many metamorphic rocks of the early Precambrian.

It is, however, a question of geochemical interest in how far mineral transformations did in actual fact take place on a submicroscopical scale in spite of their apparently good preservation both megascopically and even microscopically. The question hence arises whether and how far the chemical composition in fact corresponds to the mineral content determined by the microscopical Rosiwal method. More detailed investigations in this problem especially by microprobe analysis are desirable in order to discuss this question of fundamental importance.

Internal retrogressive alterations can likewise be of geological interest for instance with respect to absolute age determinations of minerals and rocks. Radiogenic elements, especially the gases He and Ar, could have escaped, at least partially, from the minerals whence they originated. Thereby it is possible that the extent of migration was sufficiently small to be still included within the geochemical sample. Hence it is very important for all geochemical investigations that the average distance of possible migration of mobile substances investigated be known.

Retrograde metamorphism can proceed geochemically conservatively but there may also be involved essential geochemical changes in the bulk composition of the rocks. Angel (1965) suggested only this latter type of retrograde metamorphism be called *diaphthoresis*. Yet most petrologists use both terms as synonyms, irrespective of whether chemical changes in the bulk composition did or did not take place.

III. Allochemical Series of Metamorphism (Metasomatism)

Amongst all possible allochemical reactions occurring during metamorphism the concept of *metasomatism* can be defined as the special process of practically simultaneous solution and re-deposition by which a mineral or mineral aggregate is replaced by new minerals of partly or wholly differing composition.

This process of immediate replacement of one mineral by another is evidently of great importance during regional metamorphism. The change in composition may be of variable amount, but numerous cases exist where a complete change of all elements has to be assumed because no relic whatever of the previous minerals remains. In other cases certain geochemically *immobile* elements remain behind whilst others, *mobile* ones were exchanged.

Which of these elements behave in a mobile or immobile way largely depends on the PTX-conditions prevailing. For this reason it is hardly possible to establish a sequence of geochemical mobility which would equally be valid for all PTX-conditions. Generally those components are mobile that are easily soluble in hydrothermal solutions, like Na and K. On the other hand, Al can under most conditions of the crust be regarded as rather immobile. KORZHINSKY (1955) who thoroughly investigated this question of geochemical mobility proposed on the average the following sequence from high to low mobility:

H_2O, CO_2, S, SO_3, Cl, K_2O, Na_2O, F, CaO, O_2, Fe, MgO, SiO_2, P_2O_5, Al_2O_3, TiO_2.

As already mentioned, this sequence can be valid only as a general principle. Many deviations occur dependent on the transporting medium, the degree of oxidation, the state of hydration of ions, etc. (see for instance FERSMAN, 1929; BARKER, 1964).

The first important factor of the effectiveness of metasomatic processes is based on the penetrability of the parent rock (paleosome). In porous rocks liquids and gases can diffuse easily. In most metamorphic rocks, however, the pore spaces are largely closed due to recrystallization so that any transport by diffusion can only take place along the intergranular spaces ("intergranular film"). This kind of migration is obviously furthered by deformation. Cataclastic, i.e. broken, minerals and rocks within zones of deformation, like faults, are preferred ways for the transport of migrating matter.

A second factor is the *reactivity* of the paleosome relative to the active neosomatic liquid or gas. For instance reaction proceeds quickly, if the paleosome has a more or less *basic* character, i.e. if it is rich in Ca, Mg and Fe, whilst the neosomatic substances can be of *acid* character if they are rich in HCl, HF, etc. Under these circumstances the reaction can result in a considerable local enrichment of the reaction products. If, however, the chemical contrasts are small, the reaction products are often widely distributed and the local concentration of neosomatic minerals is generally small. This is of great importance in the prospection of metasomatic ore deposits.

If the reactivity of the paleosome is great but its penetrability is small, a *metasomatic front* is formed. In this case the border between the paleosome and the neosome is more or less continuous. It is mostly unaffected by grain borders or other heterogeneities of the paleosome and often consists of parallel zones of varying mineral content.

If the reactivity is small but the penetrability is great, then the metasomatic reaction proceeds as a sort of *infiltration* or *impregnation* process. Here the border between the paleosome and the neosome is rather irregular and frequently remnants of the former are surrounded by a network of the latter.

Especially in the case of metasomatic impregnation all inhomogeneities of the paleosome are often very exactly portrayed by the neosome. In this way *pseudomorphs* of the younger minerals after the older ones are formed and it is not rarely possible to reconstruct the nature of the paleosomatic mineral simply from its former crystal shape even if no substantial relic has been preserved.

It has however to be emphasized that complete metasomatism, even of chemically extremely different paleosomes, can finally lead to an almost or entirely uniform neosome. In such cases a considerable *excess* of added substances over the paleosome, has to be assumed. Likewise an extensive removal of the dissolved substances of the paleosome has to be postulated.

Such a complete exchange of a rather considerable amount of substances added and removed over a distance up to several hundred meters is admittedly rather difficult to understand, if we have to assume migration by diffusion. Nevertheless such reactions according to the counter-current principle must have taken place as demonstrated by many natural occurrences. Presumably they can only be explained by the action of migrations in the course of geological times.

The *classification* of metasomatic rocks and ore deposits has to be based on the geochemical character of the neosome since the paleosome as a rule has more or less

Table 9-4. *Examples of alkali metasomatism through albitization (1 and 2) and potash feldspathization (3 and 4)*

	Albitization		Potash feldspathization	
	1	2	3	4
SiO_2	40.31	51.92	62.20	59.82
TiO_2	1.13	0.63	0.92	0.77
Al_2O_3	22.52	20.47	13.41	15.72
Fe_2O_3	2.31	1.51	1.55	1.80
FeO	11.49	6.98	4.73	3.15
MnO	0.11	0.10	0.26	0.14
MgO	6.80	4.67	3.55	3.46
CaO	1.85	1.19	3.20	2.74
Na_2O	2.84	4.16	3.62	3.86
K_2O	6.87	4.70	4.70	6.26
H_2O^+	2.99	2.78	1.05	1.33
H_2O^-	0.20	0.18	0.16	0.03
P_2O_5	0.11	0.44	0.21	0.58
CO_2	—	tr	0.65	0.26
	99.53	99.73	100.21	99.92

Albitization:

1: Biotite-schists with some albite porphyroblasts from Ben Ledi Beds, Portnahorna, Co. Antrim.

2: Albite-schist from the same horizon as specimen 1 and collected less than 2 feet away from it. Both analyses after REYNOLDS, 1942/43, p. 49.

Potash feldspathization:

3: Biotite-plagioclase hornfels from Geschwend, southern Black Forest.

4: The same rock rich in porphyroblasts of potash feldspar. Average of several specimens. Bot analyses after HOENES, 1940, p. 199.

changed its original composition. The number of geochemical modes of occurrence is indeed very large because every chemical element or combination of them can be present in a neosomatic rock (for a summary see GOLDSCHMIDT, 1920 or HIETANEN, 1954).

The most frequent types of metasomatism are:

Alkali Metasomatism. The alkalis Na and K belong to the most mobile components of the crust and hence frequently occur as major constituents of migrating solutions. As is well-known, for instance, NaCl is a very common component of microscopic and submicroscopic inclusions of most minerals.

Amongst rocks of the epizone alkali metasomatism in the form of *albitization* is of widespread occurrence. Table 9-4, columns 1+2 show examples of a biotite schist and its corresponding albitization product.

At generally greater depth and in the immediate vicinity of granite massifs *potash feldspathization* is a common process which results in porphyroblasts of orthoclase or microcline. In Table 9-4, columns 3 and 4 typical examples are represented. This reaction finally leads to the processes of granitization (p. 289).

Calc-silicate Metasomatism. Limestones and dolomites because of their composition react as a rule rather intensely with migrating solutions of an acid character. The latter contain beside natural acids appreciable amounts of dissolved SiO_2 and hence different types of calc-silicates can be formed by reaction with the carbonates of the paleosome, e.g.

$$\underset{\text{Calcite}}{CaCO_3} + SiO_2 \rightleftharpoons \underset{\text{Wollastonite}}{CaSiO_3} + CO_2$$

$$\underset{\text{Dolomite}}{CaMg(CO_3)_2} + 2\,SiO_2 \rightleftharpoons \underset{\text{Diopside}}{CaMg(SiO_3)_2} + 2\,CO_2$$

As CO_2 is liberated during these processes and dissipates by way of capillaries out of the rock system, the reaction proceeds entirely to the right side of the equation. These transformations are often associated with the contacts of granite massifs but also occur in regionally metamorphic terrains.

Iron-magnesia-silicate Metasomatism. This type of metasomatism often occurs in conjunction with the foregoing type. Fe-Mg-silicates originate here together with Ca-silicates. This typical paragenesis is called *skarn* and can for instance be composed of the following minerals:

Hedenbergite	$CaFe(SiO_3)_2$
Andradite	$Ca_3Fe_2(SiO_4)_3$
Cordierite	$Mg_2Al_4Si_5O_{18}$
Anthophyllite	$(Mg, Fe)_7(Si_4O_{11})_2(OH)_2$
Cummingtonite	$(Fe, Mg)_7(Si_4O_{11})_2(OH)_2$

These silicates are often associated with oxidic or sulphidic ore deposits. The former are predominantly composed of magnetite or hematite, locally also Mn-oxides and -hydroxides, whilst the latter are composed of sulphides of Fe, Cu, Pb, Zn and others.

The origin of skarn ore deposits is still debated. It is possible that the ores, at least in part, represent primary concentrations which subsequently were transformed by metamorphism and metasomatism. Anatexis and granitization presumably had some additional influence.

Hydrothermal ore Metasomatism. Under hydrothermal conditions carbonates are replaced by reaction with ore-forming solutions. Metasomatic deposits of sulphidic ore minerals can thus originate, for instance of galena and sphalerite, but also of pyrite, marcasite, chalcopyrite, chalcocite or bornite. Under certain conditions (see p. 282) deposits of siderite or magnesite can be formed.

Boron Metasomatism. Supply of boron generally leads to the formation of tourmaline or, less frequently, of axinite or other rare B-silicates. Tourmalinization is a common process in the immediate vicinity of granite stocks, often occurring as an auto-metasomatic replacement of the granite by its own "pneumatolytic" agents rich in volatiles. It is generally associated with ore deposits of cassiterite and wolframite.

It should however be pointed out that the primary content of boron in pelitic sediments is often quite appreciable. In addition, it could be locally concentrated through secondary mobilization and then simulate metasomatic enrichment.

Fluorine Metasomatism. This type of metasomatism which is often connected with boron metasomatism leads to the formation of F-containing minerals, mainly fluorite besides silicates like topaz or lepidolite and zinnwaldite. Rocks of this type which predominantly consist of quartz beside F-minerals and often cassiterite are called *greisen.* The process leading to their formation is correspondingly called "*greisenization*".

Carbon Dioxide Metasomatism. In some respects this process of carbonatization of silicates is the opposite one to that of calc-silicate metasomatism originating from carbonates as quoted above. In the former process CO_2 is added to silicate rocks leading finally to the formation of carbonates of Mg, Fe, Ca, etc. and their corresponding mixed-crystal series.

Whilst calc-silicate metasomatism generally occurs at medium to high temperatures, carbon dioxide metasomatism takes place under relatively low temperature, i.e. in the hydrothermal range, where CO_2 as a dissolved gas phase is of great importance, as is instanced by the presence of carbonates of various composition in hydrothermal veins.

Carbon dioxide metasomatism is often met with in the hydrothermal transformation of peridotites and serpentinites according to the following equation:

$$\underset{\text{Serpentine}}{2\,Mg_3Si_2O_5(OH)_4} + 3\,CO_2 \rightleftharpoons \underset{\text{Talc}}{Mg_3Si_4O_{10}(OH)_2} + \underset{\text{Magnesite}}{3\,MgCO_3} + 3\,H_2O$$

At low temperatures and sufficiently high partial pressure of CO_2 even talc can be replaced by magnesite and quartz:

$$\underset{\text{Talc}}{Mg_3Si_4O_{10}(OH)_2} + 3\,CO_2 \rightleftharpoons \underset{\text{Magnesite}}{MgCO_3} + \underset{\text{Quartz}}{4\,SiO_2} + H_2O$$

In this way ultrabasic rocks are often superficially intersected and partially replaced by siliceous magnesite.

IV. Contact Metamorphism

All transformations that occur at or near the contact of an intruding magma are called contact metamorphism. Indeed this is but a special case of general metamorphism as hitherto described. Yet it is a very important one since it portrays the action of metamorphism in all its details within the very smallest domains.

At rather shallow depth of magmatic activity only minor transformations take place. Here the solidification of magma proceeds rapidly so that the physical influence on the country rock is restricted to but a few cm or mm in width. The chemical influence is frequently but very insignificant, too, since magmas near the surface are usually *degassed* and therefore there is no possibility of reaction of the country rocks with volatiles.

At medium depth of some km below the surface the proper region of contact metamorphism is reached. Granite massifs intruding at this depth as a rule produce contact aureoles 1—2 km in width where the country rocks are altered by recrystallization and mineral blastesis.

In pelitic rocks, for instance, the following zones are normally developed:

a) Spotted Slates

In the outermost zone of an aureole generally round or oval spots of a few mm in diameter make their appearance which can either be lighter or darker than their surroundings. Dark spots consist of accretions of the previously dispersed dark pigment of slates, i.e. very fine-grained Fe- or C-bearing compounds. Light spots consist mostly of extremely poikiloblastic cordierite and, in the case of Al-rich rocks, also of andalusite.

b) Spotted Mica-schists

In an intermediate zone the contactmetamorphic rocks generally become more and more coarse-grained. Depending on the primary composition of the rocks, biotite, cordierite, garnet or hornblende together with recrystallized quartz are formed. The structure is still predominantly schistose, however less marked than in the spotted slates.

c) Hornfels

In the innermost zone the schistosity disappears entirely, the structure becoming more or less massive. The newly formed minerals are often relatively large but interspersed with fine-grained relics of other minerals so that a "sieve-like" (poikiloblastic) fabric results. In spite of the large poikiloblasts the megascopic structure of hornfelses is hence relatively fine-grained. As a rule cordierite, garnet, hypersthene, sillimanite, spinel and corundum are the main constituents beside orthoclase, plagioclase and quartz.

Geochemically contact metamorphism can progress up to the very highest grades without any change of chemical composition (Goldschmidt, 1911; Niggli, 1950). It is termed thermometamorphism if the influence of pressure was insignificant. Table 9-5 shows a series of analyses of pelitic rocks in the sequence of increasing grade of contact metamorphism after Compton (1960). The characteristic trend of these analyses is the decrease of the H_2O content. The average content in H_2O of the normal rocks outside the aureole is 3.3%, in the outer zone of the aureole it is 2.5% and in its inner zone but 1.2%. In order to detect possible differences of other components the analyses were recalculated on the basis of no water content. It can thus be seen that almost all components show relatively good agreement. Solely K_2O exhibits a somewhat greater variation which, however, can still be accepted as incidental. Similarly the components FeO and Fe_2O_3 are rather variable and a moderate increase in total iron can probably be deduced.

On the other hand, internal migrations of mobile chemical components within the rocks of the contact aureole can of course occur. For instance, carbonaceous and siliceous layers often react with each other thus forming the so-called reaction skarns purely by "topochemical" reaction, i.e. reaction *in situ* without any addition of matter from outside (KENNEDY, 1959; ITO, 1962). Hence the distance of migration in these mainly solid rocks is generally but a few cm or dm. More extensive migrations occur by the action of aqueous solutions or gases issuing from the intruding magma and more or less intensely penetrating the contact aureole (LEAKE and SKIRROW, 1960; EVANS, 1964; FLOYD, 1965).

Summing up it can thus be said that contact metamorphism is governed by the same principles geochemically as regional metamorphism. The essential difference can be seen in the fact that during contact metamorphism mineral facies are formed

Table 9-5. *Average compositions (calculated free of H_2O, CO_2 and S) and standard deviations (s) of 13 samples of normal phyllites (1), of 13 metamorphic rocks from the outer zones of a contact aureole (2), and of 13 corresponding rocks from the inner parts of the aureole, Santa Rosa Range, north-central Nevada.* (After COMPTON, 1960.)

	1 Normal phyllites (13 samples)		2 Contact metamorphic rocks of the outer zones (13 samples)		3 Contact metamorphic rocks of the inner zones (13 samples)	
	mean	*s*	mean	*s*	mean	*s*
SiO_2	64.3	2.6	63.7	1.7	64.0	1.3
TiO_2	1.0	0.05	1.05	0.1	0.99	0.03
Al_2O_3	20.0	0.7	20.3	1.6	19.5	1.2
Fe_2O_3	1.2	5.34	1.7	0.67	0.93	0.50
FeO	5.0	1.1	4.8	1.2	5.9	0.9
MnO	0.08	0.02	0.08	0.02	0.08	0.01
MgO	2.2	0.8	2.2	0.7	2.4	0.7
CaO	1.3	1.3	1.1	0.7	1.5	1.1
Na_2O	1.4	0.4	1.5	0.4	1.5	0.3
K_2O	3.5	0.7	3.2	0.5	3.1	0.3
P_2O_5	0.17	0.03	0.18	0.03	0.15	0.04
Total Fe as Fe_2O_3	6.7	1.1	7.0	0.9	7.5	0.9
H_2O	3.3		2.5		1.2	

that are characterized by relatively low pressure but high temperature (see Fig. 9-1). Hence the most frequently occurring facies are the pyroxene-hornfels facies and the hornblende-hornfels facies. At very low pressure but high temperature the sanidinite facies is formed. It is mainly restricted to xenoliths in lavas and local contact phenomena in the immediate vicinity of near-surface intrusions.

At high pressure and high temperature the contact facies merge with the corresponding facies of regional metamorphism (PITCHER and READ, 1960). The petrographical and geochemical differences between the country rocks formed by regional metamorphism and the contact metamorphic rocks hence become less and less perceptible. Finally no difference exists at all due to increasing alkali metasomatism and feldspar blastesis. Even granitic rocks can merge imperceptibly into their erstwhile country rocks now transformed to migmatites by granitization, anatexis, etc.

V. Migmatites, Anatexis, Granitization

At high temperature combined with high pressure, so that volatiles like H_2O etc. are sufficiently retained, rocks of suitable composition begin to show phenomena of partial melting. According to the experimental results of BOWEN, GORANSON, TUTTLE, YODER, WINKLER, KHITAROV, WYART rocks of granitic composition melt at temperatures and pressures very probably attained within the crust during progressive regional metamorphism (Fig. 9-1, curve "minimum melting temperature"). Thus the eutectic melting temperature of the system quartz-albite-potash feldspar-H_2O at 4 kb pressure was found to be about 650° C. If other volatiles, especially HF, are present as well, the minimum melting temperature, depending on the amount of volatiles added, is further reduced even to 550° C.

These conditions are realized in the high-temperature range of the amphibolite facies. It is interesting to note that other facies marked by high temperature and high pressure, e.g. the granulite and eclogite facies, generally exhibit but slight evidence of melting in natural rocks. The reason for this presumably has to be sought in the relatively small H_2O content or in the often basic or even ultrabasic composition of the rocks involved. Of course rocks rich in Mg-Fe-Al components have an appreciably higher melting interval than rocks rich in alkali silicates (feldspar) and quartz.

In natural rocks metamorphosed under the conditions of the high-temperature subfacies of the amphibolite facies, i.e. the sillimanite-almandine-orthoclase subfacies, rock portions of pegmatitic character often occur that cannot be introduced from outside but must have been formed *in situ*. Mixed rocks consisting of such pegmatitic or granitic portions beside their former parent rock are termed *migmatites*. The forms of interpenetration of paleosomes and their pegmatoid or granitoid neosomes are very variable. Hence many different terms, generally of Greek origin, have been introduced to describe the complex megascopic appearance of migmatites, e.g. agmatites, diktyonites, embrechites, ophthalmites, nebulites, etc. The differences between these types of penetration are mainly due to the varying tectonic action on the heterogeneous rock system and hence are geochemically insignificant.

In Table 9-6 some examples of chemical analyses of rocks from the Black Forest after MEHNERT (1951, 1962/63) are given. No. 1 is the parent rock from which the migmatites originated, a rather homogeneous biotite-plagioclase gneiss. No. 2 shows the composition of the newly formed pegmatoid portions consisting mainly of acid plagioclase and quartz. Along with these pegmatoid veins and patches there are biotite-rich rims accompanying the former so closely that a simultaneous origin is evident (No. 3). The megascopic proportion between the pegmatoid "leucosomes" and the biotite-rich "melanosomes" is on the average about 3:1. If the respective chemical compositions of both portions are added according to this ratio, the bulk composition of the whole migmatite is obtained (Nr. 4 = 2+3, in the ratio of 3:1). Comparison of the sum (No. 4) with the composition of the parent rock (No. 1) demonstrates only minor differences.

From this evidence it can be concluded that the pegmatoid leucosomes were in fact "mobilized" from the parent rock and that the biotite-rich melanosomes remained in situ as non-mobilized components. Hence the pegmatoid portions can generally be referred to as *mobilizates* whilst the biotite-rich portions can be termed *restites*. This process of *anatectic differentiation* into mobilizates and restites is evidently widely distributed in the deep parts of the crust and it can safely be assumed that many pegmatites, granites or related rocks originated in this way.

On the whole the composition of pegmatoid mobilizates closely approximates to the cotectic ratio of the minerals involved. It can thus be postulated that the mobilization process in fact was caused by initial melting. It has however to be taken into account that fusion in natural rocks will always permit a mechanical mixing of unmolten material with the adjacent melt. Hence it is often difficult to determine the unadulterated composition of mobilizates in natural rocks after reconsolidation.

Table 9-6. *Chemical analyses of a migmatite composed of leucocratic pegmatoid mobilizates (1) and melanocratic biotite-rich restites (2). The bulk composition of the migmatite (3) is compared with that of the parent rock, a biotite-plagioclase gneiss from the Black Forest*[a]. (After MEHNERT, 1951, p. 181, calculated free of P_2O_5 and H_2O^-.)

No.		SiO_2	TiO_2	Al_2O_3	Fe_2O_3	FeO	MnO	MgO	CaO	Na_2O	K_2O	H_2O^+	Σ
1	Leucocratic, pegmatoid mobilizates	72.72	0.05	16.78	0.28	0.33	0.04	0.0	3.48	4.48	1.09	0.75	100.0
2	Melanocratic, biotite-rich restites	38.06	2.50	18.80	4.10	14.52	0.23	9.79	2.28	1.20	5.51	3.01	100.0
3	Bulk composition of the migmatite (1 + 2) in the ratio of 3:1	64.05	0.66	17.30	1.25	3.89	0.08	2.44	3.18	3.64	2.19	1.32	100.0
4	Parent rock: Biotite-plagioclase gneiss	63.36	0.65	18.22	1.83	4.04	0.04	1.25[b]	2.86	3.96	1.90	1.89	100.0
	Difference between migmatite (3) and parent rock (4)												
	excess	0.69	0.01	—	—	—	0.04	1.19	0.32	—	0.29	—	—2.54
	deficit	—	—	0.92	0.58	0.15	—	—	—	0.32	—	0.57	—2.54

[a] Location: Abandoned quarry 0.5 km south of Urenkopf near Haslach, central Black Forest.
[b] This figure is obviously too low.

The corresponding non-mobilized portions, the restites, are generally enriched in biotite, cordierite, sillimanite, garnet, spinel and corundum. This paragenesis of Mg-Fe-Al-rich minerals can also result from the transformation of pelitic sediments and hence genuine restites have sometimes been interpreted as such. Their occurrence next to mobilizates, however, clearly points to their origin. Nevertheless in the deep parts of the crust restites may occur that have lost their corresponding mobilizates by some sort of squeezing-out process.

The close association of pegmatitic or granitic rocks with rocks of generally mafic composition was interpreted in a slightly different way by HOLMES and REYNOLDS (1947), REYNOLDS (1946, 1947, 1958) and LAPADU-HARGUES (1945). On the basis of geochemical investigations the authors emphasized that essentially two processes can be distinguished:

a) *Granitization*, i.e. the formation of granites from non-granitic rocks by the action of an "acid" front, viz. the metasomatic enrichment of SiO_2, Na_2O and K_2O by migrating "emanations".

b) *Basification*, i.e. the formation of melanocratic rocks by the action of a "basic" front, viz. the metasomatic enrichment of MgO, FeO resp. Fe_2O_3, MnO, TiO_2 and P_2O_5.

This was for instance demonstrated on the basis of inclusions in the granites of Flamanville and of Dartmoor (REYNOLDS, 1946). Compared with the normal country rocks the inclusions have undergone a basification, whilst the immediately adjoining granitic rocks presumably derived from them show strong granitization geochemically.

Regarding the origin of acid and basic fronts both are closely related on principle. During granitization the "basic" components are supposed to be liberated by replacement and therefore precede the front of the somewhat slower migrating granitic components. Hence rocks as a rule are at first basified and subsequently granitized.

In contrast to this, RAMBERG (1952) pointed out that the elements Ca, Fe and Mg are generally less mobile than are granitic elements. In consequence he assumed that "basification" as demonstrated above originates by the emigration of the rather mobile granitic elements, the so-called "granitophile" elements (RANKAMA and SAHAMA, 1950). The non-granitic elements, correspondingly called "granitophobe" elements, are hence enriched in the remaining rock portion, called "restites" above.

If no geochemical change took place at all in rocks due to their chemical resistance against attacking substances, these inert rock types are called "resisters" according to READ (1944, 1948). Often quartzites, amphibolites or other ultrabasic rocks remain completely unchanged amidst highly mobilized migmatites.

A detailed geochemical investigation concerning the migration of elements during metamorphism and granitization was carried out by ENGEL and ENGEL (1958/60) on a rock series of the Adirondacks. The primary metamorphic rock consists of a biotite-oligoclase gneiss of very even composition (Table 9-7). With increasing metamorphism (Nr. 2→4) it is enriched in Al, Fe^{+2}, total Fe, Mg and Ca as well as in the trace elements Cr, Ga, Ni and V. In contrast to this, the concentrations of K, Si, Fe^{+3}, H_2O and Ba decrease. This is interpreted as a metamorphic basification or "degranitization" in which Si, K and H_2O were mobilized whilst the more basic constituents were retained.

Granitization (Nr. 5→6) took place with an increase mainly of K and of the trace elements Ba and Pb as well as decreases of Ti, Fe^{+3}, Fe^{+2}, Mg, Ca and H_2O together with the trace elements Co, Cr, Mn, Ni, Sr, and V.

Mineralogically granitization is usually characterized by potash feldspar blastesis (see Table 9-4, No. 3 and 4). Frequently biotite is formed at the expense of other mafic minerals, for instance hornblende, garnet, etc. (see Table 9-8).

Chemically the process is characterized by K-(Na)-metasomatism, as instanced by BILLINGS (1938), CHAO (1951), ESKOLA (1956), HAYAMA (1962), HÄRME (1955—65). In this way rocks can finally be formed that resemble magmatic or anatectic granites very closely.

In principle granites can therefore originate in three ways:

a) *Magmatic granites* by crystallization from a granitic magma.

b) *Anatectic granites* by partial re-melting of a pre-existing rock whereby mafic restites can be formed.

Table 9-7. *Average chemical compositions of major and minor elements of paragneisses of the Adirondack Mts. during progressive metamorphism including basification and granitization.* (After Engel and Engel, 1958)

	Series showing increasing basification (degranitization)				Exemples of increasing granitization	
	1	2	3	4	5	6
SiO_2	70.25	69.61	69.16	65.97	69.25	72.31
TiO_2	0.67	0.74	0.70	0.64	0.63	0.31
Al_2O_3	14.14	14.39	15.09	16.83	14.63	14.76
Fe_2O_3	0.55	0.47	0.33	0.36	0.70	0.33
FeO	3.83	4.35	4.24	4.92	3.45	1.76
MnO	0.05	0.06	0.05	0.07	0.05	0.02
MgO	1.76	1.75	1.74	2.25	1.74	0.79
CaO	2.20	2.56	2.85	3.41	1.97	1.40
Na_2O	3.43	3.45	3.62	3.70	3.06	3.57
K_2O	2.40	2.06	1.70	1.46	3.79	4.27
H_2O^+	0.72	0.56	0.52	0.39	0.71	0.44
B	10	10	10	12	22	12
Ba	612	600	650	492	750	910
Co	8	10	8	7	7	4
Cr	35	25	40	56	16	4
Cu	16	12	15	12	4	19
Ga	11	12	12	14	10	12
Mn	356	340	350	341	434	260
Ni	15	10	12	21	12	4
Pb	12	10	12	15	21	38
Sr	310	550	500	304	320	270
Ti	2,800	3,000	2,200	3,000	1,800	750
V	56	50	65	81	53	24
Zr	171	180	220	176	172	150

1: Mean values of 24 specimens of para-biotite-plagioclase gneiss from the Emeryville area.

2: Mean values of 19 specimens of garnet-bearing biotite-plagioclase gneiss from the Edwards area.

3: Mean values of 16 specimens of garnet-bearing biotite-plagioclase gneiss in an advanced stage from the Russel area.

4: Mean values of 14 specimens of garnet-biotite-plagioclase gneiss from the Colton area.

5: Mean values of 4 specimens of incipiently granitized gneiss from the Emeryville area.

6: Mean values of 3 specimens of strongly granitized gneiss (granitic gneiss) from the Emeryville area.

c) *Metasomatic granites* by migration of granitophile components (see p. 291) and their reaction with the pre-existing rock. Granitophobe elements may be removed and be concentrated elsewhere.

It must be emphasized that the term "magma" can be applied to any melt of geological extent, irrespective of its mode of formation. Anatectic granites can thus merge petrographically and genetically into magmatic ones, if re-melting has progressed sufficiently, and the magmatic phase has become dominant. At this stage relics disappear more or less completely, rendering any genetic interpretation that might have been possible rather difficult.

Table 9-8. *Mineralogical and chemical composition of a successively granitized gabbro inclusion in microcline granite from Karlholm, southern Finland.* (After HÄRME and LAITALA, 1955)

	1	2	3		1	2	3
Plagioclase	48.9	45.7	34.6	SiO_2	49.24	47.21	50.69
Amphibole	32.3	—	—	TiO_2	2.42	2.67	2.62
Chlorite (from amphibole)	0.2	33.2	—	Al_2O_3	16.38	16.64	15.77
Biotite	7.4	8.5	40.9	Fe_2O_3	1.23	1.32	1.41
Quartz	9.6	10.7	19.1	FeO	10.85	8.96	10.85
Garnet	—	—	3.6	MnO	0.18	0.22	0.15
Accessories	1.6	1.9	1.8	MgO	5.51	4.96	5.38
				CaO	7.24	6.12	3.84
				Na_2O	0.56	0.43	0.37
				K_2O	1.68	2.87	4.20
				P_2O_5	0.38	0.37	0.37
				CO_2	1.06	2.81	0.87
				H_2O^+	2.59	4.55	2.86
				H_2O^-	0.49	1.08	0.40
					99.81	100.21	99.78

1: Original gabbro from the center of the inclusion.
2: Transitional rock between center and rim.
3: Dark rim surrounding the inclusion.

There exist granitic massifs where all the three modes of origin are in fact realized, for instance from the core towards the periphery according to the modes 1→3. With regard to the granite massifs now exposed at the Earth's surface the quantitative amount of the three types of origin is yet thoroughly debated and certainly will remain unclarified in many cases due to multiple geological implications in space and time. It can however be assumed that on principle the proportion of the three genetic types increases from 3→1 with increasing depth of the crust.

References

ANGEL, F.: Retrograde Metamorphose und Diaphthorese. Neues Jahrb. Mineral., Abhandl. **102**, 123—176 (1965).

BARKER, F.: Reaction between mafic magmas and pelitic schist, Cortlandt, New York. Am. J. Sci. **282**, 614—634 (1964).

BILLINGS, M.: Regional metamorphism of the Littleton-Moosilauke area, New Hampshire. Bull. Geol. Soc. Am. **48**, 2, 463—565 (1937).

— Introduction of potash during regional metamorphism in western New Hampshire. Bull. Geol. Soc. Am. **49**, 1, 289—302 (1938).

BIRCH, F.: Physics of the crust. Geol. Soc. Am., Spec. Papers **62**, 101—118 (1955).

BUTLER, B. C.: Metamorphism and metasomatism of rocks of the Moine series by a dolerite plug in the Glenmore, Ardnamurchan. Mineral. Mag. **32**, 866—897 (1961).

CHAO, E. C. T.: Granitization and basification. Norsk Geol. Tidsskr. **29**, 84—107 (1951).

CLARK, S. P.: Temperatures in the continental crust. Rept. Geophys. Lab., 187—190 (1960/61).

CLARKE, F. W.: The data of geochemistry. U.S. Geol. Survey, Bull. **770**, 1—841 (1924).

Compton, R. R.: Contact metamorphism in Santa Rosa Range, Nevada. Bull. Geol. Soc. Am. **71**, 1383—1416 (1960).

Den Tex, E.: A commentary on the correlation of metamorphism and deformation in space and time. Geol. en Mijnbouw **42**, 170—176 (1963).

Engel, A. E. J., and C. G. Engel: Progressive metamorphism and granitization of the major paragneiss, nothwest Adirondack Mountains, New York. Bull. Geol. Soc. Am. **69**, 1369—1414 (1958); **71**, 1—57 (1960).

Eskola, P.: The mineral facies of rocks. Norsk Geol. Tidsskr. **6**, 143—194 (1921).

— Postmagmatic potash metasomatism of granite. Compt. rend. soc. géol. Finlande **29**, 85—100 (1956).

Eugster, H. P.: Reduction and oxidation in metamorphism. Research in Geochemistry. New York 1959.

— Spurenelemente in einigen metamorphen Gesteinen des Aarmassivs. Eclogae Geol. Helv. **52**, 421—434 (1959).

Evans, B. W.: Fractionation of elements in the pelitic hornfelses of the Cashel-Lough Wheelaun intrusion, Connemora, Eire. Geochim. et Cosmochim. Acta **28**, 127—156 (1964).

Fersman, A.: Geochemische Migration der Elemente. Abhandl. prakt. Geol. **18**, 1—116; **19**, 1—86 (1929).

Floyd, P. A.: Metasomatic hornfelses of the Land's End Aureole at Tater-du, Cornwall. J. Petrology **6**, 223—245 (1965).

Goldschmidt, V. M.: Die Kontaktmetamorphose im Kristianiagebiet. Skrifter Videnskap. Kristiania, Math.-nat. Kl. **1**, 1—483 (1911).

— Die Injektionsmetamorphose im Stavangergebiet. Videnskapsselskapets-Skrifter Oslo, Nr. 10, 1—142 (1920).

— Geochemistry. Oxford 1954.

Green, J. C.: High level metamorphism of pelitic rocks in northern New Hampshire. Amer. Mineralogist **48**, 991—1023 (1963).

Grubenmann, U., u. P. Niggli: Die Gesteinsmetamorphose, S. 1—538. Zürich 1924.

Härme, M.: Examples of the granitization of gneisses. Bull. comm. géol. Finlande **184**, 41—58 (1959).

— An example of anatexis. Bull. comm. géol. Finlande **204**, 113—125 (1962).

— On the potassium migmatites of southern Finland. Bull. comm. géol. Finlande **212**, 1—43 (1965).

—, and M. Laitala: An example of granitization. Comp. rend. soc. géol. Finlande **28**, 95—99 (1955).

Harry, W. T.: The migmatites and feldspar-porphyroblast rock of Glen Dessarry, Inverness-shire. Quart. J. Geol. Soc. London **107**, 137—158 (1951).

— The composite granitic gneiss of western Ardgour, Argyll. Quart. J. Geol. Soc. London **109**, 285—309 (1953).

Hayama, Y.: Metasomatic transfer of potassium and aluminium in the Ryoke regional metamorphism of the Komogane district, Nagano Pref., Central Japan. Japan J. Geol. and Geography **33**, 79—86 (1962)

Hietanen, A.: On the geochemistry of metamorphism. J. Tenn. Acad. Sci. **29**, 286—296 (1954).

Higazy, R. A.: A geochemical study of the regional metamorphic zones of the Scottish Highlands. Compt. rend. intern. geol. congr. Alger, Sect. 13, 415—430 (1954).

Hoenes, D.: Magmatische Tätigkeit, Metamorphose und Migmatitbildung im Grundgebirge des südwestlichen Schwarzwaldes. Neues Jahrbuch Mineral., Abhandl. A **76**, 153—256 (1940).

Holmes, A., and D. L. Reynolds: A front of metasomatic metamorphism in the Dalradian of Co. Donegal. Compt. rend. Soc. géol. Finlande **20**, 25—65 (1947)

Ito, K.: Zoned skarn of the Fujigatani Mine, Yamaguchi Prefecture. Japan. J. Geol. and Geography **33**, 169—190 (1962).

Kennedy, W. Q.: The formation of a diffusion reaction skarn by pure thermal metamorphism. Mineral. Mag. **32**, 26—31 (1959).

KORZHINSKY, D. S.: Outline of metasomatic processes, 2nd ed. Moskwa: Akad. Nauk. U.S.S.R. 1955 [Russian].
LAPADU-HARGUES, P.: Sur l'existence et la nature de l'apport chimique dans certaines séries cristallophylliennes. Bull. soc. géol. France, Ser. XV, **5**, 255—310 (1945).
LEAKE, B. E., and G. SKIRROW: The pelitic hornfelses of the Cashel-Lough Wheelaun intrusion, County Galway, Eire. J. Geol. **68**, 23—40 (1960).
LUTH, W. C., R. H. JAHNS, and O. F. TUTTLE: The granite system at pressures of 4 to 10 kilobars. J. Geophys. Research **69**, 759—773 (1964).
MARMO, V.: On granites. Bull. comm. géol. Finlande **201**, 3—77 (1962).
MASON, B.: Principles of geochemistry. New York: John Wiley & Sons 1952—1958.
MEHNERT, K. R.: Zur Frage des Stoffhaushalts anatektischer Gesteine. Neues Jahrb. Mineral., Abhandl. **82**, 155—198 (1951).
— Neue Ergebnisse zur Geochemie der Metamorphose. Geol. Rundschau **51**, 384—394 (1961).
— Petrographie und Abfolge der Granitisation im Schwarzwald I—IV. Neues Jahrb. Mineral., Abhandl. **85**, 59—140 (1953); **90**, 39—90 (1957); **98**, 208—249 (1962); **99**, 161—199 (1963).
MIYASHIRO, A.: Evolution of metamorphic belts. J. Petrology **2**, 277—311 (1961).
NANZ, R. H.: Chemical composition of pre-Cambrian slates with notes on the geochemical evolution of lutites. J. Geol. **61**, 51—64 (1953).
NIGGLI, P.: Some hornfelses from Saxony and the problem of metamorphic facies. Amer. Mineralogist **35**, 867—876 (1950).
PETERLONGO, J.: Etude des phénomènes métasomatiques dans les amphibolites des monts du Lyonnais. Bull. soc. géol. France, Ser. VI, **5**, 361—374 (1955).
PITCHER, W. S., and G. W. FLINN: Controls of metamorphism. Edinburgh and London 1965, 368 p.
—, and H. H. READ: The aureole of the Main Donegal Granite. Quart. J. Geol. Soc. London **116**, 1—36 (1960).
POLDERVAART, A.: Petrological calculations in metasomatic processes. Am. J. Sci. **251**, 481—503 (1953).
RAMBERG, H.: The origin of metamorphic and metasomatic rocks. Chicago: Chicago University Press 1952.
RANKAMA, K., and TH. G. SAHAMA: Geochemistry: Chicago University Press (1960).
READ, H. H.: Meditations on granite. Proc. Geologists' Assoc. (Engl.) **55**, 45—93 (1944).
— Granites and granites. Geol. Soc. Am., Mem. **28**, 1—19 (1948).
REYNOLDS, D. L.: The albite-schists of Antrim and their petrogenetic relationship to Caledonian orogenesis. Proc. Roy. Irish Acad. **48**, 43—66 (1942/43).
— The sequence of geochemical changes leading to granitization. Quart. J. Geol. Soc. London **102**, 389—446 (1946).
— The granite controversy. Geol. Mag. **84**, 209—223 (1947).
— Granite: Some tectonic, petrological, and physico-chemical aspects. Geol. Mag. **95**, 378—396 (1958).
ROSENBUSCH, H.: Elemente der Gesteinslehre, 1. Ausg. Stuttgart 1898, 4. Ausg. (1923).
SCHUILING, R. D.: Thermal gradient of the earth. Koninkl. Ned. Akad. Wetenschap., Ser. B, **60**, No .3, 212—219 (1957).
SEDERHOLM, J. J.: On migmatites and associated pre-Cambrian rocks of southwestern Finland. Bull. comm. géol. Finlande, I: **58**, 1—153 (1923); II: **77**, 1—143 (1926); III: **107**, 1—68 (1934).
— The average composition of the earth's crust in Finland. Bull. comm. géol. Finlande **70**, 1—20 (1925).
SHAW, D. M.: Trace elements in pelitic rocks. Part I: Variations during metamorphism. Part II: Geochemical variations. Bull. Geol. Soc. Am. **65**, 1151—1182 (1954).
— Geochemistry of pelitic rocks. Part III: Major elements and general geochemistry. Bull. Geol. Soc. Am. **67**, 919—934 (1956).
SIMONEN, A.: Stratigraphy and sedimentation of the Svecofennidic, early Archean supracrustal rocks in southwestern Finland. Bull. comm. géol. Finlande **160**, 1—64 (1953).

TAYLOR, S. R.: The origin of some New Zealand metamorphic rocks as shown by their major and trace element composition. Geochim. et Cosmochim. Acta **8**, 182—197 1955).

TUOMINEN, H. V., and T. MIKKOLA: Metamorphic Mg-Fe-enrichment in the Orijärvi Region as related to folding. Compt. rend. soc. géol. Finlande **23**, 67—92 (1950).

TUTTLE, O. F.: Degré géothermique et magmas granitiques. Sciences de la terre, Nancy, 87—103 (1955).

—, and N. L. BOWEN: Origin of granite in the light of experimental studies in the system $NaAlSi_3O_8$—$KAlSi_3O_8$—H_2O. Geol. Soc. Am. ,Mem. **74**, 1—153 (1958).

WENK, E.: Berechnung von Stoffaustauschvorgängen. Schweiz. mineral. petrog. Mitt. **34**, 309—318 (1954).

WINKLER, H. G. F.: Die Genese der metamorphen Gesteine, S. 1—218. Berlin-Heidelberg-New York: Springer 1965, 2. Aufl. 1967.

—, u. H. v. PLATEN: Experimentelle Gesteinsmetamorphose I—VI. Geochim. et Cosmochim. Acta **13**, 42—69 (1957); **15**, 91—112 (1958); **18**, 294—316 (1960); **24**, 48—69, 250—259 (1961); **26**, 145—180 (1962).

YODER, H. S.: Role of water in metamorphism. Geol. Soc. Am. Spec. Papers **62**, 505—524 (1955).

ZWART, H. J.: Metamorphic history of the Central Pyrenees. Part I: Arize, Trois Seigneurs and Saint-Barthelemy Massifs. Leidse Geol. Mededel. **22**, 351—490 (1959).

— On the determination of polymetamorphic mineral associations, and its application to the Bosost area (Central Pyrenees). Geol. Rundschau **52**, 38—65 (1962).

Chapter 10

K. K. Turekian

THE OCEANS, STREAMS, AND ATMOSPHERE

In this chapter an attempt is made to bring together the most recent data on the chemical properties of the ocean, streams and the atmosphere. In addition to the chemical data, estimates of the physical properties of the systems are included, such as the dimensions of the ocean basins and the mean annual runoff of the world, since most geochemical calculations involving the cycles of the elements require flux rates and reservoir dimensions for meaningful material balances.

I. The Oceans

a) General Features of the Ocean Basins

Until about 1920 soundings were made exclusively by means of *sounding lines*. These were leaded hemp lines until the latter part of the nineteenth century after which metal wire was used. Kossina in 1921 (see Sverdrup, Johnson and Fleming, 1942) made a catalog of ocean depth distribution using these early soundings.

After 1920 more and more soundings were made by sonic devices and since World War II the topography of the ocean floor has been thoroughly investigated by means of continuously operating precision depth recorders. The limiting accuracy for ocean floor mapping is set by the accuracy of navigation devices. More and more, land controlled radio- and micro-wave frequency triangulation and satellite fixes are assisting in this problem.

Recently Menard and Smith (1966), using the most current sounding data have redone the earlier work of Kossina. The results are shown in Table 10-1.

The major geomorphic provinces of the ocean basins as typified by the Atlantic Ocean according to Heezen, Tharp and Ewing (1959) are shown in Fig. 10-1. The oceanic ridge system is a world-wide feature as is seen in Fig. 10-2 and is an area of intense seismic and thermal activity. Recent results indicate that the symmetric patterns of magnetic anomolies about the ridge axes are due to ocean floor spreading (Pitman and Heirtzler, 1966; Vine, 1966).

b) Salinity and Temperature of Ocean Water

The salinity of ocean water is approximately 35 parts per thousand. The proportions of the cations and anions do not vary greatly in the open ocean although significant departures from mean oceanic composition may occur in coastal waters. The relative concentrations of the various major components are listed in Table 10-2.

Table 10-1. *Areas, volumes, and mean depths of the ocean basins* (MENARD and SMITH, 1966)

Oceans and adjacent seas	Area 10^6 km^2	Volume 10^6 km^3	Mean depth m
Pacific	166.241	696.189	4,188
Asiatic Mediterranean	9.082	11.366	1,252
Bering Sea	2.261	3.373	1,492
Sea of Okhotsk	1.392	1.354	973
Yellow and East China Seas	1.202	0.327	272
Sea of Japan	1.013	1.690	1,667
Gulf of California	0.153	0.111	724
Pacific and adjacent seas, total	181.344	714.410	3,940
Atlantic	86.557	323.369	3,736
American Mediterranean	4.357	9.427	2,164
Mediterranean	2.510	3.771	1,502
Black Sea	0.508	0.605	1,191
Baltic Sea	0.382	0.038	101
Atlantic and adjacent seas, total	94.314	337.210	3,575
Indian	73.427	284.340	3,872
Red Sea	0.453	0.244	538
Persian Gulf	0.238	0.024	100
Indian and adjacent seas, total	74.118	284.608	3,840
Arctic	9.485	12.615	1,330
Arctic Mediterranean	2.772	1.087	392
Arctic and adjacent seas, total	12.257	13.702	1,117
Totals and mean depths	362.033	1,349.929	3,729

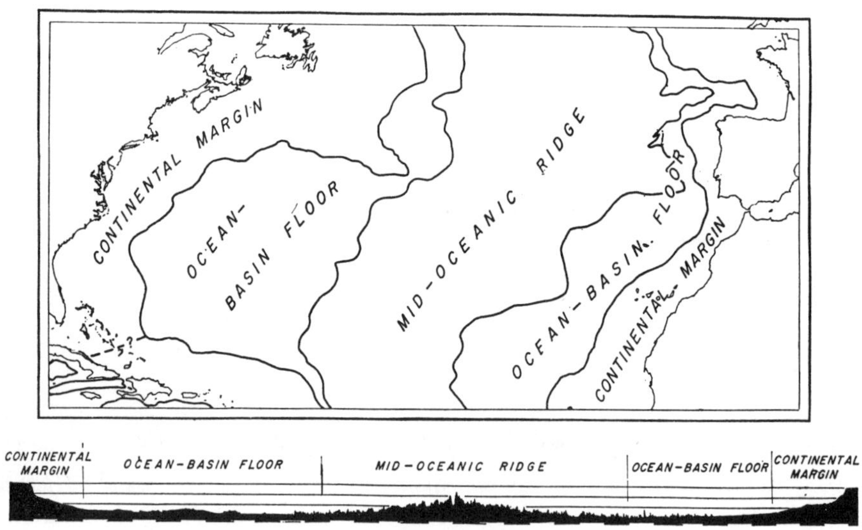

Fig. 10-1. The major geomorphic provinces of the Atlantic Ocean Basin (HEEZEN, THARP and EWING, 1959)

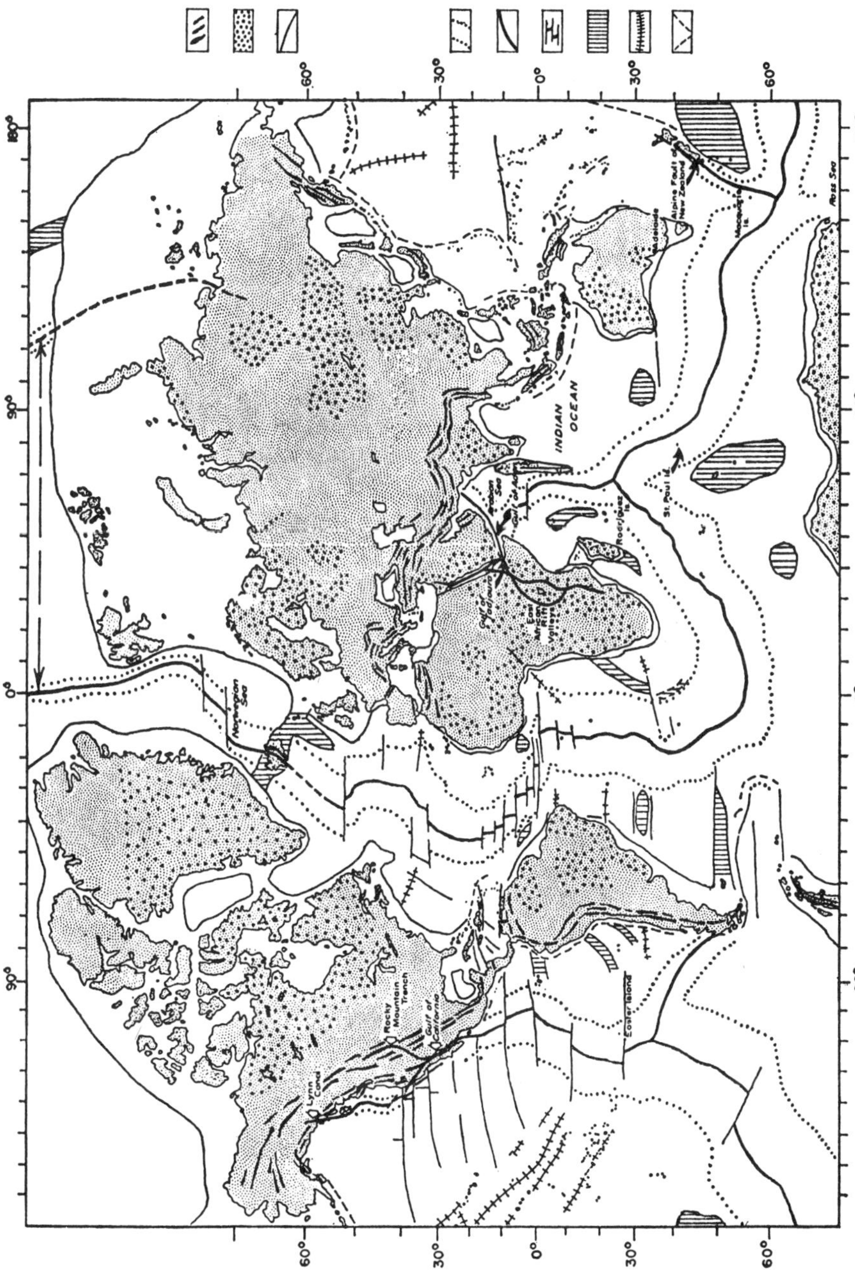

Fig. 10-2. The world-wide oceanic ridge system (HEEZEN, 1963)

Salinity will increase if the loss of water from sea water exceeds the gain. Water can be removed from sea water by evaporation and by the formation of pack ice. Water is supplied to sea water by atmospheric precipitation, surface runoff and the melting of pack ice.

The chloride ion is the most important anion in sea water and on the basis of the constancy of ionic proportions in sea water a measure of the chloride content

Table 10-2. *The concentrations of the major components of sea water (for a salinity of 35‰)* (Culkin 1965)

	g/kg		g/kg
Chloride	19.353	Bicarbonate	0.142
Sodium	10.76	Bromide	0.067
Sulfate	2.712	Strontium	0.008
Magnesium	1.294	Boron	0.004
Calcium	0.413	Fluoride	0.001
Potassium	0.387		

can be related to the salinity by the approximate equation: $S‰ = 1.805\ Cl‰ + 0.30$, where the salinity, S, and chlorinity, Cl, are in g/kg (‰) of sea water. The most efficient and accurate measure of salinity however, is made with a precision salinometer. The conductivity of a saline solution is measured and, after correction for temperature, converted to salinity based on a standard sample.

The temperature of the surface of the ocean reflects the latitude and time of year at which it is sampled. Imposed on this is the transport of water north or south as the result of the wind-driven gyre systems of the surface waters. Hence the Gulf Stream, for example, has a high temperature even at high latitudes and the southward travelling currents along the California coast, as an opposite example, transport cold water almost to the tropics.

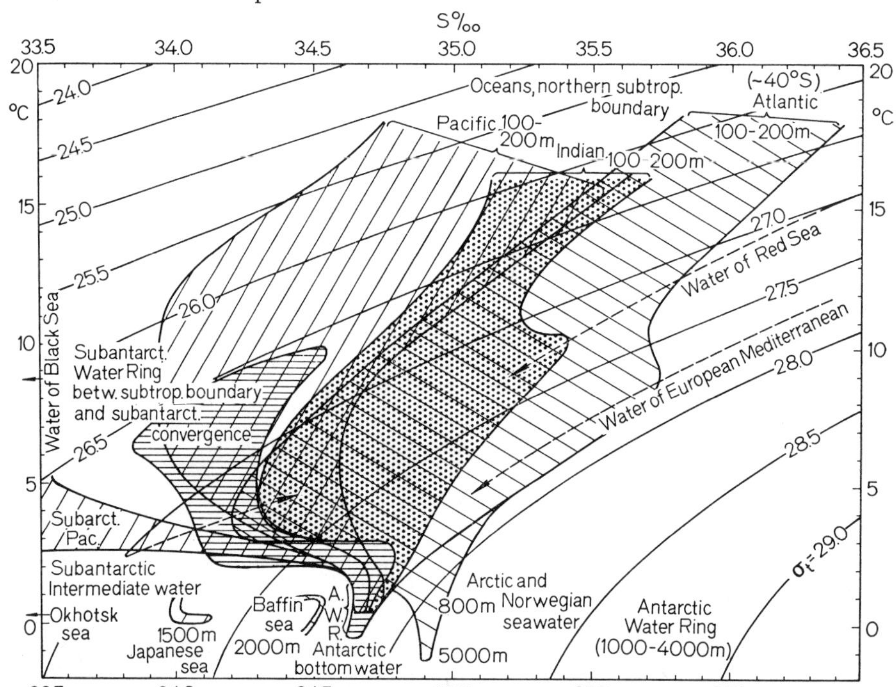

Fig. 10-3. Salinity-temperature diagram showing lines of equal density (σ_T). σ_T is equal to the density of the ocean waters at 1 atmosphere pressure minus one then multiplied by one thousand. (a density of 1.028 thus has a $\sigma_T = 28.0$.) (Dietrich, 1963)

The deep waters of the oceans have their origins in high latitudes — around Antarctica and around Greenland — hence the temperature of ocean water below about 1,000 meters is about 2—4° C.

The thermal profile of a column of water shows the following features: Down to a depth of about 100 meters, the depth of mixing, the temperature remains approximately constant. Between 100 meters and about 1,000 meters the temperature decreases monotonically to the low value typical of deep water as a result of mixing of the mixed layer and the deep layer by eddy diffusion and possibly advection. This is called the region of the thermocline.

The temperature and salinity of a volume of sea water determines its density. Since the eddy diffusion and advection rates are slow enough to permit the observation of gradients it is evident that water acquiring a certain combination of salinity and temperature will have a density which forces it to be distributed in a stratified sequence as a result of the Earth's gravity field. Lines of equal density on a temperature-salinity diagram are shown in Fig. 10-3. The displacement caused by sinking dense water at the high latitudes drives the deep circulation of the oceans.

c) Water Types and Oceanic Mixing

A particular *water type* originates in a specific part of the ocean as the result of local conditions which determine the temperature and salinity. The bottom waters of most of the ocean basins, for instance, are the result of mixing of a water type formed by processes occurring in the Antarctic winter in the Weddell sea with other adjacent waters. The freezing out of pack ice results in a cold brine which on sinking through the water mixes with water supplied to the Antarctic from other sources to produce a distinctive bottom water.

The mixing of water types produces a continuous distribution on a temperature-salinity diagram and from this *water masses* are delineated. Source water from the Mediterranean, for example, penetrates into the North Atlantic and can be recognized by a continuous plot in a temperature-salinity diagram as it mixes with surrounding North Atlantic water.

The formation and modification of seawater types can be seen in the distribution of the oxygen and hydrogen isotopes as shown by Redfield and Friedman (1965) and Craig and Gordon (1965). Fig. 10-4 is taken from the paper by Craig and Gordon, who use an oxygen isotope-salinity plot to define the path of sea water types and masses. The variations observed are based on three major fractionation processes: (1) the ratio of evaporation to precipitation results in a relationship between salinity and oxygen isotopes in surface waters for the region of the trade winds and the equator; (2) the region of dilution of sea water by surface run-off, especially in high latitudes, results in a strong lightening in oxygen isotopes with decrease in salinity; and (3) freezing out of pack ice or melting of pack ice results in no isotopic fractionation despite strong salinity changes, because the conversion of water from one condensed phase to another does not involve as large a change in bonding energetics as does vaporization hence a minimum of oxygen isotope fractionation.

The rates of oceanic circulation and mixing can be phrased in one of two models which can be treated with geochemical data — the "box" model and the "tube" model. In the box model we divide the oceans into a series of boxes which interconnect and we ask the question: "What is the mean residence time of water molecules and associated dissolved species in each box?" This question requires clocks to be tied to the water molecules as they move about among the boxes. The most useful clock has proven to be carbon-14 although silicon-32 or radium-226, in principle, might also be used. Fig. 10-5 is one such carbon-14 box model used by BROECKER *et al.* (1961). It can be seen that the longest mean residence time for water

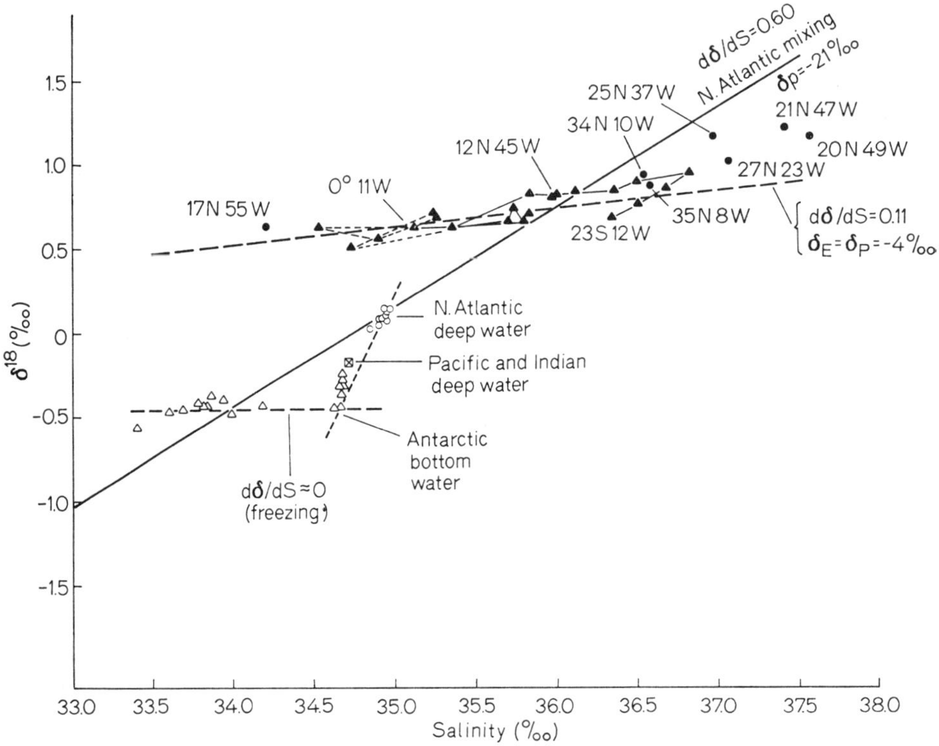

Fig. 10-4. Oxygen isotope-salinity plot of Atlantic Ocean waters (CRAIG and GORDON, 1965).

molecules in the boxes is about 1,000 years while the mean residence time for water in the oceans relative to evaporation and transport to the continents resulting in rainfall and runoff from the continents is 40,000 years.

In contrast to the box models the tube models regard the oceans as made up of columns of water. Transfer of water and associated properties occur by eddy diffusion and advection. The lateral sources and sinks are assumed to be present to supply the necessary inputs to and exits from the system.

MUNK (1966) has tried to evaluate the eddy diffusion coefficient and the vertical advection velocity using the temperature, salinity and carbon-14 distribution in the Pacific Ocean between 1—4 km. BIERI, KOIDE and GOLDBERG (1966) used dissolved rare-gas data for the depth interval 100 to 1,100 m. They both obtain the same advection velocities but MUNK's analysis requires an eddy diffusion five times greater below 1 km compared to BIERI et al.'s analysis for water between 100 to 1,000 m. These results are all very tentative and await more detailed measurements. They show, however, the potential uses of geochemical measurements on sea water.

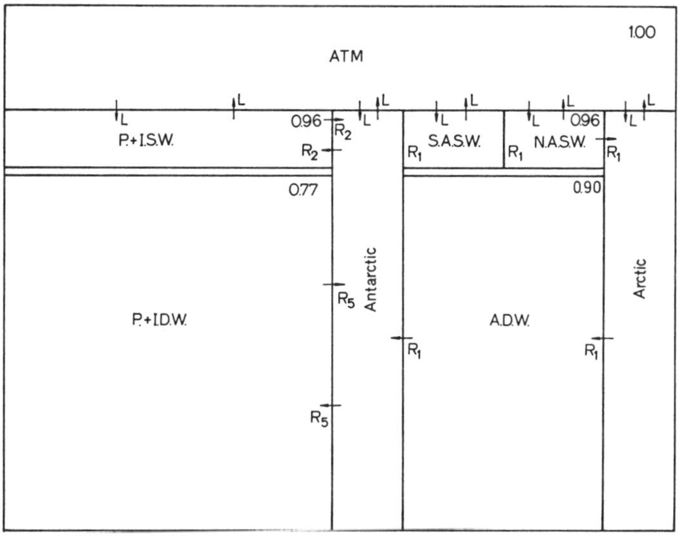

Fig. 10-5. Box model for oceanic mixing using carbon-14 (BROECKER *et al.*, 1961) P.+I.S.W.: Pacific and Indian O. surface water. P.+I.D.W.: Pacific and Indian O. deep water. S.A.S.W.: South Atlantic surface water. N.A.S.W.: North Atlantic surface water. A.D.W.: Atlantic deep water. ATM: atmosphere. L: specific rate of transfer of carbon across the ocean atmosphere boundary. R: rates of transport of carbon between oceanic reservoirs. Numbers in boxes indicate carbon-14 to carbon-12 ratios as compared to an atmospheric value of 1.0

d) Composition of Sea Water

The acidity and the oxidation — reduction potential: The carbon dioxide in the atmosphere is, to a first approximation, in equilibrium with dissolved species in sea water. The reactions and the related equilibrium constants at 25° C are:

K_P $CO_2(g)=CO_2(aq)$ $K_P^0=3.72\times10^{-2}$

K_1 $CO_2(aq)+H_2O=H^+(aq)+HCO_3^-(aq)$ $K_1^0=4.45\times10^{-7}$

K_2 $HCO_3^-(aq)=H^+(aq)+CO_3^=(aq)$ $K_2^0=4.69\times10^{-11}$

Because the ionic strength of sea water is about 0.7 instead of near zero the activities rather than concentrations of the different species must be used. The activity coefficients must be determined empirically. Alternatively the standard state can be defined as that of sea water composition and the equilibrium constants determined especially in this medium. For a detailed discussion see SKIRROW (1965). If the atmosphere is maintained at a $P_{CO_2}=3\times10^{-4}$ atm, then the equilibrium pH$\simeq$6, if only the CO_2—H_2O system is involved. For the relatively rapid mixing rates observed for sea water we must consider the fact that sea water also contains calcium ions and is receiving calcium and bicarbonate ions from streams and is in contact with calcium carbonate deposits. The equilibrium pH for the system $CaCO_3$—CO_2—H_2O is close to 8 which is what is observed in sea water.

We have assumed that the pressure of carbon dioxide is independently determined. This could only be the consequence of the biological flux on the Earth's surface. There is the possibility however, that the steady state CO_2 concentration in the atmosphere is not determined biologically but is a reflection of the pH established in sea water by reactions in sea water. SILLÉN (1961) has suggested that silicates react in sea water to stabilize the ratio of cations including the hydrogen ion by reactions of the sort:

$$\text{K-feldspar} + H^+ = \text{"H-feldspar"} + K^+.$$

The question is not settled to the satisfaction of all so that both possibilities may still be entertained. But despite the choice of model, pH=8 is the end result.

The concentration of molecular oxygen in most parts of the ocean is very close to saturation relative to the atmosphere for the temperature of the water. Its actual range, however, is from 120% of saturation in surface waters because of heating and oxygen production by phytoplankton to about 70% of saturation and sometimes less at depth because of utilization by organisms in respiration. At around 1,000 meters there is a minimum oxygen concentration layer throughout most of the oceans.

The half cell that probably determines the oxidation-reduction potential in aerated sea water is:

$$2H_2O = 4H^+ + O_2 + 4e^-; \qquad E_h^0(25°\,C) = 1.23 \text{ volts}.$$

For $P_{O_2} = 0.2$ atm and pH=8 at 25° C the E_h of sea water is then 0.85. Values this high have not actually been measured indicating either a perturbing reaction on the electrodes or E_h control by some other related half cell.

Controls on the composition of sea water: It was evident to GOLDSCHMIDT (1954) and others following him that although the cations in the oceans could be derived from the weathering of silicates, the chloride, the other halides, bicarbonate and borate could not be balanced and required an independent origin. It is not clear whether the "excess volatiles" as RUBEY (1951) has called them are being derived continuously from the Earths' interior or were supplied to a major extent in the early history of the Earth. There is no unambiguous proof of either model.

With weathered silicates deposited in sea water it is not unreasonable that ion exchange of the cations is responsible for the resultant relative concentrations of the major ions. Probably the limiting control on calcium concentration will be equilibrium with calcium carbonate, but sodium, potassium and probably magnesium apparently are controlled by silicate reactions.

The sulfate ion concentration is the direct consequence of the relative efficiency of supply by streams (or possibly volcanoes) and removal as sulfides in reducing muds and as sulfates in evaporites. Although large variations in the sulfur isotope ratio with time have been observed in evaporites (THODE and MONSTER, 1965; NIELSEN and RICKE, 1964; HOLSER and KAPLAN, 1966) this can give us no information of itself on the sulfate ion concentration in ancient seas. Barium is one of the few trace elements which may be present in sea water in concentrations approaching that in equilibrium with an observed solid phase, barite (CHOW and GOLDBERG, 1960; TUREKIAN and JOHNSON, 1966).

The concentrations of many of the trace elements, however, are not controlled by the solubility product constant of the least soluble salt in normal sea water as KRAUS-

KOPF (1956) and GOLDBERG (1965) have pointed out. Silver, gold, mercury and lead, for instance form highly soluble chloride complexes which would predict much higher concentrations of these elements than are actually found. Similarly the solubility product constants of the insoluble hydroxides or carbonates of the ferrides predict much higher concentrations of cobalt and nickel than are observed. Some metals such as Fe and Mn may be present in higher concentrations in sea water than predicted because of the formation of colloids or undissociated oxides.

Table 10-3. *Solubility controls on concentration of trace elements in sea water*

Element	Insoluble salt in normal sea water	Expected concentration (log moles/l)[a]	Oberved concentration (log moles/l)	Expected concentration in sulfide-rich sea water or mud (log moles/l)[a]
Lanthanum	$LaPO_4$	—11.1	—10.7	—
Thorium	$Th_3(PO_4)_4$	—11.8	—11.7	—
Cobalt	$CoCO_3$	— 6.5	— 8.2	—12.1
Nickel	$Ni(OH)_2$	— 3.2	— 6.9	—10.7
Copper	$Cu(OH)_2$	— 5.8	— 7.3	—26.0
Silver[b]	AgCl	— 4.2	— 8.5	—19.8
Zinc	$ZnCO_3$	— 3.7	— 6.8	—14.1
Cadmium	$CdCO_3$	— 5.0	— 9.0	—16.2
Mercury[b]	$Hg(OH)_2$	+ 1.9	— 9.1	—43.7
Lead[b]	$PbCO_3$	— 5.6	— 9.8	—16.6

[a] The expected concentrations at 25° C are calculated on the basis of the following thermodynamic concentrations, *a* ("activities") of the anions: $\log a_{PO_4^{\equiv}} = -9.3$, $\log a_{CO_3^{=}} = -5.3$, $\log a_{OH^-} = -6$, $\log a_{S^{=}} = -9$.

[b] Form strong chloride complexes.

Stability constants from the compilation of SILLÉN and MARTELL (1964).

Table 10-3 lists the compounds and the expected concentrations for several elements in aerated sea water and in strongly reducing mud in which hydrogen sulfide is being generated by sulfate-reducing bateria. It can be seen clearly that the observed concentrations in sea water of many elements indicate undersaturation relative to aerated water, but supersaturation in the context of a stinking hydrogen sulfide-rich near-shore mud.

Trace elements are also extracted from sea water by organisms either as integral parts of biologically important compounds or adsorbed on the phosphatic skeletal material or organic soft tissue. Table 10-4 shows some representative concentrations in planktonic ash.

In deep water these metals may be transported downward by particle settling but they are regenerated and returned to the surface as in the case of phosphate. Consequently in deep water, metals are not removed into the sediments. In relatively shallow water the particles reach the bottom, and while undergoing regeneration by biological utilization of the fallen debris, the ever present reducing conditions in near-shore muds may result in the strong trapping of the metals. These two processes are much more important in controlling the trace-element composition of sea water than any other.

Adsorption on clay particles, manganese oxides, and iron oxides has been invoked as a device for chelating trace elements and removing them from sea water into sediments. It is intuitively obvious, however, that unless the trace element concentrations are very much greater in sea water than in streams, a clay or oxide particle will have chelated a maximal amount of metal from the relatively pure water of streams. This adsorbed material will be released on encountering sea water, if anything happens at all, because of the strong competition of Mg^{+2} and Na^{+} for adsorption sites.

Table 10-4. *Spectrographic analyses of undifferentiated planktonic ash from the North Atlantic Ocean*

	% Ca	% Mg	ppm Sr	Parts per million				
				Pb	Sn	Cr	Ni	Ag
SF 1	—	—	—	176	530	740	~10,000	—
SF 2	—	—	—	52	14	236	80	—
SF 3	26.7	9.70	5,800	340	22	370	430	0.4
SF 4	20.6	10.51	4,600	64	24	860	610	0.3
SF 5	21.2	7.75	3,500	48	7	38	25	2.8
SF 6	18.8	5.95	2,700	15	19	52	59	3.0
SF 7	18.8	5.86	2,800	240	10	78	94	2.0
SF 8	23.6	4.29	2,400	450	16	122	46	1.8
SF 9	38.6	3.88	3,300	530	5	19	52	1.8
SF 10	23.4	5.60	6,200	376	<5	24	15	0.2
SF 11	21.8	5.86	6,500	222	<5	10	14	0.3
SF 12	9.4	4.43	1,100	38	14	90	70	2.4

SF 3 is composed mainly of crustacea.
SF 10 and 11 are composed mainly of *Sargassum*.
SF 4—7 are sub-arctic species.
SF 8—9 are warm species.

Hence metals may actually be supplied to ocean water by stream-borne detritus rather than removing metals from sea water.

The high concentrations of cobalt, nickel, and manganese, for example, in deep-sea sediments probably are not the result of extraction from sea water but represent the settling out of very fine material with high surface to mass rations in which are corporated the trace elements at the source.

e) Distribution of Elements in Sea Water

All the elements involved in the biological cycle of the oceans should show variations in concentration with depth. The primary life processes of photosynthesis occur in the top 200 meters of the oceans. Phosphorous and nitrogen especially are removed from the surface waters as well as silicon and calcium where these latter elements compose the skeletal parts of plankton. As we have seen above it is not unlikely that many of the trace elements are also extracted from sea water biologically at the surface of the oceans.

During descent of the biologically derived particles they are destroyed by organisms which return most of the elements to sea water. If sea water were not continually mixing in the upper parts of the water column to return the phosphorous and other elements to the surface the oceans would become virtually lifeless in a few years, so efficient is the removal rate of nutrient elements.

Fig. 10-6 shows the observed distribution of phosphate, nitrate, and silica in the major oceans as a function of depth. It will be evident that these chemicals are more concentrated in the deep waters of the Pacific than the Atlantic. This is related mainly to the observation that the mean residence time of water is greater in the deep Pacific than in the deep Atlantic permitting a build-up of these elements in the deeper waters.

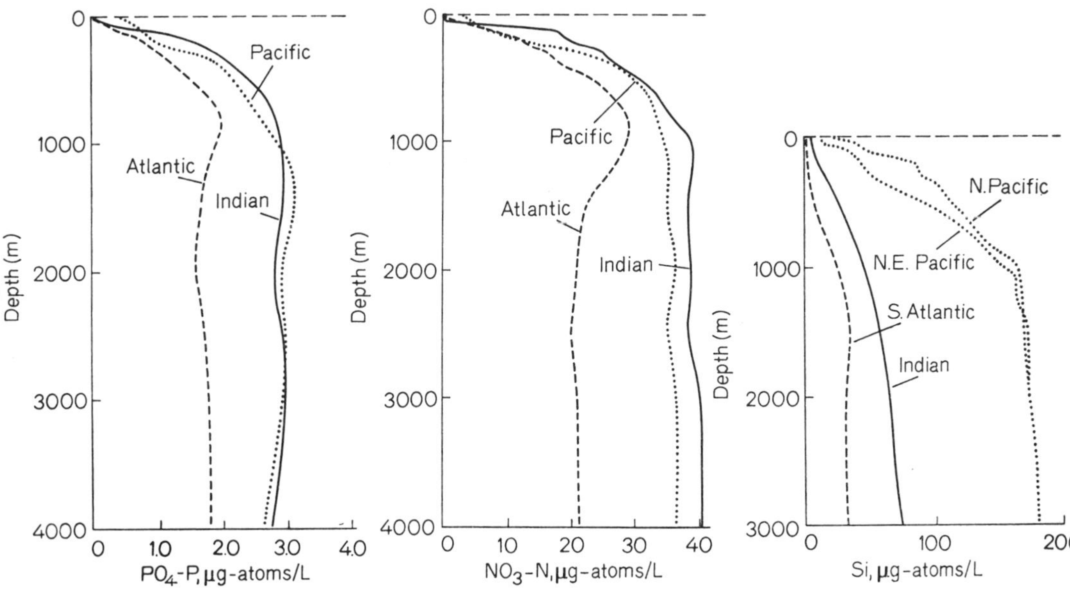

Fig. 10-6. Ocean profiles of phosphate, nitrate and silica concentration (SVERDRUP, JOHNSON and FLEMING, 1942)

In areas of upwelling and high biological productivity a monotonic increase in the concentration of some trace elements has been observed through the thermocline (Fig. 10-7) as expected by local cycling. In other parts of the oceans such as the Indian Ocean (SCHUTZ and TUREKIAN, 1965a) and the Gulf of Mexico (SLOWEY, 1966) some of the highest concentrations occur at shallow depths while deeper waters are low in elements such as silver, cobalt and manganese. The reasons for these observations are not clear but the results seem to imply a decoupling of the deeper waters from the surface waters which are being enriched in trace elements from land sources or from transport of metal-rich waters to the site.

Some trace elements also show strong regional variations while some others either are reasonably constant with depth and location or show random distribution patterns. These can best be seen when data from one investigator is used because of possible systematic variations in analytical accuracy among investigators. In Table 10-5, information on a variety of elements is listed under one of three categories of observed distribution patterns: relatively constant concentration, random varia-

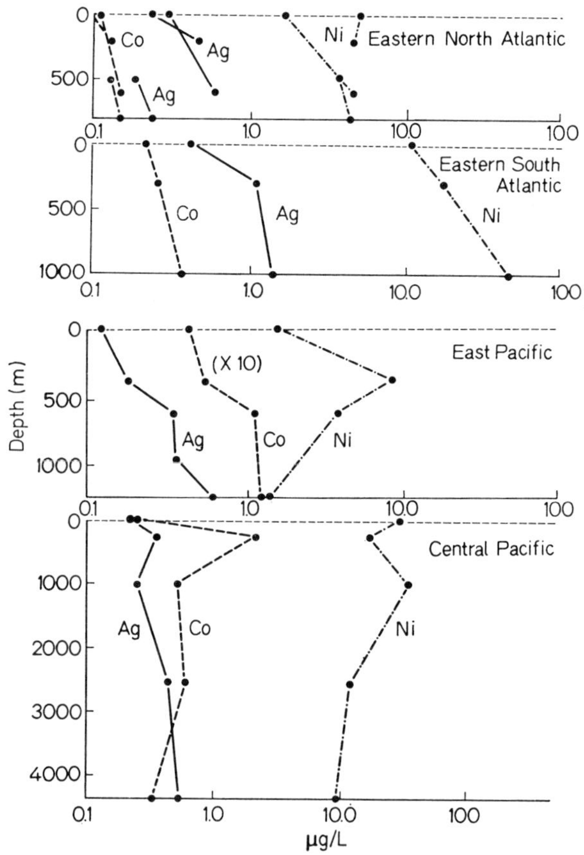

Fig. 10-7. Silver, cobalt and nickel concentration profiles in regions of upwelling (SCHUTZ and TUREKIAN, 1965a) µg/L = ppb

Table 10-5. *Categorization of some trace elements in sea water on the basis of their distribution patterns*

Relatively constant	Random variation	Regionally variable
Sr	Sb	Co
Ba	Se	Ag
Cs	Au	Ni
Rb	(Pb)	
U		
Mo		

tions in concentration; and regional differences in concentration. It is important to note that trace elements do not necessarily follow phosphate or any other strongly variable biologically associated component. The Antarctic waters for example, although high in phosphate and silica are the lowest of all the oceans in Co, Ni and Ag (SCHUTZ and TUREKIAN, 1965b).

In the open oceans most of the concentration of elements in a sea water sample are found in the fraction that passes through 0.45 μ filters. It is not certain that some of this material might not be very finely particulate or colloidal. For most purposes an analysis of open sea water, whether filtered or not, gives a good representation of the composition of the sea water. In coastal areas there is a marked trace element component in the $>$ 0.45 μ fraction of sea water (Table 10-6) hence filtering of samples may be required.

Table 10-6. *Average concentration of various forms of copper, manganese and zinc in the Gulf of Mexico* (SLOWEY, 1966)

Location	Cu (ppb)	Mn (ppb)	Zn (ppb)	No. of samples
		Ionic		
Coastal	0.73	1.5	2.5	8
Open	0.90	0.26	2.6	43
		Particulate ($>$0.45 μ)		
Coastal	0.57	2.3	1.1	5
Open	0.24	0.03	0.27	42

Despite all these difficulties of interpretation, Table 10-7 has been constructed to give an idea of the concentrations of the elements in the oceans. I have selected a particular source of data for each element and included the citation to the authors. An attempt was made to choose the most reliable results coupled with the largest number of observations. Where equally acceptable techniques were employed the more representative sampling, in general, was given preference. In this sense the compilation is different from that of RICHARDS (1957) which listed all values up to the time of his writing and from that of GOLDBERG (1965) who presents a single value for each element without documentation. The risk of appearing arbitrary is more than offset by the fact that the reader can evaluate any new data as it appears in terms of the specific data quoted.

Table 10-7. *The composition of sea water at 35‰ salinity*

Element	ppb = μg/l	Source
Hydrogen	1.10×10^8	
Helium	0.0072	BIERI, KOIDE and GOLDBERG (1964)
Lithium	170	CHOW and GOLDBERG (1964)
Beryllium	0.0006	MERRILL et al. (1960)
Boron	4,450	GREENHALGH et al. in CULKIN (1965)
Carbon (inorganic)	28,000	SVERDRUP, JOHNSON and FLEMING (1942)
(dissolved organic)	500	DUURSMA (1961)
Nitrogen (dissolved N_2)	15,500	KÖNIG et al. (1964)
(as NO_2^-, NO_3^-, NH_4^+ and dissolved organic)	670	EMERY, ORR and RITTENBERG (1955)
Oxygen (dissolved O_2)	6,000	SVERDRUP, JOHNSON and FLEMING (1942)
(as H_2O)	8.83×10^8	
Fluorine	1,300	RILEY (1965)
Neon	0.120	BIERI, KOIDE and GOLDBERG (1964)

Table 10-7. (Continued)

Element	ppb = μg/l	Source
Sodium	1.08×10^7	CULKIN (1965)
Magnesium	1.29×10^6	CULKIN (1965)
Aluminum	1	SACKETT and ARRHENIUS (1962)
Silicon	2,900	EMERY, ORR and RITTENBERG (1955)
Phosphorus	88	EMERY, ORR and RITTENBERG (1955)
Sulfur	9.04×10^5	RILEY in CULKIN (1965)
Chlorine	1.94×10^7	CULKIN (1965)
Argon	450	BIERI, KOIDE and GOLDBERG (1964)
Potassium	3.92×10^5	JENTOFT and ROBINSON (1956)
Calcium	4.11×10^5	CARPENTER (1957)
Scandium	<0.004	SCHUTZ and TUREKIAN (1965a)
Titanium	1.0	GRIEL and ROBINSON (1952)
Vanadium	1.9	SUGAWARA, NAITO and YAMADA (1956)
Chromium	0.2	LOVERIDGE et al. (1960)
Manganese	0.4	SLOWEY (1966)
Iron	3.4	LEWIS and GOLDBERG (1954)
Cobalt	0.39	SCHUTZ and TUREKIAN (1965a, b)
Nickel	6.6	SCHUTZ and TUREKIAN (1965a, b)
Copper	0.9	SLOWEY (1966)
Zinc	5	SLOWEY (1966)
Gallium	0.03	CULKIN and RILEY (1958)
Germanium	0.06	BURTON and RILEY (1958)
Arsenic	2.6	SMALES and PATE (1952)
Selenium	0.090	SCHUTZ and TUREKIAN (1965a)
Bromine	6.73×10^4	MORRIS and RILEY in CULKIN (1965)
Krypton	0.21	BIERI, KOIDE and GOLDBERG (1964)
Rubidium	120	BOLTER, TUREKIAN and SCHUTZ (1964)
Strontium	8,100	CHOW and THOMPSON (1955)
Yttrium	0.013	HØGDAHL (1967)
Zirconium	0.026	SHIGEMATSU et al. (1964)
Niobium	0.015	CARLISLE and HUMMERSTONE (1958)
Molybdenum	10	SUGAWARA and OKABE (1960)
Ruthenium	0.0007	DIXON, SLOWEY and HOOD (1966)
Rhodium		
Palladium		
Silver	0.28	SCHUTZ and TUREKIAN (1965a, b)
Cadmium	0.11	MULLIN and RILEY (1956)
Indium		
Tin	0.81	HAMAGUCHI et al. (1964)
Antimony	0.33	SCHUTZ and TUREKIAN (1965a)
Tellurium		
Iodine	64	BARKLEY and THOMPSON (1960)
Xenon	0.047	BIERI, KOIDE and GOLDBERG (1964)
Cesium	0.30	BOLTER, TUREKIAN and SCHUTZ (1964)
Barium	21	TUREKIAN and JOHNSON (1966)
Lanthanum	0.0034	HØGDAHL (1967)
Cerium	0.0012	HOGDAHL (1967)
Praesodymium	0.00064	HØGDAHL (1967)
Neodymium	0.0028	HØGDAHL (1967)
Samarium	0.00045	HØGDAHL (1967)
Europium	0.000130	HØGDAHL (1967)
Gadolinium	0.00070	HØGDAHL (1967)
Terbium	0.00014	HØGDAHL (1967)

Table 10-7. (Continued)

Element	ppb = µg/l	Source
Dysprosium	0.00091	HØGDAHL (1967)
Holmium	0.00022	HØGDAHL (1967)
Erbium	0.00087	HØGDAHL (1967)
Thulium	0.00017	HØGDAHL (1967)
Ytterbium	0.00082	HØGDAHL (1967)
Lutetium	0.00015	HØGDAHL (1967)
Hafnium	<0.008	SCHUTZ and TUREKIAN (1965a)
Tantalum	<0.0025	SCHUTZ and TUREKIAN (1965a)
Tungsten	<0.001	approximation
Rhenium	0.0084	SCADDEN (1968)
Osmium		
Iridium		
Platinum		
Gold	0.011	SCHUTZ and TUREKIAN (1965a)
Mercury	0.15	HAMAGUCHI, KURODA and HOSOHARA (1961), HOSOHARA (1961)
Thallium		
Lead	0.03	TATSUMOTO and PATTERSON (1963)
Bismuth	0.02	BROOKS (1960)
Radium	1×10^{-7}	KOCZY (1958)
Thorium	0.0004	SOMAYAJULU and GOLDBERG (1966)
Protactinium	2×10^{-10}	MOORE and SACKETT (1964)
Uranium	3.3	RONA, GILPATRICK and JEFFREY (1956)

II. Streams

a) Discharge, Dissolved Load and Sediment Load

Table 10-8 lists two recent estimates of surface runoff into the oceans and the dissolved load of streams. For comparison purposes ALEKIN and BRAZHNIKOVA's (1961) results on dissolved load based on $CO_3^=$ have been converted to a HCO_3^- basis using their factor of 1.364 based on the average composition of river water. Nevertheless it can be seen that except for South America, where both compilations undoubtedly used the same source of information, there are some disparities in both the discharge and dissolved load concentration.

ALEKIN and BRAZHNIKOVA used observed dissolved load data from the U.S.S.R. as a function of different climatic and geomorphic types and extended their results for each type (except tropical) to the other parts of the world using the geographic distribution of vegetation types as an index of dissolved load.

In light of these uncertainties it may be useful as a first approximation for material balance calculations to note the following relationships:

area of the oceans: 3.6×10^{18} cm^2;
total discharge: $\sim 3.6 \times 10^{16}$ liters/year;
total dissolved load supply rate: $\sim 3.6 \times 10^{15}$ grams/year;
discharge per unit area of ocean: ~ 10 liters/cm^2/1,000 years;
supply of dissolved load per unit area of ocean: ~ 1 gram/cm^2/1,000 years.

Table 10-8. *The dissolved load of streams*

	LIVINGSTONE (1963)			ALEKIN and BRAZHNIKOVA (1961)		
	×10^{15} liters/ year	dissolved solids (ppm)	10^{15} g/ year	×10^{15} liters/ year	dissolved solids (ppm) (including organic)	10^{15} g/ year
North America	4.55	142	0.646	6.43	89	0.572
Europe	2.50	182	0.455	3.00	101	0.303
Asia	11.05	142	1.570	12.25	111	1.360
Africa	5.90	121	0.715	6.05	96	0.581
Australia	0.32	59	0.019	0.61	176	0.107
South America	8.01	69	0.552	8.10	71	0.575
Total	32.33	120	3.957	36.45	88	3.498

Fig. 10-8. Drainage basins as used in Table 10-9 (JUDSON and RITTER, 1964)

The sediment load of streams is even more difficult to estimate than the dissolved load. The sediment load of a particular stream varies with the discharge of that stream in a non-linear fashion. It is thus difficult to estimate the total supply of solid material to the oceans by streams, unless the major streams are monitored over long periods of time.

A recent attempt at determining the sediment transport by streams and the regional denudation rates in the United States has been made by JUDSON and RITTER (1964). They used continuous data from gauging stations on the rivers of major regions of the U.S. as delineated in Fig. 10-8. Their summary is presented in Table 10-9.

If we take the ratio of dissolved load to suspended load typical of the United States and assume that it is approximated by the Earth as a whole the average world-

Table 10-9. *Sediment load of streams in various regions of the United States* (JUDSON and RITTER, 1964)

Drainage region	Drainage area $\times 10^3$ km^2	Runoff $\times 10^{15}$ liters/year	Average concentration suspended load (mg/l)	Average concentration dissolved load (mg/l)	Average annual suspended load 10^{15} g/year	Average annual dissolved load 10^{15} g/year
Colorado	637	0.021	13,930	760		
Pacific Slopes, Calif.	303	0.072	970	167		
Western Gulf	829	0.049	1,880	770		
Mississippi	3,240	0.555	604	248		
S. Atlantic and eastern Gulf	735	0.291	131	171		
N. Atlantic	383	0.188	156	128		
Columbia	679	0.309	106	138		
Total	6,807	1.485	602	214	0.894	0.318

wide supended load is 330 mg/l. JUDSON and RITTER (1964) believe that the correspondence of the accumulation rate of sediment in man-made Lake Mead, behind Hoover Dam on the Colorado River, and the suspended sediment supply by the Colorado River indicates that the traction or bed load is not very important in most rivers. They suggest 10% of the suspended load as the maximum for the bed-load. This raises the total supply of detrital load by streams to 365 mg/l. We will round this off to 400 mg/l for the world average.

b) Composition of Streams

The composition of the major components of the dissolved material in river waters from the compilation of LIVINGSTONE (1963) is presented in Table 10-10. It is doubtful if the iron listed as Fe really exists in true solution in the amount listed.

Table 10-10. *Average composition of strems* (LIVINGSTONE, 1963)

	ppm		ppm
HCO_3^-	58.4	Na^+	6.3
$SO_4^=$	11.2	K^+	2.3
Cl^-	7.8	(Fe)	(0.67)
NO_3^-	1.0	SiO_2	13.1
Ca^{++}	15.0		
Mg^{++}	4.1	Total	120

The role of suspended sediment as an additional transport agent for cations has been evaluated by KENNEDY (1965) for some U.S. streams. The average is 28 milliequivalents of adsorbed cations per 100 grams of sediment.

The trace element composition of streams is even more difficult to ascertain than it is for sea water. Not only is it probable that there are real regional variations depending on rock type being weathered and degree of weathering but also the likelihood of human contamination is no longer trivial. The results in Table 10-11, compiled in a manner similar to that for sea water, should not be taken more seriously than as a first estimate. We may never know what the "true" amounts of some of these elements supplied to the oceans are.

Table 10-11. *The trace element composition of streams*

	ppb = μg/l	Approximate estimate (ppb)	Region	Reference
Lithium	3.3	3	North America	Durum and Haffty (1961)
Beryllium				
Boron	13	10	U.S.S.R.	Konovalov (1959)
Fluorine	88	100	U.S.S.R.	Konovalov (1959)
	150		Japan	Sugawara (1967)
Aluminum	360	400	Japan	Sugawara (1967)
Phosphorus	19	20	Columbia R.	Silker (1964)
Scandium	0.004	0.004	Columbia R.	Silker (1964)
Titanium	2.7	3	Maine (U.S.A.) lakes and streams	Turekian and Kleinkopf (1956)
Vanadium	0.9	0.9	Japan	Sugawara, Naito and Yamada (1956)
Chromium	1.4	1	U.S. Streams, Rhone, Amazon	Kharkar, Turekian and Bertine (1968)
	0.3		Maine (U.S.A.) lakes and streams	Turekian and Kleinkopf (1956)
Manganese	12	7	U.S.S.R.	Konovalov (1959)
	4.0		Maine (U.S.A.) lakes and streams	Turekian and Kleinkopf (1956)
	4.8		Columbia R.	Silker (1964)
Cobalt	0.19	0.2	U.S. streams, Rhone, Amazon	Kharkar, Turekian and Bertine (1968)
Nickel	0.3	0.3	Maine (U.S.A.) lakes and streams	Turekian and Kleinkopf (1956)
Copper	0.9	7	Japan	Sugawara (1967)
	10		U.S.S.R.	Konovalov (1956)
	12		Maine (U.S.A.) lakes and streams	Turekian and Kleinkopf (1956)
	4.4		Columbia R.	Silker (1964)
Zinc	5.0	20	Japan	Sugawara (1967)
	45		U.S.S.R.	Konovalov (1956)
	16		Columbia R.	Silker (1964)
Gallium	0.089	0.09	Saale and Elbe (Germany)	Heide andKödderitzsch (1964)
Germanium				
Arsenic	1.7	2	Japan	Sugawara (1967)
	1.6		Columbia R.	Silker (1964)
Selenium	0.20	0.2	U.S. streams, Rhone, Amazon	Kharkar, Turekian and Bertine (1968)

Table 10-11. (Continued)

	ppb = μg/l	Approximate estimate (ppb)	Region	Reference
Bromine	19	20	U.S.S.R.	KONOVALOV (1959)
Rubidium	1.1	1	U.S. streams, Rhone, Amazon	KHARKAR, TUREKIAN and BERTINE (1968)
Strontium	46	50	Eastern U.S.	TUREKIAN (1966)
Yttrium		0.7		estimate[a]
Zirconium				
Niobium				
Molybdenum	0.6	1	Japan	SUGAWARA (1967)
	1.8		U.S. streams, Amazon	KHARKAR, TUREKIAN and BERTINE (1968)
Ruthenium				
Rhodium				
Palladium				
Silver	0.39	0.3	U.S. streams, Rhone, Amazon	KHARKAR, TUREKIAN and BERTINE (1968)
Cadmium				
Indium				
Tin				
Antimony	1.1	1	U.S. streams, Rhone, Amazon	KHARKAR, TUREKIAN and BERTINE (1968)
Tellurium				
Iodine	7.1	7	U.S.S.R.	KONOVALOV (1959)
Cesium	0.020	0.02	U.S. streams, Rhone, Amazon	KHARKAR, TUREKIAN and BERTINE (1968)
Barium	11	10	Eastern U.S.	TUREKIAN (1966)
Lanthanum	0.2	0.2	Sweden	LANDSTRÖM and WENNER (1965)
		0.19	Columbia R.	SILKER (1964)
Cerium		0.06		estimate[a]
Praseodymium		0.03		estimate[a]
Neodymium		0.2		estimate[a]
Samarium		0.03		estimate[a]
Europium		0.007		estimate[a]
Gadolinium		0.04		estimate[a]
Terbium		0.008		estimate[a]
Dysprosium		0.05		estimate[a]
Holmium		0.01		estimate[a]
Erbium		0.05		estimate[a]
Thulium		0.009		estimate[a]
Ytterbium		0.05		estimate[a]
Lutetium		0.008		estimate[a]

[a] Estimate based on prorating rare-earth values in streams using La concentration in streams and the relative proportions of rare-earths found in the oceans as given in Table 10-7.

Table 10-11. (Continued)

	ppb = μg/l	Approximate estimate (ppb)	Region	Reference
Hafnium				
Tantalum				
Tungsten	0.03	0.03	Sweden	Landström and Wenner (1965)
Rhenium				
Osmium				
Iridium				
Platinum				
Gold	0.002	0.002	Sweden	Landström and Wenner (1965)
Mercury	0.074	0.07	Saale and Elbe (Germany)	Heide, Lerz, and Böhm (1957)
Thallium	?			
Lead	3.9	3	Saale and Elbe (Germany)	Heide, Lerz, and Böhm (1957)
	2.3		Maine (U.S.A.) lakes and streams	Turekian and Kleinkopf (1956)
Bismuth	?			
Thorium	0.096	0.1	Amazon	Moore (1967)
Uranium	0.06	0.04	Sweden	Landström and Wenner (1965)
	0.043		Amazon	Moore (1967)
	0.026		North America	Rona and Urry (1952)

III. The Atmosphere

a) The Structure of the Atmosphere

The Earth's atmosphere is divided into several regions on the basis of particle density, temperature and reactivity with photons.

The sun emits radiation which corresponds roughly with the spectrum of a black body at a temperature of 6,000° K. The main component is thus in the visible with lesser numbers of photons in the ultraviolet and in the infrared. The mean flux at the top of the atmosphere is about 1.4×10^6 ergs/cm²/sec. About 40% of the radiation is lost by reflection back onto space, the remaining 60% reacts with the Earth's surface and atmosphere and ultimately is reradiated off at different wavelengths. The temperature of the Earth's surface due to the incident radiation reaching the surface is formally 245° K but some of the emitted radiation in the infrared is reabsorbed so that the true mean surface temperature of the Earth is closer to 290° K.

The regions of the atmosphere, classified on the basis of temperature primarily, are the following:

The *troposphere* is the region in which the infrared radiation is absorbed mainly by water vapor to raise the surface temperature. Since the particle density or the pressure of each gas diminishes away from the surface of the Earth, the absorption decreases with height both because of the continually attenuating flux of infrared radiation upward and the decrease in the water molecule density. The point at which the temperature decrease is reserved is called the *tropopause* and the region above, which contains very little water vapor, is called the *stratosphere*. The temperature decrease in the troposphere is about 6.5° K/km up to the tropopause at a height of about 12 km. The mean temperature at the tropopause is about 210° K.

The temperature rises in the stratosphere with increasing elevation because of the absorption of ultraviolet light by ozone. Temperature is a maximum at about 50 km and is marked as the top of the stratosphere. Between 50 and 80 km is a transition region of decreasing temperature with height called the *mesosphere*. Above 80—100 km a region of strong heating occurs with increasing altitude due to direct interaction of the suns rays with the oxygen and nitrogen molecules to produce photodissociation and photoionization. This is called the *thermosphere*. At about 700 km the highest temperature is attained, approximately 1,500° K and the high temperature plateau above this height is called the *exosphere*. Beyond these layers are the regions of magnetic and electrical interaction typified by the van Allen belt.

b) Composition of the Atmosphere

In the troposphere there is fairly rapid mixing due to convection so that there is no separation of the different gases comprising the atmosphere. This results in a relatively homogenous atmosphere except for water which varies with location and season as well as with elevation. The composition of water-free air at sea level is given in Table 10-12.

Nitrogen in the presence of oxygen at the surface of the oceans should combine to form nitrate in solution as the stable form (SILLÉN, 1966). Both nitrogen and oxygen are maintained at their levels by biological processes. Oxygen is more obviously biologically controlled than nitrogen but both are dependent on the chemical actions of life.

The argon in the atmosphere is almost completely argon-40 which has been produced by the radioactive decay of potassium-40 in the Earth and released to the atmosphere by degassing of the Earth. It is not certain whether most of the argon was supplied from the argon produced in the Earth by potassium-40 decay at some major degassing epoch in the Earth's early history of by the continuous loss of the continuously generated argon in the Earth's crust and mantle. It has been shown that radiogenic argon is released from most minerals at about 300° C which corresponds to depths in the Earth of not much greater than 7 km.

Methane and carbon dioxide are closely tied to biological activity. Methane is oxidized to carbon dioxide and its presence is directly sustained by production by bacteria and animals.

There are also present the man-made impurities such as sulfur dioxide and carbon monoxide in high concentration in urban areas which have been responsible for the physical discomforts of smog.

Table 10-12. *Composition of clean, dry air near sea level*

Component	Content, per cent by volume	Molecular weight
Nitrogen	78.084	28.0134
Oxygen	20.9476	31.9988
Argon	0.934	39.948
Carbon dioxide	0.0314	44.00995
Neon	0.001818	20.183
Helium	0.000524	4.0026
Krypton	0.000114	83.80
Xenon	0.0000087	131.30
Hydrogen	0.00005	2.01594
Methane	0.0002	16.04303
Nitrous oxide	0.00005	44.0128
Ozone		47.9982
Summcr	0 to 0.000007	
Winter	0 to 0.000002	
Sulfur dioxide	0 to 0.0001	64.0628
Nitrogen dioxide	0 to 0.000002	46.0055
Ammonia	0 to trace	17.03061
Carbon monoxide	0 to trace	28.01055
Iodine	0 to 0.000001	253.8088

c) Atmospheric Precipitation

Atmospheric precipitation in the form of rain and snow, primarily, is responsible for the transfer of a considerable amount of ions from the oceans to the continents as well as from the atmosphere to the Earth's surface.

The water balance between the continents and the oceans by means of atmospheric precipitation and evaporation has been evaluated by Kalle (1945) and his results are presented in Table 10-13. Since the stream runoff from continents is 36×10^{15} liters per year, it is obvious that much of the rainfall on the continents is recycled on the continents.

Table 10-13. *Water balance of oceans and continents.* (After Kalle, 1945 in Hutchinson, 1957)

	Water balance	
	$gr. cm^{-2} yr.^{-1}$	$10^{20} gr. yr.^{-1}$
Evaporation from ocean surfaces	106	3.83
Precipitation on ocean surfaces	96	3.47
Evaporation from land surfaces	42	0.63
Precipitation on land surfaces	67	0.99

The composition of atmospheric precipitation varies from location to location. The major influence along the sea coasts is the transfer of salts by means of aerosols from the sea to land (Fig. 10-9) but in the interior of continents this is modified by the effects of recycling of water which has been part of the continental ground- and surface-water system.

Most analyses of recent atmospheric precipitation show the influence of industrial contamination. This is particularly evident in the increased high sulfate concentration of rain water due to the burning of sulfur-rich fossil fuels.

Table 10-14 shows HUTCHINSON's best estimate of what typical uncontaminated rainfall might be expected to look like. It can be used as a first approximation in the evaluation of the effects of atmospheric precipitation on the geochemical cycle.

There are not many data on the trace-element composition of rain waters and snows, but SUGAWARA (1967) has summarized the work done by him and his co-

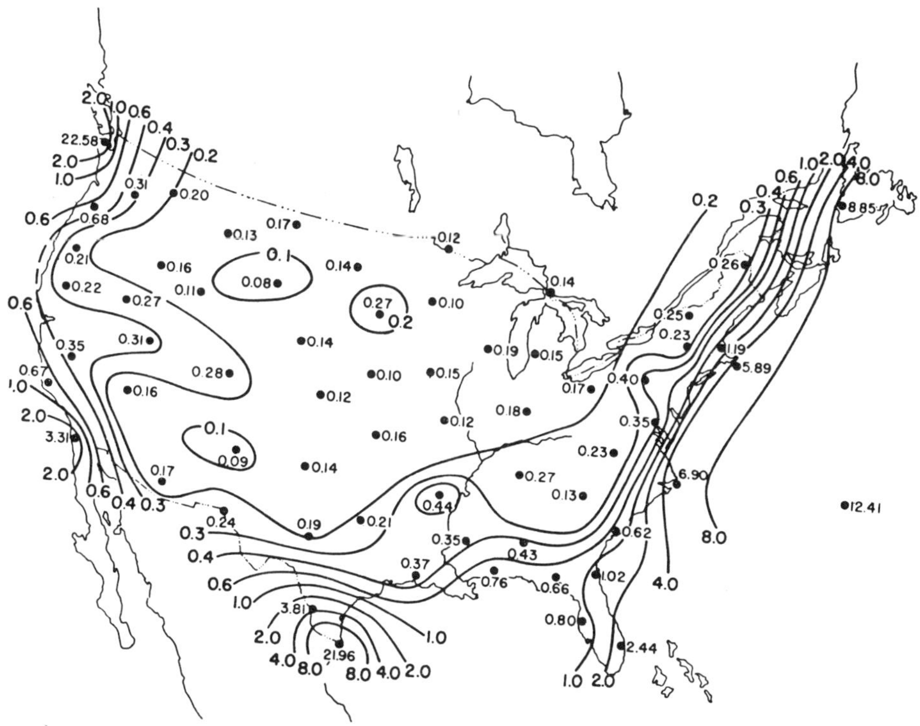

Fig. 10-9. Average Cl^- concentration (ppm) in rain over the U.S., July 1955—June 1956 (JUNGE and WERBY, 1958)

Table 10-14. *Composition of rain and of dilute lake water.* (After HUTCHINSON, 1957)

	Rain water (provisional estimate), ppm		Rain water (provisional estimate), ppm
Cl	0.5	K	0.03 or more
Br	0.03	Mg	0.1 or more
I	0.001	Ca	0.1—10
SO_4	2.0	$N \cdot NH_3$	0.5
B	0.01	$N \cdot NO_3$	0.2
Na	0.4 or more		

workers on the composition of atmospheric precipitation and river water in Japan. Table 10-15, taken from his summary, shows that some of the trace elements in the surface runoff are actually depleted relative to rain water falling on the drainage basin, indicating adsorption on particles or biological removal. Molybdenum, alone, of the trace elements studied shows a factor of two increase over the rain water supply.

Table 10-15. *Average chemical composition of precipitation and river water in Japan* (SUGAWARA, 1967)

	Precipitation (ppm)	River water (ppm)	River water relative to precipitation normalized to Cl ratio = 1
Na	1.1	5.1	1
K	0.26	1.0	0.8
Mg	0.36	2.4	0.7
Ca	0.94	6.3	1.4
Sr	0.011	0.057	1.1
Cl	1.1	5.2	1
I	0.0018	0.0022	0.3
F	0.08	0.15	0.4
S	1.5	3.5	0.5
Si	0.83	8.1	0.5
Fe	0.23	0.48	0.4
Al	0.11	0.36	0.7
P	0.014		
Mo	0.00006	0.0006	2.1
V	0.0014	0.0010	0.2
Cu	0.0008	0.0014	0.4
Zn	0.0042	0.0050	0.2
As	0.0016	0.0017	0.2

Although it is evident that for the Japanese Islands as well as the maritime provinces of the continents the trace element as well as the major element concentration is controlled by the supply by atmospheric precipitation, it seems likely that continental weathering should have a major part in determining the trace element composition of large rivers.

References

ALEKIN, O. A., and L. V. BRAZHNIKOVA: The discharge of soluble matter from dry land of the Earth. Gidrokhim. Materialy **32**, 12—24 (1961). (Translated into English by M. FLEISCHER, U.S.G.S.).

BARKLEY, R. A., and T. G. THOMPSON: The total iodine and iodate-iodine content of sea-water. Deep-Sea Res. **7**, 24—34 (1960).

BIERI, R., M. KOIDE, and E. D. GOLDBERG: Noble gases contents of Pacific seawaters. J. Geophys. Res. **71**, 5243—5265 (1966).

BOLTER, E., K. K. TUREKIAN, and D. F. SCHUTZ: The distribution of rubidium, cesium and barium in the oceans. Geochim. Cosmochim. Acta **28**, 1459—1466 (1964).

BROECKER, W. S., R. D. GERARD, M. EWING, and B. C. HEEZEN: Geochemistry and physics of ocean circulation. In: Oceanography (ed. M. SEARS), p. 301—322, Publ. 67. Washington, D.C.: Am. Assoc. Adv. Sci. 1961.

BROOKS, R. R.: The use of ion-exchange enrichment in the determination of trace elements in sea water. Analyst **85**, 745—748 (1960).

BURTON, J. D., and J. P. RILEY: Germanium in sea-water. Nature **181**, 179—180 (1958).

CARLISLE, D. B., and L. G. HUMMERSTONE: Niobium in sea-water. Nature **181**, 1002—1003 (1958).

CARPENTER, J. H.: The determination of calcium in natural waters. Limnol. and Oceanog. **2**, 271—280 (1957).

CHOW, T. J., and E. D. GOLDBERG: On the marine geochemistry of barium. Geochim. Cosmochim. Acta **20**, 192—198 (1960).

— — Mass spectrometric determination of lithium in seawater. J. Marine Res. **20**, 163—167 (1962).

—, and T. G. THOMPSON: Flame photometric determination of strontium in seawater. Anal. Chem. **27**, 18—21 (1955).

CRAIG, H., and L. I. GORDON: Isotopic oceanography. Deuterium and oxygen 18 variations in the ocean and marine atmosphere. Narragansett Mar. Lab. Occ. Publ. No. 3 277—374 (1965), and in Stable isotopes in oceanographic studies and paleotemperatures, Spoleto, Italy, July 26—30, 122 p. (1965).

CULKIN, F.: The major constituents of sea water. In: Chemical oceanography (ed. J. P. RILEY and G. SKIRROW), vol. 1, p. 121—161. London: Academic Press 1965.

—, and J. P. RILEY: Gallium in sea-water. Nature **181**, 179—180 (1958).

DIETRICH, G.: General oceanography. Interscience, 588 p. New York: John Wiley & Sons 1963.

DIXON, B. W., J. F. SLOWEY, and D. W. HOOD: In: The chemistry and analysis of trace metals in sea water. Final Report AEC Contract AT (40-1)-2799, Texas A + M Project 276, D. W. HOOD, Director (1966).

DURUM, W. H., and J. HAFFTY: Occurrence of minor elements in water. U.S. Geol. Surv. Circ. **445**, 11 p. (1960).

DUURSMA, E. K.: Dissolved organic carbon, nitrogen and phosphorus in the sea. Neth. J. Sea Res. **1**, 1—148 (1961).

EMERY, K. O., W. L. ORR, and S. C. RITTENBERG: Nutrient budgets in the ocean. In: Essays in the natural sciences in honor of Captain ALLAN HANCOCK, p. 299—309. Los Angeles: Univ. of South. Calif. Press 1955.

GOLDBERG, E. D.: Minor elements in sea water. In: Chemical oceanography (ed. J. P. RILEY and G. SKIRROW), vol. 1, p. 163—196. London: Academic Press 1965.

GOLDSCHMIDT, V. M.: Geochemistry. 730 p. Oxford: Clarendon Press 1954.

GRIEL, J. V., and R. J. ROBINSON: Titanium in sea water. J. Marine Res. **11**, 173—179 (1952).

HAMAGUCHI, H., R. KURODA, and K. HOSOHARA: Photometric determination of mercury in seawater. Nippon Kagaku Zasshi **82**, 347—349 (1961).

— — N. ONUMA, K. KAWABUCHI, T. MITSUBAYASHI, and K. HOSOHARA: The geochemistry of tin. Geochim. Cosmochim. Acta **28**, 1039—1053 (1964).

HEEZEN, B. C., and M. EWING: The mid-oceanic ridge. In: The sea (ed. M. N. HILL), vol. 3, p. 388—410. New York: Interscience (1963).

— M. THARP, and M. EWING: The floors of the oceans I. The North Atlantic. Geol. Soc. Am., Spec. Papers **65**, 122 p. (1959).

HEIDE, F., u. H. KODDERITZSCH: Der Galliumgehalt des Saale- und Elbewassers. Naturwissenschaften **51**, 104 (1964).

— H. LERZ u. G. BOHM: Gehalt des Saalewassers an Blei und Quecksilber. Naturwissenschaften **44**, 441—442 (1957).

HØGDAHL, O.: Distribution of the rare-earth elements in sea water. Central Institute for Ind. Res., Oslo, NATO Grant 203, Semiannual Prog. Rept. 4, 1967.

HOLSER, W. T., and I. R. KAPLAN: Isotope geochemistry of sedimentary sulfates. Chem. Geol. **1**, 93—135 (1966).

HOSOHARA, K.: Mercury content of deep-sea water. Nippon Kagaku Zasshi **82**, 1107—1108 (1961). CA **56**,4535.

HUTCHINSON, G. E.: Treatice on limnology, vol. 1. John Wiley & Sons 1957.

JENTOFT, R. E., and R. J. ROBINSON: The potassium-chlorinity ratio of ocean water. J. Marine Res. **15**, 170—180 (1956).

JUDSON, S., and D. F. RITTER: Rates of regional denudation in the United States: J. Geophys. Res. **49**, 3395—3401 (1964).

JUNGE, C. E., and R. T. WERBY: The concentration of chloride, sodium, potassium, calcium and sulfate in rain water over the United States: J. Meteorology **15**, 417—425 (1958).

KENNEDY, V. C.: Mineralogy and cation-exchange capacity of sediments from selected streams: U.S. Geol. Surv. Profess. Paper **433**-D (1965).

KHARKAR, D. P., K. K. TUREKIAN, and K. K. BERTINE: Stream supply of dissolved silver, molybdenum, antimony, selenium, chromium, cobalt, rubidium and cesium to the oceans. Geochim. Cosmochim. Acta **32**, 285—298 (1968).

KOCZY, F. F.: Natural radium as a tracer in the ocean: Proceedings of the Second Int. Conf. on the Peaceful Uses of Atomic Energy, Geneva, vol. 18, p. 351—357. New York: United Nations 1958.

KÖNIG, H., H. WÄNKE, G. S. BIEN, N. W. RAKESTRAW, and H. E. SUESS: Helium, neon and argon in the oceans. Deep-Sea Res. **11**, 243—247 (1964).

KONOVAL, G. S.: The transport of microelements by the most important rivers of the U.S.S.R. Dokl Akad. Nauk USSR **129**, No. 4, 912—915 (1959). (Translated into English by M. FLEISCHER, U. S. G. S.).

KRAUSKOPF, K. B.: Factors controlling the concentrations of thirteen rare metals in sea-water. Geochim. Cosmochim. Acta **9**, 1—32b (1956).

LANDSTRÖM, O., and C. G. WENNER: Neutron-activation analysis of natural water applied to hydrogeology. Aktiebolaget Atomenergi (Sweden) AE-204 1965.

LEWIS, G. J., and E. D. GOLDBERG: Iron in marine waters. J. Marine Res. **13**, 183—197 (1954).

LIVINGSTONE, D.: Chemical composition of rivers and lakes. U.S. Geol. Surv. Profess. Paper **440**-G (1963).

LOVERIDGE, B. A., G. W. C. MILNER, G. A. BARNETT, A. M. THOMAS, and W. M. HENRY: The determination of copper, chromium, lead, and manganese in sea water. At. Energy Res. Estab. (Gt. Brit.), Rep. R 3323 1960.

MENARD, H. W., and S. M. SMITH: Hypsometry of ocean basin provinces. J. Geophys. Res. **71**, 4305—4325 (1966).

MERRILL, J. R., E. F. X. LYDEN, M. HONDA, and J. R. ARNOLD: The sedimentary geochemistry of the beryllium isotopes. Geochim. Cosmochim. Acta **18**, 108—129 (1960).

MOORE, W. S.: Amazon and Missisisippi River concentrations of uranium, thorium, and radium isotopes. Earth and Planetary Science Letters **2**, 231—234 (1967).

—, and W. M. SACKETT: Uranium and thorium series inequilibrium in sea water. J. Geophys. Res. **69**, 5401—5405 (1964).

MULLIN, J. B., and J. P. RILEY: The occurrence of cadmium in seawater and in marine organisms and sediments. J. Marine Res. **15**, 103—122 (1956).

MUNK, W. H.: Abyssal recipes. Deep-Sea Res. **13**, 707—730 (1966).

NIELSEN, H., u. W. RICKE: Schwefel-Isotopenverhältnisse von Evaporiten aus Deutschland. Geochim. Cosmochim Acta **28**, 577—591 (1964).

PITMAN, W. C., and J. R. HEIRTZLER: Magnetic anomolies over the Pacific-Antarctic Ridge. Science **154**, 1164—1171 (1966).

REDFIELD, A. C., and I. FRIEDMAN: Factors affecting the distribution of deuterium in the

RICHARDS, F. A.: Some current aspects of chemical oceanography. In Physics and Chemistry of the Earth **2**, 77—128 (1957).

ocean. Narragansett Mar. Lab. Occ. Publ. No. 3-1965, 149—168 (1965).

RILEY, J. P.: The occurrence of anomolously high fluoride concentrations in the North Atlantic. Deep-Sea Res. **12**, 219—220 (1965).

—, and P. SINHASENI: The determination of copper in sea water, silicate rocks and biological materials. Analyst **83**, 299—304 (1958).

RONA, E., L. O. GILPATRICK, and L. M. JEFFREY: Uranium determination in sea water. Trans. Am. Geophys. Union **37**, 697—701 (1956).

—, and W. D. URRY: Radioactivity of ocean sediments. VIII Radium and Uranium content of ocean and river waters. Am. J. Sci **250**, 241 (1952).

RUBEY, W. W.: Geologic history of sea water. Bull. Geol. Soc. Am. **62**, 1111—1147 (1951).
SACKETT, W., and G. ARRHENIUS: Distribution of aluminum species in the hydrosphere — I Aluminum in the ocean. Geochim. Cosmochim. Acta **26**, 955—968 (1962).
SCADDEN, E. M.: The concentration of rhenium in Pacific Ocean surface waters. Geochim. Cosmochim. Acta (in press).
SCHUTZ, D. F., and K. K. TUREKIAN: The investigation of the geographical and vertical distribution of several trace elements in sea water using neutron activation analyses. Geochim. Cosmochim. Acta **29**, 259—313 (1965a).
SCHUTZ, D. F., and K. K. TUREKIAN: The distribution of cobalt, nickel, and silver in ocean water profiles around Pacific Antarctica. J. Geophys. Res. **70**, 5519—5528 (1965b).
SHIGEMATSU, T., Y. NISHIKAWA, K. HIRAKI, and H. NAKAGAWA: Determination of zirconium in sea water. Nippon Kagaku Zasshi **85**, 490—493 (1964). CA **61**, 13043.
SILKER, W. B.: Variations in elemental concentrations in the Columbia River. Limnol. and Oceanog. **9**, 540—545 (1964).
SILLÉN, L. G.: The physical chemistry of sea water. In: Oceanography (ed. M. SEARS), p. 549—581, Publ. 67. Washington, D.C.: Am. Assoc. Adv. Sci., 1961.
— Regulation of O_2, N_2 and CO_2 in the atmosphere; thoughts of a laboratory chemist. Tellus **18**, 198—206 (1966).
—, and A. E. MARTELL: Stability constants of metal-ion complexes. Chem. Soc. (London), Spec. Publ. No. 17, 754 p. (1964).
SKIRROW, G.: The dissolved gases-carbon dioxide. In: Chemical oceanography (ed. J. P. RILEY and G. SKIRROW), vol. 1, p. 227—322. London: Academic Press 1965.
SLOWEY, J. I.: Studies on the distribution of copper, manganese and zinc in the ocean using neutron activation analysis. In: The chemistry and analysis of trace metals in sea water. Final Report AEC Contract AT (40-1)-2799, Texas A & M Project 276, D. W. HOOD, Director (1966).
SMALES, A. A., and B. D. PATE: The determination of sub-microgram quantities of arsenic by radioactivation. Part II. The determination of arsenic in seawater. Analyst **77**, 188—195 (1952).
SOMAYAJULU, B. L. K., and E. D. GOLDBERG: Thorium and uranium isotopes in sea water and sediments. Earth and Planetary Science **1**, 102—106 (1966).
SUGAWARA, K.: Personal communication (1967).
— H. NAITO, and S. YAMADA: Geochemistry of vanadium in natural waters. J. Earth Sci. Nagoya Univ. **4**, 44—61 (1956).
—, and S. OKABE: Geochemistry of molybdenum in natural water (1): J. Earth Sci. Nagoya Univ. **8**, 93—107 (1960).
SVERDRUP, H. U., M. W. JOHNSON, and R. H. FLEMING: The oceans. New York: Prentice-Hall, Inc. 1942, 1087 p.
TATSUMOTO, M., and C. C. PATTERSON: The concentration of common lead in sea water. In: Earth Science and Meteorites, p. 74—89. Amsterdam: North Holland Publ. Co. 1963.
THODE, H. G., and J. MONSTER: Sulfur-isotope geochemistry of petroleum, evaporites and ancient seas. Bull. Am. Assoc. Petrol. Geologists, Mem. **4**, 367—377 (1965).
TUREKIAN, K. K.: Trace elements in sea water and other natural waters: Annual Report AEC Contract AT (30-1)-2912, Publ. Yale-2912-12. 59 p. (1966).
—, and D. G. JOHNSON: The barium distribution in sea water: Geochim. Cosmochim. Acta **30**, 1153—1174 (1966).
—, and M. D. KLEINKOPF: Estimates of the average abundance of Cu, Mn, Pb, Ti, Ni, and Cr in surface waters of Maine. Bull. Geol. Soc. Am. **67**, 1129—1132 (1956).
VINE, F. J.: Spreading of the ocean floor: new evidence. Science **154**, 1405—1415 (1966).

CHAPTER 11

D. M. SHAW

EVALUATION OF DATA

I. Analytical Errors and Related Topics in Geochemistry

All measurements are erroneous to some degree. It is customary to recognise two main components to the errors which accompany chemical analyses:

a) Statistical error, precision (Ger: zufälliger Fehler, Reproduzierbarkeit; Fr; reproductibilité).

b) Systematic error, accuracy (Ger: systematischer Fehler; Genauigkeit; Fr: précision).

These are theoretically quite different in nature. Although they are usually treated as being independent, this is far from the truth and in most instrumental analyses they are intimately related.

These two forms of error are more often independent of the absolute abundance of the element sought in physical than in chemical methods, especially when expressed in terms of log concentrations. This relationship however breaks down at the lower end of the working range, where it is convenient to recognise the intervention of a factor producing a third source of error:

c) Sensitivity (Ger: Nachweisempfindlichkeit; Fr: sensibilité).

a) Precision

1. This error arises from random, unexpected fluctuations in the analytical procedure, coming from instrumental, environmental and personal sources. This error ε differs with each analysis carried out. If the quantity to be measured is x, then it may be supposed that a particular analysis j made by the method i yields the value x_{ij}: this value is made up of the statistical error ε_{ij}, plus the quantity sought, μ_i:

$$x_{ij} = \mu_i + \varepsilon_{ij}.$$

We shall first assume the magnitude of the error to be unrelated to the magnitude of the quantity sought.

Neither ε nor μ can ever be known: using certain assumptions however they can be estimated. Most of the statistical discussion which follows may be found in Section II, which also gives the necessary calculation formulae.

If n replicate analyses are made we may obtain the mean value:

$$\bar{x}_{i.} = \frac{1}{n} \sum_j x_{ij} = \mu_i + \bar{\varepsilon}_{i.}.$$

If the errors ε_{ij} are small, then it is reasonable to suppose that their average $\bar{\varepsilon}_{i.}$ will be much smaller, and $\bar{x}_{i.}$ will closely approximate to μ_i.

Individuals deviations $(x_{ij}-\bar{x}_{i.})$ give an indication of the magnitude of the errors ε_{ij}, since if

$$\bar{\varepsilon}_{i.} \cong 0$$

then

$$(x_{ij}-\bar{x}_{i.}) \cong \varepsilon_{ij}.$$

In practice, a more useful measure is given by the *standard deviation* s_i, where

$$s_i^2 = \frac{1}{n-1} \sum_j (x_{ij}-\bar{x}_{i.})^2.$$

If n is large, s_i^2 is approximately equal to the *average value of the square of the random error*, and s_i may also be called the *root mean square* (*RMS*) *error*. It is often desirable to compare the standard deviations at different levels of elemental abundance: for this, the *relative standard deviation* or *coefficient of variation* is c_i or C_i (in per cent), where

$$c_i = \frac{s_i}{\bar{x}_i}.$$

Example: 9 analyses of the standard rock W-1 for Co by an optical spectrographic method (Shaw, 1964; Table 4-1) gave the values 35, 32, 37, 33, 38, 33, 39, 38, 36 ppm, from which the following statistics may be obtained:

$$\bar{x}_i = 35.7 \text{ ppm}, \qquad s_i = 2.6 \text{ ppm}, \qquad c_i = 0.073, \qquad C_i = 7.3\%.$$

The group of n analyses referred to can be considered to be one sample or n samples, as desired: in this case they will be taken as being one sample, and the first result from the Example presented is that the best estimate of the Co content of W-1 given by the sample of 9 analyses is 35.7 ppm. This may be rounded if so desired, to 36 ppm (rules for rounding are given in Section II).

2. In order to interpret the meaning of the standard deviation it is necessary to make some assumption about the future behaviour of the errors. It is usual to assume that these are *distributed normally* (by the law of Gauss), with mean value zero and variance σ_i^2 (this is also the variance of x). Once again the use of a Greek symbol reminds us that this quantity can not be known (in some cases it can however be closely approximated by theoretical calculation) and the statistic s_i is its best estimator. For a large sample little error is to be expected by using s_i in the place of σ_i, and in such circumstances the probably value of the next analysis is expressed in statements such as the following:

$$P\{\bar{x}_{i.} - 1.96\, s_i < x_{ik} < \bar{x}_{i.} + 1.96\, s_i\} = 0.95$$

$$P\{\bar{x}_{i.} - 2.58\, s_i < x_{ik} < \bar{x}_{i.} + 2.58\, s_i\} = 0.99$$

where x_{ik} is the next analytical result and P is probability.

Example: suppose that in the previous example $\bar{x}_i$ and s_i had been obtained from a sample of 999 analyses. Then there is a probability of approximately 95% that the next analysis will lie in the range

$$35.7 \pm 1.96 \times 2.6 \qquad \text{or} \qquad 30.6 \text{ to } 40.8 \text{ ppm}$$

there is a 5% probability of this forecast being wrong.

For a small sample (e.g. $n < 100$) it is necessary to multiply s_i by the different factor STUDENT t. This is tabulated (Appendix 2) for different values of $(n-1)$, the *number of degrees of freedom*, and for different probability levels. The table of t is entered for the probability $(1-(\alpha/2))$, when the level desired is $(1-\alpha)$[1].

Example: we will use the original data. For a sample of 9 analyses and probability level of 95% we find $t_{8,0.975} = 2.31$.

There is 95% probability that the next analysis will lie in the range

$$35.7 \pm 2.31 \times 2.6 \qquad \text{or} \qquad 29.7 \text{ to } 41.7 \text{ ppm.}$$

It is more usual to make these predictions about the mean value, thereby obtaining a *confidence interval* for the predicted value of the mean, which is of course μ_i, the "true value" as obtained by method i. The procedure is the same, except that s_i is divided by $\sqrt{n}$ to obtain the *standard error of the mean*. The formal expression of the confidence interval for a *significance level* (accepted probability of being wrong) equal to α is:

$$P\left\{\bar{x}_{i.} - t_{n-1,1-(\alpha/2)} \frac{s_i}{\sqrt{n}} < \mu_i < \bar{x}_{i.} + t_{n-1,1-(\alpha/2)} \frac{s_i}{\sqrt{n}}\right\} = 1-\alpha.$$

Example: the values used previously give us

$$\bar{x}_{i.} = 35.7, \qquad \frac{s_i}{\sqrt{n}} = 0.87, \qquad t_{8,0.975} = 2.31$$

whence the confidence interval is

$$35.7 \pm 2.31 \times 0.87 \qquad \text{or} \qquad 33.7 \text{ to } 37.7 \text{ ppm.}$$

3. If a number of sets of analyses for a given element have been obtained under identical circumstances it is possible to calculate the standard deviation from the *pooled variance*. For k sets, where set p contained n_p analyses, we calculate

$$s_i^2 = \frac{\sum_{p=1}^{k} (n_p - 1)\, s_p^2}{\sum_{p=1}^{k} (n_p - 1)}.$$

4. If the magnitudes of s_i and $\bar{x}_i$ are similar (which is commonly the case in trace element geochemistry) then confidence intervals may extend into regions of negative concentration. Such results indicate that the normal law assumption does not apply: it is recommended that the calculations be repeated using log concentrations.

5. It is often possible to recognise several components to analytical error, such as sample preparation, weighing, electronic circuitry, calculation procedure etc. If the respective variance contributions $\sigma_{i1}^2, \sigma_{i2}^2 \ldots \sigma_{ip}^2$ are each known, then the overall precision is given by

$$\sigma_i^2 = \sum_p \sigma_{ip}^2.$$

Where the variance contributions have been estimated by values such as $s_{i1}^2, s_{i2}^2 \ldots s_{ip}^2$, an analogous addition will give an overall variance estimate $s_i'^2$. Whether or not this value provides a satisfactory estimate of the variance σ_i^2 will depend very much on the circumstances of determination of the separate variance components. It is generally

[1] For definitions and symbols see page 332.

much more satisfactory to measure these in a carefully planned experimental design, using analysis of variance (see Section II).

6. Acceptable values of the relative standard deviation for geochemical research depend on the nature of the problem being studied. Much of the early data used for sketching in the major distribution features of the trace elements had values of C_i of perhaps 50 to 200% : modern methods have improved the precision by a factor of about 10×, and many trace element analyses now have values of C_i between 5 and 15%. Present-day major element (oxide) values of C_i fall usually in the range of 0.5 to 5% : low values are essential for SiO_2 and Al_2O_3 in silicate rocks, and for CaO, MgO and CO_2 in carbonatic ones. The most precise analyses at the present time are often found with measurements at trace levels, using isotopic techniques, and obtaining relative standard deviations of 0.1 to 0.5%. Most of the methods currently in use are capable of high precision, but of course the necessary expenditure of time and labour may not be justified by the nature of the problem. What is of the utmost importance is that every article includes a statement of the precision attained, and how it was measured.

7. It is still common to see analytical precision expressed in terms such as 10.21± 0.52%. This is to be deplored, unless it is clearly indicated what significance to give to such a statement, which as it stands is meaningless.

b) Accuracy

1. In the previous section it was taken that an analysis result x_{ij}, for a given element, estimates the quantity sought, μ_i, when using a particular method i.

The implication of this is that any given method of analysis[1] imparts a certain bias (α_i) which is always present, in effect "camouflaging" the true value μ:

$$x_{ij} = \mu_i + \varepsilon_{ij} = \mu + \alpha_i + \varepsilon_{ij}.$$

Whatever manipulations of the results are made, all that can be estimated by method i is the combined magnitude of $\mu + \alpha_i$.

It should be noticed that this approach assumes the analytical bias or accuracy to be constant, unaffected by random components: in practice the relation is more complex. It should also be borne in mind that analytical bias is theoretically negligible in so-called *absolute methods* (e.g. gravimetric analysis): when absolute methods are used by a skilled analyst this is undoubtedly the case, but the study of G-1 and W-1 (q.v.) indicated clearly the scarcity of skilled analysts. Moreover the competence required by such methods renders them very liable to errors of precision.

To judge the accuracy of an analysis it is highly desirable to have *reference* samples (composition not known from preparation) or *standard* samples whose composition is known from preparation. This can not be perfectly achieved in any branch of science, but in some it is possible to prepare useful synthetic materials (e.g. metallurgical alloy standards) whose composition is known to a higher degree of accuracy than can be expected in their analysis. This is rarely the case in geochemistry because of (a) losses of constituents at the high temperatures usually necessary for reaction, and (b) difficulties of homogenisation of reaction products. It is consequently necessary to

[1] "Method of analysis" is understood to mean a given sequence of chemical operation, as practised in a particular laboratory, at a particular period, by a particular analyst etc.

use natural materials for standards, and their "standardisation" or certification must itself rely on analytical procedures.

Since any analysis will err both in accuracy and in precison, it is evident that the former can only be studied by comparing the results of different methods, each highly precise. In such a case, assuming the bias is constant in each of the methods u and v, we have

$$\bar{x}_{u.} \text{ estimates } \mu + \alpha_u$$

$$\bar{x}_{v.} \text{ estimates } \mu + \alpha_v$$

so that $(\bar{x}_{u.} - \bar{x}_{v.})$ estimates $(\alpha_u - \alpha_v)$.

If, for a number of different methods, it is found that $\bar{x}_{1.} \cong \bar{x}_{2.} \cong \bar{x}_{3.}$ etc. then it may be accepted that the bias (α) in each is negligible and all are of equal accuracy. The best estimate of μ will then be the grand average $\bar{x}_{..}$.

The difficulty of applying these considerations will be examined using examples drawn from the history of attempts to certificate the well-known American samples G-1 and W-1.

2. The history of these two samples has been recorded in a number of publications (see References), beginning with Bulletin 980 of the U.S. Geological Survey in 1951, and continuing until the most recent compilation of new results in 1969. These should be read by all geochemists.

Table 1 in a recent paper (FLEISCHER, 1965) lists three successive attempts to choose recommended values for the 14 major constituents of these rocks, each attempt made 9 years or more after Bulletin 980 (which presented the results of about 30 analyses of each sample in laboratories all over the world). It is fair to say that this Table 1 represents the aggregate results of nearly all of the world's best silicate analysts, yet it is clear that G-1 and W-1 can still not be regarded as standards. For only 4 constituents in G-1 and 2 constituents in W-1, in both cases at abundance levels of < 0.5 per cent, do the three columns agree in the second decimal place. This means that, assuming the second decimal to be meaningful, the analytical precision is superior to the accuracy in most laboratories (i.e. $\bar{\varepsilon}_{i.} < \alpha_i$). Only when the errors in accuracy have been reduced below random statistical error can a substance be considered to be a standard.

Example: in the first 30 analyses reported for W-1, FAIRBAIRN et al. (1951) listed SiO_2 values in the range 51.28 to 53.01 per cent: the mean was 52.69 and the relative standard deviation 0.6 per cent. In 1960 STEVENS and NILES recommended the average 52.46. This was amended by FLEISCHER and STEVENS (1962) to 52.64 (this figure is not given explicitly, but is so attributed in a later article). INGAMELLS and SUHR (1963) preferred the value 52.58. These three values were reproduced by FLEISCHER (1965, Table 1); 11 new analyses for SiO_2 were also included, showing variations between 51.7 and 52.67.

The situation is worse for trace elements, but in the absence of other well-calibrated standards, G-1 and W-1 have provided some much-needed overall control on reported results (in the country of the blind, the one-eyed man is king). Some indication of the results for one well-studied element follow.

Example: among the earliest values reported (up to 1955) for Sr in G-1 were the figures 900, 120, 250, 280, 200, 450 ppm; the range thus extends for a factor of nearly $10\times$ the smallest value. During 1955—56 the values 395, 225, 500, 440, 300, 218 and 233 ppm were reported. Between 1957 and 1959 eleven values in the range 200 to 320 ppm were found, and the authors of Bulletin 1113 suggested either 280 or 250 ppm as the most representative value.

The 1965 article by FLEISCHER contains an additional 13 estimates, representing the results obtained on at least 54 individual analyses, but still ranging from 191 to 382 ppm. In Dr. FLEISCHER's words (op. cit., p. 1275) — "Discrepancies are still serious. Values of 250 and 180[1] appear to be reasonable compromises".

Inhomogeneities in the powdered samples of G-1 and W-1 had been suspected from the beginning. Such effects are difficult to disentangle from other analytical errors, as shown by the case of Pb in G-1: the correct assignation of various errors was only possible after a carefully designed analysis of variance programme described by FLANAGAN in Bulletin 1113. Another and briefer example follows.

Example: the Cs content of G-1 had only been determined 3 times by 1962, giving 1.5, 2.0 and 2.2 ppm (*in* FLEISCHER and STEVENS, 1962). Using samples taken from 2 different bottles of G-1, BUTLER and THOMPSON (1962) found values of 2.3, 2.4 ppm for one bottle, and 1.6, 1.6 ppm for the other. Simultaneous control analyses (neutron activation was used) of other rocks show that beyond any doubt the Cs content was different in the two bottles. Additional determinations given by FLEISCHER in 1965 are 1.5, 4.7 and 2.1 ppm.

Bearing in mind the intensive efforts brought to bear by so many analysts on analysing these two rocks, the results suggest that few published geochemical data can be regarded as well-established. Numerous additional standard rocks are now available or in preparation: once the problems of their certification have been resolved, these controls will no doubt help ensure much more reliable data in the future.

3. The foregoing remarks indicate that the demonstration of inaccuracy, and its subsequent control are difficult.

In many cases this does not interfere with the results of a particular study, made in one laboratory. Differences in the estimates $\bar{x}_{i.}$ and $\bar{y}_{i.}$ of different amounts of an element in two different rocks will give an estimate of the differences between the two amounts:

$$\bar{x}_{i.} = \mu_x + \alpha_{xi} + \bar{\varepsilon}_{xi.}, \qquad \bar{y}_{i.} = \mu_y + \alpha_{yi} + \bar{\varepsilon}_{yi.}$$

and

$$(\bar{x}_{i.} - \bar{y}_{i.}) \cong (\mu_x - \mu_y) + (\alpha_{xi} - \alpha_{yi}).$$

Usually it can safely be assumed for physical methods that the errors in accuracy are independent of the amount being determined: in this case the last term of the equation drops out, and $(\bar{x}_{i.} - \bar{y}_{i.})$ is an unbiased estimate of the real difference.

If a well-calibrated standard is available, it should be included in the series of analyses. Knowing its composition μ_s, the inaccuracy can be determined from its analysis $\bar{z}_{i.}$:

$$\alpha_i = \mu_s - \bar{z}_i \ .$$

In some cases, notably emission spectrography, these procedures are best carried out on log concentrations. If two standards are available it is possible to verify whether the analytical bias varies with the absolute amount present.

[1] For W-1.

Example: analysis of G-1 and W-1 for Ga (SHAW, 1964) gave 25 and 18 ppm, in the course of analysing other materials. Comparison with the then recommended values (20 and 15 ppm) showed nearly constant differences in log concentration. This indicated that a parallel displacement of the spectrographic working curve was desirable, or alternatively that $\log \alpha(\text{G-1}) \cong \log \alpha(\text{W-1})$.

Results for Ti however were 1,450 and 5,100 ppm respectively. Comparison with the recommended values of 1,400 and 6,700 ppm showed that in this case the values of $\log \alpha$ were not the same: the method gave acceptable results at low concentrations but not at higher ones, and a rotational adjustment of the working curve was indicated.

When adequate standards are not available, it is strongly recommended to make inter-laboratory comparisons: if the results from two institutions agree on one or more samples, the confidence in both analysts is fortified:

Example: the Rb content of basaltic and ultramafic rocks is very low and results might consequently be expected to be unreliable: analyses by GAST (1965) of three oceanic tholeiitic basalts gave 1.42, 1.14 and 0.35 ppm, whereas TATSUMOTO et al. (1965) obtained 1.14, 1.06 and 0.45 ppm. At these abundance levels such agreement is excellent and helps confirm the validity of the unusually high K/Rb ratios in such rocks.

If a geochemical article contains no indication of the value of the results in absolute terms (i.e. an assessment of the ability to reproduce the figures obtained by good analysts elsewhere) it is not to be expected that the paper will be widely accepted. This is particularly true of fields of great analytical difficulty, but also still applies even in major element petrological studies

4. It will no doubt have been noticed that in the preceding paragraphs the determination of the characteristic concentrations in a standard has been described as a "choice", a "recommendation", the result of a "good" analyst etc. This was not only customary usage, but deliberate too. There is no way, as indicated several times already, of knowing the concentration of an element in a substance: in the final analysis a subjective choice must be made, among the best of the available data. Only when the accuracy exceeds the precision in all the results obtained does the question of the "best" value resolve itself.

c) Sensitivity

1. The lower limit to an analytical method is determined by the point where a weak *signal* (e.g. faint spectral line, low scintillometer count, small balance needle deflection etc.) cannot be readily distinguished from *background* or *noise*. Background effects have two main components.

The first is *structureless* or *continuous* background, and comprises the aggregate effect of many different kinds of random fluctuations which release energy that can accumulate with the quantity being measured. The zeropoint of a balance, for example, will be subject to variations in temperature, air-currents, dust, mechanical effects such as friction, electrostatic attraction etc.; these will together determine the minimum needle deflection (weight) which can be detected. Sometimes a single specific effect can be identified in the background (e.g. graininess of a photographic emulsion), but not always. The history of advances in scientific instrumentation has always been one of identifying and then suppressing background interferences.

Structured background is the second component, usually referred to as *interference* or *coincidence*. It may occur in any spectral method, consisting as it does of a second, spectral signal superimposed on the first: there will be of course a continuous background as well. Correction for the effects of coincidences can be extremely difficult and three main approaches are used. The first is to prevent the coincidence from appearing: e.g. since the emission line Fe 2497.82 Å interferes with B 2497.73 Å, then spectrographic analyses for boron would best be made after preliminary separation of iron from the samples. The second approach is an improvement of the instrumental *resolution*, relying on the fact that no two line wavelengths or energies are ever quite identical. Such a procedure may be feasible in emission spectroscopy, by using a higher diffraction order, but is impossible in mass-spectrocopy, where the differences in masses such as e.g. $^{32}S^{-1}$ and $2^{16}O^{-1}$ are quite negligible. The third method is to measure another line produced by the interfering substance, take this values as the background and apply a proportional correction (measured in advance) to the analysis line.

Correction for the effect of continuous background may be simple or difficult, depending on the particular instrumental method. It will not be discussed further here, since many excellent specialised articles and texts are available (e.g. AHRENS and TAYLOR, 1961; ADLER, 1966).

2. The determination of the *sensitivity limit* of an analytical method consists essentially of finding out how weak a signal can still be profitably corrected for background. Down to this limit a numerical result is given; below the limit, the signal can still be detected, and is usually worth recording as a *trace* (*tr*); where no signal can be detected at all the symbol *n.d.* or * is usually used. It is important to make clear whether an element has been sought but not found (*n.d.* or *), or not sought (usually given as –); unfortunately the English *n.d.* is used by some for *not detected* and by some for *not determined* (in the sense of *not sought*).

Determination of the sensitivity limit is quite simple for an instrumental method depending on a linear response scale starting at zero. The limit here is simply zero plus the rounding interval (see Section II). If, for example, the method permits discrimination of the values 0.21 and 0.22 per cent, then the limit of sensitivity will be 0.01 per cent. Unfortunately this is of little practical value, for most linear scales do not remain linear close to the zero point: this is the reason why one frequently encounters tables of analyses containing e.g. wt. per cent MnO, in which values such as 0.06, 0.07 occur, but in which the lowest number recorded might be 0.03, accompanied by tr for the lower concentrations.

A practical approach for reporting a sensitivity limit by use of an agreed *convention* is as follows. Suppose the scale reading u is converted to x wt. per cent or ppm, using a working-curve. Let u be measured on a sample of p blanks, containing none of the element in question. These will give an average $\bar{u}_b$ and standard deviation s_b, leading to the following approximate confidence statement (Section II):

$$P\{\bar{u}_b - 2.58\, s_b < \mu_u < \bar{u}_b + 2.58\, s_b\} = 0.99\,.$$

If now u_{s1} is defined by

$$u_{s1} = \bar{u}_b + 2.58\, s_b$$

then it is exceedingly unlikely that a higher value of u ($\geq u_{s1}$) could occur when the element is totally absent. The value of x_{s1} corresponding to u_{s1} may be arbitrarily taken as the sensitivity limit. If it is customary to make n replicates of each analysis, then s_b in the above formula should be replaced by $s_b/\sqrt{n}$.

Analyses along these lines have been made in several articles by H. KAISER (see e.g. 1965, which contains references to earlier papers). A modification of this theory is used in X-ray fluorescence analysis (see e.g. BIRKS, 1959; ADLER, 1966).

II. Statistical Procedures

a) Introduction

1. Intention

The intention of this part of Ch. 11 is to present a summary of the most commonly used statistical procedures in geochemical problems.

It is assumed that the reader is generally conversant with the language and the conceptual framework in which statistical theory is applied, and for this reason the statements which follow are presented with stress on the model and the computational procedure, omitting both mathematical derivation and explanatory discussion. A text in applied statistics (e.g. BENNETT and FRANKLIN, 1954; OSTLE, 1954) may be consulted if amplification is needed.

Many useful topics, e.g. trend surface analysis, multiple regression, partial correlation, discriminatory analysis, factor analysis, are omitted. The geochemist intending to use more advanced procedures such as these will find geological applications and mathematical references in MILLER and KAHN (1962), and in KRUMBEIN and GRAYBILL (1965).

The writer is grateful for critical review of the manuscript by G. V. MIDDLETON and J. G. PAYNE.

2. General Definitions

x_{ij} — the measured value of a variable, classified in both of categories (sets) i and j

$\bar{x}_{i.}$ — the mean value of the variable, with reference to all values in the set j, but classified in another set i

$\sum_i = \sum_{i=1}^{n_i}$ — summation over all the (n_i) given values in the set i

$P\{\ \} = 1 - \alpha$ — the probability that the statement in parentheses is true is equal to $1 - \alpha$

α — the significance level of a test

d.f. — degrees of freedom

$z_{\alpha/2}$ — the abscissa of the normal probability function such that the area under the curve to the right of this point is $\alpha/2$

$t_{\alpha/2,\nu}$ — the abscissa of the Student-t function with ν d.f. such that the area under the curve to the right of this point is $\alpha/2$

$\chi^2_{\alpha/2,\nu}$ — the abscissa of the χ^2 function with ν d.f. such that the area under the curve to the right of this point is $\alpha/2$

$F_{\alpha/2,\nu_1,\nu_2}$ the abscissa of the F function with ν_1 (numerator) and ν_2 (denominator) d.f. such that the area under the curve to the right of this point is $\alpha/2$.

Tables of z, t, χ^2 and F are in Appendices 1 to 4 respectively.

3. Definitions of Moments

1. For a probability distribution $y=f(x)$, the moment of order p about the origin is μ'_p where

$$\mu'_p=\int_R x^p f(x)\,dx, \tag{1}$$

and R indicates the range of values of x.

The first moment μ'_1 is usually written as μ and defines the *mean*.

2. The central moment of order p is μ_p, where

$$u_p=\int_R (x-\mu)^p f(x)\,dx. \tag{2}$$

It may be noted that

$$\mu_0=\mu'_0=1, \tag{3}$$

$$\mu_1=0, \tag{4}$$

$$\mu_2=\sigma^2=\text{variance}. \tag{5}$$

3. For sample moments substitute m_p for μ_p, a summation $\sum_i$ over all the items for the integration, and observed frequencies for probability density. Thus

$$m'_p=\frac{1}{n}\sum_i x_i^p, \tag{6}$$

$$m_p=\frac{1}{n}\sum_i (x_i-\bar{x})^p. \tag{7}$$

For data grouped in k classes, where each of the f_j values falling in class j is represented by the class mid-point x_j, the moments are

$$m'_{pk}=\frac{1}{\sum_j f_j}\sum_j f_j x_j^p, \tag{8}$$

$$m_{pk}=\frac{1}{\sum_j f_j}\sum_j f_j(x_j-\bar{x})^p. \tag{9}$$

b) Propagation of Error

If F is a function of several independent variables x_i, then small errors Δx_i in the variables give rise to an error ΔF by

$$\Delta F=\sum_i \frac{\partial F}{\partial x_i}\Delta x_i$$

neglecting second-order derivatives in the Taylor series expansion. If the errors in variables x_i are expressed as sample variances s_i^2, then the variance of F is

$$s_F^2=\sum_i\left(\frac{\partial F}{\partial x_i}s_i\right)^2.$$

c) Rules for Rounding Sample Statistics

1. In a sample of n items, let the maximum number of digits necessary to include all reported values be m: e.g. if the item-values range between 0.01 and 98.5 then $m=4$. Let us write $n=f\times 10^k$, where f is between 0 and 10, and where k is an integer. The following rules are semi-empirical (SHAW, 1962) and are conservative: they have been criticised by AGTERBERG (1963).

2. For samples where $n>100$ the maximum useful number of figures is $(m+k)$ or $(m+k+1)$ for the sum and the mean; $(2m+k)$ or $(2m+k+1)$ or $(2m+k+2)$ for the variance; one less for the standard deviation than for the variance.

3. For smaller samples the numbers are m or $(m+1)$ for the sum and the mean; $2m$ or $(2m+1)$ or $(2m+2)$ for the variance; one less for the standard deviation than for the variance.

4. If the maximum number of item-value digits m includes p digits (or zeros) to the left of the decimal-point, then these p digits (or zeros) must be counted with the permissible number of digits in the standard deviation: similarly, $2p$ left-hand digits (or zeros) should be included in the variance.

5. When a statistic is to be used in further calculations it is preferable not to round it.

6. In rules 2 and 3 above the choice of how many digits to retain is dictated by the final arithmetic division by which the statistic is obtained, using the principle that the quotient obtained contains enough figures to recompute the numerator

e.g. Suppose 7 values of wt. % Al_2O_3 range from 15.89 to 21.22. Then $m=4$, $k=0$. Further suppose that

$\sum x_i^2 = 2{,}478.0834$ with $8=2m$ figures

and $(n-1)\,s^2 = 0{,}019.020542\ldots$ retaining 4 decimals (8 figures)

$n-1=6$

then $s^2 = 0{,}003.1700904\ldots$ retaining $9=2m+1$ figures

since $0{,}003.17009\times 6 = 0{,}019.02054$,

but $0{,}003.1701 \times 6 = 0{,}019.0206$, which does not equal $(n-1)\,s^2$.

Finally $s = 01.780475$ retaining $9-1=8$ figures.

d) Summary Statistics — Single Variable

1. Central Measures

1. *Arithmetic mean or average* ($\bar{x}$) *or first moment about zero* (m_1').

Continuous variable[1]:

$$\bar{x} = m_1' = \frac{1}{n}\sum_{i=1}^{n} x_i. \tag{1}$$

Grouped variable: the n items have been grouped in k classes, replacing each of the f_j values falling in class j by the class midpoint x_j.

$$\bar{x}_k = m_{1k}' = \frac{1}{\sum_j f_j}\sum_{j=1}^{k} f_j x_j \tag{2}$$

[1] "Continuous variable" will be used to refer to numerical data which have not been grouped into classes, other than by rounding.

the first moment about an origin a is given by

$$\bar{x}_a = m_1' = \frac{1}{n} \sum_i (x_i - a), \tag{3}$$

$$\bar{x}_{ka} = m_{1ka}' = \frac{1}{\sum_j f_j} \sum_j f_j (x_j - a) \tag{4}$$

the *first central moment*, m_1, is calculated about the mean

$$m_1 = \frac{1}{n} \sum_i (x_i - \bar{x}) = 0, \tag{5}$$

$$m_{1k} = \frac{1}{\sum_j f_j} \sum_j f_j (x_j - \bar{x}_k) = 0. \tag{6}$$

2. *Geometric mean*, $\bar{x}_g$:

Continuous variable:

$$\bar{x}_g = (x_1 . x_2 . \dots x_i . \dots x_n)^{1/n} \tag{7}$$

or if $y_i = \ln x_i$, then $\ln \bar{x}_g = \bar{y}$.

Grouped data:

$$\ln \bar{x}_{gk} = \frac{1}{\sum_j f_j} \sum_j f_j \ln x_j. \tag{8}$$

3. *Median*.

The middle value of the sample data, when ranked in order of magnitude: the fiftieth percentile.

The median is indeterminate except for an odd value of n (but may be interpolated).

4. *Mode*.

The most frequent value in a sample: for a continuous variable the mode is the abscissa-value at the maximum probability density. With grouped or rounded data its magnitude is a function of the grouping interval, and more than one mode may occur. It is common practice to extend the definition to *a commonly occurring value*, in order to refer to polymodal samples.

5. *Linear transformation of variable*.

For a linear transformation of the kind $z_i = a + b x_i$

$$\bar{z} = a + b\bar{x} \tag{9}$$

for continuous variable and grouped data similar transformations apply to the median and mode, but not to the geometric mean $\bar{z}_g$. However, for the transformation $u_i = b x_i$, it follows that

$$\bar{u}_g = b\,\bar{x}_g.$$

2. Variance, s^2 or V

1. $$V = s^2 = \frac{1}{n-1} \sum_i (x_i - \bar{x})^2 = \frac{1}{n^2 - n} \left[n \sum_i x_i^2 - \left(\sum_i x_i\right)^2 \right], \tag{1}$$

$$V_k = s_k^2 = \frac{1}{\left(\sum_j f_j\right) - 1} \sum_j f_j (x_j - \bar{x}_k)^2 = \frac{\sum_j f_j \left(\sum_j f_j x_j^2\right) - \left(\sum_j f_j x_j\right)^2}{\left(\sum_j f_j\right)^2 - \sum_j f_j}. \tag{2}$$

For computation on a desk calculator the second forms of these expressions are preferable. The square root of the variance, s, is the *standard deviation* or *root-mean square deviation.* The *coefficient of variation* c (or C in per cent) is defined as

$$c = s/\bar{x}. \tag{3}$$

2. The *standard error of the mean* is $s_{\bar{x}}$ where

$$s_{\bar{x}} = \frac{s}{\sqrt{n}}. \tag{4}$$

3. For grouped data the variance s_k^2 is biased, and it may be desirable to apply SHEPPARD's correction to obtain s_c^2 where

$$s_c^2 = s_k^2 - \frac{h^2}{12} \tag{5}$$

h is the width of the class-interval. This correction is seldom worth-while in small samples ($n < 1{,}000$) or where h is less than 1/20 of the range (YULE and KENDALL, 1950, p. 134).

4. The *second central moment* is

$$m_2 = \frac{1}{n} \sum_i (x_i - \bar{x})^2 = \frac{n-1}{n} s^2 \tag{6}$$

or

$$m_{2k} = \frac{1}{\sum_j f_j} \sum_j f_j (x_j - \bar{x}_k)^2. \tag{7}$$

5. For the linear transformation $z_i = a + b x_i$ then

$$s_z = b s_x. \tag{8}$$

Computations of mean and variance may be simplified in many cases by linear transformations.

6. The *pooled variance* s^2 of p different samples, of which sample i contains n_i items and has variance s_i^2 is given by

$$s^2 = \frac{\sum_{i=1}^{p} (n_i - 1)\, s_i^2}{\sum_{i=1}^{p} n_i - p} = \frac{\sum_i \sum_j x_{ij}^2 - \sum_i (T_{i.}^2 / n_i)}{\sum_i n_i - p} \tag{9}$$

where x_{ij} is the value for item j in sample i and $T_{i.} = \sum_{j=1}^{n_i} x_{ij}$.

3. Moments, Semi-invariants

1. Moments may be used to measure the *skewness* (deviation from symmetry about the mode) and *kurtosis* (sharpness or flatness of the peak) for a sampling frequency distribution.

2. More useful are the *semi-invariants* or *k-statistics* of R. A. FISHER, the first four of which are:

$$k_1 = \frac{S_1}{n} = \bar{x}, \tag{10}$$

$$k_2 = \frac{n S_2 - S_1^2}{n(n-1)} = s^2, \tag{11}$$

$$k_3 = \frac{n^2 S_3 - 3n S_2 S_1 + 2 S_1^3}{n(n-1)(n-2)}, \tag{12}$$

$$k_4 = \frac{(n^3 + n^2) S_4 - 4(n^2 + n) S_3 S_1 - 3(n^2 - n) S_2^2 + 12 n S_2 S_1^2 - 6 S_1^4}{n(n-1)(n-2)(n-3)} \tag{13}$$

where $S_p = \sum_i x_i^p$. FISHER's *g-statistics* are convenient measure of skewness (g_1) and kurtosis (g_2):

$$g_1 = \frac{k_3}{k_2^{3/2}} \quad \text{and} \quad g_2 = \frac{k_4}{k_2^2}. \tag{14}$$

A symmetrical distribution gives $g_1 = 0$: positive values correspond to a distribution leaning to the left (mode at smaller value than the mean): this is called *positive skewness*: negative values indicate *negative skewness*. The normal distribution has $g_2 = 0$: a distribution more peaked than a normal curve is *leptokurtic* and has $g_2 > 0$: a distribution flatter than a normal curve is *platykurtic* and has $g_2 < 0$.

4. Standardized Variable

1. The variable x_i may be *standardized* (normalised) by the linear transformation to z_i where

$$z_i = \frac{x_i - \bar{x}}{s}$$

z_i then has the properties that

$$\bar{z} = 0 \quad \text{and} \quad s_z^2 = 1.$$

2. For many geochemical purposes it is preferable to use logarithmic units such that

$$y_i = \log x_i$$

and

$$z_i = \frac{\log x_i - \bar{y}}{s_y}.$$

e) Summary Statistics — Two or More Variables

1. Correlation Coefficient, Covariance

A sample of n items consists of measurements of two variables, which on item j may be denoted by x_{1j}, x_{2j}. The *correlation coefficient* is r, where

$$r = \frac{\sum_j (x_{1j} - \bar{x}_{1.})(x_{2j} - \bar{x}_{2.})}{\left[\sum_j (x_{1j} - \bar{x}_{1.})^2 \cdot \sum_j (x_{2j} - \bar{x}_{2.})^2\right]^{\frac{1}{2}}}. \tag{1}$$

Using the notation

$$S_1(x_1x_2)=\sum_j x_{1j}\,x_{2j}$$
$$S_1(x_1)=\sum_j x_{1j}$$
$$S_2(x_1)=\sum_j x_{1j}^2 \text{ etc.}$$

the best form for computation by desk calculator is

$$r=\frac{nS_1(x_1x_2)-S_1(x_1)S_2(x_2)}{\{(nS_2(x_1)-[S_1(x_1)]^2)\,(nS_2(x_2)-[S_1(x_2)]^2)\}^{\frac{1}{2}}}. \tag{2}$$

The *covariance* is $s(x_1x_2)$ where

$$s(x_1x_2)=\frac{1}{n-1}\sum_j(x_{1j}-\bar{x}_{1.})(x_{2j}-\bar{x}_{2.}). \tag{3}$$

Using previous definitions it follows that

$$r=\frac{s(x_1x_2)}{s(x_1)\,s(x_2)} \tag{4}$$

where $s(x_1)$ and $s(x_2)$ are the standard deviations.

If the variables had been standardised to z_{1j} and z_{2j} then

$$r=r_z=\frac{s(z_1z_2)}{s(z_1)\,s(z_2)}=s(z_1z_2) \tag{5}$$

i.e. the correlation coefficient is the covariance for standardised variables.

2. Sample and Dispersion Matrices

1. Measurements of variables 1, 2, ... i, ... m on each of 1, 2, ... j, ... n items of a sample define the sample matrix of order $(m\times n)$ of values x_{ij}, given by

$$\mathbf{X}=\mathbf{X}_{ij}=[x_{ij}]=\begin{bmatrix} x_{11} & x_{12}\cdots x_{1j}\cdots x_{1n}\\ x_{21} & x_{22}\cdots x_{2j}\cdots x_{2n}\\ \vdots & \vdots\\ x_{i1} & \cdots\cdots\; x_{ij}\cdots x_{in}\\ \vdots & \vdots\\ x_{ml} & \cdots\cdots\cdots\cdots\; x_{mn}\end{bmatrix}. \tag{1}$$

The m measurements on each item j constitute the column-vector $\mathbf{X}_j$: the n measurements of any variable i on all the items constitutes the row-vector $\mathbf{X}_i$.

The *sum of squares and cross-products matrix* is a $(m\times m)$ matrix given by the sum of the product of each column-vector $\mathbf{X}_j$ by its transpose $\mathbf{X}_j'$

$$\sum_j \mathbf{X}_j\cdot\mathbf{X}_j'=\Big[\sum_j x_{ij}\cdot x_{i'j}\Big]=\begin{bmatrix}\sum_j x_{1j}^2 & \sum_j x_{1j}x_{2j}\cdots & \sum_j x_{1j}x_{mj}\\ \vdots & & \vdots\\ \sum_j x_{mj}x_{1j} & \cdots\cdots & \sum_j x_{mj}^2\end{bmatrix}. \tag{2}$$

Addition of the values in each row of the sample matrix gives a column-vector

$$\left[\sum_j x_{ij}\right] = [T_i.] = n\overline{\mathbf{X}}. \tag{3}$$

The product of n, $\overline{\mathbf{X}}$ and its transpose $\overline{\mathbf{X}}'$ gives the $(m \times m)$ matrix

$$n\overline{\mathbf{X}} \cdot \overline{\mathbf{X}}' = \begin{bmatrix} \frac{T_1^2.}{n} & \frac{T_1.\,T_2.}{n} \ldots & \frac{T_1.\,T_m.}{n} \\ \vdots & & \vdots \\ \frac{T_m.\,T_1.}{n} & \ldots\ldots\ldots\ldots & \frac{T_m^2.}{n} \end{bmatrix}. \tag{4}$$

Subtraction of corresponding elements in the matrices (2) and (4), followed by division by $(n-1)$ gives the dispersion matrix, $\mathbf{W}$:

$$\mathbf{W} = \frac{1}{n-1}\left[\sum_j \mathbf{X}_j \cdot \mathbf{X}_j' - n\overline{\mathbf{X}} \cdot \overline{\mathbf{X}}'\right] = [s_{ii'}^2] = [w_{ii'}]. \tag{5}$$

Diagonal elements of $\mathbf{W}$ are *variances* of the kind $s_{ii'}^2 = s_i^2$: other elements are covariances, e.g. s_{12}^2. The *correlation matrix* $\mathbf{R}$ may be readily obtained as

$$\mathbf{R} = [R_{ii'}] \qquad \text{where} \qquad R_{ii'} = \frac{w'_{ii}}{\sqrt{w_{ii} \cdot w'_{i'i'}}}. \tag{6}$$

The matrices $\mathbf{W}$ and $\mathbf{R}$ are symmetrical.

A useful check on computations is obtained as follows. If the sum of values in the column vector $\mathbf{X}_j$ is $T_{.j}$ then

$$\sum_i \sum_{i'} \left[\sum_j x_{ij} \cdot x_{i'j}\right] = \sum_j (T_{.j})^2 \tag{7}$$

also if the grand total of the sample-matrix is $T_{..} = \sum_j T_{.j}$ then

$$\sum_i \sum_{i'} \left[\frac{T_i.\,T_{i'}.}{n}\right] = T_{..}^2. \tag{8}$$

2. Computation of the dispersion and correlation matrices is usually the first step in any multivariate statistical analysis. Several kinds of multivariate analysis can be chosen, and the detailed procedures must usually be adapted to each problem encountered according to its nature. For this reason it was deemed inadvisable to present such procedures here in a synoptic form, in case the reader might falsely conclude that the kinds of analysis are stereotyped. For further details see Kendall (1957), Rao (1952), Fisher (1950).

3. Closed Array Data

In a sample of size n, the value of variable i in item j is x_{ij} and the sample matrix is $\mathbf{X} = [x_{ij}]$. If the sum of the variables in each item is equal to a constant k, that is if

$$\sum_{i=1}^{m} x_{ij} = k \qquad \text{for all } j$$

then the matrix is a *closed array*. In geochemistry this occurs most frequently with chemical analyses which sum to 100 per cent or 10^6 ppm. It has been shown (Chayes,

1960, 1962) that the following algebraic relations hold true:

$$\sum_i \sum_k w_{ik} = 0 \qquad \text{where } [w_{ik}] \text{ is the dispersion matrix,} \tag{1}$$

$$s_i^2 + \sum_{k=1}^{m} w_{ik} = 0 \qquad \text{for } k \neq i, \tag{2}$$

$$s_i + \sum_{k=1}^{m} r_{ik} \cdot s_k = 0 \qquad \text{for } k \neq i. \tag{3}$$

Certain consequences of these relations are important when interpreting correlations and variances of closed array data:

1. Since the s_i^2 are all positive, then at least one w_{ik} in each row must be negative$_3$
2. for the variable of maximum variance at least two covariances must be negative$_3$
3. if any s_i is greater than the sum of any p of the other standard deviations, then at least $p+1$ of the covariances formed by i must be negative$_3$
4. the expected value of any r_{ik} for variables associated as randomly as possible is $1/(1-m)$.

Other consequences are discussed by CHAYES (op. cit.).

Caution should therefore be exercised when interpreting correlations.

4. Ratio Correlation

1. It is commonly necessary to study the relative behaviour of two variables x, y, which are themselves ratios of primary data-variables u_1, u_2, u_3, u_4:

$$x = \frac{u_1}{u_2} \qquad \text{and} \qquad y = \frac{u_3}{u_4}.$$

The correlation coefficient r_{xy} is an algebraic function of the correlation coefficients $r_{u_1u_2} = r_{12}$ and coefficients of variation $c_{u_1} = c_1$ of the primary variables:

$$r_{xy} = \frac{r_{13}c_1c_3 - r_{14}c_1c_4 - r_{23}c_2c_3 + r_{24}c_2c_4}{(c_1^2 + c_2^2 - 2r_{12}c_1c_2)^{\frac{1}{2}}\,(c_3^2 + c_4^2 - 2r_{34}c_3c_4)^{\frac{1}{2}}}. \tag{1}$$

This relation has been discussed by CHAYES (1949), and is subject to the assumption that the c_i are suffiiciently small that powers greater than two are negligible. Among the many consequences of this general equation, two may be presented:

a) r_{xy} can only equal zero, if each $r_{ij} = 0$ or if the numerator is zero;

b) r_{xy} can equal unity (perfect linear correlation) for a variety of combinations of r_{ij} and c_i.

It is consequently unwise to draw any inferences concerning the variables u_i from correlation studies of their ratios.

2. An interesting situation arises where $u_2 = u_4$, or the ratios have the same denominator:

$$x = \frac{u_1}{u_2} \qquad \text{and} \qquad y = \frac{u_3}{u_2}.$$

In the case where u_1, u_2 and u_3 are uncorrelated ($r_{12} = r_{23} = r_{31} = 0$) then

$$r_{xy} = \frac{c_2^2}{(c_1^2 + c_2^2)^{\frac{1}{2}}\,(c_2^2 + c_3^2)^{\frac{1}{2}}} \tag{2}$$

r_{xy} will therefore differ from zero, and represents *spurious correlation;* for example if $c_1 = c_2 = c_3$ then $r_{xy} = 0.5$.

f) Properties of Some Probability Density Functions

1. Normal (Gauss-Laplace) Law

1. Equation:

$$y=\frac{1}{\sigma\sqrt{2\pi}}\,e^{-\frac{1}{2}\left(\frac{x-\mu}{\sigma}\right)^2} \qquad \text{for variable } x, \text{ where } -\infty<x<+\infty \tag{1}$$

and

$$y=\frac{1}{\sqrt{2\pi}}\,e^{-\frac{z^2}{2}} \qquad \text{for transformed variable } z=\frac{x-\mu}{\sigma}. \tag{2}$$

These equations may be abbreviated to

$$N(\mu,\sigma^2\,|\,x) \quad \text{or} \quad x \text{ is } N(\mu,\sigma^2)$$

and

$$N(0,1\,|\,z) \quad \text{or} \quad z \text{ is } N(0,1).$$

2. For $N(\mu,\sigma^2)\,|\,x)$:

$$P\{\mu-\sigma<x<\mu+\sigma\}=0.6827$$
$$P\{\mu-2\sigma<x<\mu+2\sigma\}=0.9545$$
$$P\{\mu-3\sigma<x<\mu+3\sigma\}=0.9973.$$

3. If x_1 is $N(\mu_1,\sigma_1^2)$ and x_2 is $N(\mu_2,\sigma_2^2)$ then x_1+x_2 is $N(\mu_1+\mu_2,\sigma_1^2+\sigma_2^2)$, and similarly any linear combination $a_1+b_1x_1+a_2+b_2x_2$ is $N(a_1+b_1\mu_1+a_2+b_2\mu_2, b_1^2\sigma_1^2+b_2^2\sigma_2^2)$, so long as x_1, x_2 are independent.

4. If the central moment of degree n is μ_n, then

$$\mu_n=0 \quad \text{for } n \text{ odd,} \quad \text{and} \quad \mu_n=\frac{n!\,\sigma^n}{2^{n/2}\left(\frac{n}{2}!\right)} \quad \text{for } n \text{ even}$$

thus

$$\mu_1=0, \quad \mu_2=\sigma^2, \quad \mu_3=0, \quad \mu_4=3\sigma^4.$$

5. For $x=N(\mu,\sigma^2)$ then

$$\mu=\text{mean}=\text{mode}=\text{median}=\text{geometric mean.}$$

2. Lognormal (Galton) Law

1. Equation:

$$y=\frac{1}{x\sigma\sqrt{2\pi}}\,e^{-\frac{1}{2}\left(\frac{\log x-\mu}{\sigma}\right)^2} \qquad \text{for variable } x, \text{ where } 0<x<+\infty. \tag{1}$$

This may be abbreviated to $\Lambda(\mu,\sigma^2\,|\,x)$ or x is $\Lambda(\mu,\sigma^2)$.

If we write $u=\ln x$, then u is $N(\mu,\sigma^2)$ and any of the properties of the normal distribution apply to u.

2. If x is $\Lambda(\mu,\sigma^2)$ and if g, α, β^2 are respectively the geometric mean, the arithmetic mean and the variance of x, then:

$$g=e^{\mu}=\text{median }(x), \tag{2}$$

$$g>\text{mode }(x)=e^{\mu-\sigma^2}, \tag{3}$$

$$g < \alpha = e^{\left(\mu + \frac{\sigma}{2}\right)^2} = \mu_1' \quad \text{(first moment of } x\text{)}, \tag{4}$$

$$\beta^2 = e^{(2\mu + \sigma^2)} (e^{\sigma^2} - 1) = \mu_2 \quad \text{(second central moment of } x\text{)}, \tag{5}$$

$$\mu = \ln \frac{\alpha^2}{\sqrt{\alpha^2 + \beta^2}}, \tag{6}$$

$$\sigma^2 = \ln \frac{\alpha^2 + \beta^2}{\alpha^2}. \tag{7}$$

Estimation of α and β is troublesome and the reader is referred to AITCHISON and BROWN (1957).

3. The moment of degree n is

$$\mu_n' = e^{\left(n\mu + \frac{n^2\sigma^2}{2}\right)} \tag{8}$$

hence this expression defines

$$\alpha = \mu_1' = e^{\left(\mu + \frac{\sigma^2}{2}\right)} = g\, e^{\frac{\sigma^2}{2}}.$$

4. If the coefficient of variation of x is γ, then

$$\gamma^2 = e^{\sigma^2} - 1 \tag{9}$$

and if two populations have the same γ then they have the same σ^2.

5. If x is $\Lambda(\mu, \sigma^2)$ then cx^b is $\Lambda(a + b\mu, b^2 \sigma^2)$, where $c = e^a$: thus e.g. $\frac{1}{x}$ is $\Lambda(-\mu, \sigma^2)$.

If x_1 is $\Lambda(\mu_1, \sigma_1^2)$ and x_2 is $\Lambda(\mu_2, \sigma_2^2)$ and x_1 and x_2 are independent then $x_1 x_2$ is $\Lambda(\mu_1 + \mu_2, \sigma_1^2 + \sigma_2^2)$: similarly x_1/x_2 is $\Lambda(\mu_1 - \mu_2, \sigma_1^2 + \sigma_2^2)$.

3. Binomial Law

1. If the probability of a desired event is p and of any other event is q, where $p + q = 1$, the probability of x desired events in a sample of size n is P_x, where

$$P_x = \frac{n!}{x!\,(n-x)!}\, p^{n-x} q^x \tag{1}$$

where $0 < x < \infty$.

2. Moments of the distribution include

$$\mu_1' = \mu = np, \tag{2}$$

$$\mu_2 = \sigma^2 = np\,q. \tag{3}$$

3. Where n is large the calculation of P_x is facilitated by STIRLING's approximation:

$$n! \cong \sqrt{2n\pi} \cdot n^n e^{-n}. \tag{4}$$

4. Poisson Law

1. Where p is small and n large, the binomial probability distribution approximates to the Poisson law. If $np = m$, the probability of x desired events in a sample of size n is P_x, where

$$P_x = e^{-m} \frac{m^x}{x!}. \tag{1}$$

2. Moments of the distribution include

$$\mu_1' = \mu = m, \tag{2}$$

$$\mu_2 = \sigma^2 = m. \tag{3}$$

g) Assumptions of Randomness and Normality

1. General

1. Many statistical tests and procedures are founded on the twin assumptions that samples have (a), been randomly drawn and (b), come from a normally-distributed parent population.

2. To ensure that 1 (a) is satisfied it is necessary to establish with great care the overall plan and objective of the investigation, to understand thoroughly the conceptual and actual populations from which samples are taken, and to make every effort to randomise the actual sampling. It should be remembered that random sampling is *not* the same as haphazard sampling.

3. Testing for normality 1 (b), or for any other frequency law, is difficult and inconclusive, especially with small samples. Four points should be considered:

a) plotting a cumulative sampling frequency distribution on normal probability paper will often give an assurance that an assumption of normality is reasonably acceptable;

b) transformation of the variable (especially $y = \log x$) commonly improves the confidence that deviations from normality are not serious;

c) many tests and procedures are known to be rather *robust*, i.e. not unduly influenced by deviations from normality;

d) non-parametric tests are available if needed.

A specific test for normality follows.

2. Hypothesis that a Sample has a Normally Distributed Parent Population

1. $H: x$ is $N(\mu, \sigma^2)$.
2. Compute g_1 and g_2 (see § 4.3) and then

$$t_1 = g_1 \sqrt{\frac{n}{6}}, \qquad t_2 = g_2 \sqrt{\frac{n}{24}}.$$

3. If H is true, then

$$P\left\{ t_{1-\alpha/2} < \begin{matrix} t_1 \\ \text{or} \\ t_2 \end{matrix} < t_{\alpha/2} \right\} = 1 - \alpha, \qquad \text{where } t_{\alpha/2} = -t_{1-\alpha/2} \text{ and has } (n-1) \text{ d.f.}$$

4. For samples where n is small (<500) a modification of this test is available (Bennett and Franklin, 1954, p. 96).

h) Estimation

It will be assumed in the following that the probability level desired is $P = 1 - \alpha$, where α is the significance level. It is further assumed that sampling is random.

1. Confidence Interval for Mean μ

1. Assume x is $N(\mu, \sigma^2)$, and random sample n.

2. Compute $c = t_{\alpha/2} \frac{s}{\sqrt{n}}$, where $t_{\alpha/2}$ has $(n-1)$ d.f. in tables of the t-distribution.

3. The confidence interval is defined by

$$P\{\bar{x} - c < \mu < \bar{x} + c\} = 1 - \alpha.$$

2. Confidence Interval for Sum (or Difference) of Means μ_1 and μ_2

1. Assume samples of n_1 and n_2 items drawn randomly from $x_1 = N(\mu_1, \sigma_1^2)$ and $x_2 = N(\mu_2, \sigma_2^2)$ respectively, where $\sigma_1^2 = \sigma_2^2 = \sigma^2$.

2. Compute $c = t_{\alpha/2} \cdot s^2 \left(\frac{1}{n_1} + \frac{1}{n_2}\right)^{\frac{1}{2}}$, where s^2 is the pooled variance [see Eq. (d.2.9.)] and $t_{\alpha/2}$ has $(n_1 + n_2 - 2)$ d.f.

3. Then

$$P\{(\bar{x}_1 \pm \bar{x}_2) - c < \mu_1 \pm \mu_2 < (\bar{x}_1 \pm \bar{x}_2) + c\} = 1 - \alpha.$$

For cases where it cannot be assumed that $\sigma_1^2 = \sigma_2^2$, then use s and $t_{\alpha/2}$ as given in § i.5.7.

3. Confidence Interval for Variance σ^2

1. Assume x is $N(\mu, \sigma^2)$ and random sample n.

2. Compute $c_1 = \frac{(n-1)s^2}{\chi^2_{\alpha/2}}$ and $c_2 = \frac{(n-1)s^2}{\chi^2_{1-\alpha/2}}$, where χ^2 has $(n-1)$ d.f. in tables of the χ^2 distribution.

3. The confidence interval is defined by

$$P\{c_2 < \sigma^2 < c_1\} = 1 - \alpha.$$

4. Confidence Interval for Correlation Coefficient r

1. Assume a sample n drawn randomly from a bivariate normal distribution in $x_1 x_2$ with parameters $\mu_1, \mu_2, \sigma_1^2, \sigma_2^2, \varrho$.

2. Compute $z = \frac{1}{2} \ln \frac{1+r}{1-r}$ and $t = \frac{t_{\alpha/2}}{(n-3)^{\frac{1}{2}}}$ where $t_{\alpha/2}$ has an infinite number of d.f. (i.e. is an abscissa of the standard normal distribution).

3. Then

$$P\{z - t < \varrho < z + t\} = 1 - \alpha.$$

4. The limits $(z - t)$ and $(z + t)$ may be retransformed by § h.4.2., to give two values of r.

5. The foregoing applies to cases where $n > 50$ and r is small. Otherwise graphs prepared by F. N. DAVID may be used (see BENNETT and FRANKLIN, 1954, pp. 273—80).

5. Confidence Interval for Mean of Binomial Distribution

1. Assume x is randomly drawn from a binomial population with parameter π, which is estimated by the number of successful events p in the sample n. The following is applicable where $np > 5$.

2. Compute

$$m = \frac{n}{n + z^2_{\alpha/2}} \left(p + \frac{z^2_{\alpha/2}}{2n}\right)$$

and

$$c = \frac{n z_{\alpha/2}}{n + z^2_{\alpha/2}} \left[p \frac{(1-p)}{n} + \frac{z^2_{\alpha/2}}{4n}\right]^{\frac{1}{2}}$$

where $z_{\alpha/2}$ is the abscissa of $z = N(0, 1)$ at the quantile $\alpha/2$.

3. Then

$$P\{m - c < \pi < m + c\} = 1 - \alpha.$$

i) Tests of Hypothesis

1. General

1. A hypothesis will be symbolised by H and usually refers to the magnitude of a parameter:

e.g. $H: \mu = c$, means the hypothesis that the mean of the population is equal to a numerical value c.

2. A test leads to a statement of probability, P_t, that the hypothesis is true. Depending on the prior establishment of a probability level $P = 1 - \alpha$ for judgement (where α is known as the *significance level of the test*), one of three outcomes is usual:

a) if $P_t < P$: reject H;
b) if $P_t > P$: accept H;
c) if $P_t \simeq P$: defer decision.

Neglecting outcome 3, two kinds of error may ensue:

d) Type I error: rejection of H when H is true, the probability of this error is α;

e) Type II error: acceptance of H when H is false; the probability of this error is a function of the true magnitude of the parameter in question.

2. Hypothesis that μ is Equal to Some Numerical Value (Population Variance Unknown)

1. $H: \mu = c$.

2. Assume a random sample n from $x = N(\mu, \sigma^2)$.

3. Compute $t = \frac{\bar{x} - c}{s} \cdot n^{\frac{1}{2}}$.

4. If H is true, then $P\{t_{1-\alpha/2} < t < t_{\alpha/2}\} = 1 - \alpha$ where $t_{\alpha/2} = -t_{1-\alpha/2}$ and has $(n-1)$ d.f. This is sometimes called a two-tailed test.

5. The related hypotheses, $H: \mu \geqq c$ and $H: \mu \leqq c$ are treated in the same fashion, except in one respect, namely the choice of the critical value of t:

i.e. for $H: \mu \leqq c$ then

$$P\{t < t_\alpha\} = 1 - \alpha.$$

This is a one-tailed test.

3. Hypothesis that σ^2 is Equal to some Numerical Value

1. $H: \sigma^2 = c$.

2. Assume a random sample n from $x = N(\mu, \sigma^2)$.

3. Compute $\chi^2 = \frac{(n-1)\, s^2}{c}$.

4. If H is true, then

$$P\{\chi^2_{1-\alpha/2} < \chi^2 < \chi^2_{\alpha/2}\} = 1-\alpha, \qquad \text{where } \chi^2_{\alpha/2} \text{ has } (n-1) \text{ d.f.}$$

5. The related one-tailed test is similar to above:

e.g. If $H: \sigma^2 \leqq c$ is true then

$$P\{\chi^2 < \chi^2_\alpha\} = 1-\alpha, \qquad \text{where } \chi^2_\alpha \text{ has } (n-1) \text{ d.f.}$$

6. In the case where s^2 is a pooled variance from p samples [see Eq. (d.2.9)] the procedure is the same except that:

a) compute $\chi^2 = \frac{\left(\sum_i n_i - p\right) s^2}{c}$;

b) $\chi^2_{\alpha/2}$ (or χ^2_α) has $\left(\sum_i n_i - p\right)$ d.f.

4. Hypothesis that Two Samples have the same Population Variance

1. Assume random samples n_1 and n_2 from $x_1 = N(\mu_1, \sigma_1^2)$ and $x_2 = N(\mu_2, \sigma_2^2)$.

2. $H: \sigma_1^2 = \sigma_2^2 = c$.

3. Compute $F = \frac{s_1^2}{s_2^2}$.

4. If H is true, then

$$P\{F_{1-\alpha/2} < F < F_{\alpha/2}\} = 1-\alpha,$$

where

$$F_{\alpha/2} \text{ and } F_{1-\alpha/2} \text{ each have } v_1 = n_1 - 1 \quad \text{and} \quad v_2 = n_2 - 1 \text{ d.f.}$$

5. If F is computed by choosing to put the larger sample variance in the numerator, then the test should be modified to the one-tailed case:

$$P\{F < F_\alpha\} = 1-\alpha.$$

5. Hypothesis that Two Samples have the same Population Mean

1. Assume random samples n_1 and n_2 from $x_1 = N(\mu_1, \sigma_1^2)$ and $x_2 = N(\mu_2, \sigma_2^2)$.

2. First test $H_\sigma: \sigma_1^2 = \sigma_2^2$.

3. If H_σ is accepted proceed as follows.

4. $H: \mu_1 = \mu_2$.

5. Compute $t = \frac{\bar{x}_1 - \bar{x}_2}{s}$, where $s^2 = s^2_{(\bar{x}_1 - \bar{x}_2)} = s_p^2 \left(\frac{1}{n_1} + \frac{1}{n_2}\right)$ and $s_p^2 =$ pooled sample variance [Eq. (d.2.9.)].

6. If H is true, then

$$P\{t_{1-\alpha/2} < t < t_{\alpha/2}\} = 1-\alpha, \qquad \text{where } t_{\alpha/2} = -t_{1-\alpha/2} \text{ and has } (n_1 + n_2 - 2) \text{ d.f.}$$

7. If H_σ is rejected proceed as in 4., 5., and 6., with the following changes:

a) $w_1 = s_1^2/n_1$ and $w_2 = s_2^2/n_2$ where w_1 and w_2 have $v_1 = (n_1 - 1)$ and $v_2 = (n_2 - 1)$ d.f. respectively;

b) $s^2 = w_1 + w_2$;

c) compute the d.f. v of t as follows:

$$\frac{1}{v} = \frac{1}{v_1}\left(\frac{w_1}{w_1+w_2}\right)^2 + \frac{1}{v_2}\left(\frac{w_2}{w_1+w_2}\right)^2.$$

d) compare t with $t_{\alpha/2}$ and $t_{1-\alpha/2}$, with v d.f.

6. Hypothesis that Several Samples have the same Population Variance (Bartlett's Test)

1. Assume random samples $n_1, n_2 \ldots n_i \ldots n_p$ from a set of variables x_i which are $N(\mu_i, \sigma_i^2)$.
2. $H: \sigma_1^2 = \sigma_2^2 = \cdots = \sigma_i^2 = \cdots = \sigma_p^2$.
3. Compute:
a) s^2 = pooled variance;
b) $C = 1 + \frac{1}{3(p-1)}\left(\sum_i \left(\frac{1}{n_i-1}\right) - \frac{1}{\sum_i (n_i-1)}\right)$;
c) $B = \frac{1}{C}\left(\sum_i (n_i-1) \ln s^2 - \sum_i (n_i-1) \ln s_i^2\right)$.
4. If H is true, then

$$P\{B < \chi_\alpha^2\} = 1-\alpha, \qquad \text{where } \chi_\alpha^2 \text{ has } (p-1) \text{ d.f.}$$

5. Note:
a) It is recommended that each $n_i > 6$;
b) since $1/C$ is < 1 it is unnecessary to compute C, unless B appears significant when C is taken as unity.

7. Hypothesis that the Population Correlation Coefficient ϱ is Equal to some Numerical Value

1. Assume a random sample n from a bivariate normal population in $x_1\, x_2$ with parameters $\mu_1, \mu_2, \sigma_1^2, \sigma_2^2, \varrho$.
2. Assume $n \geq 50$ and ϱ_0 not large.
3. $H: \varrho = \varrho_0 \neq 0$.
4. Compute:

$$z = \frac{1}{2} \ln \frac{1+r}{1-r}$$

$$z_0 = \frac{1}{2} \ln \frac{1+\varrho_0}{1-\varrho_0} + \frac{\varrho_0}{2(n-1)}$$

and

$$t = \frac{z-z_0}{(n-3)^{\frac{1}{2}}}.$$

5. If H is true, then

$$P\{t_{1-\alpha/2} < t < t_{\alpha/2}\} = 1-\alpha, \qquad \text{where } t_{\alpha/2} \text{ and } t_{1-\alpha/2} \text{ are}$$

abscissa points for a standard normal deviate (or t-distribution with ∞ d.f.).

6. For the hypothesis $H: \varrho = 0$ compute

$$t = \frac{r}{(1-r^2)^{\frac{1}{2}}}(n-2)^{\frac{1}{2}}$$

and if H is true then

$$P\{t_{1-\alpha/2} < t < t_{\alpha/2}\} = 1-\alpha, \quad \text{where } t_{\alpha/2} \text{ and } t_{1-\alpha/2} \text{ have } (n-2) \text{ d.f.}$$

8. Hypothesis that a Sample of Enumeration Data is from a Binomial Population of Probability π_0

1. Assume that a random sample n, from binomial population with parameter π, shows d successes.

2. $H: \pi = \pi_0$.

3. Compute: $\chi^2 = \dfrac{(d - n\,\pi_0)^2}{n\,\pi_0(1-\pi_0)}$.

4. If H is true, then

$$P\{\chi^2_{1-\alpha/2} < \chi^2 < \chi^2_{\alpha/2}\} = 1 - \alpha, \quad \text{where } \chi^2_{\alpha/2} \text{ has 1 d.f.}$$

5. This test applies when n is large.

k) Analysis of Variance

1. General

1. Analysis of variance (ANOVA) is used to obtain information about population means and variances of categories, classes or sub-sets in a set of measurements of a variable.

2. $x_{ij\alpha}$ will denote a measurement of x which is replicate number α, in class j and also in class i (the two are jointly referred to as *cell ij*), where

$$i = 1, 2, \ldots p$$

$$j = 1, 2, \ldots q$$

$$\alpha = 1, 2, \ldots n_{ij}.$$

Also it is convenient to define

$$T_{ij} = \sum_{\alpha=1}^{n_{ij}} x_{ij\alpha}, \quad \bar{x}_{ij} = \frac{T_{ij}}{n_{ij}}, \quad T_{i.} = \sum_j \sum_\alpha x_{ij\alpha}, \quad \bar{x}_{i.} = \frac{T_{i.}}{\sum_i n_{ij}}$$

$$T = T_{..} = \sum_i \sum_j \sum_\alpha x_{ij\alpha} = \sum_i T_{i.} = \sum_j T_{.j}, \qquad \bar{x}_{..} = \frac{T_{..}}{\sum_i \sum_j n_{ij}} = \bar{x}\,.$$

Both computation and theory is simplified if n_{ij} is constant ($= n$) for all i, j and this will be assumed in the following. Then

$$\bar{x}_{ij} = \frac{T_{ij}}{n}, \qquad \bar{x}_{i.} = \frac{T_i}{qn}, \qquad \bar{x}_{..} = \frac{T_{..}}{pq\,n} = \bar{x}\,.$$

3. The model to be analysed or examined takes the form of a statement such as the following:

$$x_{ij\alpha} = \mu + \beta_i + \gamma_j + \delta_{ij} + \varepsilon_{\alpha(ij)}$$

in which μ = population mean

- β_i, γ_j numerical constants characteristic of every measurement in sub-sets i, j respectively
- δ_{ij} a numerical quantity called *interaction*, associated with the quantities β_i and γ_j in the cell ij only
- $\varepsilon_{\alpha(ij)}$ random error associated with each measurement in the cell ij.

4. The minimum assumptions are that the $x_{ij\alpha}$ are sampled randomly, in a fashion appropriate to the particular model, that the $\varepsilon_{\alpha(ij)}$ are uncorrelated, and that ε is $N(0, \sigma^2)$. The analysis of variance takes the form of examining various estimates of the variance σ^2 in relation to the variances associated with other terms in the model.

5. There are three principal types of ANOVA, each of which may be adapted to a variety of classifications and sampling plans.

Type 1. Parameters are mean values of sub-sets.

Variances of the kind $\sigma_\beta^2, \sigma_\gamma^2, \sigma_\delta^2$ can be estimated but are constants, with no meaning outside the sub-sets. In addition, estimates of the means of sub-sets, their differences etc., can be made.

Type 2: Components of variance.

The sub-sets are samples from infinite populations of sub-sets. Their variances are therefore population parameters and can be used for prediction as well as estimation.

Type 3: An extension of Type 2 to the case where the sub-sets are drawn from finite populations.

Sampling and testing procedures may differ for the different types, but the algebraic procedures are basically similar, varying rather with the hierarchical relations of the sub-sets and the model.

6. There are two principal kinds of classification, sampling plan or model (each of which must be further identified as to whether it is Type 1, 2 or 3). In the first of these the kinds of classification (factors) are different in kind and are said to be *crossed*:

e.g. the distribution of Ce in igneous rocks might be related both to the type of accessory mineral (apatite, sphene, allanite etc.) and to the type of rock (pegmatite, granite, syenite etc.): analyses made of each mineral separated from each rock, if suitably randomised, might be amenable to a two-factor crossed analysis of variance.

In the second kind of model, the factors represent successive sub-divisions or sampling procedures: this is a *hierarchical* or *nested* classification:

e. g. the apparent chemical homogeneity of a rock-type with respect to some element might be affected by the variability introduced by collection of specimens, crushing and quartering, chemical analysis: if each specimen were crushed and quartered, and each quarter analysed several times, the whole sampling operation being randomised, the analyses could be examined by a two-factor nested analysis of variance.

In other cases a *mixed classification* may be appropriate, where two or more factors may be crossed and others nested. Such analysis rapidly increases in complexity as factors of both kinds are added, and will not be discussed here. (There is a thorough discussion in BENNETT and FRANKLIN, 1954, Chapter 7.)

7. Hypotheses concerning the equality of different variance estimates s_1^2 and s_2^2 may be tested by comparing their ratio, F, with values of the F-distribution at the appropriate significance-level (see § i.4) using the d.f. of s_1^2 and s_2^2.

Confidence limits and comparisons of mean values are made by using the t-statistic as in § h and § i.

8. The ANOVA tables which follow give E.M.S. values applicable to Type 2 models. In other cases modification may be necessary, which will of course affect any tests to be used.

9. The following symbols will be used
S.S. sum of squares
d.f. degree of freedom
M.S. mean square
E.M.S. expected value of the mean square
F ratio of mean squares.

2. Single Factor with Replication

Source	S.S.	d.f.	M.S.	E.M.S.	F
Between classes	$S_i = n\sum_i (\bar{x}_{i.} - \bar{x})^2$	$p-1$	$\frac{S_i}{p-1} = s_1^2$	$\sigma^2 + n\sigma_\beta^2$	$F = s_1^2/s^2$
Within classes	$S_{\alpha(i)} = \sum_i \sum_\alpha (x_{i\alpha} - \bar{x}_{i.})^2$	$p(n-1)$	$S_{\alpha(i)} = s^2$	σ^2	
Total	$S = \sum_i \sum_\alpha (x_{i\alpha} - \bar{x})^2$	$np-1$			

1. Model: $x_{i\alpha} = \mu + \beta_i + \varepsilon_{i\alpha}$.
2. Assume: n is the same in each class.
3. Computation:

$$S = \sum_i \sum_\alpha x_{i\alpha}^2 - T^2/np$$

$$S_i = \sum_i T_i^2/n - T^2/np$$

$$S_{\alpha(i)} = S - S_i$$

$$F = s_i^2/s^2, \qquad \text{with } \nu_1 = p-1 \text{ and } \nu_2 = p(n-1) \text{ d.f.}$$

4. Hypothesis $H: \sigma_\beta^2 = 0$ is accepted at significance level α, if

$$P\{F < F_{\alpha, \nu_1, \nu_2}\} = 1 - \alpha.$$

5. If the hypothesis is rejected then
a) the estimator of σ^2 is s^2;
b) the estimator of σ_β^2 is $s_\beta^2 = (s_i^2 - s^2)/n$;
c) the estimator of μ is $\bar{x} = T/np$, with variance estimate $s_{\bar{x}}^2 = s^2/np$;
d) the estimator of $\mu + \beta_i$ is $\bar{x}_{i.} = T_{i.}/n$ with variance estimate $s_{\bar{x}_i}^2 = s^2/n$.

6. On the basis of the estimators obtained in § k.2.5 it is possible to obtain confidence limits and to test hypotheses concerning the various statistics: these mostly follow the lines indicated earlier (§ h and § i), but since care is needed on some points of procedure (especially defining d.f.) it is recommended that a text be consulted (e.g. BENNETT and FRANKLIN, 1954, Chapter 7).

7. If the hypothesis is accepted, then the data may be treated as a sample of np items from $x = N(\mu, \sigma^2)$.

3. Two Factor Crossed with Replication

1. Model:

$$x_{ij\alpha} = \mu + \beta_i + \gamma_j + \delta_{ij} + \varepsilon_{ij\alpha} \qquad \text{where } i = 1, 2 \ldots p$$
$$j = 1, 2 \ldots q$$
$$\alpha = 1, 2 \ldots n.$$

Source	S.S.	d.f.	M.S.	E.M.S.	F
Between rows	$S_i = nq \sum_i (\bar{x}_{i.} - \bar{x})^2$	$p-1$	$s_i^2 = S_i/(p-1)$	$\sigma^2 + n\sigma_\delta^2 + nq\sigma_\beta^2$	
Between columns	$S_j = np \sum_j (\bar{x}_{.j} - \bar{x})^2$	$q-1$	$s_j^2 = S_j/(q-1)$	$\sigma^2 + n\sigma_\delta^2 + np\sigma_\gamma^2$	
Inter-action	$S_{ij} = n \sum_i \sum_j (\bar{x}_{ij} - \bar{x}_{i.} - \bar{x}_{.j} + \bar{x})^2$	$(p-1)(q-1)$	$s_{ij}^2 = S_{ij}/(p-1)(q-1)$	$\sigma^2 + n\sigma_\delta^2$	$F_1 = s_{ij}^2/s^2$
Within cells (replicates)	$S_{\alpha(ij)} = \sum_i \sum_j \sum_\alpha (x_{ij\alpha} - \bar{x}_{ij})^2$	$pq(n-1)$	$s^2 = S_{\alpha(ij)}/pq(n-1)$	σ^2	
Total	$S = \sum_i \sum_j \sum_\alpha (x_{ij\alpha} - \bar{x})^2$	$pqn-1$			

2. Computations:

$$S = \sum_i \sum_j \sum_\alpha x_{ij\alpha}^2 - T^2/npq$$

$$S_i = \sum_i T_{i.}^2/nq - T^2/npq$$

$$S_j = \sum_j T_{.j}^2/np - T^2/npq$$

$$S_{ij} = \sum_i \sum_j T_{ij}^2/n - \sum_i T_{i.}^2/nq - \sum_j T_{.j}^2/np + T^2/npq$$

$$S_{\alpha(ij)} = S - S_i - S_j - S_{ij}$$

$$F_1 = s_{ij}^2/s^2 \qquad \text{with } \nu_1 = (p-1)(q-1) \text{ and } \nu_2 = pq(n-1) \text{ d.f.}$$

3. Hypothesis $H_1: \sigma_\delta^2 = 0$ is first tested (see § k.2.4.) using F_1.

4. If H_1 is rejected:

a) Hypotheses $H_2: \sigma_\gamma^2 = 0$ and $H_3: \sigma_\beta^2 = 0$ may be tested using the ratios $F_2 = s_j^2/s_{ij}^2$ and $F_3 = s_i^2/s_{ij}^2$ respectively. If either is rejected the estimator of variance σ_γ^2 or σ_β^2 may be computed as before.

b) The estimator of σ^2 is s^2.

5. If H_1 is accepted:

a) Pool the Interaction and Within Cells sums of squares to obtain

$$s'^2 = (S_{ij} + S_{\alpha(ij)})/(pqn - p - q + 1).$$

b) Test hypotheses H_2 and H_3 using the ratios

$$F_2' = s_j^2/s'^2 \qquad \text{and} \qquad F_3' = s_i^2/s'^2.$$

c) If both H_2 and H_3 are rejected the estimators of σ_γ^2 and σ_β^2 may be computed.

d) If both H_2 and H_3 are rejected the estimator of σ^2 is s'^2.

e) If one of H_2 or H_3 is accepted a new pooled variance s''^2 may be computed to estimate σ^2.

6. If H_1, H_2 and H_3 are all accepted then the data may be treated as a sample of npq items from $x = N(\mu, \sigma^2)$.

7. Estimators of μ, β_i, γ_j, δ_{ij} are obtained as before (where the appropriate hypothesis has been rejected). Their sample variances are obtained by dividing s^2 (or s'^2 or s''^2) by npq, nq, np and n respectively.

8. In the case where there is only one observation in each cell, or $n=1$ *(two factor crossed without replication)* the Within Cells variance disappears and it is not possible to distinguish interaction from residual (replication) variance. Otherwise the procedure follows § k.2.5, omitting sections 3 and 4.

4. Two Factor Nested, with Replication

1. Model:

$$x_{ij\alpha}=\mu+\beta_i+\gamma_{j(i)}+\varepsilon_{ij\alpha} \quad \text{where} \quad \begin{aligned} i&=1,2\ldots p\\ j&=1,2\ldots q\\ \alpha&=1,2\ldots n.\end{aligned}$$

Source	S.S.	d.f.	M.S.	E.M.S.	F
Between classes	$S_i=nq\sum_i(\bar{x}_{i.}-\bar{x})^2$	$p-1$	$s_i^2=S_i/(p-1)$	$\sigma^2+n\sigma_\gamma^2+nq\sigma_\beta^2$	
Between subclasses	$S_{j(i)}=n\sum_i\sum_j(\bar{x}_{ij}-\bar{x}_{i.})^2$	$p(q-1)$	$s_{ij}^2=S_{j(i)}/(pq-p)$	$\sigma^2+n\sigma_\gamma^2$	$F_1=s_{ij}^2/s^2$
Within subclasses (replicates)	$S_{\alpha(ij)}=\sum_i\sum_j\sum_\alpha(x_{ij\alpha}-\bar{x}_{ij})^2$	$pq(n-1)$	$s^2=S_{\alpha(ij)}/(pqn-pq)$		
Total	$S=\sum_i\sum_j\sum_\alpha(x_{ij\alpha}-\bar{x})^2$	$pqn-1$			

2. Computations:

$$S=\sum_i\sum_j\sum_\alpha x_{ij\alpha}^2-T^2/npq$$

$$S_i=\sum_i T_{i.}^2/np-T^2/npq$$

$$S_{j(i)}=\sum_i\sum_j T_{ij}^2/n-\sum_i T_{i.}^2/nq$$

$$S_{\alpha(ij)}=S-S_i-S_{j(i)}$$

$$F_1=s_{ij}^2/s^2 \quad \text{with} \quad \nu_1=p(q-1) \text{ and } \nu_2=pq(n-1) \text{ d.f.}$$

3. Hypothesis H_1: $\sigma_\gamma^2=0$ is first tested using F_1.

4. If H_1 is rejected:

a) the estimator of σ_γ^2 is $(s_{ij}^2-s^2)/n$;

b) the estimator of σ^2 is s^2;

c) Hypothesis H_2:$\sigma_\beta^2=0$ is tested, using $F_2=s_i^2/s_{ij}^2$;

d) if H_2 is rejected, then σ_β^2 can be estimated by $(s_i^2-s_{ij}^2)/nq$.

5. If H_1 is accepted:

a) the estimator of σ^2 is $s'^2=(S_{j(i)}+S_{\alpha(ij)})/pq\,n-p)$;

b) Hypothesis H_2 is tested, using $F_2=s_i^2/s'^2$.

6. If both H_1 and H_2 are accepted the data may be treated as a sample of npq items from $x=N(\mu,\sigma^2)$.

7. Estimators of μ, β_i, $\gamma_{j(i)}$ are obtained as before (where the appropriate hypothesis has been rejected). Their sample variances are obtained by dividing s^2 (or s'^2) by npq, nq or n respectively.

l) Linear Regression

1. General

1. Assume a random sample of n pairs of values y_i and x_i where

a) the x_i are fixed values, subject to no error;

b) the y_i are $N(\alpha+\beta x_i, \sigma^2)$.

2. Compute:

$$b=\frac{n\sum x_i y_i-\sum x_i \sum y_i}{n\sum x_i^2-(\sum x_i)^2}$$

$$a=\frac{\sum y_i-b\sum x_i}{n}=\bar{y}-b\bar{x}$$

$$s_{y|x}^2=\frac{\sum y_i^2-a\sum y_i-b\sum x_i y_i}{n-2}.$$

3. Then b and a estimate the parameters β and α respectively and define the *linear regression* of y on x: $s_{y|x}^2$ estimates the variance σ^2.

4. Hypothesis $H:\beta=\beta_0$:

a) Compute

$$s_b^2=\frac{s_{y|x}^2}{\sum x_i^2-(\sum x_i)^2/n}$$

and

$$t=\frac{b-\beta_0}{s_b};$$

b) if H is true, then

$P\{t_{1-\alpha/2}< t< t_{\alpha/2}\}=1-\alpha$ where $t_{1-\alpha/2}$ and $t_{\alpha/2}$ have $(n-2)$ d.f.

5. The confidence limits for β are given by

$P\{b-t_{\alpha/2}\, s_b<\beta< b+t_{\alpha/2}\cdot s_b\}=1-\alpha$ where $t_{\alpha/2}$ has $(n-2)$ d.f.

6. If there are a priori reasons for expecting that y and x are not linearly related, it may be possible to transform one or both variables so that a linear relationship obtains: a common transformation is $u=\log y$ and/or $v=\log x$. Alternatively, if y is linearly related to $f(x)$ where $f(x)$ is a polynomial or a power series in x, *multiple linear regression* may be employed (see BENNETT and FRANKLIN, 1954, Chapter 6).

7. The following relationships may be useful:

$$b_{y|x}=r\frac{s_y}{s_x}$$

$$b_{x|y}=r\frac{s_x}{s_y}$$

$$s_{y|x}^2=\frac{n-1}{n-2}s_y^2(1-r^2)$$

where r is the correlation coefficient and $b_{x/y}$ is the slope for the regression of x on y.

2. Analysis of Regression (Test of Independence of Two Variables)

1. The hypothesis $H: \beta = 0$ may alternatively be tested by the methods of analysis of variance, using the following table:

Source	S.S.	d.f.	M.S.	E. M. S.	F
Linear regression	$S_r = b^2 \sum_i (x_i - \bar{x})^2$	1	$s_r^2 = S_r$	$\sigma^2 + \beta^2 \sum_i (x_i - \bar{x})^2$	$F = s_r^2 / s^2$
Residual	$S_\alpha = \sum_i (y_i - a - b x_i)^2$	$n-2$	$s^2 = S_\alpha/(n-2)$	σ^2	
Total	$S = \sum_i (y_i - \bar{y})^2$	$n-1$			

2. Computations:

$$S_r = b^2 \sum x_i^2 - b^2 (\sum x_i)^2 / n$$

$$S = \sum y_i^2 - (\sum y_i)^2 / n$$

$$S_\alpha = S - S_r$$

$$F = s_r^2 / s^2, \quad \text{with} \quad v_1 = 1 \quad \text{and} \quad v_2 = (n-2) \text{ d.f.}$$

3. If H is true then

$$P\{F < F_{\alpha, v_1, v_2}\} = 1 - \alpha.$$

4. The estimator of σ^2 is s^2, which is identical with $s_{y|x}^2$ as previously computed.

3. Analysis of Regression (Test of Linearity)

1. Assume:

a) n_i random values of y_{ij} are associated with each fixed value of x_i, where $i = 1, 2 \ldots p$ and $j = 1, 2, \ldots n_i$;

b) y_{ij} is $N(\mu_i, \sigma^2)$ and $\mu_i = \alpha + \beta x_i$.

2. Model: $y_{ij} = \alpha + \beta x_i + \delta_i + \varepsilon_{ij}$, where the δ_i are deviations from linearity.

3. Compute: b and a as before (see § 1.1.2. and use $n = \sum_i n_i$)

$$S_r = b^2 \sum_i n_i x_i^2 - b^2 \sum_i (n_i x_i)^2 / \sum_i n_i$$

$$S = \sum_i \sum_j y_{ij}^2 - \left(\sum_i \sum_j y_{ij}\right)^2 / \sum_i n_i$$

$$S_\alpha = \sum_i \sum_j y_{ij}^2 - \sum_i T_{i.}^2(y) / n_i$$

$$S_d = S - S_r - S_\alpha$$

$$F_1 = s_d^2 / s^2, \quad \text{with } v_1 = (p-2) \text{ and } v_2 = \sum_i (n_i - 1) \text{ d.f.}$$

4. Hypothesis H_1: $\delta_i = 0$ expresses the condition of linearity $\mu_i = \alpha + \beta x_i$ and may be tested by F_1 in the usual way.

5. If H_1 is accepted then compute the pooled residual variance $s'^2 = (S_\alpha + S_d) / \left(\sum_i n_i - 2\right)$ and obtain the ratio $F_2 = s_r^2 / s'^2$ to test the hypothesis H_2: $\beta = 0$.

6. If H_1 is rejected, then test the hypothesis H_2: $\beta = 0$ by using the ratio $F_2 = s_r^2 / s_d$.

Source	S.S.	d.f.	M.S.	E.M.S.	F
Linear regression	$S_r = b^2 \sum_i n_i(x_i - \bar{x})^2$	1	$s_r^2 = S_r$	$\sigma^2 + \beta^2 \sum_i n_i(x_i - \bar{x})^2$	
Deviations from linear regression	$S_d = \sum_i n_i(\bar{y}_{i.} - a - b x_i)^2$	$p-2$	$s_d^2 = S_d/(p-2)$	$\sigma^2 + \sum_i n_i \delta_i^2/(p-2)$	$F_1 = s_d^2/s^2$
Within group (residual)	$S_\alpha = \sum_i \sum_j (y_{ij} - \bar{y}_{i.})^2$	$\sum_i n_i - p$	$s^2 = S_\alpha / \sum_i n_i - p$	σ^2	
Total	$S = \sum_i \sum_j (y_{ij} - \bar{y})^2$	$\sum_i n_i - 1$			

III. Interpretation of Chemical Analysis of Silicates

Analyses of minerals and rocks must commonly be interpreted by recalculations of various kinds, appropriate to the situation. Many procedures have been devised, but only a few will be discussed here.

a) Calculation of Atomic Proportion

Analyses are given in wt. per cent metallic oxides, except for certain minor elements and anions. The first step in most interpretative procedures is to calculate the atomic proportions of the metal and/or oxygen, by dividing by the molecular weight, obtaining the molecular proportion, and then multiplying by the number of element atoms in the formula (conversion tables are in Appendix 11-5) e.g.:

	Wt. %	Form. Wt.	Mol. Prop.	Cation. Prop.	Atom. Prop. of O
SiO_2	54.6	60.06	0.9091	0.9091	1.8182
Al_2O_3	15.3	101.94	0.1501	0.3002	0.4503

Analyses are seldom reliable enough to attach any real meaning to the fourth decimal in atomic proportions, but it is advisable to retain them throughout the computations, rounding at the end.

It is sometimes useful to remember that such atomic *proportions* are identical with *gram-atoms per 100 grams*.

b) Significance of Oxygen

It is important to realize that oxygen is never determined in most conventional analytical procedures, and various consequences arise from this fact:

1. The reporting of metallic elements as oxides is a convention, adopted on the assumption that the sum of the constituent oxides in a silicate rock will equal 100 per cent.

2. In cases where other anions are present, too much 0 will be included in the sum of the metallic oxides, and it is necessary to subtract the anion-equivalent oxygen from the sum:

e.g. in the mica analysis below, the sum of constituents reported is 99.05, this including 0.66% *F*: it is necessary therefore to substract 0.28% O_2, which is the fluorine equivalent, to obtain a corrected total of 98.77%: this correction does not apply, of course, for any anion reported as an oxide (e.g. *S* as SO_2).

3. Metals of variable valency may not be determined or reported in the valency states appropriate to the specimen: much care is commonly devoted to the correct determination of Fe^{+2} and Fe^{+3}, but this is not the case for titanium and manganese (to choose two common examples), which are usually reported as TiO_2 and MnO without regard to their particular oxidation states.

4. The analysis will not distinguish between H_2O and OH^{-1}, both of which will be determined as H_2O^+: the O content of a rock is the same in both cases, but the structural role of the O associated with H may differ.

It would be preferable in many ways to report silicate analyses only as elemental abundances actually determined, but the great disadvantage of this would be the absence of the quick check on accuracy provided by the summation of the oxides (99.5—100.5 per cent) which applies in most cases.

c) Interpretation of the Analysis of a Complex Silicate

The method used to verify that an analysis conforms to the expected formula depends on the circumstances. The following indicates a common approach:

Example: The structural formula of phlogopite can be written $(OH,O,F)_4$ $W_2\,Y_{4-6}\,Z_8O_{20}$ where Z includes Si, Al, Ti,

Y includes Al, Ti, Fe^{+2}, Fe^{+3}, Mn, Mg
W includes K, Na, Ca.

The following analysis of a phlogopite (SHAW, SCHWARCZ and SHEPPARD, 1965) is recomputed to verify this formula (see Table p. 357).

Adsorbed water, P_2O_5 and trace constituents are neglected. Atom proportions of cations and of O and other anions are then computed.

Next, the sum of atomic proportions of (O+F) are divided into 24, to give a multiplication factor for the cations. The multiplied cations appear in the last column, ready for substitution in the formula. At this point it is decided arbitrarily to use all the Si and enough Al to give $Z = 8.000$. Residual Al and all the Ti goes into the octahedral Y group. The final formula agrees with the expected one and can be approximated by the following:

$$(O_{0.4}OH_{3.3}F_{0.3})\,(K_{1.3}Na_{0.3}Ca_{0.3})\,Mg_{5.5}Al_{0.4})\,(Si_{6.1}Al_{1.9})\,O_{20}.$$

d) Calculation of the Igneous Norm of a Rock

A combination of the methods of the CIPW authors and of Niggli will be used here. From the weight percentages are calculated the metal-atom or *cation proportions* as before: the sum of the cations is taken, and each is re-calculated as a percentage.

	Wt. %	Atom. Prop. of O	Atom. Prop. of metals, F, H	Atom. Prop. to 24 (O+F)		
SiO_2	43.74	1.4566	0.7283	6.102		$Z = 8.000$
Al_2O_3	13.74	0.4044	0.2696	2.259	1.898 0.361	
TiO_2	0.20	0.0050	0.0025	0.021		
Fe_2O_3	0.38	0.0072	0.0048	0.040		
FeO	0.10	0.0014	0.0014	0.012		$Y = 5.893$
MnO	0.00	—	—	—		
MgO	26.27	0.6515	0.6515	5.459		
CaO	1.89	0.0337	0.0337	0.282		
Na_2O	1.00	0.0162	0.0323	0.271		$W = 1.809$
K_2O	7.06	0.0749	0.1499	1.256		
H_2O^+	3.52	0.1953	0.3907	3.274		
H_2O^-	0.47	—				
F	0.66	—	0.0347	0.291		
P_2O_5	0.02	(0.0007)	—			
CO_2	0.00	—				
Cl	tr	—				
S	0.00	—				
BaO	tr	—				
Sum	99.05	2.8469				
Less O ≡ F	0.28	−0.0173				
Total	98.77	2.8296				
+F		0.0347				
Total O + F		2.8643				

Anions = 3.274 OH
0.291 F
(0.435 O) + 20.000 O

Total 4.000 + 20.000 = 24.000

$$\text{Multiplication factor} = \frac{24}{2.8643} = 8.379$$

The analysis is then rearranged from these figures to give the proportions of the simple mineral molecules, by allotting the cations in the following sequence:

1. 3P:5Ca for *Apatite*.
2. 1 Ti:1 Fe^{+2} for *Ilmenite*.

3. 2 S:1 Fe^{+2} for *Pyrite.*
4. 1 C:1 Ca for *Calcite* (only if the C is expressed as CO_2 in the analysis).
5. 1 K:1 Al for *Orthoclase* or *Leucite.*
6. 1 Na:1 Al for *Albite* or *Nepheline.*
7. Residual 1 Na:1 Fe^{+3} for *Acmite.*
8. 2 Fe^{+3}: 1 Fe^{+2} for *Magnetite.*
9. Residual Fe^{+3} as Hematite.
10. Residual 2 Al:1 Ca for *Anorthite.*
11. Residual 2 Al as *Corundum.*
12. At this stage the ratio of Mg to residual Fe^{+2} (adding Mn^{+2}) is found, and this ratio preserved for diopside, hypersthene, olivine. Combine Mg, Fe^{+2} and Mn^{+2} to give Mg′.
13. 1 Ca:1 Mg′ for *Diopside.*
14. Residual 1 Ca for *Wollastonite.*
15. Residual Mg′ for *Hypersthene* or *Olivine.*

Then allot silicon in accordance with the following formulas:

1. Anorthite, $CaAl_2Si_2O_8$.
2. Diopside, $CaMg'Si_2O_6$.
3. Acmite, $NaFe^{+3}Si_2O_6$.
4. Olivine, Mg'_2SiO_4.
5. Wollastonite, $CaSiO_3$.
6. Nepheline, $NaAlSiO_4$ (see 9).
7. Orthoclase, $KAlSi_3O_8$ (see 8).

8. If available Si is not sufficient for orthoclase, distribute between x units orthoclase and y units leucite, given by $x+y=$ available K and $3x+2y=$ available Si.

9. If residual Si remains after 8, convert part or all the nepheline to albite in respective proportions u and v, where $u+v=$ available Na and $3u+v=$ available Si.

10. If residual Si remains after 9, convert olivine partially or wholly to hypersthene, in respective proportions p and q, where $p+q=\text{Mg}'$ and $p+\frac{q}{2}=$ available Si.

11. Residual Si is reckoned as quartz.

Note the Following.

1. The sum of the normative minerals should still be 100.

2. If small amounts of zirconium, manganese, fluorine and water are present in the analysis, they may usually be ignored in calculating the cation proportions.

3. This system may be adapted for interpretation of the actual *mode* of the rock very easily, and is somewhat superior to the CIPW norm in this regard.

Example: An altered basic tuff (VAN DE KAMP, 1964; Appendix 1, No. 50) has the composition given below. In computing the igneous norm H_2O and CO_2 will be ignored as a product of alteration. Details are given in the tabular form, which permits a check on arithmetic by summation.

	Wt. %	Cation Prop.	Cation %
SiO_2	41.50	6,910	44.81
TiO_2	3.51	439	2.85
Al_2O_3	12.64	2,480	16.08
Fe_2O_3	6.92	867	5.62
FeO	5.22	727	4.71
MnO	0.11	16	0.10
MgO	7.61	1,887	12.24
CaO	5.70	1,016	6.59
Na_2O	1.98	639	4.14
K_2O	1.76	374	2.43
H_2O^+	5.23	(5,805)	—
H_2O^-	6.11	—	—
P_2O_5	0.46	65	0.42
CO_2	1.50	(341)	—
Total	100.25	15,420	99.99

The results of the norm calculation may be summarized as follows:

	Cation norm %
Quartz	2.22
Orthoclase	12.15
Plagioclase ($Ab_{47}An_{53}$)	44.48
Clinopyroxene ($Di_{100}Hed_0$)	4.52
Orthopyroxene ($En_{100}Fs_0$)	22.22
Magnetite	5.88
Ilmenite	5.70
Hematite	1.70
Apatite	1.12
Total	99.99

It should be remembered that the units are not in wt. per cent, but can be readily converted if so desired by using formula weights.

e) Metamorphic and Plutonic Norm Calculation

1. Introduction

The CIPW norm corresponds to magmatic conditions and its hypothetical minerals are characteristic of the catazone or the granulite and sanidinite facies. BARTH recommends the calculation of the *catanorm* for metamorphic rocks in these facies, of the *mesonorm* for amphibolite facies rocks and of the *epinorm* for low-grade rocks. The mesonorm is the most useful and applies also to most granitic rocks.

The catanorm is simply the CIPW norm calculated using the Niggli-Barth cation method (see previous section). The mesonorm is calculated on the same basis but introduces some different hypothetical minerals, as does the epinorm.

2. Minerals of the Mesonorm

Biotite and amphibole are used as standard (hypothetical) minerals in place of certain assemblages containing olivine, orthopyroxene and clinopyroxene. It will be recalled that the Niggli-Barth cationic system takes the unit of any mineral equal to the normal formula divided by the total cation content (ignoring hydrogen):

$$\text{e.g.} \qquad 1\ \mathrm{Fo} = \tfrac{1}{3}\ \mathrm{Mg_2SiO_4}.$$

The biotite and amphibole minerals used are Bi, Ho, Ac and Bk (barkevikite) where

$$1\ \mathrm{Bi} = \tfrac{1}{8} \cdot \mathrm{KAlMg_3Si_3O_{10}(OH)_2}$$

$$1\ \mathrm{Ho} = \tfrac{1}{15} \cdot \mathrm{Ca_2Mg_4Al_2Si_7O_{22}(OH)_2}$$

$$1\ \mathrm{Ac} = \tfrac{1}{15} \cdot \mathrm{Ca_2Mg_5Si_8O_{22}(OH)_2}$$

$$1\ \mathrm{Bk} = \tfrac{1}{16} \cdot \mathrm{Na_2CaMg_4Al_2Si_7O_{22}(OH)_2}.$$

		Si	Ti	Al	Fe^{+3}	Fe^{+2} +Mn	Mg	Ca	Na	K	P	Cation Norm
Cation %		44.81	2.85	16.08	5.62	4.81	12.24	6.59	4.14	2.43	0.42	
Quartz	Si	2.22										2.22 Q
Orthoclase	K · Al · 3Si	7.29		2.43						2.43		12.15 Or
Albite	Na · Al · 3Si	12.42		4.14					4.14			20.70 Ab
Anorthite	Ca · 2Al · 2Si	9.51		9.51				4.755				23.78 An
Leucite	K · Al · 2Si											Lc
Nepheline	Na · Al · Si											Ne
Corundum	2Al											Cor
Acmite	Na · Fe^{+3} · 2Si											Acm
Diopside	Ca · (Mg + Fe^{+2}) · 2Si	2.26					1.13	1.13				4.52 Di
Wollastonite	Ca · Si											Wo
Hypersthene	(Mg + Fe^{+2}) · Si	11.11					11.11					22.22 Hy
Olivine	2(Mg + Fe^{+2}) · Si											Or
Magnetite	Fe^{+2} · 2Fe^{+3}				3.92	1.96						5.88 Mt
Haematite	2Fe^{+3}				1.70							1.70 Hm
Ilmenite	Fe^{+2} · Ti		2.85			2.85						5.70 Il
Pyrite	Fe^{+2} · 2S											Py
Apatite	5Ca · 3P							0.70			0.42	1.12 Ap
Calcite	Ca · C											Ct
Total												99.99

The last is used for silica-poor rocks. It is understood that ferrous iron will be included with magnesia in these minerals. Edenite was later added (1962) for rocks over-saturated with silica:

$$1\ \mathrm{Ed} = \tfrac{1}{16} \cdot \mathrm{NaCa_2Mg_5AlSi_7O_{22}(OH)_2}.$$

If so desired, these minerals can readily be calculated from the CIPW norm, using equations such as the following:

$$10\ \mathrm{Or} + 9\ \mathrm{Fo} = 16\ \mathrm{Bi} + 3\ \mathrm{Q}$$

$$8\ \mathrm{Di} + 6\ \mathrm{En} + \mathrm{Q} = 15\ \mathrm{Ac}$$

$$5\ \mathrm{An} + 4\ \mathrm{Di} + 6\ \mathrm{En} = 15\ \mathrm{Ho}$$

$$10\ \mathrm{Ab} + 4\ \mathrm{Di} + 6\ \mathrm{En} = 16\ \mathrm{Bk} + 4\ \mathrm{Q}$$

Another useful mineral is sphene:

$$1\ \mathrm{Sph} = \tfrac{1}{3}\ \mathrm{CaSiTiO_5}.$$

Sphene should be calculated in place of ilmenite.

Other normative minerals are in common with the catanorm. It will be noticed that common metamorphic minerals such as muscovite, sillimanite, kyanite, andalusite, garnet, staurolite and cordierite are not used. The reason for this is the complex set of factors which determine their formation, and the desire to avoid these factors in the calculations. They can in any case be readily computed from the norm.

3. Principles of Calculation

The detailed principles will be found in BARTH's paper. In general however, the calculation procedure adopts a sequence which prevents the appearance of incompatible assemblages, according to the bulk composition. This is accomplished by considering the relative amounts of Si, Al, Ca, Mg, Fe and K. The different steps are outlined in the following.

4. Calculation Procedure (Mesonorm)

1. Accessories such as Mt, Sph, Ap, Ct, etc. are calculated as usual.
2. Calculate Ab and (provisionally) Or.
3. Combine residual Fe^{+2} with Mg to obtain Mg′; remaining Ca and Al become Ca′ and Al′.
4. Different steps now follow according to the mutual relations in the tetrahedron Al′—Ca′—Mg′—K.
5. If Al′ > 2 Ca′.
   a) This is characteristic of granitic gneisses;
   b) use Ca′ for An; state Al′ as Cor;
   c) form Bi according to the equation

$$3\ \mathrm{Mg}' + 5\ \mathrm{Or} = 8\ \mathrm{Bi}.$$

6. If Al′ < 2 Ca′ and Al′ > 2 Ca′ − Mg′/2.
   a) Various hornblende gneisses;
   b) Ca′ will be distributed between An and Ho:
   if x is Ca′ in An and y is Ca′ in Ho then

$$x + y = \mathrm{Ca}',$$

$$2x + y = \mathrm{Al}';$$

   c) form Bi as in 5.3.
7. If Al′ < 2 Ca′ − Mg′/2.
   a) More basic rocks than in 6;
   b) Ca′ must be distributed between An, Ho, Di: let x, y, z be Ca′ in An, Ho, Di:

$$x + y + z = \mathrm{Ca}',$$

$$2x + y = \mathrm{Al}',$$

$$2y + z = \mathrm{Mg}',$$

   thus Ca′ in An $= x = 1/3\ (\mathrm{Al}' + \mathrm{Ca}' - \mathrm{Mg}')$;

c) Bi does not appear here.

8. If Ca′ > Al′/2+Mg′.

a) This condition would lead to excess Ca going in Wo, but in the mesofacies it is better to use Ct.

9. If Al′+Mg′/2>2 Ca′+3/2 K and if Mg′>3 K then potash feldspar will be absent.

10. If Mg′>2 Ca′+3 K then anorthite will be absent.

11. If the rock is deficient in Si the calculation will end with a deficiency in quartz. In the mesonorm this deficiency is taken up in the barkevikite molecule, according to one of the following equations:

$$15\ \mathrm{Ho}+10\ \mathrm{Ab}=16\ \mathrm{Bk}+5\ \mathrm{An}+4\ \mathrm{Q}$$

$$10\ \mathrm{Ab}+7\tfrac{1}{2}\mathrm{Ho}+4\ \mathrm{Hy}=16\ \mathrm{Bk}+4\tfrac{1}{2}\ \mathrm{Q}+\mathrm{C}$$

$$10\ \mathrm{Ab}+15\ \mathrm{Ac}=16\ \mathrm{Bk}+4\ \mathrm{Di}+5\ \mathrm{Q}$$

$$10\ \mathrm{Ab}+7\tfrac{1}{2}\ \mathrm{Ac}+3\ \mathrm{Hy}=16\ \mathrm{Bk}+4\tfrac{1}{2}\ \mathrm{Q}$$

In rare cases it may also be necessary to convert Hy to Ol and Ab to Ne.

Example: The mesonorm for the tuff used in the last example has been calculated in the following table.

	Si	Ti	Al	Fe^{+3}	Fe^{+2} +Mn	Mg	Ca	Na	K	P	Total	Cation Mesonorm
	44.81	2.85	16.08	5.62	4.81	12.24	6.59	4.14	2.43	0.42	99.99	
Ap 5Ca·3P							0.70			0.42		1.12 Ap
Sph Ca·Ti·Si	2.85	2.85					2.85					8.55 Sph
Mt Fe^{+2}·2Fe^{+3}				5.62	2.81							8.43 Mt
Ab Na·Al·3Si	12.42		4.14					4.14				20.70 Ab
Or K·Al·3Si	7.29		2.43						2.43			12.15 Or
	22.25		9.51		2.00	12.24	3.04					
	Ca′ = 3.04 Al′ = 9.51 Mg′ = 14.24	Since Al′ > 2 Ca′, proceed to compute anorthite and biotite										
An Ca·2Al·2Si	6.08		6.08				3.04					15.20 An
Cor 2Al			3.43									3.43 Cor
Bi 5Or·3Mg′						7.29			(+12.15 Or)			19.44 Bi
Hyp Mg′·Si	6.95					6.95						13.90 Hyp
Qtz Si	9.22											9.22 Qtz
											Sum	112 14
											Less Or	−12.15
											Total	99.99

The procedure is very flexible. If, for example, it is undesirable to include Cor in the norm, knowing that the rock has been metamorphosed and contains garnet, the following reaction may be used:

$$6\ \mathrm{Hyp} + 2\ \mathrm{C} = 8\ \mathrm{Alm},$$

$$10.29 + 3.43 = 13.72.$$

Only realistic reactions should however be considered: in this case, the high Mg/Fe ratio in Mg′ would lead to a high pyrope fraction in the garnet, which is not appropriate for the mesonorm.

5. Epinorm

Details of the procedure will not be elaborated here. Important additional mineral species are:

$$\text{Zoisite}\quad \mathrm{Zo} = \tfrac{1}{8}\, Ca_2Al_3Si_3O_{12}(OH)$$

$$\text{Antigorite}\quad \mathrm{Ant} = \tfrac{1}{5}\, Mg_3Si_2O_5(OH)_4$$

$$\text{Amesite}\quad \mathrm{Ame} = \tfrac{1}{5}\, Mg_2Al_2SiO_5(OH)_4.$$

Note added in proof:

The method of mesonorm calculation given on pp. 360—363 follows the original scheme (1959) proposed by Barth, and is satisfactory for many common rocks. With silica-deficient rocks it is preferable to use Act and Ed in place of Ho and Bk, following the desilification steps explained in the second reference (Barth, 1962). I am obliged to Dr. Brenda B. Jensen, Mineralogisk-Geologisk Museum, Oslo, for drawing my attention to this.

Appendix

Appendix 11-1. *Areas of the normal curve.* (From DIXON and MASSEY, 1951, Table 4)

z	X	Area	z	X	Area
−3.0	$\mu-3.0\ \sigma$	0.0013	0.1	$\mu+0.1\ \sigma$	0.5398
−2.9	$\mu-2.9\ \sigma$	0.0019	0.2	$\mu+0.2\ \sigma$	0.5793
−2.8	$\mu-2.8\ \sigma$	0.0026	0.3	$\mu+0.3\ \sigma$	0.6179
−2.7	$\mu-2.7\ \sigma$	0.0035	0.4	$\mu+0.4\ \sigma$	0.6554
−2.6	$\mu-2.6\ \sigma$	0.0047	0.5	$\mu+0.5\ \sigma$	0.6915
−2.5	$\mu-2.5\ \sigma$	0.0062	0.6	$\mu+0.6\ \sigma$	0.7257
−2.4	$\mu-2.4\ \sigma$	0.0082	0.7	$\mu+0.7\ \sigma$	0.7580
−2.3	$\mu-2.3\ \sigma$	0.0107	0.8	$\mu+0.8\ \sigma$	0.7881
−2.2	$\mu-2.2\ \sigma$	0.0139	0.9	$\mu+0.9\ \sigma$	0.8159
−2.1	$\mu-2.1\ \sigma$	0.0179	1.0	$\mu+1.0\ \sigma$	0.8413
−2.0	$\mu-2.0\ \sigma$	0.0228	1.1	$\mu+1.1\ \sigma$	0.8643
−1.9	$\mu-1.9\ \sigma$	0.0287	1.2	$\mu+1.2\ \sigma$	0.8849
−1.8	$\mu-1.8\ \sigma$	0.0359	1.3	$\mu+1.3\ \sigma$	0.9032
−1.7	$\mu-1.7\ \sigma$	0.0446	1.4	$\mu+1.4\ \sigma$	0.9192
−1.6	$\mu-1.6\ \sigma$	0.0548	1.5	$\mu+1.5\ \sigma$	0.9332
−1.5	$\mu-1.5\ \sigma$	0.0668	1.6	$\mu+1.6\ \sigma$	0.9452
−1.4	$\mu-1.4\ \sigma$	0.0808	1.7	$\mu+1.7\ \sigma$	0.9554
−1.3	$\mu-1.3\ \sigma$	0.0968	1.8	$\mu+1.8\ \sigma$	0.9641
−1.2	$\mu-1.2\ \sigma$	0.1151	1.9	$\mu+1.9\ \sigma$	0.9713
−1.1	$\mu-1.1\ \sigma$	0.1357	2.0	$\mu+2.0\ \sigma$	0.9772
−1.0	$\mu-1.0\ \sigma$	0.1587	2.1	$\mu+2.1\ \sigma$	0.9821
−0.9	$\mu-0.9\ \sigma$	0.1841	2.2	$\mu+2.2\ \sigma$	0.9861
−0.8	$\mu-0.8\ \sigma$	0.2119	2.3	$\mu+2.3\ \sigma$	0.9893
−0.7	$\mu-0.7\ \sigma$	0.2420	2.4	$\mu+2.4\ \sigma$	0.9918
−0.6	$\mu-0.6\ \sigma$	0.2741	2.5	$\mu+2.5\ \sigma$	0.9938
−0.5	$\mu-0.5\ \sigma$	0.3085	2.6	$\mu+2.6\ \sigma$	0.9953
−0.4	$\mu-0.4\ \sigma$	0.3446	2.7	$\mu+2.7\ \sigma$	0.9965
−0.3	$\mu-0.3\ \sigma$	0.3821	2.8	$\mu+2.8\ \sigma$	0.9974
−0.2	$\mu-0.2\ \sigma$	0.4207	2.9	$\mu+2.9\ \sigma$	0.9981
−0.1	$\mu-0.1\ \sigma$	0.4602	3.0	$\mu+3.0\ \sigma$	0.9987
0	μ	0.5000			
−3.090	$\mu-3.090\ \sigma$	0.001	+3.090	$\mu+3.090\ \sigma$	0.999
−2.576	$\mu-2.576\ \sigma$	0.005	+2.576	$\mu+2.576\ \sigma$	0.995
−2.326	$\mu-2.326\ \sigma$	0.010	+2.326	$\mu+2.326\ \sigma$	0.990
−1.960	$\mu-1.960\ \sigma$	0.025	+1.960	$\mu+1.960\ \sigma$	0.975
−1.645	$\mu-1.645\ \sigma$	0.050	+1.645	$\mu+1.645\ \sigma$	0.950
−1.282	$\mu-1.282\ \sigma$	0.100	+1.282	$\mu+1.282\ \sigma$	0.900
−1.036	$\mu-1.036\ \sigma$	0.150	+1.036	$\mu+1.036\ \sigma$	0.850
−0.842	$\mu-0.842\ \sigma$	0.200	+0.842	$\mu-0.842\ \sigma$	0.800
−0.674	$\mu-0.674\ \sigma$	0.250	+0.674	$\mu+0.674\ \sigma$	0.750
−0.524	$\mu-0.524\ \sigma$	0.300	+0.524	$\mu+0.524\ \sigma$	0.700
−0.385	$\mu-0.385\ \sigma$	0.350	+0.385	$\mu+0.385\ \sigma$	0.650
−0.253	$\mu-0.253\ \sigma$	0.400	+0.253	$\mu+0.253\ \sigma$	0.600
−0.126	$\mu-0.126\ \sigma$	0.450	+0.126	$\mu+0.126\ \sigma$	0.550
0	μ	0.500			

Appendix 11-2. *Values of t. For significance level* $\frac{\alpha}{2}$, *enter the table at percentile* $100\left(1-\frac{\alpha}{2}\right)$.
(From DIXON and MASSEY, 1951, Table 5)

d.f.	$t_{0.95}$	$t_{0.975}$	$t_{0.9875}$	$t_{0.995}$	$t_{0.9975}$
1	6.31	12.7	25.5	63.7	127
2	2.92	4.30	6.21	9.92	14.1
3	2.35	3.18	4.18	5.84	7.45
4	2.13	2.78	3.50	4.60	5.60
5	2.01	2.57	3.16	4.03	4.77
6	1.94	2.45	2.97	3.71	4.32
7	1.89	2.36	2.84	3.50	4.03
8	1.86	2.31	2.75	3.36	3.83
9	1.83	2.26	2.69	3.25	3.69
10	1.81	2.23	2.63	3.17	3.58
11	1.80	2.20	2.59	3.11	3.50
12	1.78	2.18	2.56	3.05	3.43
13	1.77	2.16	2.53	3.01	3.37
14	1.76	2.14	2.51	2.98	3.33
15	1.75	2.13	2.49	2.95	3.29
16	1.75	2.12	2.47	2.92	3.25
17	1.74	2.11	2.46	2.90	3.22
18	1.73	2.10	2.45	2.88	3.20
19	1.73	2.09	2.43	2.86	3.17
20	1.72	2.09	2.42	2.85	3.15
21	1.72	2.08	2.41	2.83	3.14
22	1.72	2.07	2.41	2.82	3.12
23	1.71	2.07	2.40	2.81	3.10
24	1.71	2.06	2.39	2.80	3.09
25	1.71	2.06	2.38	2.79	3.08
26	1.71	2.06	2.38	2.78	3.07
27	1.70	2.05	2.37	2.77	3.06
28	1.70	2.05	2.37	2.76	3.05
29	1.70	2.05	2.36	2.76	3.04
30	1.70	2.04	2.36	2.75	3.03
40	1.68	2.02	2.33	2.70	2.97
60	1.67	2.00	2.30	2.66	2.91
120	1.66	1.98	2.27	2.62	2.96
∞	1.64	1.96	2.24	2.58	2.81
d.f.	$t_{0.05}$	$t_{0.025}$	$t_{0.0125}$	$t_{0.005}$	$t_{0.0025}$

When the table is read from the foot, the tabled values are to be prefixed with a negative sign. Interpolation should be performed using the reciprocals of the degrees of freedom The values in the above table were computed from percentiles of the F distribution.

Appendix 11-3. *Percentiles of the χ^2 distribution. For significance level $\frac{\alpha}{2}$, enter the table at percentile $100\left(1-\frac{\alpha}{2}\right)$.* (From DIXON and MASSEY, 1951, Table 6a)

d.f.	Percentiles									
	0.5	1	2.5	5	10	90	95	97.5	99	99.5
1	0.000039	0.00016	0.00098	0.0039	0.0158	2.71	3.84	5.02	6.63	7.88
2	0.0100	0.0201	0.0506	0.1026	0.2107	4.61	5.99	7.38	9.21	10.60
3	0.0717	0.115	0.216	0.352	0.584	6.25	7.81	9.35	11.34	12.84
4	0.207	0.297	0.484	0.711	1.064	7.78	9.49	11.14	13.28	14.86
5	0.412	0.554	0.831	1.15	1.61	9.24	11.07	12.83	15.09	16.75
6	0.676	0.872	1.24	1.64	2.20	10.64	12.59	14.45	16.81	18.55
7	0.989	1.24	1.69	2.17	2.83	12.02	14.07	16.01	18.48	20.28
8	1.34	1.65	2.18	2.73	3.49	13.36	15.51	17.53	20.09	21.96
9	1.73	2.09	2.70	3.33	4.17	14.68	16.92	19.02	21.67	23.59
10	2.16	2.56	3.25	3.94	4.87	15.99	18.31	20.48	23.21	25.19
11	2.60	3.05	3.82	4.57	5.58	17.28	19.68	21.92	24.73	26.76
12	3.07	3.57	4.40	5.23	6.30	18.55	21.03	23.34	26.22	28.30
13	3.57	4.11	5.01	5.89	7.04	19.81	22.36	24.74	27.69	29.82
14	4.07	4.66	5.63	6.57	7.79	21.06	23.68	26.12	29.14	31.32
15	4.60	5.23	6.26	7.26	8.55	22.31	25.00	27.49	30.58	32.80
16	5.14	5.81	6.91	7.96	9.31	23.54	26.30	28.85	32.00	34.27
18	6.26	7.01	8.23	9.39	10.86	25.99	28.87	31.53	34.81	37.16
20	7.43	8.26	9.59	10.85	12.44	28.41	31.41	34.17	37.57	40.00
24	9.89	10.86	12.40	13.85	15.66	33.20	36.42	39.36	42.98	45.56
30	13.79	14.95	16.79	18.49	20.60	40.26	43.77	46.98	50.89	53.67
40	20.71	22.16	24.43	26.51	29.05	51.81	55.76	59.34	63.69	66.77
60	35.53	37.48	40.48	43.19	46.46	74.40	79.08	83.30	88.38	91.95
120	83.85	86.92	91.58	95.70	100.62	140.23	146.57	152.21	158.95	163.64

For large values of degrees of freedom the approximate formula

$$\chi^2_\alpha = \tfrac{1}{2}(z_\alpha + \sqrt{2n-1})^2$$

where z_α is the normal deviate and n is the number of degrees of freedom, may be used. For example $\chi^2_{0.99} = \frac{1}{2}(2.326 + 10.909)^2 = 87.6$ for the 99th percentile for 60 degrees of freedom.

Appendix 11-4. *(a) F distribution, upper 5% points*[a]. (From DIXON and MASSEY, 1951, Table 7a)

Degrees of freedom for denominator	Degrees of freedom for numerator																		
	1	2	3	4	5	6	7	8	9	10	12	15	20	24	30	40	60	120	∞
1	161	200	216	225	230	234	237	239	241	242	244	246	248	249	250	251	252	253	254
2	18.5	19.0	19.2	19.2	19.3	19.3	19.4	19.4	19.4	19.4	19.4	19.4	19.4	19.5	19.5	19.5	19.5	19.5	19.5
3	10.1	9.55	9.28	9.12	9.01	8.94	8.89	8.85	8.81	8.79	8.74	8.70	8.66	8.64	8.62	8.59	8.57	8.55	8.53
4	7.71	6.94	6.59	6.39	6.26	6.16	6.09	6.04	6.00	5.96	5.91	5.86	5.80	5.77	5.75	5.72	5.69	5.66	5.63
5	6.61	5.79	5.41	5.19	5.05	4.95	4.88	4.82	4.77	4.74	4.68	4.62	4.56	4.53	4.50	4.46	4.43	4.40	4.37
6	5.99	5.14	4.76	4.53	4.39	4.28	4.21	4.15	4.10	4.06	4.00	3.94	3.87	3.84	3.81	3.77	3.74	3.70	3.67
7	5.59	4.74	4.35	4.12	3.97	3.87	3.79	3.73	3.68	3.64	3.57	3.51	3.44	3.41	3.38	3.34	3.30	3.27	3.23
8	5.32	4.46	4.07	3.84	3.69	3.58	3.50	3.44	3.39	3.35	3.28	3.22	3.15	3.12	3.08	3.04	3.01	2.97	2.93
9	5.12	4.26	3.86	3.63	3.48	3.37	3.29	3.23	3.18	3.14	3.07	3.01	2.94	2.90	2.86	2.83	2.79	2.75	2.71
10	4.96	4.10	3.71	3.48	3.33	3.22	3.14	3.07	3.02	2.98	2.91	2.85	2.77	2.74	2.70	2.66	2.62	2.58	2.54
11	4.84	3.98	3.59	3.36	3.20	3.09	3.01	2.95	2.90	2.85	2.79	2.72	2.65	2.61	2.57	2.53	2.49	2.45	2.40
12	4.75	3.89	3.49	3.26	3.11	3.00	2.91	2.85	2.80	2.75	2.69	2.62	2.54	2.51	2.47	2.43	2.38	2.34	2.30
13	4.67	3.81	3.41	3.18	3.03	2.92	2.83	2.77	2.71	2.67	2.60	2.53	2.46	2.42	2.38	2.34	2.30	2.25	2.21
14	4.60	3.74	3.34	3.11	2.96	2.85	2.76	2.70	2.65	2.60	2.53	2.46	2.39	2.35	2.31	2.27	2.22	2.18	2.13
15	4.54	3.68	3.29	3.06	2.90	2.79	2.71	2.64	2.59	2.54	2.48	2.40	2.33	2.29	2.25	2.20	2.16	2.11	2.07
16	4.49	3.63	3.24	3.01	2.85	2.74	2.66	2.59	2.54	2.49	2.42	2.35	2.28	2.24	2.19	2.15	2.11	2.06	2.01
17	4.45	3.59	3.20	2.96	2.81	2.70	2.61	2.55	2.49	2.45	2.38	2.31	2.23	2.19	2.15	2.10	2.06	2.01	1.96
18	4.41	3.55	3.16	2.93	2.77	2.66	2.58	2.51	2.46	2.41	2.34	2.27	2.19	2.15	2.11	2.06	2.02	1.97	1.92
19	4.38	3.52	3.13	2.90	2.74	2.63	2.54	2.48	2.42	2.38	2.31	2.23	2.16	2.11	2.07	2.03	1.98	1.93	1.88
20	4.35	3.49	3.10	2.87	2.71	2.60	2.51	2.45	2.39	2.35	2.28	2.20	2.12	2.08	2.04	1.99	1.95	1.90	1.84
21	4.32	3.47	3.07	2.84	2.68	2.57	2.49	2.42	2.37	2.32	2.25	2.18	2.10	2.05	2.01	1.96	1.92	1.87	1.81
22	4.30	3.44	3.05	2.82	2.66	2.55	2.46	2.40	2.34	2.30	2.23	2.15	2.07	2.03	1.98	1.94	1.89	1.84	1.78
23	4.28	3.42	3.03	2.80	2.64	2.53	2.44	2.37	2.32	2.27	2.20	2.13	2.05	2.01	1.96	1.91	1.86	1.81	1.76
24	4.26	3.40	3.01	2.78	2.62	2.51	2.42	2.36	2.30	2.25	2.18	2.11	2.03	1.98	1.94	1.89	1.84	1.79	1.73
25	4.24	3.39	2.99	2.76	2.60	2.49	2.40	2.34	2.28	2.24	2.16	2.09	2.01	1.96	1.92	1.87	1.82	1.77	1.71
30	4.17	3.32	2.92	2.69	2.53	2.42	2.33	2.27	2.21	2.16	2.09	2.01	1.93	1.89	1.84	1.79	1.74	1.68	1.62
40	4.08	3.23	2.84	2.61	2.45	2.34	2.25	2.18	2.12	2.08	2.00	1.92	1.84	1.79	1.74	1.69	1.64	1.58	1.51
60	4.00	3.15	2.76	2.53	2.37	2.25	2.17	2.10	2.04	1.99	1.92	1.84	1.75	1.70	1.65	1.59	1.53	1.47	1.39
120	3.92	3.07	2.68	2.45	2.29	2.18	2.09	2.02	1.96	1.91	1.83	1.75	1.66	1.61	1.55	1.50	1.43	1.35	1.25
∞	3.84	3.00	2.60	2.37	2.21	2.10	2.01	1.94	1.88	1.83	1.75	1.67	1.57	1.52	1.46	1.39	1.32	1.22	1.00

Interpolation should be performed using reciprocals of the degrees of freedom.

[a] This table is reproduced with the permission of Professor E. S. PEARSON from MERRINGTON, M., THOMPSON, C. M., "Tables of percentage points of the inverted beta (F) distribution", *Biometrika*, vol. 33 (1943), p. 73.

Appendix 11-4. *(b) F distribution, upper 1% points*[a]. (From DIXON and MASSEY, 1951, Table 7c)

Degrees of freedom for denominator	Degrees of freedom for numerator																		
	1	2	3	4	5	6	7	8	9	10	12	15	20	24	30	40	60	120	∞
1	4,052	5,000	5,403	5,625	5,764	5,859	5,928	5,982	6,023	6,056	6,106	6,157	6,209	6,235	6,261	6,287	6,313	6,339	6,366
2	98.5	99.0	99.2	99.2	99.3	99.3	99.4	99.4	99.4	99.4	99.4	99.4	99.4	99.5	99.5	99.5	99.5	99.5	99.5
3	34.1	30.8	29.5	28.7	28.2	27.9	27.7	27.5	27.3	27.2	27.1	26.9	26.7	26.6	26.5	26.4	26.3	26.2	26.1
4	21.2	18.0	16.7	16.0	15.5	15.2	15.0	14.8	14.7	14.5	14.4	14.2	14.0	13.9	13.8	13.7	13.7	13.6	13.5
5	16.3	13.3	12.1	11.4	11.0	10.7	10.5	10.3	10.2	10.1	9.89	9.72	9.55	9.47	9.38	9.29	9.20	9.11	9.02
6	13.7	10.9	9.78	9.15	8.75	8.47	8.26	8.10	7.98	7.87	7.72	7.56	7.40	7.31	7.23	7.14	7.06	6.97	6.88
7	12.2	9.55	8.45	7.85	7.46	7.19	6.99	6.84	6.72	6.62	6.47	6.31	6.16	6.07	5.99	5.91	5.82	5.74	5.65
8	11.3	8.65	7.59	7.01	6.63	6.37	6.18	6.03	5.91	5.81	5.67	5.52	5.36	5.28	5.20	5.12	5.03	4.95	4.86
9	10.6	8.02	6.99	6.42	6.06	5.80	5.61	5.47	5.35	5.26	5.11	4.96	4.81	4,73	4.65	4.57	4.48	4.40	4.31
10	10.0	7.56	6.55	5.99	5.64	5.39	5.20	5.06	4.94	4.85	4.71	4.56	4.41	4.33	4.25	4.17	4.08	4.00	3.91
11	9.65	7.21	6.22	5.67	5.32	5.07	4.89	4.74	4.63	4.54	4.40	4.25	4.10	4.02	3.94	3.86	3.78	3.69	3.60
12	9.33	6.93	5.95	5.41	5.06	4.82	4.64	4.50	4.39	4.30	4.16	4.01	3.86	3.78	3.70	3.62	3.54	3.45	3.36
13	9.07	6.70	5.74	5.21	4.86	4.62	4.44	4.30	4.19	4.10	3.96	3.82	3.66	3.59	3.51	3.43	3.34	3.25	3.17
14	8.86	6.51	5.56	5.04	4.70	4.46	4.28	4.14	4.03	3.94	3.80	3.66	3.51	3.43	3.35	3.27	3.18	3.09	3.00
15	8.68	6.36	5.42	4.89	4.56	4.32	4.14	4.00	3.89	3.80	3.67	3.52	3.37	3.29	3.21	3.13	3.05	2.96	2.87
16	8.53	6.23	5.29	4.77	4.44	4.20	4.03	3.89	3.78	3.69	3.55	3.41	3.26	3.18	3.10	3.02	2.93	2.84	2.75
17	8.40	6.11	5.19	4.67	4.34	4.10	3.93	3.79	3.68	3.59	3.46	3.31	3.16	3.08	3.00	2.92	2.83	2.75	2.65
18	8.29	6.01	5.09	4.58	4.25	4.01	3.84	3.71	3.60	3.51	3.37	3.23	3.08	3.00	2.92	2.84	2.75	2.66	2.57
19	8.19	5.93	5.01	4.50	4.17	3.94	3.77	3.63	3.52	3.43	3.30	3.15	3.00	2.92	2.84	2.76	2.67	2.58	2.49
20	8.10	5.85	4.94	4.43	4.10	3.87	3.70	3.56	3.46	3.37	3.23	3.09	2.94	2.86	2.78	2.69	2.61	2.52	2.42
21	8.02	5.78	4.87	4.37	4.04	3.81	3.64	3.51	3.40	3.31	3.17	3.03	2.88	2.80	2.72	2.64	2.55	2.46	2.36
22	7.95	5.72	4.82	4.31	3.99	3.76	3.59	3.45	3.35	3.26	3.12	2.98	2.83	2.75	2.67	2.58	2.50	2.40	2.31
23	7.88	5.66	4.76	4.26	3.94	3.71	3.54	3.41	3.30	3.21	3.07	2.93	2.78	2.70	2.62	2.54	2.45	2.35	2.26
24	7.82	5.61	4.72	4.22	3.90	3.67	3.50	3.36	3.26	3.17	3.03	2.89	2.74	2.66	2.58	2.49	2.40	2.31	2.21
25	7.77	5.57	4.68	4.18	3.86	3.63	3.46	3.32	3.22	3.13	2.99	2.85	2.70	2.62	2.53	2.45	2.36	2.27	2.17
30	7.56	5.39	4.51	4.02	3.70	3.47	3.30	3.17	3.07	2.98	2.84	2.70	2.55	2.47	2.39	2.30	2.21	2.11	2.01
40	7.31	5.18	4.31	3.83	3.51	3.29	3.12	2.99	2.89	2.80	2.66	2.52	2.37	2.29	2.20	2.11	2.02	1.92	1.80
60	7.08	4.98	4.13	3.65	3.34	3.12	2.95	2.82	2.72	2.63	2.50	2.35	2.20	2.12	2.03	1.94	1.84	1.73	1.60
120	6.85	4.79	3.95	3.48	3.17	2.96	2.79	2.66	2.56	2.47	2.34	2.19	2.03	1.95	1.86	1.76	1.66	1.53	1.38
∞	6.63	4.61	3.78	3.32	3.02	2.80	2.64	2.51	2.41	2.32	2.18	2.04	1.88	1.79	1.70	1.59	1.47	1.32	1.00

Interpolation should be performed using reciprocals of the degrees of freedom.

[a] This table is reproduced with the permission of Professor E. S. PEARSON from MERRINGTON, M., THOMPSON, C. M., "Tables of percentage points of the inverted beta (*F*) distribution", *Biometrika*, vol. 33 (1943), p. 73.

Appendix 11-5. *Conversion tables, weight per cent to gram-atoms × 10^4*

SiO_2

Conversion of Wt. % to Si × 10^4 gr. atoms

M.W. 60.06: Divisor 0.006006: Factor 166.50

Wt. %	0.0	0.1	0.2	0.3	0.4	0.5	0.6	0.7	0.8	0.9
0	0	17	33	50	67	83	100	117	133	150
1	167	183	200	216	233	250	266	283	300	316
2	333	350	366	383	400	416	433	450	466	483
3	500	516	533	549	566	583	599	616	633	649
4	666	683	699	716	733	749	766	783	799	816
5	833	849	866	882	899	916	932	949	966	982
6	999	1,016	1,032	1,049	1,066	1,082	1,099	1,116	1,132	1,149
7	1,166	1,182	1,199	1,215	1,232	1,249	1,265	1,282	1,299	1,315
8	1,332	1,349	1,365	1,382	1,399	1,415	1,432	1,449	1,465	1,482
9	1,499	1,515	1,532	1,548	1,566	1,582	1,598	1,615	1,632	1,648

40	6,660	50	8,325	60	9,990	70	11,655
41	6,827	51	8,492	61	10,157	71	11,822
42	6,993	52	8,658	62	10,323	72	11,988
43	7,160	53	8,825	63	10,490	73	12,155
44	7,326	54	8,991	64	10,656	74	12,321
45	7,493	55	9,158	65	10,823	75	12,488
46	7,659	56	9,324	66	10,989	76	12,654
47	7,826	57	9,491	67	11,156	77	12,821
48	7,992	58	9,657	68	11,322	78	12,987
49	8,159	59	9,824	69	11,489	79	13,154

Al_2O_3

Conversion of Wt. % to Al × 10^4 gr. atoms

M.W. 101.94: Divisor 0.005097: Factor 196.19

Wt. %	0.0	0.1	0.2	0.3	0.4	0.5	0.6	0.7	0.8	0.9
0	0	20	39	59	78	98	118	137	157	177
1	196	216	235	255	275	294	314	334	353	373
2	392	412	432	451	471	490	510	530	549	569
3	589	608	628	647	667	687	706	726	746	765
4	785	804	824	844	863	883	902	922	942	961
5	981	1,001	1,020	1,040	1,059	1,079	1,099	1,118	1,138	1,158
6	1,177	1,197	1,216	1,236	1,256	1,275	1,295	1,314	1,334	1,354
7	1,373	1,393	1,413	1,432	1,452	1,471	1,491	1,511	1,530	1,550
8	1,570	1,589	1,609	1,628	1,648	1,668	1,687	1,707	1,726	1,746
9	1,766	1,785	1,805	1,825	1,844	1,864	1,883	1,903	1,923	1,942

10	1,962	15	2,943	20	3,924	25	4,905
11	2,158	16	3,139	21	4,120	26	5,101
12	2,354	17	3,335	22	4,316	27	5,297
13	2,550	18	3,531	23	4,512	28	5,493
14	2,747	19	3,728	24	4,709	29	5,690

Appendix 11-5 (Continued)

TiO_2

Conversion of Wt. % to Ti × 10^4 gr. atoms

M.W. 79.90 : Divisor 0.007990 : Factor 125.16

Wt. %	0.0	0.1	0.2	0.3	0.4	0.5	0.6	0.7	0.8	0.9
0	0	13	25	38	50	63	75	88	100	113
1	125	138	150	163	175	188	200	213	225	238
2	250	263	275	288	300	313	325	338	350	363
3	375	388	401	413	426	438	451	463	476	488
4	501	513	526	538	551	563	576	588	601	613
5	626	638	651	663	676	688	701	713	726	738
6	751	763	776	789	801	814	826	839	851	864
7	876	889	901	914	926	939	951	964	976	989
8	1,001	1,014	1,026	1,039	1,051	1,064	1,076	1,089	1,101	1,114
9	1,126	1,139	1,151	1,164	1,177	1,189	1,202	1,214	1,227	1,239

MnO

Conversion of Wt. % to Mn × 10^4 gr. atoms

M.W 70.93 : Divisor 0.007093 : Factor 140.98

Wt. %	0.0	0.1	0.2	0.3	0.4	0.5	0.6	0.7	0.8	0.9
0	0	14	28	42	56	71	85	99	113	127
1	141	155	169	183	197	212	226	240	254	268
2	282	296	310	324	338	353	367	381	395	409
3	423	437	451	465	480	494	508	522	536	550
4	564	578	592	606	621	635	649	663	677	691
5	705	719	733	748	762	776	790	804	818	832

Fe_2O_3

Conversion of Wt. % to Fe × 10^4 gr. atoms

M.W. 159.70 : Divisor 0.007985 : Factor 125.23

Wt. %	0.0	0.1	0.2	0.3	0.4	0.5	0.6	0.7	0.8	0.9
0	0	13	25	38	50	63	75	88	100	113
1	125	138	150	163	175	188	200	213	225	238
2	250	263	276	288	301	313	326	338	351	363
3	376	388	401	413	426	438	451	463	476	488
4	501	513	526	538	551	564	576	589	601	614
5	626	639	651	664	676	689	701	714	726	739
6	751	764	776	789	801	814	827	839	852	864
7	877	889	902	914	927	939	952	964	977	989
8	1,002	1,014	1,027	1,039	1,052	1,064	1,077	1,090	1,102	1,115
9	1,127	1,140	1,152	1,165	1,177	1,190	1,202	1,215	1,227	1,240

Appendix 11-5 (Continued)

FeO

Conversion of Wt. % to Fe × 10^4 gr. atoms

M.W. 71.85 : Divisor *0.007185 : Factor 139.18*

Wt. %	0.0	0.1	0.2	0.3	0.4	0.5	0.6	0.7	0.8	0.9
0	0	14	28	42	56	70	84	97	111	125
1	139	153	167	181	195	209	223	237	251	264
2	278	292	306	320	334	348	362	376	390	404
3	418	431	445	459	473	487	501	515	529	543
4	557	571	585	598	612	626	640	654	668	682
5	696	710	724	738	752	765	779	793	807	821
6	835	849	863	877	891	905	919	933	946	960
7	974	988	1,002	1,016	1,030	1,044	1,058	1,072	1,086	1,100
8	1,113	1,127	1,141	1,155	1,169	1,183	1,197	1,211	1,225	1,239
9	1,253	1,267	1,280	1,294	1,308	1,322	1,336	1,350	1,364	1,378

MgO

Conversion of Wt. % to Mg × 10^4 gr. atoms

M.W. 40.32 : Divisor 0.004032 : Factor 248.02

Wt. %	0.0	0.1	0.2	0.3	0.4	0.5	0.6	0.7	0.8	0.9
0	0	25	50	74	99	124	149	174	198	223
1	248	273	298	322	347	372	397	422	446	471
2	496	521	546	570	595	620	645	670	694	719
3	744	769	794	818	843	868	893	918	942	967
4	992	1,017	1,042	1,066	1,091	1,116	1,141	1,166	1,190	1,215
5	1,240	1,265	1,290	1,315	1,339	1,364	1,389	1,414	1,439	1,463
6	1,488	1,513	1,538	1,563	1,587	1,612	1,637	1,662	1,687	1,711
7	1,736	1,761	1,786	1,811	1,835	1,860	1,885	1,910	1,935	1,959
8	1,984	2,009	2,034	2,059	2,083	2,108	2,133	2,158	2,183	2,207
9	2,232	2,257	2,282	2,307	2,331	2,356	2,381	2,406	2,431	2,455

CaO

Conversion of Wt. % to Ca × 10^4 gr. atoms

M.W. 56.08 : Divisor 0.005608 : Factor 178.32

Wt. %	0.0	0.1	0.2	0.3	0.4	0.5	0.6	0.7	0.8	0.9
0	0	18	36	53	71	89	107	125	143	160
1	178	196	214	232	250	267	285	303	321	339
2	357	374	392	410	428	446	464	481	499	517
3	535	553	571	588	606	624	642	660	678	695
4	713	731	749	767	785	802	820	838	856	874
5	892	909	927	945	963	981	999	1,016	1,034	1,052
6	1,070	1,088	1,106	1,123	1,141	1,159	1,177	1,195	1,213	1,230
7	1,248	1,266	1,284	1,302	1,320	1,337	1,355	1,373	1,391	1,409
8	1,427	1,444	1,462	1,480	1,498	1,516	1,534	1,551	1,569	1,587
9	1,605	1,623	1,641	1,658	1,676	1,694	1,712	1,730	1,748	1,765

Wt. %	Value	Wt. %	Value
10	1,783	15	2,675
11	1,962	16	2,853
12	2,140	17	3,031
13	2,318	18	3,210
14	2,496	19	3,388

Appendix 11-5 (Continued)

Na_2O

Conversion of Wt. % to Na $\times 10^4$ gr. atoms

M.W. 62.00 : Divisor 0.003100 : Factor 322.58

Wt. %	0.0	0.1	0.2	0.3	0.4	0.5	0.6	0.7	0.8	0.9
0	0	32	65	97	129	161	194	226	258	290
1	323	355	387	419	452	484	516	548	581	613
2	645	677	710	742	774	806	839	871	903	935
3	968	1,000	1,032	1,065	1,097	1,129	1,161	1,194	1,226	1,258
4	1,290	1,323	1,355	1,387	1,419	1,452	1,484	1,516	1,548	1,581
5	1,613	1,645	1,677	1,710	1,742	1,774	1,806	1,839	1,871	1,903
6	1,935	1,968	2,000	2,032	2,065	2,097	2,129	2,161	2,194	2,226
7	2,258	2,290	2,323	2,355	2,387	2,419	2,452	2,484	2,516	2,548
8	2,581	2,613	2,645	2,677	2,710	2,742	2,774	2,806	2,839	2,871
9	2,903	2,935	2,968	3,000	3,032	3,065	3,097	3,129	3,161	3,194

K_2O

Conversion of Wt. % to K $\times 10^4$ gr. atoms

M.W. 94.19 : Divisor 0.004710 : Factor 212.31

Wt. %	0.0	0.1	0.2	0.3	0.4	0.5	0.6	0.7	0.8	0.9
0	0	21	42	64	85	106	127	149	170	191
1	212	234	255	276	297	318	340	361	382	403
2	425	446	467	488	510	531	552	573	594	616
3	637	658	679	701	722	743	764	786	807	828
4	849	870	892	913	934	955	977	998	1,019	1,040
5	1,062	1,083	1,104	1,125	1,146	1,168	1,189	1,210	1,231	1,253
6	1,274	1,295	1,316	1,338	1,359	1,380	1,401	1,422	1,444	1,465
7	1,486	1,507	1,529	1,550	1,571	1,592	1,614	1,635	1,656	1,677
8	1,698	1,720	1,741	1,762	1,783	1,805	1,826	1,847	1,868	1,890
9	1,911	1,932	1,953	1,974	1,996	2,017	2,038	2,059	2,081	2,102

H_2O

Conversion of Wt. % to H $\times 10^4$ gr. atoms

M. W. 18.02 : Divisor 0.000901 : Factor 1109.88

Wt. %	0.0	0.1	0.2	0.3	0.4	0.5	0.6	0.7	0.8	0.9
0	0	111	222	333	444	555	666	777	888	999
1	1,110	1,221	1,332	1,443	1,554	1,665	1,776	1,887	1,998	2,109
2	2,220	2,331	2,442	2,553	2,664	2,775	2,886	2,997	3,108	3,219
3	3,330	3,441	3,552	3,663	3,774	3,885	3,996	4,107	4,218	4,329
4	4,440	4,551	4,661	4,772	4,883	4,994	5,105	5,216	5,327	5,438
5	5,549	5,660	5,771	5,882	5,993	6,104	6,215	6,326	6,437	6,548

Appendix 11-5 (Continued)

CO_2

Conversion of Wt. % to $C \times 10^4$ gr. atoms

M.W. 44.01 : Divisor 0.004401 : Factor 227.22

Wt. %	0.0	0.1	0.2	0.3	0.4	0.5	0.6	0.7	0.8	0.9
0	0	23	45	68	91	114	136	159	182	204
1	227	250	273	295	318	341	364	386	409	432
2	454	477	500	523	545	568	591	613	636	659
3	682	704	727	750	773	795	818	841	863	886
4	909	932	954	977	1,000	1,022	1,045	1,068	1,091	1,113
5	1,136	1,159	1,182	1,204	1,227	1,250	1,272	1,295	1,318	1,341

SO_3

Conversion of Wt. % to $S \times 10^4$ gr. atoms

M.W. 80.07 : Divisor 0.008007 : Factor 124.89

Wt. %	0.0	0.1	0.2	0.3	0.4	0.5	0.6	0.7	0.8	0.9
0	0	12	25	37	50	62	75	87	100	112
1	125	137	150	162	175	187	200	212	225	237
2	250	262	275	287	300	312	325	337	350	362
3	375	387	400	412	425	437	450	462	475	487
4	500	512	525	537	550	562	574	587	599	612
5	624	637	649	662	674	687	699	712	724	737

P_2O_5

Conversion of Wt. % to $P \times 10^4$ gr. atoms

M.W. 141.96 : Divisor 0.007098 : Factor 140.88

Wt. %	0.0	0.1	0.2	0.3	0.4	0.5	0.6	0.7	0.8	0.9
0	0	14	28	42	56	70	85	99	113	127
1	141	155	169	183	197	211	225	239	254	268
2	282	296	310	324	338	352	366	380	394	409
3	423	437	451	465	479	493	507	521	535	549
4	564	578	592	606	620	634	648	662	676	690
5	704	718	733	747	761	775	789	803	817	831

Cl

Conversion of Wt. % to $Cl \times 10^4$ gr. atoms

A.W. 35.46 : Divisor 0.003546 : Factor 282.01

Wt. %	0.0	0.1	0.2	0.3	0.4	0.5	0.6	0.7	0.8	0.9
0	0	28	56	85	113	141	169	197	226	254
1	282	310	338	367	395	423	451	479	508	536
2	564	592	620	649	677	705	733	761	790	818
3	846	874	902	931	959	987	1,015	1,043	1,072	1,100
4	1,128	1,156	1,184	1,213	1,241	1,269	1,297	1,325	1,354	1,382
5	1,410	1,438	1,466	1,495	1,523	1,551	1,579	1,607	1,636	1,664

Appendix 11-5 (Continued)

F

Conversion of Wt. % to F × 10^4 gr. atoms
A.W. 19.00 : Divisor 0.001900 : Factor 526.32

Wt. %	0.0	0.1	0.2	0.3	0.4	0.5	0.6	0.7	0.8	0.9
0	0	53	105	158	211	263	316	368	421	474
1	526	579	632	684	737	789	842	895	947	1,000
2	1,053	1,105	1,158	1,211	1,263	1,316	1,368	1,421	1,474	1,526
3	1,579	1,632	1,684	1,737	1,789	1,842	1,895	1,947	2,000	2,053
4	2,105	2,158	2,211	2,263	2,316	2,368	2,421	2,474	2,526	2,579
5	2,632	2,684	2,737	2,789	2,842	2,895	2,947	3,000	3,053	3,105

S

Conversion of Wt. % to S × 10^4 gr. atoms
A.W. 32.07 : Divisor 0.003207 : Factor 311.82

Wt. %	0.0	0.1	0.2	0.3	0.4	0.5	0.6	0.7	0.8	0.9
0	0	31	62	94	125	156	187	218	249	281

ZrO_2

Conversion of Wt. % to Zr × 10^4 gr. atoms
M.W. 123.22 : Divisor 0.012322 : Factor 81.16

Li_2O

Conversion of Wt. % to Li × 10^4 gr. atoms
M.W. 29.88 : Divisor 0.001494 : Factor 669.34

Rb_2O

Conversion of Wt. % to Rb × 10^4 gr. atoms
M.W. 186.96 : Divisor 0.009348 : Factor 106.97

BaO

Conversion of Wt. % to Ba × 10^4 gr. atoms
M.W. 153.36 : Divisor 0.015336 : Factor 65.21

SrO

Conversion of Wt. % to Sr × 10^4 gr. atoms
M.W. 103.63 : Divisor 0.010363 : Factor 96.50

References: Section I

ADLER, I.: X-ray emission spectrography in geology, p. 1—258. Amsterdam: Elsevier 1966.

AHRENS, L. H., and S. R. TAYLOR: Spectrochemical analysis, 2nd ed. p. 1—454. Reading, Mass.: Addison-Wesley 1961.

BIRKS, L. S.: X-ray spectrochemical analysis, p. 1—137. New York: Interscience 1959.

BUTLER, J. R., and A. J. THOMPSON: Different values for Cs in G-1. Geochim. Cosmochim. Acta **26**, 1349—1350 (1962).

CALDER, A. B.: Evaluation and presentation of spectro-analytical results. London: Hilger & Watts Ltd. 1959.

DOERFEL, K.: Beurteilung von Analysenverfahren und -ergebnissen. Berlin-Göttingen-Heidelberg: Springer 1962.

FAIRBAIRN, H. W., and others: A cooperative investigation of precision and accuracy in chemical, spectrochemical and modal analysis of silicate rocks. U.S., Geol. Survey., Bull. **980**, 1—71 (1951).
FLEISCHER, M.: Summary of new data on rock samples G-1 and W-1, 1962—1965. Geochim. Cosmochim. Acta **29**, 1263—1283 (1965).
— U.S. Geological Survey standards-I-additional data on rodes G-1 and W-1 1965—1967. Geochim. Cosmochim. Acta **33**, 65—80 (1969).
—, and R. E. STEVENS: Summary of new data on rock samples G-1 and W-1. Geochim. Cosmochim. Acta **26**, 525—543 (1962).
GAST, P. W.: Terrestrial ratio of K to Rb and the composition of the earth's mantle. Science **147**, 858—860 (1965).
INGAMELLS, C. O., and N. H. SUHR: Chemical and spectrographic analysis of standard silicate samples. Geochim. Cosmochim. Acta **27**, 897—910 (1963).
KAISER, H.: Zum Problem der Nachweisgrenze. Z. Anal. Chem. **209**, 1—18 (1965).
SHAW, D. M.: Interprétation géochimique des éléments en traces dans les roches cristallines, p. 1—237. Paris: Masson & Cie. 1964.
SMALES, A. A., and L. R. WAGER: Methods in geochemistry, p. 1—464. New York and London: Interscience 1960.
STEVENS, R. E., and others: Second report on a cooperative investigation of the composition of two silicate rocks. U.S., Geol. Surv., Bull. **1113**, 1—126 (1960).
TATSUMOTO, M., C. E. HEDGE, and A. E. J. ENGEL: K, Rb, Sr, Th, U and the ratio of Sr^{87} to Sr^{86} in oceanic tholeiitic basalt. Science **150**, 886—888 (1965).

References: Sections II and III

AGTERBERG, F. P.: Rounding off in geochemistry. Letter to the editor. Can. Mineralogist **7**, part 4, 655—662 (1963).
AITCHISON, J., and J. A. C. BROWN: The lognormal distribution, p. 1—176. Cambridge 1957.
BARTH, T. F. W.: Principles of classification and norm calculations of metamorphic rocks. J. Geol. **67**, 135—152 (1959).
— A final proposal for calculating the mesonorm of metamorphic rocks. J. Geol. **70**, 497—498 (1962).
BENNETT, C. A., and N. L. FRANKLIN: Statistical analysis in chemistry and the chemical industry. New York: John Wiley & Sons. 1954.
CHAYES, F.: On ratio correlation in petrography. J. Geol. **57**, 239—254 (1949).
— On correlation between variables of constant sum. J. Geophys. Res. **65**, 4185—4193 (1960).
— Numerical correlation and petrographic variation. J. Geol. **70**, 440—452 (1962).
DIXON, W. J., and F. J. MASSEY JR.: Introduction to statistical analysis, p. 1—370. New York: McGraw-Hill Book Co. 1951.
FISHER, R. A.: Statistical methods for research workers, 11th ed. Edinburgh: Oliver & Boyd 1950.
KAMP, P. C. VAN DE: Geochemistry and classification of amphibolites and related rocks. McMaster University, M.Sc. Thesis 1964 (unpublished).
KENDALL, M. G.: A course in multivariate analysis. New York: Hafner 1957.
KRUMBEIN, W. C., and F. A. GRAYBILL: An introduction to statistical models in geology, p. 1—475. New York: McGraw-Hill Book Co. 1965.
MILLER, R. L., and J. S. KAHN: Statistical analysis in the geological sciences, p. 1—483. New York: John Wiley & Sons 1962.
OSTLE, B.: Statistics in research, p. 1—487. Iowa State College Press 1954.
RAO, C. R.: Advanced statistical methods in biometric research. New York: John Wiley & Sons 1952.
SHAW, D. M.: Rounding errors and Chayes' paradox. Can. Mineralogist **7**, 236—244 (1962).
— H. P. SCHWARCZ, and S. M. F. SHEPPARD: The petrology of two zoned scapolite skarns. Can. J. Earth Sci. **2**, 577—595 (1965).
YULE, G. V., and M. G. KENDALL: An introduction to the theory of statistics. New York: Hafner 1950.

Chapter 12

A. Heydemann

TABLES

Table 12-1. *Physical constants.* (From D'Ans-Lax, 1967)

Where pertinent, values are based on the $^{12}C = 12$ amu scale		
Avogadro's number	N_0	6.02252×10^{23} (g mole)$^{-1}$
Bohr magneton	$\mu_B = \frac{h \cdot e}{4 \cdot \pi \cdot m}$	9.273×10^{-21} erg gauss^{-1}
Boltzmann constant	$k = \frac{R_0}{N_0}$	1.38054×10^{-16} erg deg^{-1}
Electronic charge	$e = \frac{F}{N_0}$	1.60203×10^{-20} emu 4.8030×10^{-10} esu
Faraday's constant	F	9.6487×10^{3} emu (g mole)$^{-1}$
Gas constant	R_0	8.3143×10^{7} erg deg^{-1} (g mole)$^{-1}$ 1.98585 cal deg^{-1} (g mole)$^{-1}$
Loschmidt's number	$n_0 = \frac{N_0}{V_0}$	2.68699×10^{19} cm^{-3}
Mass of atom of unit atomic weight	$M_0 = \frac{1}{N_0}$	1.66043×10^{-24} g
Mass of α-particle	M_α	6.6442×10^{-24} g
Mass of electron	m	9.10891×10^{-28} g
Mass of proton	M_p	1.67252×10^{-24} g
Planck's constant	h	6.6256×10^{-27} erg s
Speed of light	c	2.99793×10^{10} cm s^{-1}
Volume of ideal gas	V_0	2.24136×10^{4} cm^{3} (g mole)$^{-1}$

Table 12-2. *Periodic system*

	Ia	a II b	a III b	b IV a	b V a	b VI a	b VII a	VIIIb			Ib	O
1	1 **H** 1.00797											2 **He** 4.0026
2	3 **Li** 6.939	4 **Be** 9.0122	5 **B** 10.811	6 **C** 12.0111	7 **N** 14.0067	8 **O** 15.9994	9 **F** 18.9984					10 **Ne** 20.183
3	11 **Na** 22.9898	12 **Mg** 24.312	13 **Al** 26.9815	14 **Si** 28.086	15 **P** 30.9738	16 **S** 32.064	17 **Cl** 35.453					18 **Ar** 39.948
4	19 **K** 39.102	20 **Ca** 40.08 30 **Zn** 65.37	21 **Sc** 44.956 31 **Ga** 69.72	22 **Ti** 47.90 32 **Ge** 72.59	23 **V** 50.942 33 **As** 74.9216	24 **Cr** 51.996 34 **Se** 78.96	25 **Mn** 54.938 35 **Br** 79.909	26 **Fe** 55.847	27 **Co** 58.9332	28 **Ni** 58.71	29 **Cu** 63.54	36 **Kr** 83.80
5	37 **Rb** 85.47	38 **Sr** 87.62 48 **Cd** 112.40	39 **Y** 88.905 49 **In** 114.82	40 **Zr** 91.22 50 **Sn** 118.69	41 **Nb** 92.906 51 **Sb** 121.75	42 **Mo** 95.94 52 **Te** 127.60	43 **Tc** 99 53 **J** 126.904	44 **Ru** 101.07	45 **Rh** 102.905	46 **Pd** 106.4	47 **Ag** 107.87	54 **Xe** 131.30
6	55 **Cs** 132.905	56 **Ba** 137.34 80 **Hg** 200.59	57 **La** 138.91 59 … 70[a] 71 **Lu** 174.97 81 **Tl** 204.37	58 **Ce** 140.12 72 **Hf** 178.49 82 **Pb** 207.19	73 **Ta** 180.948 83 **Bi** 208.98	74 **W** 183.85 84 **Po** 210	75 **Re** 186.2 85 **At** 210	76 **Os** 190.2	77 **Ir** 192.2	78 **Pt** 195.09	79 **Au** 196.967	86 **Rn** 222
7	87 **Fr** 223.05	88 **Ra** 226.05	89 **Ac** 227	90 **Th** 232.038	91 **Pa** 231	92 **U**[b] 238.03						

[a] 59 **Pr** 140.907, 60 **Nd** 144.24, 61 **Pm** 145, 62 **Sm** 150.35, 63 **Eu** 151.96, 64 **Gd** 157.25, 65 **Tb** 158.924, 66 **Dy** 162.50, 67 **Ho** 164.93, 68 **Er** 167.26, 69 **Tm** 168.934, 70 **Yb** 173.04

[b] 93 **Np** 237, 94 **Pu** 242, 95 **Am** 243, 96 **Cm** 247, 97 **Bk** 249, 98 **Cf** 251, 99 **Es**, 100 **Fm**, 101 **Md**, 102 **No**

Table 12-3. *Table of relative atomic weights.* (From IUPAC Comptes Rendus XXII Conference, 1963)

Based on the atomic mass of $^{12}C = 12$.

The values for atomic weights given in the table apply to elements as they exist in nature, without artificial alteration of their isotopic composition and, further, to natural mixtures that do not include isotopes of radiogenic origin.

Name	Symbol	Atomic number	Atomic weight
Actinium	Ac	89	...
Aluminium	Al	13	26.9815
Americum	Am	95	...
Antimony, stibium	Sb	51	121.75
Argon	Ar	18	39.948
Arsenic	As	33	74.9216
Astatine	At	85	...
Barium	Ba	56	137.34
Berkelium	Bk	97	...
Beryllium	Be	4	9.0122
Bismuth	Bi	83	208.980
Boron	B	5	10.811[a]
Bromine	Br	35	79.909[b]
Cadmium	Cd	48	112.40
Caesium	Cs	55	132.905
Calcium	Ca	20	40.08
Californium	Cf	98	...
Carbon	C	6	12.01115[a]
Cerium	Ce	58	140.12
Chlorine	Cl	17	35.453[b]
Chromium	Cr	24	51.996[b]
Cobalt	Co	27	58.9332
Copper, cuprum	Cu	29	63.54
Curium	Cm	96	...
Dysprosium	Dy	66	162.50
Einsteinium	Es	99	...
Erbium	Er	68	167.26
Europium	Eu	63	151.96
Fermium	Fm	100	...
Fluorine	F	9	18.9984
Francium	Fr	87	...
Gadolinium	Gd	64	157.25
Gallium	Ga	31	69.72
Germanium	Ge	32	72.59
Gold, aurum	Au	79	196.967
Hafnium	Hf	72	178.49
Helium	He	2	4.0026
Holmium	Ho	67	164.930
Hydrogen	H	1	1.00797[a]
Indium	In	49	114.82
Iodine	I	53	126.9044
Iridium	Ir	77	192.2
Iron, ferrum	Fe	26	55.847[b]
Krypton	Kr	36	83.80
Lanthanum	La	57	138.91
Lawrencium	Lr	103	...
Lead, plumbum	Pb	82	207.19
Lithium	Li	3	6.939
Lutetium	Lu	71	174.97
Magnesium	Mg	12	24.312
Manganese	Mn	25	54.9380
Mende-levium	Md	101	...
Mercury, hydrar-gyrum	Hg	80	200.59
Molyb-denum	Mo	42	95.94
Neodymium	Nd	60	144.24
Neon	Ne	10	20.183
Neptunium	Np	93	...
Nickel	Ni	28	58.71
Niobium (colum-bium)	Nb	41	92.906
Nitrogen	N	7	14.0067
Nobelium	No	102	...
Osmium	Os	76	190.2
Oxygen	O	8	15.9994[a]
Palladium	Pd	46	106.4
Phosphorus	P	15	30.9738
Platinum	Pt	78	195.09
Plutonium	Pu	94	...
Polonium	Po	84	...
Potassium, kalium	K	19	39.102
Praseodym	Pr	59	140.907
Promethium	Pm	61	...
Protactinium	Pa	91	...
Radium	Ra	88	...
Radon	Rn	86	...
Rhenium	Re	75	186.2
Rhodium	Rh	45	102.905
Rubidium	Rb	37	85.47
Ruthenium	Ru	44	101.07
Samarium	Sm	62	150.35
Scandium	Sc	21	44.956
Selenium	Se	34	78.96
Silicon	Si	14	28.086[a]

Table 12-3 (Continued)

Name	Symbol	Atomic number	Atomic weight	Name	Symbol	Atomic number	Atomic weight
Silver, argentum	Ag	47	107.870[b]	Thulium	Tm	69	168.934
				Tin, stanum	Sn	50	118.69
Sodium, natrium	Na	11	22.9898	Titanium	Ti	22	47.90
				Tungsten, wolfram	W	74	183.85
Strontium	Sr	38	87.62				
Sulfur	S	16	32.064[a]	Uranium	U	92	238.03
Tantalum	Ta	73	180.948	Vanadium	V	23	50.942
Technetium	Tc	43	...	Xenon	Xe	54	131.30
Tellurium	Te	52	127.60	Ytterbium	Yb	70	173.04
Terbium	Tb	65	158.924	Yttrium	Y	39	88.905
Thallium	Tl	81	204.37	Zinc	Zn	30	65.37
Thorium	Th	90	232.038	Zirconium	Zr	40	91.22

[a] Atomic weights so designated are known to be variable because of natural variations in isotopic composition. The observed ranges are:

Hydrogen	± 0.00001	Oxygen	± 0.0001
Boron	± 0.003	Silicon	± 0.001
Carbon	± 0.00005	Sulfur	± 0.003

[b] Atomic weights so designated are believed to have the following experimental uncertainties:

Chlorine	± 0.001	Bromine	± 0.002
Chromium	± 0.001	Silver	± 0.003
Iron	± 0.003		

Table 12-4. *Table of isotopic abundance and relative atomic masses.* (From Handbook of Chemistry and Physics 1967/68; STROMINGER, HOLLANDER and SEABORG, 1958)

Based on the atomic mass of $^{12}C = 12$.

Atomic No.	Symbol	Mass No.	Abundance (per cent)	Atomic weight
1	H	1	99.99	1.007825
		2	0.01	2.01410
2	He	3	0.0001	3.01603
		4	99.99	4.0026
3	Li	6	7.4	6.01513
		7	92.6	7.01601
4	Be	9	100	9.01219
5	B	10	19.6	10.01294
		11	80.4	11.00931
6	C	12	98.9	12.00000
		13	1.1	13.00335
7	N	14	99.6	14.00307
8		15	0.4	15.00011
8	O	16	99.8	15.99491
		17	0.04	16.99914
		18	0.2	17.99916
9	F	19	100	18.99840
10	Ne	20	90.9	19.99244
		21	0.3	20.99395
		22	8.8	21.99138
11	Na	23	100	22.98977
12	Mg	24	78.7	23.98504
		25	10.1	24.98584
		26	11.2	25.98259
13	Al	27	100	26.98153
14	Si	28	92.2	27.97693
		29	4.7	28.97649
		30	3.1	29.97376
15	P	31	100	30.97376
16	S	32	95.0	31.97207
		33	0.8	32.97146
		34	4.2	33.96786
		36	0.014	35.96709
17	Cl	35	75.5	34.96885
		37	24.5	36.96590
18	Ar	36	0.3	35.96755
		38	0.06	37.96272
		40	99.6	39.96238
19	K	39	93.1	38.96371
		40	0.01	
		41	6.9	40.96184
20	Ca	40	97.0	39.96259
		42	0.6	41.95863
		43	0.1	42.95878
		44	2.1	43.95549
		46	0.003	45.9537
		48	0.2	47.9524
21	Sc	45	100	44.95592
22	Ti	46	8.0	45.95263

Table 12-4 (Continued)

Atomic No.	Symbol	Mass No.	Abundance (per cent)	Atomic weight
22	Ti	47	7.3	46.9518
		48	73.9	47.94795
		49	5.5	48.94787
		50	5.3	49.9448
23	V	50	0.2	49.9472
		51	99.8	50.9440
24	Cr	50	4.3	49.9461
		52	83.8	51.9405
		53	9.6	52.9407
		54	2.4	53.9389
25	Mn	55	100	54.9381
26	Fe	54	5.8	53.9396
		56	91.7	55.9349
		57	2.2	56.9354
		58	0.3	57.9333
27	Co	59	100	58.9332
28	Ni	58	67.8	57.9353
		60	26.2	59.9332
		61	1.2	60.9310
		62	3.7	61.9283
		64	1.1	63.9280
29	Cu	63	69.1	62.9298
		65	30.9	64.2978
30	Zn	64	48.9	63.9291
		66	27.8	65.9260
		67	4.1	66.9271
		68	18.6	67.9249
		70	0.6	69.9253
31	Ga	69	60.4	68.9257
		71	39.6	70.9249
32	Ge	70	20.5	69.9243
		72	27.4	71.9217
		73	7.8	72.9234
		74	36.5	73.9212
		76	7.8	75.9214
33	As	75	100	74.9216
34	Se	74	0.9	73.9225
		76	9.0	75.9192
		77	7.6	76.9199
		78	23.5	77.9173
		80	49.8	79.9165
		82	9.2	81.9167
35	Br	79	50.5	78.9183
		81	49.5	80.9163
36	Kr	78	0.4	77.9204
		80	2.3	79.9164
		82	11.6	81.9135
		83	11.5	82.9141
		84	56.9	83.9115
		86	17.4	85.9106
37	Rb	85	72.2	84.9117
		87	27.8	

Table 12-4 (Continued)

Atomic No.	Symbol	Mass No.	Abundance (per cent)	Atomic weight
38	Sr	84	0.5	83.9134
		86	9.9	85.9094
		87	7.0	86.9089
		88	82.6	87.9056
39	Y	89	100	88.9054
40	Zr	90	51.5	89.9043
		91	11.2	90.9053
		92	17.1	91.9046
		94	17.4	93.9061
		96	2.8	95.9082
41	Nb	93	100	92.9060
42	Mo	92	15.9	91.9063
		94	9.1	93.9047
		95	15.7	94.9046
		96	16.5	95.9046
		97	9.5	96.9058
		98	23.7	97.9055
		100	9.6	99.9076
44	Ru	96	5.5	95.9076
		98	1.9	97.9055
		99	12.7	98.9061
		100	12.6	99.9030
		101	17.1	100.9041
		102	31.6	101.9037
		104	18.6	103.9055
45	Rh	103	100	102.9048
46	Pd	102	1.0	101.9049
		104	11.0	103.9036
		105	22.2	104.9046
		106	27.3	105.9032
		108	26.7	107.9030
		110	11.8	109.9045
47	Ag	107	51.8	106.9041
		109	48.2	108.9047
48	Cd	106	1.2	105.907
		108	0.9	107.9050
		110	12.4	109.9030
		111	12.7	110.9042
		112	24.1	111.9028
		113	12.3	112.9046
		114	28.9	113.9036
		116	7.6	115.905
49	In	113	4.3	112.9043
		115	95.7	114.9041
50	Sn	112	1.0	111.9040
		114	0.6	113.9030
		115	0.3	114.9035
		116	14.3	115.9021
		117	7.6	116.9031
		118	24.0	117.9018
		119	8.6	118.9034
		120	32.8	119.9021

Table 12-4 (Continued)

Atomic No.	Symbol	Mass No.	Abundance (per cent)	Atomic weight
50	Sn	122	4.9	121.9034
		124	5.9	123.9052
51	Sb	121	57.3	120.9038
		123	42.7	122.9041
52	Te	120	0.1	119.9045
		122	2.5	121.9030
		123	0.9	122.9042
		124	4.6	123.9028
		125	7.0	124.9044
		126	18.7	125.9032
		128	31.8	127.9047
		130	34.5	129.9067
53	I	127	100	126.9044
54	Xe	124	0.1	123.9061
		126	0.1	125.9042
		128	1.9	127.9035
		129	26.4	128.9048
		130	4.1	129.9035
		131	21.2	130.9051
		132	26.9	131.9042
		134	10.4	133.9054
		136	8.9	135.9072
55	Cs	133	100	132.9041
56	Ba	130	0.1	129.9062
		132	0.1	131.9057
		134	2.4	133.9043
		135	6.6	134.9056
		136	7.8	135.9044
		137	11.3	136.9056
		138	71.7	137.9050
57	La	138	0.1	137.9068
		139	99.9	138.9061
58	Ce	136	0.2	135.907
		138	0.2	137.9057
		140	88.5	139.9053
		142	11.1	141.9090
59	Pr	141	100	140.9074
60	Nd	142	27.1	141.9075
		143	12.2	142.9096
		144	23.9	143.9099
		145	8.3	144.9122
		146	17.2	145.9127
		148	5.7	147.9165
		150	5.6	149.9207
62	Sm	144	3.1	143.9117
		147	15.0	146.9146
		148	11.2	147.9146
		149	13.8	148.9169
		150	7.5	149.9170
		152	26.7	151.9195
		154	22.7	153.9220
63	Eu	151	47.8	150.9196

Table 12-4. (Continued)

Atomic No.	Symbol	Mass No.	Abundance (per cent)	Atomic weight
63	Eu	153	52.2	152.9209
64	Gd	152	0.2	151.9195
		154	2.2	153.9207
		155	14.7	154.9226
		156	20.5	155.9221
		157	15.7	156.9339
		158	24.9	157.9241
		160	21.9	159.9071
65	Tb	159	100	158.9250
66	Dy	156	0.1	155.9238
		158	0.1	157.9240
		160	2.3	159.9248
		161	19.0	160.9266
		162	25.5	161.9265
		163	24.9	162.9284
		164	28.1	163.9288
67	Ho	165	100	164.9303
68	Er	162	0.1	161.9288
		164	1.6	163.9293
		166	33.4	165.9304
		167	22.9	166.9320
		168	27.1	167.9324
		170	14.9	169.9355
69	Tm	169	100	168.9344
70	Yb	168	0.1	167.9339
		170	3.1	169.9349
		171	14.4	170.9365
		172	21.9	171.9366
		173	16.2	172.9283
		174	31.6	173.9390
		176	12.6	175.9427
71	Lu	175	97.4	174.9409
		176	2.6	
72	Hf	174	0.2	173.9403
		176	5.2	175.9435
		177	18.5	176.9435
		178	27.1	177.9439
		179	13.8	178.9460
		180	35.2	179.9468
73	Ta	180	0.01	179.9475
		181	99.99	180.9480
74	W	180	0.1	179.9470
		182	26.4	181.9483
		183	14.4	182.9503
		184	30.6	183.9510
		186	28.4	185.9543
75	Re	185	37.1	184.9530
		187	62.9	186.9560
76	Os	184	0.02	183.9526
		186	1.6	185.9539
		187	1.6	186.9560
		188	13.3	187.9560

Table 12-4 (Continued)

Atomic No.	Symbol	Mass No.	Abundance (per cent)	Atomic weight
76	Os	189	16.1	188.9586
		190	26.4	189.9586
		192	41.0	191.9612
77	Ir	191	37.3	190.9609
		193	62.7	192.9633
78	Pt	190	0.01	189.9600
		192	0.8	191.9614
		194	32.9	193.9628
		195	33.8	194.9648
		196	25.3	195.9650
		198	7.2	197.9675
79	Au	197	100	196.9666
80	Hg	196	0.1	195.9658
		198	10.0	197.9668
		199	16.8	198.9683
		200	23.1	199.9683
		201	13.2	200.9703
		202	29.8	201.9706
		204	6.9	203.9735
81	Tl	203	29.5	202.9723
		205	70.5	204.9745
82	Pb	204	1.5	203.973
		206	23.6	205.9745
		207	22.6	206.9759
		208	52.3	207.9766
83	Bi	209	100	208.9804
90	Th	232	100	232.0382
92	U	234	0.006	234.0409
		2235	0.72	235.0439
		238	99.27	238.0508

Table 12-5. *Naturally occurring radioactive elements.* (From HAMILTON, 1965; WETHERILL, 1966)

Parent	% Abundance	Daughter and mode of decay	Half-life $T_{\frac{1}{2}}$ [year]	Decay constant λ [year^{-1}]	Decay energies [cal g^{-1} year^{-1}]
H^3	$\sim 4\times 10^{-18}$	β-^{3}He	1.226×10	5.653×10^{-2}	
C^{14}	$\sim 10^{-11}$	β-^{14}N	5.568×10^3	1.245×10^{-4}	
K^{40}	0.0119	e ^{40}Ar β-^{40}Ca [a]	1.33×10^9	5.21×10^{-10}	0.21
Rb^{87}	27.85	β-^{87}Sr	5.0×10^{10}	1.39×10^{-11}	
In^{115}	95.77	β-^{115}Sn	6.0×10^{14}	1.16×10^{-15}	
La^{138}	0.089	e ^{138}Ba β-^{135}Ce [a]	$\sim 7.0 \times 10^{10}$	9.9×10^{-12}	
Sm^{147}	15.09	α ^{143}Nd	1.25×10^{16}	5.54×10^{-17}	
Lu^{176}	2.59	e ^{176}Tb β-^{176}Hf [a]	2.4×10^{10}	2.89×10^{-11}	
Re^{187}	62.93	β-^{187}Os	$\sim 5 \times 10^{10}$	1.39×10^{-11}	
Th^{232}	100.00	complex ^{208}Pb	1.39×10^{10}	4.99×10^{-11}	0.20
U^{235}	0.72	complex ^{207}Pb	7.1×10^8	9.76×10^{-10}	4.3
U^{238}	99.27	complex ^{206}Pb	4.5×10^9	1.54×10^{-10}	0.71

[a] Branching mode of decay $\left(\text{branching ratio for } ^{40}\text{K} \rightarrow {}^{40}\text{Ar}, {}^{40}\text{Ca } \frac{\lambda_e}{\lambda_\beta} = 0.11\right)$.

Table 12-6. *Electronic configuration of the elements in their normal states.* (From PAULING, 1960)

		He	Neon		Argon		Krypton			Xenon			Radon				Eka-radon				Term
		1*s*	2*s*	2*p*	3*s*	3*p*	3*d*	4*s*	4*p*	4*d*	5*s*	5*p*	4*f*	5*d*	6*s*	6*p*	5*f*	6*d*	7*s*	7*p*	symbol
H	1	1																			$^2S_{1/2}$
He	2	2																			$^1S_0$
Li	3	2	1																		$^2S_{1/2}$
Be	4	2	2																		$^1S_0$
B	5	2	2	1																	$^2P_{1/2}$
C	6	2	2	2																	$^3P_0$
N	7	2	2	3																	$^4S_{3/2}$
O	8	2	2	4																	$^3P_2$
F	9	2	2	5																	$^2P_{3/2}$
Ne	10	2	2	6																	$^1S_0$
Na	11	10 Neon			1																$^2S_{1/2}$
Mg	12	core			2																$^1S_0$
Al	13				2	1															$^2P_{1/2}$
Si	14				2	2															$^3P_0$
P	15				2	3															$^4S_{3/2}$
S	16				2	4															$^3P_2$
Cl	17				2	5															$^2P_{3/2}$
Ar	18	2	2	6	2	6															$^1S_0$
K	19	18 Argon core						1													$^2S_{1/2}$
Ca	20							2													$^1S_0$
Sc	21						1	2													$^2D_{3/2}$
Ti	22						2	2													$^3F_2$
V	23						3	2													$^4F_{3/2}$
Cr	24						5	1													$^7S_3$
Mn	25						5	2													$^6S_{5/2}$
Fe	26						6	2													$^5D_4$
Co	27						7	2													$^4F_{9/2}$
Ni	28						8	2													$^3F_4$
Cu	29						10	1													$^2S_{1/2}$
Zn	30						10	2													$^1S_0$
Ga	31						10	2	1												$^2P_{1/2}$
Ge	32						10	2	2												$^3P_0$
As	33						10	2	3												$^4S_{3/2}$
Se	34						10	2	4												$^3P_2$
Br	35						10	2	5												$^2P_{3/2}$
Kr	36	2	2	6	2	6	10	2	6												$^1S_0$
Rb	37	36 Krypton core									1										$^2S_{1/2}$
Sr	38										2										$^1S_0$
Y	39									1	2										$^2D_{3/2}$
Zr	40									2	2										$^3F_2$
Nb	41									4	1										$^6D_{1/2}$
Mo	42									5	1										$^7S_3$
Tc	43									5	2										$^6S_{5/2}$
Ru	44									7	1										$^5F_5$
Rh	45									8	1										$^4F_{9/2}$
Pd	46									10											$^1S_0$
Ag	47									10	1										$^2S_{1/2}$

Table 12-6 (Continued)

		He 1s	Neon 2s	2p	Argon 3s	3p	Krypton 3d	4s	4p	Xenon 4d	5s	5p	Radon 4f	5d	6s	6p	Eka-radon 5f	6d	7s	7p	Term symbol
Cd	48	36 Krypton core								10	2										$^1S_0$
In	49									10	2	1									$^2P_{1/2}$
Sn	50									10	2	2									$^3P_0$
Sb	51									10	2	3									$^4S_{3/2}$
Te	52									10	2	4									$^3P_2$
I	53									10	2	5									$^2P_{3/2}$
Xe	54	2	2	6	2	6	10	2	6	10	2	6									$^1S_0$
Cs	55	54 Xenon core													1						$^2S_{1/2}$
Ba	56														2						$^1S_0$
La	57													1	2						$^2D_{3/2}$
Ce	58												1	1	2						$^3H_4$
Pr	59												2	1	2						$^4K_{11/2}$
Nd	60												3	1	2						$^5L_6$
Pm	61												4	1	2						$^6L_{9/2}$
Sm	62												5	1	2						$^7K_4$
Eu	63												6	1	2						$^8H_{3/2}$
Gd	64												7	1	2						$^9D_2$
Tb	65												8	1	2						$^8H_{17/2}$
Dy	66												9	1	2						$^7K_{10}$
Ho	67												10	1	2						$^6K_{19/2}$
Er	68												11	1	2						$^5L_{10}$
Tm	69												12	1	2						$^4K_{17/2}$
Yb	70												13	1	2						$^3H_6$
Lu	71												14	1	2						$^2D_{3/2}$
Hf	72												14	2	2						$^3F_2$
Ta	73												14	3	2						$^4F_{3/2}$
W	74												14	4	2						$^5D_0$
Re	75												14	5	2						$^6S_{5/2}$
Os	76												14	6	2						$^5D_4$
Ir	77												14	7	2						$^4F_{9/2}$
Pt	78												14	9	1						$^3D_3$
Au	79												14	10	1						$^2S_{1/2}$
Hg	80												14	10	2						$^1S_0$
Tl	81												14	10	2	1					$^2P_{1/2}$
Pb	82												14	10	2	2					$^3P_0$
Bi	83												14	10	2	3					$^4S_{3/2}$
Po	84												14	10	2	4					$^3P_2$
At	85												14	10	2	5					$^2P_{3/2}$
Rn	86	2	2	6	2	6	10	2	6	10	2	6	14	10	2	6					$^1S_0$
Fr	87	86 Radon core																	1		$^2S_{1/2}$
Ra	88																		2		$^1S_0$
Ac	89																	1	2		$^2D_{3/2}$
Th	90																	2	2		$^3F_2$
Pa	91																	3	2		$^4F_{3/2}$
U	92																	4	2		$^5D_0$
Eka-Rn	118	2	2	6	2	6	10	2	6	10	2	6	14	10	2	6	14	10	2	6	$^1S_0$

Table 12-7. *The effective radii of atoms.* (From PAULING, 1960; ZEMANN, 1966)

(The values of the effective radii are given for ligancy 12; correction for ligancy 8: -3%; for ligancy 6: -4%; for ligancy 4: -12%.)

Symbol	Atomic No.	PAULING 1960 [Å]	ZEMANN 1966 [Å]	Symbol	Atomic No.	PAULING 1960 [Å]	ZEMANN 1966 [Å]
Ac	89	—	1.88	Nb	41	1.46	1.48
Ag	47	1.44	1.44	Nd	60	1.82	1.82
Al	13	1.43	1.43	Ne	10	(1.60)	—
Am	95	—	1.73	Ni	28	1.24	1.24
Ar	18	(1.92)	—	Np	93	—	1.58
As	33	1.48	~1.50	O	8	(0.60)	—
Au	79	1.44	1.44	Os	76	1.35	1.34
B	5	0.98	—	P	15	1.28	—
Ba	56	2.22	2.24	Pa	91	—	1.61
Be	4	1.12	1.13	Pb	82	1.70	1.75
Bi	83	1.78	~1.70	Pd	46	1.37	1.37
Br	35	(1.11)	—	Pm	61	1.83	—
C	6	(0.86)	—	Po	84	(1.41)	~1.70
Ca	20	1.97	1.97	Pr	59	1.82	1.83
Cd	48	1.51	1.57	Pt	78	1.39	1.39
Ce	58	1.82	1.82	Pu	94	—	1.64
Cl	17	(0.99)	—	Rb	37	2.48	2.50
Co	27	1.25	1.26	Re	75	1.37	1.37
Cr	24	1.28	1.27	Rh	45	1.34	1.34
Cs	55	2.67	2.71	Ru	44	1.34	1.32
Cu	29	1.28	1.28	S	16	1.27	—
Dy	66	1.78	1.77	Sb	51	1.66	~1.60
Er	68	1.76	1.76	Sc	21	1.62	1.64
Eu	63	2.08	2.04	Se	34	1.40	—
F	9	(0.64)	—	Si	14	1.38	—
Fe	26	1.26	1.27	Sm	62	1.80	1.80
Ga	31	1.40	~1.40	Sn	50	1.62	1.58
Gd	64	1.80	1.80	Sr	38	2.15	2.15
Ge	32	1.44	~1.40	Ta	73	1.46	1.48
H	1	(0.46)	—	Tb	65	1.77	1.78
He	2	(1.45)	—	Tc	43	1.36	1.35
Hf	72	1.59	1.59	Te	52	1.60	—
Hg	80	1.51	1.62	Th	90	1.80	1.80
Ho	67	1.76	1.77	Ti	22	1.47	1.45
I	53	(1.36)	—	Tl	81	1.60	1.73
In	49	1.58	1.66	Tm	69	1.76	1.75
Ir	77	1.36	1.36	U	92	1.52	1.55
K	19	2.35	2.34	V	23	1.34	1.35
Kr	36	(1.98)	—	W	74	1.39	1.41
La	57	1.87	1.87	Xe	54	(2.18)	—
Li	3	1.55	1.56	Y	39	1.80	1.80
Lu	71	1.74	1.73	Yb	70	1.93	1.94
Mg	12	1.60	1.60	Zn	30	1.34	1.39
Mn	25	1.27	1.32	Zr	40	1.60	1.60
Mo	42	1.39	1.40				
N	7	(0.8)	—				
Na	11	1.90	1.91				

Table 12-8. *Table of the effective radii of ions.* (From AHRENS, 1952; GOLDSCHMIDT, 1926, 1934; PAULING, 1927; WYCKOFF, 1948; ZEMANN 1966
(Most values for the effective radii are given for ions in 6 fold co-ordination.)

Symbol	Atomic No.	Charge	GOLDSCHMIDT, 1926/34 [Å]	PAULING, 1927 [Å]	WYCKOFF, 1948 [Å]	AHRENS, 1952 [Å]	ZEMANN, 1966 [Å]
Ac	89	+3	—	—	—	1.18	1.11
Ag	47	+1	1.13	1.26	0.97	1.26	1.13
		+2	—	—	—	0.89	—
Al	13	+3	0.57	0.50	0.55	0.51	0.57
Am	95	+3	—	—	—	1.07	—
		+4	—	—	—	0.92	1.01
As	33	−3	—	2.22	1.91	—	—
		+3	0.69	—	0.69	0.58	—
		+5	—	0.47	—	0.46	0.46
At	85	+7	—	—	—	0.62	—
Au	79	+1	—	1.37	—	1.37	—
		+3	—	—	—	0.85	—
B	5	+3	—	0.20	—	0.23	—
Ba	56	+2	1.43	1.35	1.38	1.34	1.43
Be	4	+2	0.34	0.31	0.30	0.35	0.34
Bi	83	−3	—	—	2.13	—	—
		+3	—	—	1.20	0.96	0.96
		+5	—	0.74	—	0.74	0.74
Br	35	−1	1.96	1.95	1.96	—	1.96
		+5	—	—	—	0.47	—
		+7	—	0.39	0.39	0.39	~0.40
C	6	−4	—	2.60	2.60	—	—
		+4	0.2	0.15	0.15	0.16	—
Ca	20	+2	1.06	0.99	1.05	0.99	1.06
Cd	48	+2	1.03	0.97	0.99	0.97	1.03
Ce	58	+3	1.18	—	1.10	1.07	1.09
		+4	1.02	1.01	1.01	0.94	1.03
Cl	17	−1	1.81	1.81	1.80	—	1.81
		+5	—	—	—	0.34	—
		+7	—	0.26	0.26	0.27	~0.30
Co	27	+2	0.82	—	0.78	0.72	0.82
		+3	—	—	0.65	0.63	0.64
Cr	24	+3	0.64	—	0.70	0.63	~0.64
		+6	~0.35	0.52	0.52	0.52	~0.35
Cs	55	+1	1.65	1.69	1.70	1.67	1.65
Cu	29	+1	—	0.96	0.58	0.96	—
		+2	—	—	—	0.72	~0.70
Dy	66	+3	1.07	—	—	0.92	0.97
Er	68	+3	1.04	—	—	0.89	0.94
Eu	63	+2	—	—	—	—	1.25
		+3	1.13	—	—	0.98	1.01
F	9	−1	1.33	1.36	1.33	—	1.33
		+7	—	0.07	0.07	0.08	—
Fe	26	+2	0.83	—	0.80	0.74	0.82
		+3	0.67	—	0.67	0.64	0.67
Fr	87	+1	—	—	—	1.80	—
Ga	31	+3	0.62	0.62	0.65	0.62	0.67
Gd	64	+3	1.11	—	—	0.97	1.00

Table 12-8 (Continued)

Symbol	Atomic No.	Charge	Goldschmidt, 1926/34 [Å]	Pauling, 1927 [Å]	Wyckoff, 1948 [Å]	Ahrens, 1952 [Å]	Zemann, 1966 [Å]
Ge	32	−4	—	2.72	—	—	—
		+2	—	—	0.65	0.73	—
		+4	0.44	0.53	0.55	0.53	0.44
H	1	−1	1.54	2.08	1.27	—	—
Hf	72	+1	—	—	0.72	—	—
		+4	—	—	—	0.78	0.84
Hg	80	+2	1.12	1.10	0.66	1.10	1.12
Ho	67	+3	1.05	—	—	0.91	0.95
I	53	−1	2.20	2.16	2.20	—	2.20
		+5	0.94	—	—	0.62	—
		+7	—	0.50	—	0.50	~0.50
In	49	+3	0.92	0.81	0.95	0.81	0.92
Ir	77	+4	0.66	—	0.65	0.68	0.66
K	19	+1	1.33	1.33	1.33	1.33	1.33
La	57	+3	1.22	1.15	1.15	1.14	1.12
Li	3	+1	0.78	0.60	0.70	0.68	0.78
Lu	71	+3	—	—	—	0.85	0.91
Mg	12	+2	0.78	0.65	0.75	0.67	0.78
Mn	25	+2	0.91	—	0.83	0.80	0.91
		+3	0.70	—	—	0.66	0.70
		+4	0.52	—	0.52	0.60	0.52
		+7	—	0.46	—	0.46	—
Mo	42	+4	0.68	—	0.68	0.70	0.68
		+6	—	0.62	0.65	0.62	0.62
N	7	−3	—	1.71	1.48	—	—
		+3	—	—	—	0.16	—
		+5	~0.15	0.11	—	0.13	—
NH_4		+1	1.43	—	—	—	—
Na	11	+1	0.98	0.95	1.00	0.97	0.98
Nb	41	+4	0.69	—	0.67	0.74	—
		+5	0.69	0.70	—	0.69	0.69
Nd	60	+3	1.15	—	1.07	1.04	1.06
Ni	28	+2	0.78	—	0.74	0.69	0.78
Np	93	+3	—	—	—	1.10	—
		+4	—	—	—	0.95	1.04
		+7	—	—	—	0.71	—
O	8	−2	1.32	1.40	1.35	—	1.32
		+6	—	0.09	—	0.10	—
Os	76	+4	0.67	—	0.65	—	0.67
		+6	—	—	—	0.69	—
P	15	−3	—	2.12	1.86	—	—
		+3	—	—	—	0.44	—
		+5	~0.35	0.34	0.34	0.35	~0.35
Pa	91	+3	—	—	—	1.13	—
		+4	—	—	—	0.98	1.06
		+5	—	—	—	0.89	—
Pb	82	+2	1.32	—	1.18	1.20	1.32
		+4	0.84	0.84	0.70	0.84	0.84
Pd	46	+2	—	—	—	0.80	0.93
		+4	—	—	—	0.65	—

Table 12-8 (Continued)

Symbol	Atomic No.	Charge	GOLDSCHMIDT, 1926/34 [Å]	PAULING, 1927 [Å]	WYCKOFF, 1948 [Å]	AHRENS, 1952 [Å]	ZEMANN, 1966 [Å]
Pm	61	+3	—	—	—	1.06	—
Po	84	+6	—	—	—	0.67	~1.10
Pr	59	+3	1.16	—	1.09	1.06	1.07
		+4	1.00	—	—	0.92	—
Pt	78	+2	—	—	0.52	0.80	—
		+4	—	—	0.55	0.65	—
Pu	94	+3	—	—	—	1.08	—
		+4	—	—	—	0.93	1.02
Ra	88	+2	—	—	1.42	1.43	1.52
Rb	37	+1	1.49	1.48	1.52	1.47	1.49
Re	75	+4	—	—	—	0.72	~0.75
		+6	—	—	—	—	0.55
		+7	—	—	—	0.56	—
Rh	45	+3	0.68	—	0.75	0.68	0.68
Ru	44	+4	0.65	—	0.65	0.67	0.65
S	16	−2	1.74	1.84	1.82	—	1.74
		+4	—	—	—	0.37	—
		+6	0.34	0.29	—	0.30	0.34
Sb	51	−3	—	2.45	2.08	—	—
		+3	0.90	—	0.90	0.76	—
		+5	—	0.62	—	0.62	0.62
Sc	21	+3	0.83	0.81	0.83	0.81	0.83
Se	34	−2	1.91	1.98	1.93	—	1.91
		+3	—	—	0.78	—	—
		+4	—	—	—	0.50	—
		+6	~0.35	0.42	—	0.42	~0.35
Si	14	−4	1.98	2.71	—	—	—
		+4	0.39	0.41	0.40	0.42	0.39
Sm	62	+2	—	—	—	—	0.93
		+3	1.13	—	—	1.00	1.02
Sn	50	−4	—	2.94	—	—	—
		+2	—	—	1.02	0.93	—
		+4	0.74	0.71	0.65	0.71	0.74
Sr	38	+2	1.27	1.13	1.18	1.12	1.27
Ta	73	+5	—	—	—	0.68	0.68
Tb	65	+3	1.09	—	—	0.93	0.98
		+4	0.89	—	—	0.81	—
Tc	43	+7	—	—	—	0.56	~0.55
Te	52	−2	2.11	2.21	2.12	—	2.11
		+4	0.89	—	0.84	0.70	—
		+6	—	0.56	—	0.56	0.56
Th	90	+4	1.10	—	1.10	1.02	1.10
Ti	22	+2	—	—	0.76	—	—
		+3	0.69	—	0.70	0.76	0.69
		+4	0.64	0.68	0.60	0.68	0.64
Tl	81	+1	1.49	—	1.50	1.47	1.49
		+3	1.05	0.95	—	0.95	1.05
Tm	69	+3	1.04	—	—	0.87	0.93
U	92	+4	1.05	—	1.05	0.97	1.05
		+6	—	—	—	0.80	0.83

Table 12-8 (Continued)

Symbol	Atomic No.	Charge	GOLDSCHMIDT, 1926/34 [Å]	PAULING, 1927 [Å]	WYCKOFF, 1948 [Å]	AHRENS, 1952 [Å]	ZEMANN, 1966 [Å]
V	23	+2	—	—	—	0.95	—
		+3	0.65	—	0.75	0.74	0.65
		+4	0.61	—	0.57	0.63	—
		+5	~0.4	0.59	—	0.59	~0.40
W	74	+4	0.68	—	0.68	0.70	0.68
		+6	—	—	0.65	0.62	0.62
Y	39	+3	1.06	0.93	0.95	0.92	1.06
Yb	70	+3	1.00	—	—	0.86	0.92
Zn	30	+2	0.83	0.74	0.83	0.74	0.83
Zr	40	+4	0.87	0.80	0.80	0.79	0.87

Table 12-9. *Ionization potentials for elements in the atomic state.* (From AHRENS, 1952 and Handbook of Chemistry and Physics 1967/68)

Symbol	Atomic No.	Ionization potentials, volts						
		I	II	III	IV	V	VI	VII
Ac	89							
Ag	47	7.542	21.4	35.9				
Al	13	5.96	18.74	28.31	119.37	153.4		
Ar	18	15.68	27.76	40.75	(61)	(78)		
As	33	10.5	20.1	28.0	49.9	62.5		
Au	79	9.18	19.95					
B	5	8.257	25.00	37.75	258.1	338.5		
Ba	56	5.19	9.95					
Be	4	9.28	18.12	153.1	216.6			
Bi	83	8.0	16.6	25.42	45.1	55.7		
Br	35	11.80	19.1	25.7	(50)			103.5
C	6	11.217	24.27	47.65	64.22	390.1		
Ca	20	6.09	11.82	50.96	69.7			
Cd	48	8.96	16.84	38.0				
Ce	58	6.54	14.8	(36.5)	(36.5)			
Cl	17	12.952	23.67	39.69	53.16	67.4		114
Co	27	7.81	17.3					
Cr	24	6.74	16.6	~32.1		(73)	96	
Cs	55	3.87	23.4	(35)	(51)	(58)		
Cu	29	7.68	20.34	29.5				
Dy	66	6.8						
Er	68							
Eu	63	5.64	11.4					
F	9	17.34	34.81	62.35	86.72	113.67	156.37	184.26
Fe	26	7.83	16.16					
Ga	31	5.97	20.43	30.6	63.8			
Gd	64	6.7						
Ge	32	8.09	15.86	34.07	45.5	93.0		
H	1	13.527						
He	2	24.46	54.14					
Hf	72		(14.8)		31			
Hg	80	10.39	18.65	34.3	(72)	(82)		

Table 12-9 (Continued)

Symbol	Atomic No.	Ionization potentials, volts						
		I	II	III	IV	V	VI	VII
Ho	67							
I	53	10.6	19.4					90.2
In	49	5.76	18.79	27.9	57.8			
K	19	4.318	31.66	46.5				
Kr	36	13.93	26.4	36.8	(68)			
La	57	5.6	11.4	(20.4)				
Li	3	5.363	75.26	121.8				
Mg	12	7.61	14.96	79.72	108.9			
Mn	25	7.41	15.70			(76)		~122
Mo	42	7.35				60.8	70	
N	7	14.48	29.47	47.40	77.0	97.4		
Na	11	5.12	47.06	70.72				
Nb	41			24.2		49.3		
Nd	60	6.3						
Ne	10	21.47	40.9	63.2				
Ni	28	7.61	18.2					
O	8	13.550	34.93	54.87	76.99	113	137.5	
Os	76	(8.7)						
P	15	10.9	19.56	30.012	51.106	64.698		
Pb	82	7.38	14.96	(31.9)	42.11	69.4		
Pd	46	8.3	19.8					
Pr	59	5.8						
Pt	78	8.88	19.3					
Ra	88	5.252	10.099					
Rb	37	4.159	27.36	(47)	(80)			
Rh	45	7.7						
Rn	86	10.698						
Ru	44	7.7						
S	16	10.30	23.3	34.9	47.08	63	87.65	
Sb	51	8.5	(18)	24.7	44.0	55.5		
Sc	21	6.7	12.8	24.61	(73.9)	(97.0)		
Se	34	9.70	21.3	33.9	42.72	72.8	82	
Si	14	8.12	16.27	33.35	44.93	165.6		
Sm	62	6.6	11.4					
Sn	50	7.30	14.5	30.5	39.4	80.7		
Sr	38	5.667	10.98					
Tb	65	6.7						
Tc	43							(95)
Te	52	8.96		30.5	37.7	60.0	(72)	
Th	90			29.4				
Ti	22	6.81	13.6	27.6	42.98	(99.6)		
Tl	81	6.07	20.32	29.7	50.5			
V	23	6.71	14.1	(26.4)	(48)	(65)		
W	74	8.1					61	
Xe	54	12.08	(21.1)	32.0	(46)	(76)		
Y	39	6.5	12.3	20.4				
Yb	70	7.1						
Zn	30	9.36	17.89	40.0				
Zr	40	6.92	13.97	24.00	33.8			

The degree of ionization is indicated by the numerals I, II, etc. Doubtful values are indicated by parentheses.

Table 12-10. *Electronegativity values.* (From ALLRED and ROCHOW, 1958; GORDY and THOMAS, 1956; PAULING, 1960. (The values given in the table refer to the common oxidation states of the elements. For some elements variation of the electronegativity with oxidation number is observed.)

Symbol	ALLRED, ROCHOW, 1958	PAULING, 1960	GORDY, THOMAS, 1956
Ac	—	1.1	1.1
Ag	1.42	1.9	1.8
Al	1.47	1.5	1.5
Am	—	1.3	$\sim$1.3
As	2.20	2.0	2.0
At	—	2.2	2.2
Au	—	2.4	2.3
B	2.01	2.0	2.0
Ba	0.97	0.9	0.9
Be	1.47	1.5	1.5
Bi	—	1.9	1.8
Bk	—	1.3	$\sim$1.3
Br	2.74	2.8	2.8
C	2.50	2.5	2.5
Ca	1.04	1.0	1.0
Cd	1.46	1.7	Cd^{II} 1.5
Ce	—	1.1	Ce^{III} 1.1
Cf	—	1.3	$\sim$1.3
Cl	2.83	3.0	3.0
Cm	—	1.3	$\sim$1.3
Co	1.70	1.8	1.7
Cr	1.56	1.6	Cr^{II} 1.4
			Cr^{III} 1.6
			Cr^{IV} 2.2
Cs	0.86	0.7	0.75
Cu	1.75	1.9	Cu^{I} 1.8
			Cu^{II} 2.0
Dy	—	1.2	$\sim$1.2
Er	—	1.2	$\sim$1.2
Es	—	1.3	—
Eu	—	1.1	$\sim$1.1
F	4.10	4.0	3.9_5
Fe	1.64	1.8	Fe^{II} 1.7
			Fe^{III} 1.8
Fm	—	1.3	—
Fr	—	0.7	0.7
Ga	1.82	1.6	1.5
Gd	—	1.2	$\sim$1.2
Ge	2.02	1.8	1.8
H	—	2.1	2.1_5
Hf	—	1.3	1.4
Hg	—	1.9	Hg^{II} 1.8
Ho	—	1.2	$\sim$1.2
I	2.21	2.5	2.5_5
In	1.49	1.7	1.5
Ir	—	2.2	2.1
K	0.91	0.8	0.8
La	1.08	1.1	1.1
Li	0.97	1.0	0.95
Lu	—	1.2	$\sim$1.2
Md	—	1.3	—
Mg	1.23	1.2	1.2
Mn	1.60	1.5	Mn^{II} 1.4
			Mn^{III} 1.5
			Mn^{VII} 2.5
Mo	—	1.8	Mo^{IV} $\sim$1.6
			Mo^{VI} $\sim$2.1
N	3.07	3.0	3.0
Na	1.01	0.9	0.9
Nb	—	1.6	1.7
Nd	—	1.2	$\sim$1.2
Ni	1.75	1.8	1.8
No	—	1.3	—
Np	—	1.3	$\sim$1.1
O	3.50	3.5	3.5
Os	—	2.2	2.0
P	2.06	2.1	2.1
Pa	—	1.5	Pa^{III} 1.3
			Pa^{V} 1.7
Pb	—	1.8	Pb^{II} 1.6
			Pb^{IV} 1.8
Pd	—	2.2	2.0
Pm	—	1.2	$\sim$1.2
Po	—	2.0	2.0
Pr	—	1.1	1.1
Pt	—	2.2	2.1
Pu	—	1.3	$\sim$1.3
Ra	—	0.9	0.9
Rb	0.89	0.8	0.8
Re	—	1.9	Re^{V} 1.8
			Re^{VII} 2.2
Rh	—	2.2	2.1
Ru	—	2.2	Ru^{III} 2.0
S	2.44	2.5	2.5
Sb	1.82	1.9	Sb^{III} 1.8
			Sb^{V} 2.1
Sc	1.20	1.3	1.3
Se	2.48	2.4	2.4
Si	1.74	1.8	1.8
Sm	—	1.2	$\sim$1.2
Sn	1.72	1.8	Sn^{II} 1.7
			Sn^{IV} 1.8
Sr	0.99	1.0	1.0
Ta	—	1.5	Ta^{III} 1.3
			Ta^{V} 1.7
Tb	—	1.2	$\sim$1.2

Table 12-10 (Continued)

Symbol	ALLRED, ROCHOW, 1958	PAULING, 1960	GORDY, THOMAS, 1956		Symbol	ALLRED, ROCHOW, 1958	PAULING, 1960	GORDY, THOMAS, 1956	
Tc	—	1.9	Tc^{V}	1.9	V			U^{VI}	1.9
			Tc^{VII}	2.3	V	1.45	1.6	V^{III}	1.4
Te	2.01	2.1		2.1				V^{IV}	1.7
Th	—	1.3	Th^{II}	1.0				V^{V}	1.9
			Th^{IV}	1.4	W	—	1.7	W^{IV}	1.6
Ti	1.32	1.5		1.6				W^{VI}	2.0
Tl	—	1.8	Tl^{I}	1.5	Y	1.11	1.2		1.2
			Tl^{III}	1.9	Yb	—	1.1		~1.1
Tm	—	1.2		~1.2	Zn	1.66	1.6	Zn^{II}	1.5
U	—	1.7	U^{IV}	1.4	Zr	—	1.4		1.5

Table 12-11. *Molar volumes and densities of minerals.* (From ROBIE and BETHKE, 1966)

Mineral	Formula	Molar volume cm³	Density g/cm³	Temp. °C
		Sulfides and arsenides		
Antimonite	Sb_2S_3	73.414 ± 0.05	4.696	25
Argentite	Ag_2S	34.2 ± 0.2	7.254	r
Arsenopyrite	FeAsS	26.42 ± 0.05	6.163	r
Bornite	Cu_5FeS_4	98.6 ± 0.3	5.090	r
Chalcocite	Cu_2S	27.48 ± 0.01	5.792	r
Chalcopyrite	$CuFeS_2$	44.11 ± 0.05	4.088	r
Cinnabarite	HgS	28.419 ± 0.010	8.187	r
Cobaltite	CoAsS	26.4 ± 0.4	6.28	r
Covellite	CuS	20.427 ± 0.020	4.680	r
Cubanite	$CuFe_2S_3$	67.39 ± 0.20	4.028	20
Digenite	Cu_9S_5	26.01 ± 0.01	5.605	25
Enargite	Cu_3AsS_4	88.2 ± 0.1	4.46	26
Galena	PbS	31.495 ± 0.010	7.597	r
Gersdorffite	NiAsS	27.78 ± 0.01	5.964	26
Loellingite	$FeAs_2$	27.51 ± 0.02	7.476	26
Marcasite	FeS_2	24.58 ± 0.02	4.885	25
Millerite	NiS	16.891 ± 0.008	5.374	r
Molybdenite	MoS_2	32.03 ± 0.01	4.998	26
Niccolite	NiAs	17.186 ± 0.020	7.775	r
Pentlandite	$Fe_{5.25}Ni_{3.75}S_8$	159.6 ± 0.3	4.824	r
Proustite	Ag_3AsS_3	88.4 ± 0.1	5.597	26
Pyrargyrite	Ag_3SbS_3	92.5 ± 0.3	5.85_5	26
Pyrite	FeS_2	23.942 ± 0.004	5.016	r
Rammelsbergite	$NiAs_2$	29.42 ± 0.04	7.088	26
Safflorite	$CoAs_2$	27.92 ± 0.02	7.477	26
Sphalerite	ZnS	23.834 ± 0.008	4.088_5	r
Tennantite	$CuAs_4S$	318.7 ± 0.8	4.641	26
Tetraedrite	$Cu_{12}Sb_4S_{13}$	331.7 ± 0.7	5.024	26
		Oxides and hydroxides		
Brucite	$Mg(OH)_2$	24.64 ± 0.03	2.368	26
α-Cristobalite	SiO_2	25.74 ± 0.02	2.334	25
Corundum	Al_2O_3	25.57 ± 0.01	3.988	26

Table 12-11 (Continued)

Mineral	Formula	Molar volume cm³	Density g/cm³	Temp. °C
Cuprite	Cu_2O	23.44 ±0.02	6.104	26
Diaspore	AlO(OH)	17.76 ±0.03	3.377	25
Gibbsite	$Al(OH)_3$	31.96 ±0.06	2.441	r
Goethite	α-FeO(OH)	20.82 ±0.03	4.268	r
Hausmannite	Mn_3O_4	46.96 ±0.08	4.873	r
Hematite	Fe_2O_3	30.28 ±0.02	5.274	25
Hercynite	$FeAl_2O_4$	40.82 ±0.06	4.258	25
Ilmenite	$FeTiO_3$	31.71 ±0.05	4.786	r
Magnetite	Fe_3O_4	44.50 ±0.03	5.206	25
Periclase	MgO	11.25 ±0.01	3.584	25
Perovskite	$CaTiO_3$	33.63 ±0.03	4.043	r
Pyrolusite	MnO_2	16.61 ±0.06	5.233	r
α-Quartz	SiO_2	22.690±0.005	2.648	25
Rutile	TiO_2	18.80 ±0.02	4.250	25
Spinel	$MgAl_2O_4$	39.72 ±0.03	3.582	26
Tenorite	CuO	12.22 ±0.02	6.508	26
	Carbonates, halides, phosphate and sulfates			
Anhydrite	$CaSO_4$	45.94 ±0.05	2.963	26
Apatite	$Ca_5(PO_4)_3OH$	159.66 ±0.40	3.146	r
Barytes	$BaSO_4$	52.11 ±0.05	4.480	26
Calcite	$CaCO_3$	36.94 ±0.02	2.712	26
Celestine	$SrSO_4$	46.25 ±0.05	3.972	26
Dolomite	$CaMg(CO_3)_2$	64.35 ±0.04	2.866	26±3
Fluorite	CaF_2	24.54 ±0.01	3.181	25
Gypsum	$CaSO_4 \cdot 2\,H_2O$	74.31 ±0.16	2.317	r
Halite	NaCl	27.018±0.007	2.163	25
Magnesite	$MgCO_3$	28.02 ±0.01	3.009	26±3
Siderite	$FeCO_3$	29.38 ±0.02	3.944	26±3
Strontianite	$SrCO_3$	39.01 ±0.03	3.785	26
Sylvine	KCl	37.528±0.007	1.987	25
	Silicates			
Åkermanite	$Ca_2[MgSi_2O_7]$	92.82 ±0.15	2.938	r
Albite	$Na[AlSi_3O_8]$	100.21 ±0.19	2.617	r
Almandine	$Fe_3Al_2[Si_3O_{12}]$	115.28 ±0.04	4.318	25
Analcite	$Na[AlSi_2O_6] \cdot H_2O$	97.50 ±0.10	2.258	r
Andalusite	Al_2SiO_5	51.54 ±0.01	3.144	25
Andradite	$Ca_3Fe_2[Si_3O_{12}]$	131.67 ±0.04	3.860	25
Annite	$KFe_3[AlSi_3O_{10}](OH)_2$	154.32 ±1.0	3.317	26
Anorthite	$Ca[Al_2Si_2O_8]$	100.73 ±0.15	2.762	r
Anthophyillite	$Mg_7[Si_8O_{22}](OH)_2$	274.0 ±3.5	2.850	r
Clinozoisite	$Ca_2AlAl_2OOH[Si_2O_7][SiO_4]$	136.2 ±0.30	3.336	r
Cordierite	$Mg_2Al_3[AlSi_5O_{18}]$	233.50 ±0.30	2.505	r
Diopside	$CaMg[Si_2O_6]$	66.10 ±0.10	3.277	r
Enstatite	$Mg_2[Si_2O_6]$	62.80 ±0.10	3.198	26
Fayalite	$Fe_2[SiO_4]$	46.39 ±0.08	4.393	25
Forsterite	$Mg_2[SiO_4]$	43.79 ±0.03	3.214	25
Gehlenite	$Ca_2[Al_2SiO_7]$	90.25 ±0.15	3.038	r
Glaucophane	$Na_2Mg_3Al_2[Si_8O_{22}](OH)_2$	269.7 ±0.8	2.906	r
Grossular	$Ca_3Al_2[Si_3O_{12}]$	125.32 ±0.04	3.595	25
Grunerite	$Fe_7[Si_8O_{22}](OH)_2$	278.5 ±1.0	3.597	r

Table 12-11 (Continued)

Mineral	Formula	Molar volume cm³	Density g/cm³	Temp. °C
Hedenbergite	$CaFe[Si_2O_6]$	65.97 ±0.30	3.55 ±0.01	r
Jadeite	$NaAl[Si_2O_6]$	60.98 ±0.40	3.315	r
Kaolinite	$Al_2[Si_2O_5](OH)_4$	99.31 ±0.30	2.600	r
Kyanite	Al_2SiO_5	44.11 ±0.02	3.674	25
Lawsonite	$CaAl_2[Si_2O_7](OH)_2 \cdot H_2O$	101.33 ±0.15	3.110	r
Leucite	$K[AlSi_2O_6]$	88.01 ±0.15	2.480	r
Microcline	$K[AlSi_3O_8]$	108.69 ±0.20	2.561	r
Monticellite	$CaMg[SiO_4]$	51.37 ±0.15	3.046	r
Muscovite	$KAl_2[AlSi_3O_{10}](OH)_2$	140.55 ±0.50	2.834	27
Nepheline	$Na[AlSiO_4]$	54.17 ±0.15	2.623	r
Orthoclase	$K[AlSi_3O_8]$	109.11 ±0.30	2.551	r
Phlogopite	$KMg_3[AlSi_3O_{10}](OH)_2$	149.66 ±1.0	2.788	r
Pyrope	$Mg_3Al_2[Si_3O_{12}]$	113.29 ±0.04	3.559	25
Pyrophyllite	$Al_2[Si_4O_{10}](OH)_2$	126.6 ±0.50	2.845	r
Sillimanite	Al_2SiO_5	49.91 ±0.02	3.247	25
Sphene	$CaTiO[SiO_4]$	55.70 ±0.30	3.520	r
Talc	$Mg_3[Si_4O_{10}](OH)_2$	136.7 ±0.30	2.788	r
Topaz	$Al_2[SiO_4]F_2$	51.66 ±0.10	3.563	26
Tremolite	$Ca_2Mg_5[Si_8O_{22}](OH)_2$	272.95 ±0.90	2.976	r
Wollastonite	$Ca[SiO_3]$	39.94 ±0.08	2.909	r
Zircon	$Zr[SiO_4]$	39.27 ±0.08	4.668	25

Molar volumes were calculated from measured densities or from the unit-cell dimensions.

The letter r indicates that the measurements were made at an unspecified room temperature.

Table 12-12. *The theoretical composition of some silicate minerals in cation percentage*

Mineral	Formula	Formula weight [g]	Si	Al	Fe^{3+}	Fe^{2+}	Mg	Ca	Na	K
Acmite	$NaFe[Si_2O_6]$	230.97	50.0		25.0				25.0	
Actinolite	$Ca_2Mg_4Fe[Si_8O_{22}](OH)_2$	843.79	53.4			6.6	26.7	13.3		
Åkermanite	$Ca_2[MgSi_2O_7]$	272.60	40.0				20.0	40.0		
Albite	$Na[AlSi_3O_8]$	262.15	60.0	20.0					20.0	
Almandine	$Fe_3Al_2[Si_3O_{12}]$	497.67	37.5	25.0		37.5				
Amesite	$Mg_2Al[AlSiO_5](OH)_4$	278.67	20.0	40.0			40.0			
Analcite	$Na[AlSi_2O_6]\cdot H_2O$	220.11	50.0	25.0					25.0	
Andradite	$Ca_3Fe_2[Si_3O_{12}]$	508.12	37.5		25.0			37.5		
Anorthite	$Ca[Al_2Si_2O_8]$	278.14	40.0	40.0				20.0		
Anthophyllite	$Mg_7[Si_8O_{22}](OH)_2$	780.74	53.3				46.7			
Antigorite	$Mg_3[Si_2O_5](OH)_4$	277.11	40.0				60.0			
Barkevikite	$Na_2CaMg_4Al[AlSi_7O_{22}](OH)_2$	755.72	48.6	13.9			16.7	6.9	13.9	
Biotite	$KMg_3[AlSi_3O_{10}](OH)_2$	417.22	37.5	12.5			37.5			12.5
	$KMg_2Fe[AlSi_3O_{10}](OH)_2$	448.76	37.5	12.5		12.5	25.0			12.5
Chabazite	$Ca[Al_2Si_4O_{12}]\cdot 6H_2O$	506.36	57.2	28.5				14.3		
Chlorite	$Mg_5Al[AlSi_3O_{10}](OH)_8$	555.78	30.0	20.0			50.0			
	$Mg_3Fe_2Al[AlSi_3O_{10}](OH)_8$	618.84	30.0	20.0		20.0	30.0			
Chloritoid	$Fe_2AlAl_3O_2[SiO_4]_2(OH)_4$	503.73	25.0	50.0		25.0				
Clinozoisite	$Ca_2AlAl_2OOH[Si_2O_7][SiO_4]$	454.26	37.5	37.5				25.0		
Cordierite	$Mg_2Al_3[AlSi_5O_{18}]$	584.82	45.4	36.4			18.2			
	$MgFeAl_3[AlSi_5O_{18}]$	616.35	45.4	36.4		9.1	9.1			
Cummingtonite	$Mg_{3.5}Fe_{3.5}[Si_8O_{22}](OH)_2$	891.20	53.3			23.4	23.4			
Diopside	$CaMg[Si_2O_6]$	216.52	50.0				25.0	25.0		
Edenite	$NaCa_2Mg_5[AlSi_7O_{22}](OH)_2$	754.16	50.0	7.1			21.5	14.3	7.1	
Enstatite	$Mg_2[Si_2O_6]$	200.76	50.0				50.0			
Epidote	$Ca_2FeAl_2OOH[Si_2O_7][SiO_4]$	483.14	37.5	25.0	12.5			25.0		
Fayalite	$Fe_2[SiO_4]$	203.76	33.3			66.7				

Ferro-Gedrite	$Fe_5Al_2[Al_2Si_6O_{22}](OH)_2$	941.52	40.1	26.6		33.3					
Forsterite	$Mg_2[SiO_4]$	140.70	33.3				66.7				
Gehlenite	$Ca_2[Al_2SiO_7]$	274.16	20.0	40.0				40.0			
Glaucophane	$Na_2Mg_3Al_2[Si_8O_{22}](OH)_2$	783.39	53.4	13.3			20.0		13.3		
Grossular	$Ca_3Al_2[Si_3O_{12}]$	450.36	37.5	25.0				37.5			
Hauyne	$Na_6Ca[Al_6Si_6O_{24}](SO_4)$	988.31	31.5	31.5				5.4	31.5		
Hedenbergite	$CaFe[Si_2O_6]$	248.05	50.0			25.0		25.0			
Heulandite	$Ca[Al_2Si_7O_{18}]\cdot 6H_2O$	686.54	70.0	20.0				10.0			
Hornblende	$Ca_2Mg_4Al[AlSi_7O_{22}](OH)_2$	749.81	52.2	14.9			18.0	14.9			
Jadeite	$NaAl[Si_2O_6]$	202.09	50.0	25.0					25.0		
Kaolinite	$Al_2[Si_2O_5](OH)_4$	258.09	50.0	50.0							
Laumontite	$Ca[Al_2Si_4O_{12}]\cdot 4H_2O$	470.32	57.2	28.6				14.2			
Lawsonite	$CaAl_2[Si_2O_7](OH)_2\cdot H_2O$	314.17	40.0	40.0				20.0			
Leucite	$K[AlSi_2O_6]$	218.19	50.0	25.0						25.0	
Mesolite	$Na_2Ca_2[Al_2Si_3O_{10}]_3\cdot 8H_2O$	1,164.64	47.4	31.6				10.5	10.5		
Monticellite	$CaMg[SiO_4]$	156.46	33.3				33.3	33.3			
Muscovite	$KAl_2[AlSi_3O_{10}](OH)_2$	398.21	42.8	42.8						14.3	
Natrolite	$Na_2[Al_2Si_3O_{10}]\cdot 2H_2O$	380.17	42.8	28.6					28.6		
Nepheline	$Na[AlSiO_4]$	142.03	33.3	33.3					33.3		
Nosean	$Na_8[Al_6Si_6O_{24}](SO_4)$	994.23	30.0	30.0					40.0		
Paragonite	$NaAl_2[AlSi_3O_{10}](OH)_2$	382.11	42.8	42.8					14.3		
Phillipsite	$NaKCa_{0.5}[Al_3Si_5O_{16}]\cdot 6H_2O$	667.45	47.7	28.6				4.7	9.5	9.5	
Phlogopite	$KMg_3[AlSi_3O_{10}](OH)_2$	417.23	37.5	12.5			37.5			12.5	
Pigeonite	$MgFe[Si_2O_6]$	232.29	50.0			25.0	25.0				
Potassium feldspar	$K[AlSi_3O_8]$	278.25	60.0	20.0						20.0	
Prehnite	$Ca_2Al[AlSi_3O_{10}](OH)_2$	412.30	42.9	28.5				28.5			
Pyrope	$Mg_3Al_2[Si_3O_{12}]$	403.08	37.5	25.0			37.5				
Pyrophyllite	$Al_2[Si_4O_{10}](OH)_2$	360.20	66.7	33.3							
Riebeckite	$Na_2Fe_3Fe_2[Si_8O_{22}](OH)_2$	935.74	53.2		13.5	19.9			13.5		
Scolecite	$Ca[Al_2Si_3O_{10}]\cdot 3H_2O$	392.25	50.0	33.3				16.7			
Sillimanite	$Al_2O[SiO_4]$	162.00	33.3	66.7							
Sodalite	$Na_8[Al_6Si_6O_{24}]Cl_2$	969.07	30.0	30.0					40.0		
Sphene	$CaTiO[SiO_4]$	196.04	33.3					33.3			Ti 33.3

Table 12-12 (Continued)

Mineral	Formula	Formula weight [g]	Si	Al	Fe^{3+}	Fe^{2+}	Mg	Ca	Na	K	
Spodumene	$LiAl[Si_2O_6]$	186.03	50.0	25.0							Li 25.0
Staurolite	$Fe_2Al_9O_6[SiO_4]_4OOH$	851.67	26.7	60.0		13.3					
Stilbite	$Ca[Al_2Si_7O_{18}]\cdot 7H_2O$	704.55	70.0	20.0				10.0			
Talc	$Mg_3[Si_4O_{10}](OH)_2$	379.22	57.2				42.8				
Thomsonite	$NaCa_2[Al_5Si_5O_{20}]\cdot 6H_2O$	806.41	38.5	38.5				15.4	7.6		
Tremolite	$Ca_2Mg_5[Si_8O_{22}](OH)_2$	812.26	53.4				33.3	13.3			
Vesuvianite	$Ca_{10}Mg_2Al_4[Si_2O_7]_2[SiO_4]_5(OH)_4$	1,421.89	36.0	16.0			8.0	40.0			
Wollastonite	$Ca[SiO_3)$	116.14	50.0					50.0			
Zircon	$Zr[SiO_4]$	183.28	50.0								Zr 50.0

Table 12-13. *The theoretical composition of some silicate minerals in weight percentage*

Mineral	Formula	Formula weight [g]	SiO_2	Al_2O_3	Fe_2O_3	FeO	MgO	CaO	Na_2O	K_2O	H_2O	
Acmite	$NaFe[Si_2O_6]$	230.97	52.0		34.6				13.4			
Actinolite	$Ca_2Mg_4Fe[Si_8O_{22}](OH)_2$	843.79	56.9			8.5	19.1	13.3			2.1	
Åkermanite	$Ca_2[MgSi_2O_7]$	272.60	44.0				14.8	41.2				
Albite	$Na[AlSi_3O_8]$	262.15	68.7	19.4					11.8			
Almandine	$Fe_3Al_2[Si_3O_{12}]$	497.67	36.2	20.5		43.3						
Amesite	$Mg_2Al[AlSiO_5](OH)_4$	278.67	21.6	36.6			28.9				12.9	
Analcite	$Na[AlSi_2O_6]\cdot H_2O$	220.11	54.6	23.2					14.1		8.1	
Andradite	$Ca_3Fe_2[Si_3O_{12}]$	508.12	35.5		31.4			33.1				
Anorthite	$Ca[Al_2Si_2O_8]$	278.14	43.2	36.6				20.2				
Anthophyllite	$Mg_7[Si_8O_{22}](OH)_2$	780.74	61.5				36.2				2.3	
Antigorite	$Mg_3[Si_2O_5](OH)_4$	277.11	43.3				43.7				13.0	

Barkevikite	$Na_2CaMg_4Al[AlSi_7O_{22}](OH)_2$	755.72	55.6	13.5			12.9	7.4	8.2		2.4	
Biotite	$KMg_3[AlSi_3O_{10}](OH)_2$	417.22	43.2	12.2			29.0			11.3	4.3	
	$KMg_2Fe[AlSi_3O_{10}](OH)_2$	448.76	40.1	11.4		16.0	18.0			10.5	4.0	
Chabazite	$Ca[Al_2Si_4O_{12}] \cdot 6H_2O$	506.36	47.5	20.1				11.1			21.3	
Chlorite	$Mg_5Al[AlSi_3O_{10}](OH)_8$	555.78	32.4	18.3			36.3				13.0	
	$Mg_3Fe_2Al[AlSi_3O_{10}](OH)_8$	618.84	29.1	16.5		23.2	19.6				11.6	
Chloritoid	$Fe_2AlAl_3O_2[SiO_4]_2(OH)_4$	503.73	23.8	40.5		28.5					7.2	
Clinozoisite	$Ca_2AlAl_2OOH[Si_2O_7][SiO_4]$	454.26	39.7	33.6				24.7			2.0	
Cordierite	$Mg_2Al_3[AlSi_5O_{18}]$	584.82	51.3	34.9			13.8					
	$MgFeAl_3[AlSi_5O_{18}]$	616.35	48.7	33.1		11.7	6.5					
Cummingtonite	$Mg_{3.5}Fe_{3.5}[Si_8O_{22}](OH)_2$	891.20	53.9			28.2	15.9				2.0	
Diopside	$CaMg[Si_2O_6]$	216.52	55.5				18.6	25.9				
Edenite	$NaCa_2Mg_5[AlSi_7O_{22}](OH)_2$	754.16	55.7	6.8			16.1	14.9	4.1		2.4	
Enstatite	$Mg_2[Si_2O_6]$	200.76	59.8				40.2					
Epidote	$Ca_2FeAl_2OOH[Si_2O_7][SiO_4]$	483.14	37.3	21.1	16.5			23.2			1.9	
Fayalite	$Fe_2[SiO_4]$	203.76	29.5			70.5						
Ferro-Gedrite	$Fe_5Al_2[Al_2Si_6O_{22}]\ (OH)_2$	941.52	38.3	21.6		38.2					1.9	
Forsterite	$Mg_2[SiO_4]$	140.70	42.7				57.3					
Gehlenite	$Ca_2[Al_2SiO_7]$	274.16	21.9	37.2				40.9				
Glaucophane	$Na_2Mg_3Al_2[Si_8O_{22}](OH)_2$	783.39	61.4	13.0			15.4		7.9		2.3	
Grossular	$Ca_3Al_2[Si_3O_{12}]$	450.36	40.0	22.6				37.4				
Hauyne	$Na_6Ca[Al_6Si_6O_{24}](SO_4)$	988.31	36.5	30.9				5.7	18.8			SO_3 8.1
Hedenbergite	$CaFe[Si_2O_6]$	248.05	48.4			29.0		22.6				
Heulandite	$Ca[Al_2Si_7O_{18}] \cdot 6H_2O$	686.54	61.2	14.8				8.2			15.8	
Hornblende	$Ca_2Mg_4Al[AlSi_7O_{22}](OH)_2$	749.81	56.1	13.6			13.0	14.9			2.4	
Jadeite	$NaAl[Si_2O_6]$	202.09	59.4	25.2					15.3			
Kaolinite	$Al_2[Si_2O_5](OH)_4$	258.09	46.5	39.5							14.0	
Laumontite	$Ca[Al_2Si_4O_{12}] \cdot 4H_2O$	470.32	51.1	21.7				11.9			15.3	
Lawsonite	$CaAl_2[Si_2O_7](OH)_2 \cdot H_2O$	314.17	38.2	32.4				17.9			11.5	
Leucite	$K[AlSi_2O_6]$	218.19	55.0	23.4						21.6		
Mesolite	$Na_2Ca_2[Al_2Si_3O_{10}]_3 \cdot 8H_2O$	1,164.64	46.4	26.3				9.6	5.3		12.4	
Monticellite	$CaMg[SiO_4]$	156.46	38.4				25.8	35.8				

Table 12-13 (Continued)

Mineral	Formula	Formula weight [g]	SiO_2	Al_2O_3	Fe_2O_3	FeO	MgO	CaO	Na_2O	K_2O	H_2O	
Muscovite	$KAl_2[AlSi_3O_{10}](OH)_2$	398.21	45.2	38.4						11.8	4.5	
Natrolite	$Na_2[Al_2Si_3O_{10}]\cdot 2H_2O$	380.17	47.4	26.8					16.3		9.5	
Nepheline	$Na[AlSiO_4]$	142.03	42.3	35.9					21.8			
Nosean	$Na_8[Al_6Si_6O_{24}](SO_4)$	994.23	36.3	30.8					24.9		8.0	
Paragonite	$NaAl_2[AlSi_3O_{10}](OH)_2$	382.11	47.2	40.0					8.1		4.7	
Phillipsite	$NaKCa_{0.5}[Al_3Si_5O_{16}]\cdot 6H_2O$	667.45	45.0	22.9				4.2	4.6	7.1	16.2	
Phlogopite	$KMg_3[AlSi_3O_{10}](OH)_2$	417.23	43.2	12.2			29.0			11.3	4.3	
Pigeonite	$MgFe[Si_2O_6]$	232.29	51.7			30.9	17.4					
Potassium feldspar	$K[AlSi_3O_8]$	278.25	64.8	18.3						16.9		
Prehnite	$Ca_2Al[AlSi_3O_{10}](OH)_2$	412.30	43.7	24.7				27.2			4.4	
Pyrope	$Mg_3Al_2[Si_3O_{12}]$	403.08	44.7	25.3			30.0					
Pyrophyllite	$Al_2[Si_4O_{10}](OH)_2$	360.20	66.7	28.3							5.0	
Riebeckite	$Na_2Fe_3Fe_2[Si_8O_{22}](OH)_2$	935.74	51.3		17.0	23.0			6.8		1.9	
Scolecite	$Ca[Al_2Si_3O_{10}]\cdot 3H_2O$	392.25	45.9	26.0				14.3			13.8	
Sillimanite	$Al_2O[SiO_4]$	162.00	37.1	62.9								
Sodalite	$Na_8[Al_6Si_6O_{24}]Cl_2$	969.07	37.2	31.6					19.2			NaCl 12.0
Sphene	$CaTiO[SiO_4]$	196.04	30.6					28.6				TiO_2 40.8
Spodumene	$LiAl[Si_2O_6]$	186.03	64.6	27.4								Li_2O 8.0
Staurolite	$Fe_2Al_9O_6[SiO_4]_4OOH$	851.67	28.2	53.9		16.9					1.0	
Stilbite	$Ca[Al_2Si_7O_{18}]\cdot 7H_2O$	704.55	59.6	14.5				8.0			17.9	
Talc	$Mg_3[Si_4O_{10}](OH)_2$	379.22	63.4				31.9				4.7	
Thomsonite	$NaCa_2[Al_5Si_5O_{20}]\cdot 6H_2O$	806.41	37.2	31.6				13.9	3.8		13.4	
Tremolite	$Ca_2Mg_5[Si_8O_{22}](OH)_2$	812.26	59.2				24.8	13.8			2.2	
Vesuvianite	$Ca_{10}Mg_2Al_4[Si_2O_7]_2[SiO_4]_5(OH)_4$	1,421.89	38.0	14.3			5.7	39.4			2.5	
Wollastonite	$Ca[SiO_3]$	116.14	51.7					48.3				
Zircon	$Zr[SiO_4]$	183.28	32.8									ZrO_2 67.2

Table 12-14. *The theoretical composition of some non-silicate minerals in weight percentage*

Mineral	Formula	Formula weight [g]	Al_2O_3	Fe_2O_3	FeO	MgO	CaO			CO_2	SO_3	P_2O_5	H_2O	
Anhydrite	$CaSO_4$	136.15					41.2				58.8			
Apatite	$Ca_5(PO_4)_3OH$	502.35					55.8					42.4	1.8	
Baryte	$BaSO_4$	233.43						BaO	65.7		34.3			
Brucite	$Mg(OH)_2$	58.34				69.1							30.9	
Calcite	$CaCO_3$	100.09					56.0			44.0				
Celestine	$SrSO_4$	183.70						SrO	56.4		43.6			
Chromite	$FeCr_2O_4$	223.83			32.1			Cr_2O_3	67.9					
Diaspore	AlOOH	64.98	85.0										15.0	
Dolomite	$CaMg(CO_3)_2$	184.42				21.9	30.4			47.7				
Fluorite	CaF_2	78.08						Ca	51.3					F 48.7
Gibbsite	$Al(OH)_3$	82.99	67.4										32.6	
Goethite	FeOOH	88.87		89.8									10.2	
Gypsum	$CaSO_4 \cdot 2H_2O$	172.18					32.6				46.5		20.9	
Halite	NaCl	58.46						Na	39.3					Cl 60.7
Hercynite	$FeAl_2O_4$	173.79	58.7		41.3									
Ilmenite	$FeTiO_3$	151.75			47.3			TiO_2	52.7					
Magnesite	$MgCO_3$	84.33				47.8				52.2				
Magnetite	Fe_3O_4	231.55		69.0	31.0									
Perovskite	$CaTiO_3$	135.98					41.2	TiO_2	58.8					
Siderite	$FeCO_3$	115.86			62.0					38.0				
Spinel	$MgAl_2O_4$	142.26	71.7			28.3								
Strontianite	$SrCO_3$	147.64						SrO	70.2	29.8				
Sylvine	KCl	74.56						K	52.4					Cl 47.6

Table 12-15. *The theoretical composition of some ore minerals in weight percentage*

Mineral	Formula	Formula weight [g]	Ag	Co	Cu	Fe	Ni	Pb	Zn		As	Sb	S
Antimonite	Sb_2S_3	339.72										71.7	28.3
Argentite	Ag_2S	247.83	87.1										12.9
Arsenopyrite	$FeAsS$	162.83				34.3					46.0		19.7
Bornite	Cu_5FeS_4	501.79			63.3	11.1							25.6
Boulangerite	$Pb_5Sb_4S_{11}$	1,875.82						55.2				26.0	18.8
Bournonite	$PbCuSbS_3$	488.71			13.0			42.4				24.9	19.7
Chalcocite	Cu_2S	159.15			79.9								20.1
Chalcopyrite	$CuFeS_2$	183.52			34.6	30.4							35.0
Cinnabarite	HgS	232.68								Hg 86.2			13.8
Cobaltite	$CoAsS$	165.92		35.5							45.2		19.3
Covellite	CuS	95.61			66.5								33.5
Cubanite	$CuFe_2S_3$	271.44			23.4	41.2							35.4
Digenite	Cu_9S_5	732.16			78.1								21.9
Enargite	Cu_3AsS_4	393.77			48.4						19.0		32.6
Galena	PbS	239.28						86.6					13.4
Gersdorffite	$NiAsS$	165.67					35.4				45.2		19.4
Idaite	Cu_5FeS_6	565.94			56.1	9.9							34.0
Jamesonite	$Pb_4FeSb_6S_{14}$	2,064.17				2.7		40.2				35.4	21.7
Loellingite	$FeAs_2$	205.67				27.2					72.8		
Millerite	NiS	90.76					64.7						35.3
Molybdenite	MoS_2	160.08								Mo 59.9			40.1
Niccolite	$NiAs$	133.60					43.9				56.1		
Pentlandite	$Ni_{4.5}Fe_{4.5}S_8$	771.92				32.6	34.2						33.2
Proustite	Ag_3AsS_3	494.75	65.4								15.1		19.5
Pyrargyrite	Ag_3SbS_3	541.60	59.8									22.5	17.7
Pyrite	FeS_2	119.98				46.5							53.5
Pyrrhotite	FeS	87.92				63.5							36.5
Rammelsbergite	$NiAs_2$	208.51					28.2				71.8		
Safflorite	$CoAs_2$	208.76		28.2							71.8		
Sphalerite	ZnS	97.45							67.1				32.9
Tetraedrite	$Cu_{12}Sb_4S_{13}$	1,666.37			45.8							29.2	25.0

Table 12-16. *Astronomical constants. (Reference list of recommended constants.)* (From Astronomer's Handbook, 1966)

Defining constants	
Number of ephemeris seconds in 1 tropical year (1900)	$s = 31{,}556{,}925.9747$
Gaussian gravitational constant, defining the A.U.	$k = 0.01720209895$
Primary constants	
Measure of the A.U in metres	$A = 149{,}600 \times 10^6$
Velocity of light in metres per second	$c = 299{,}792.5 \times 10^3$
Equatorial radius for Earth in metres	$a_e = 6{,}378{,}160$
Dynamical form-factor for Earth	$J_2 = 0.0010827$
Geocentric gravitational constant (units: $m^3\ s^{-2}$)	$GE = 398{,}603 \times 10^9$
Ratio of the masses of the Moon and Earth	$\mu = 1/81.30$
Sidereal mean motion of Moon in radians per second (1900)	$n_☾^* = 2.661699489 \times 10^{-6}$
General precession in longitude per tropical century (1900)	$p = 5{,}025''.64$
Obliquity of the ecliptic (1900)	$\varepsilon = 23°27'08''.26$
Constant of nutation (1900)	$N = 9''.210$
Derived constants	
Heliocentric gravitational constant (units: $m^3\ s^{-2}$)	$GS = 132{,}718 \times 10^{15}$
Ratio of masses of Sun and Earth	$S/E = 322{,}958$
Ratio of masses of Sun and Earth + Moon	$S/E\,(1+\mu) = 328{,}912$
Perturbed mean distance of Moon, in metres	$a_☾ = 384{,}400 \times 10^3$

Table 12-17. *Solar dimensions.* (From WALDMEIER, 1965)

Radius	6.960×10^{10} cm
Surface area	6.087×10^{22} cm^2
Volume	1.412×10^{33} cm^3
Mass	1.989×10^{33} g
Mean density	1.409 g cm^{-3}
Density at the center	98 g cm^{-3}
Gravitational acceleration at the solar surface	2.740×10^4 cm s^{-2}
Escape velocity at the surface	6.177×10^7 cm s^{-1}
Effective temperature	5,785° K
Temperature at the center	13.6×10^6 ° K
Radiation	3.9×10^{33} erg s^{-1}
Specific surface emission	6.41×10^{10} erg cm^{-2} s^{-1}
Specific mean energy production	1.96 erg g^{-1} s^{-1}
Solar constant = extraterrestrial energy flux at the mean distance between earth and sun	1.39×10^6 erg cm^{-2} s^{-1} 2.00 cal cm^{-2} min^{-1}

Table 12-18. *Dimensions of the planets and the moon.* (From GONDOLATSCH, 1965; KUIPER, 1965)

Name	Sym-bol	a [10^6 km]	P [a]	Diameter [km]	Mass[a] [10^{26} g]	Volume [10^{10} km^3]
Mercury	☿	57.9	0.240 85	4,840	3.333	5.958
Venus	♀	108.2	0.615 21	12,228	48.70	95.765
Earth	♁	149.6	1.000 04	12,742.06	59.76	108.332
Moon	☾	0.384[b]	0.074 80[c]	3,476	0.735	2.192
Mars	♂	227.9	1.880 89	6,770	6.443	16.250
Jupiter	♃	778	11.862 23	140,720	18,993	145,923.204
Saturn	♄	1,427	29.457 7	116,820	5,684	83,469.806
Uranus	⛢	2,870	84.015 3	47,100	867.6	5,481.599
Neptune	♆	4,496	164.788 3	44,600	1,029	4,636.610
Pluto	Pl	5,881.9 to 5,946.5	247.7	6,000	55.3	10.833

a = semi-major axis of the orbit.

P = sidereal period = true period of the planet's revolution around the Sun (with respect to the fixed star field).

[a] Mass without moons.

[b] Mean distance from the Earth.

[c] true period of the Moon's revolution around the Earth (with respect to the fixed star field).

Name	$\bar{\varrho}$ [g/cm^3]	g_{Eq} [cm/s^2]	v_e [km/s]	A	T_{max} [°K]	$T_{av.}$ [°K]	Atm. constituents
Mercury	5.62	380	4.29	0.056	625	—	((^{40}Ar))
Venus	5.09	869	10.3	0.76	324	229	CO_2, H_2O
Earth	5.517	978	11.2	0.39	349	246	N_2, O_2
Moon	3.35	162	2.37	0.067	387	274	—
Mars	3.97	372	5.03	0.16	306	216	N_2, CO_2, H_2O
Jupiter	1.30	2,301	57.5	0.67	131	93	H_2, CH_4, NH_3
Saturn	0.68	906	33.1	0.69	95	68	H_2, CH_4, NH_3
Uranus	1.58	972	21.6	0.93[a]	67[a]	47[a]	He, H_2, CH_4
Neptune	2.22	1,347	24.6	0.84[a]	53[a]	38[a]	He, H_2, CH_4
Pluto	—	—	—	0.14	60	43	uncertain

g_{Eq} = total acceleration, including centrifugal acceleration, at equator.

v_e = velocity of escape at equator.

A = Albedo = total reflectivity, wavelength λ_{eff} = 5,500 Å.

T_{max} = maximum temperature for the subsolar point of a slowly rotating planet or satellite (computed from the visual albedo).

T_{av} = average temperature of a rapidly rotating sphere.

Atm. constituents = main atmospheric constituents.

[a] Since the albedos of Uranus and Neptune are very low in the red and infrared an effective value $A = 0.7$ has been adopted for calculating the temperatures.

Table 12-19. *Earth's dimensions.* (From Astronomer's Handbook, 1966, GONDOLATSCH, 1965, and MACDONALD 1966)

	I.E.R.[a] 1924	I.A.U.[b] 1966
Equatorial radius a_e	6,378.388 km	6,378.160 km
Polar radius a_p	6,356.912 km	6,356.775 km
Flattening factor f	1/297	1/298.25

Radius of sphere of equal volume a_0	6,371 km
Area of surface	5.101×10^8 km^2
Volume	1.083×10^{12} km^3
Mass	5.976×10^{27} g
Mean density	5.517 g cm^{-3}
Gravitational constant G	6.670×10^{-8} dynes cm^{-2} g^{-2}
Normal acceleration of gravity at equator g_e (based on Potsdam standard)	978.0436 cm s^{-2}
Mean solar day d	86,400 s = 24^h
Sidereal day S	86,164.09 s = 23^h 56^m 4.09^s
Velocity of rotation at equator	465.12 m s^{-1}
Mean moment of inertia C_0	8.02×10^{44} g cm^2

[a] I.E.R. International Ellipsoid of Reference (1924).
[b] I.A.U. International Astronomical Union (1966).

Table 12-20. *Earth's interior, masses and dimensions of the principle subdivisions.* (From MACDONALD, 1966; POLDERVAART, 1955; and SCHMUCKER, 1969)

	Mass [10^{25} g]	Mean density [g/cm^3]	Surface area [10^6 km^2]	Radius or thickness [km]	Volume [10^9 km^3]	Mean moment of inertia (spherical symmetry) [10^{42} g cm^2]
Core	192	11.0	151	3,471	175	90
Mantle	403	4.5			898	705
below 1000 km	240	5.1	362	1,900	474	333
above 1000 km	163	3.9	505	970—990	424	372
Crust	2.5	2.8	510		8.9	(7)
continental	2.0	2.75	242[b]	30	7.3	
oceanic	0.5	2.9	268[b]	6	1.6	
Oceans and marginal seas	0.14	1.03	361[b]	3.8	1.4[b]	(<1)
Whole Earth	597.6	5.52	510	6,371	1,083	802
Atmosphere	0.00051	0.0013[c]	—	8[d]	—	—

[a] Table 6-5, p 217.
[b] After POLDERVAART (1955).
[c] Surface value.
[d] Scale height of the "homogeneous" atmosphere.

Table 12-21. *The surface areas of the Earth.* (From POLDERVAART, 1955)

	10^6 km²		10^6 km²
Continental shield region Region of young folded belts Volcanic islands in deep oceanic and suboceanic region	105 42 2	land about 29.2% of total	149
Shelves and continental slopes region Deep oceanic region	93 268	ocean about 70.8% of total	361
		Total surface	510

Table 12-22. *Geological time-scales.* (From AFANASSYEV *et al.*, 1964; HOLMES' Symposium, 1964; KULP, 1960; and KRAUSKOPF, 1967; ZUBAKOV, 1967; VINOGRADOV *et al.*, 1968)

Era	Period	Epoch	Time since beginning in millions of years				
			KULP[a] 1960	AFANASSYEF et al.[a] 1964	HOLMES' Symposium[a] 1964	KRAUS-KOPF[b] 1967	VINO-GRADOV 1968
Cenozoic	Quaternary	Pleistocene[c]	1	1.5—2	1.5—2	2	
	Tertiary	Pliocene	12	12±1	7	10	
		Miocene	23	26±1	26	27	
		Oligocene	35	37±2	37—38	38	
		Eocene	55	60±2	53—54	55	
		Paleocene	70	67±3	65	65—70	
Mesozoic	Cretaceous		135	137± 5	136	130	
	Jurassic		180	195± 5	190—195	180	
	Triassic		220	240±10	225	225	
Paleozoic	Permian		270	285± 10	280	260	
	Carboniferous		350	340—360	345	340	
	Devonian		400	410± 10	395	405	
	Silurian		430	440± 15	430—440	435	
	Ordovician		490	500± 20	500	480	
	Cambrian		600	570	570	550—570	570
Pre-cambrian	Upper Precambrian						1,900
	Middle Precambrian						2,700
	Lower Precambrian						3,500

[a] Used by the I.U.G.S.-Commission on Geochronology as base for discussion of a revised geological time-scale.

[b] Source: A. KNOPF, private communication (1965).

[c] Subdivisions (from ZUBAKOV, 1967)

Wurmian glaciation	10— 70	thousand years
Warthian glaciation	100— 120	thousand years
Rissian glaciation	175— 210	thousand years
Mindelian glaciation	370— 600	thousand years
Gunzian glaciation	750—1,000	thousand years
Begin of Calabrian age	1,800± 200	thousand years

Table 12-23. *Measures, units and conversion factors.* (*Metric and U.S. System.*) (From Handbook of Chemistry and Physics, 1967/68)

Prefix	Symbol	Meaning	Units
Tera	T	1,000,000,000,000	10^{12}
Giga	G	1,000,000,000	10^{9}
Mega	M	1,000,000	10^{6}
Kilo	k	1,000	10^{3}
Hecto	h	100	10^{2}
Deka	dk	10	10^{1}
Deci	d	0.1	10^{-1}
Centi	c	0.01	10^{-2}
Milli	m	0.001	10^{-3}
Micro	μ	0.000001	10^{-6}
Nano	n	0.000000001	10^{-9}
Pico	p	0.000000000001	10^{-12}

Lengths

Metric System		U.S. System
10^{-8} cm	1 Å	$3.937 \cdot 10^{-9}$ in.
10^{-4} cm	1 μ	$3.937 \cdot 10^{-5}$ in.
	1 cm	0.3937 in.
	2.540 cm	1 in.
	0.3048 m	1 ft.
	0.9144 m	1 yd.
10^{2} cm	1 m	1.09361 yd.
	1.8288 m	1 fath.
10^{5} cm	1 km	0.62137 mi.
	1.60935 km	1 mi.
	1.852 km	1 intern. nautical mile

1 A.U. (astronomical unit) = $1.49598 \cdot 10^{8}$ km

Area

Metric System	U.S. System
1 mm²	0.00155 in.² (sq. in.)
1 cm²	0.155 in.²
6.45163 cm²	1 in.²
0.0929 m²	1 ft.² (sq. ft.)
0.83613 m²	1 yd.² (sq. yd.)
1 m²	10.7639 ft.²
1 km²	0.3861 mi.² (sq. mi.).
2.58998 km²	1 mi.²

Volume

Metric System	U.S. System
1 mm³	$0.6102 \cdot 10^{-4}$ in.³ (cu. in.)
1 cm³	0.06102 in.³
16.3872 cm³	1 in.³
0.02831 m³	1 ft.³ (cu. ft.)
0.76456 m³	1 yd.³ (cu. yd.)
1 m³	1.30794 yd.³

Table 12-23 (Continued)

Liquid measures

Metric system	U.S. System	
1 ml	0.0610 in.^{3}	
0.473 L	28.875 in.^{3}	1 pt. (pint)
0.946 L	57.749 in.^{3}	1 qt. (quart)
1 L	61.0 in.^{3}	1.0567 qt.
3.7853 L	231 in.^{3}	1 gal. (gallon)

Mass

Metric system	U.S. System
1 g	0.035 oz. av. (ounce av.)
28.349 g	1 oz. av.
453. 59 g	1 lb. av. (pound av.[a])
1 kg	2.20462 lb. av.
907.1848 kg	1 tn. sh. (short ton)
1 t	1.1023 tn. sh.
1016.047 kg	1 tn. l. (long ton)

[a] 1 lb. av. = 1 pound avoirdupois is the mass of 27.692 in.^{3} of water weighed in air at 4° C, 760 mm pressure.

Density

Metric System	U.S. System
1 g/cm^3	0.036127 lb./in.^{3}
27.68 g/cm^3	1 lb./in.^{3}
0.0160 g/cm^3	1 lb./ft.^{3}

Energy

	erg	Joule$_{int}$	kW$_{int}$h	kcal$_{15}$	Liter-atmos.	B.T.U.
erg	1	0.9997×10^{-7}	2.7769×10^{-14}	2.389×10^{-11}	9.8692×10^{-10}	9.4805×10^{-11}
Joule$_{int}$	1.0002×10^{7}	1	2.7778×10^{-7}	2.390×10^{-4}	9.8722×10^{-3}	9.480×10^{-4}
kW$_{int}$h	3.6011×10^{13}	3.6000×10^{6}	1	8.6041×10^{2}	3.5540×10^{4}	3.413×10^{3}
kcal$_{15}$	4.1853×10^{10}	4.186×10^{3}	1.1622×10^{-3}	1	4.1306×10^{1}	3.9685
Liter-atmos.	1.0133×10^{9}	1.0133×10^{2}	2.8137×10^{-5}	2.421×10^{-2}	1	9.607×10^{-2}
B.T.U.	1.0548×10^{10}	1.0548×10^{3}	2.930×10^{-4}	2.5198×10^{-1}	1.0409×10^{1}	1

Pressure

	bar	Torr	atm.	at	lb./in.^{2}
1 bar (10^6 dynes/cm^2)	1	750	0.98692	1.0197	14.504
1 Torr	0.00133	1	0.00131	0.001359	0.01934
1 atm	1.0133	760	1	1.033	14.696
1 at (1 kg/cm^2)	0.98067	735.56	0.96784	1	14.223
1 lb./in.^{2}	0.06895	51.7144	0.068046	0.07031	1

Table 12-23 (Continued)

Temperature

Absolute Centigrade or Kelvin (K)	x °K $= T$ °C $+ 273.18$
Degrees Centigrade (°C)	x °C $= \frac{5}{9}$ (T °F-32)
	x °C $= \frac{5}{4}$ T °R
Degrees Fahrenheit (°F)	x °F $= \frac{9}{5}$ T °C $+32$
	x °F $= \frac{9}{4}$ T °R $+32$
Degrees Réaumur (°R)	x °R $= \frac{4}{9}$ (T °F-32)
	x °R $= \frac{4}{5}$ T °C

Centigrade to Fahrenheit

°C	°F	°C	°F	°C	F°
−200	−328	60	140	200	392
−150	−238	70	158	250	482
−100	−148	80	176	300	572
− 50	− 58	90	194	400	752
0	+ 32	100	212	500	932
10	50	110	230	600	1,112
20	68	120	248	700	1,292
30	86	130	266	800	1,472
40	104	140	284	900	1,652
50	122	150	302	1,000	1,832

Time

1 sidereal second	= 0.99727 mean solar second
1 sidereal day	= 86,164 mean solar seconds
1 solar day	= 86,400 mean solar seconds
1 mean solar year	= 365.242 mean solar days = 3.1557×10^7 mean solar seconds
1 sidereal year	= 365.256 mean solar days = 3.15581×10^7 mean solar seconds

References

AFANASSYEV, G. D., *et al.*: The project of a revised geological time-scale in absolute chronology. Contr. Geol. Soviétiques Congr. géol. inter. 22e Sess., India, 287—324 (1964).

AHRENS, L. H.: The use of ionization potentials. Part I. Ionic radii of the elements. Geochim. Cosmochim. Acta **2**, 155 (1952).

ALLRED, A. L., and E. G. ROCHOW: A scale of electronegativity based on electrostatic force. J. Inorg. Nucl. Chem. **5**, 264 (1958).

Astronomer's Handbook, Transactions of the International Astronomical Union. Vol. XIIC. London and New York: Academic Press 1966.

D'ANS-LAX: Taschenbuch für Chemiker und Physiker. Bd. 1. Berlin-Heidelberg-New York: Springer 1967.

GOLDSCHMIDT, V. M.: Kristallchemie. Handwörterbuch der Naturwissenschaften, 2. Aufl., Bd. 5. Jena: Gustav Fischer 1934.

—, T. BARTH, G. LUNDE u. W. ZACHARIASEN: Geochemische Verteilungsgesetze der Elemente. VII. Skrifter Norske Videnskaps-Akad. Oslo, I. Mat.-Naturv. Kl., No. 2 (1926).

GONDOLATSCH, F.: Mechanical data of planets and satellites. LANDOLT-BÖRNSTEIN, New series, Group VI: Astronomy, astrophysics and space research, vol. I. Berlin-Heidelberg-New York: Springer 1965.

GORDY, W., and W. J. O. THOMAS: Electronegativities of the elements. J. Chem. Phys. **24**, 439 (1956).
HAMILTON, E. I.: Applied geochronology. London and New York: Academic Press 1965.
Handbook of Chemistry and Physics, 48th ed. Cleveland, Ohio: The Chemical Rubber Co. 1967/68.
HOLMES' Symposium: Geological society phanerozoic time-scale. Quart. J. Geol. Soc. London **120** s, 260 (1964).
IUPAG (International Union of Pure and Applied Chemistry): Comptes Rendus, XXII. Conference. London: Butterworths Sci. Publ. 1963.
KRAUSKOPF, K.: Introduction to geochemistry. New York: McGraw-Hill Book Co. 1967.
KUIPER, G. P.: Physics of planets and satellites. LANDOLT-BÖRNSTEIN, New series, Group VI: Astronomy, astrophysics and space research, vol. I. Berlin-Heidelberg-New York: Springer 1965.
KULP, J. L.: The geological time-scale. Report Intern. geol. Congr. 21st Sess., Norden, part III, 18—27 (1960).
MACDONALD, G. J. F.: Geodetic data. Handbook of physical constants, revised edit. Geol. Soc. Am., Mem. 97 (1966).
PAULING, L.: The sizes of ions and the structure of ionic crystals. J. Am. Chem. Soc. **49**, 763 (1927).
— The nature of the chemical bond. Ithaca, N.Y.: Cornell University Press 1960.
POLDERVAART, A.: Chemistry of the earth crust. Geol. Soc. Am., Spec. papers 62: Crust of the earth, p. 119 (1955).
ROBIE, R. A., P. M. BETHKE, M. S. TOULMIN, and J. L. EDWARDS: X-ray crystallographic data, densities, and molar volumes of minerals. Handbook of physical constants, revised edit. Geol. Soc. Am. Mem. 97 (1966).
SCHMUCKER, U.: Chapter 6 of this Handbook.
STROMINGER, D., J. M. HOLLANDER, and G. T. SEABORG: Table of isotopes. Rev. Mod. Phys. **30**, 585 (1958).
VINOGRADOV, A. P., and A. I. TUGARINOV: Geochronological scale of the Precambrian. Report Intern. geol. Congr. 23rd Sess., Czechoslovakia, vol. 6, 205—222 (1968).
WALDMEIER, M.: The quiet sun. LANDOLT-BÖRNSTEIN, New series, Group VI: Astronomy, astrophysics and space research, vol. I. Berlin-Heidelberg-New York: Springer 1965.
WETHERILL, G. W.: Radioactive decay constants and energies. Handbook of physical constants, revised edit. Geol. Soc. Am. Mem. 97 (1966).
WYCKOFF, R. W. G.: Crystal structures, section I. New York: Interscience Publ. 1948.
ZEMANN, J.: Kristallchemie. Sammlung GÖSCHEN, Bd. 1220/1220a. Berlin: W. de Gruyter & Co. 1966.
ZUBAKOV, V. A.: Geochronology of the continental Pleistocene deposits (based on radiometric data). Geochem. Intern. **4**, 144—154 (1967).

AUTHOR INDEX

Page numbers in *italics* refer to the references

SUBJECT INDEX

Universitätsdruckerei H. Stürtz AG Würzburg